a LANGE medical book

Examination & Board Review

Biochemistry

First Edition

Walter X. Balcavage, PhD
Professor
Department of Biochemistry and Molecular Biology
Indiana University School of Medicine
Terre Haute Center for Medical Education

Michael W. King, PhD
Associate Professor
Department of Biochemistry and Molecular Biology
Indiana University School of Medicine
Terre Haute Center for Medical Education

D1547935

APPLETON & LANGE
Norwalk, Connecticut

95 96 97 98 / 10 9 8 7 6 5 4 3 2 1

Prentice Hall International (UK) Limited, *London*
Prentice Hall of Australia Pty. Limited, *Sydney*
Prentice Hall of Canada, Inc., *Toronto*
Prentice Hall Hispanoamericana, S.A., *Mexico*
Prentice Hall of India Private Limited, *New Delhi*
Prentice Hall of Japan, Inc., *Tokyo*
Simon & Schuster Asia Pte. Ltd., *Singapore*
Editora Prentice Hall do Brasil Ltda., *Rio de Janeiro*
Prentice Hall, *Englewood Cliffs*, *New Jersey*

ISBN 0-8385-0661-5

Acquisitions Editor: John Dolan
Managing Editor: Gregory R, Huth
Production Editor: Chris Langan
Senior Art Coordinator: Maggie Belis Darrow
Designer: Libby Schmitz

PRINTED IN THE UNITED STATES OF AMERICA

Table of Contents

I. CHEMICAL PROPERTIES OF BIOLOGICAL SYSTEMS _____ 1

Chapter   1. Ionic Equilibria and Acid Base Balance 1
Chapter   2. Thermodynamic Principals Applied to the Life Process 12

II. AMINO ACIDS AND PROTEINS, STRUCTURE AND FUNCTION _____ 18

Chapter   3. Amino Acids 18
Chapter   4. Protein Structure 26
Chapter   5. Introduction to Enzymes 36
Chapter   6. Kinetics 42
Chapter   7. Mechanisms and Regulation of Enzyme Activity 53
Chapter   8. Integration of Protein Structure and Function 60

III. STRUCTURE OF CARBOHYDRATES, LIPIDS, NUCLEIC ACIDS _____ 71

Chapter   9. Carbohydrates 71
Chapter 10. Lipids 81
Chapter 11. Nucleic Acids 91

IV. BIOLOGICAL MEMBRANES _____ 103

Chapter 12. Structure and Function of Biological Membranes 103

V. INTERMEDIARY METABOLISM _____ 112

Chapter 13. Introduction to Intermediary Metabolism 112
Chapter 14. Glycolysis and Gluconeogenesis 120

Chapter 15. Glycogen Metabolism and Regulation 133

Chapter 16. Pentose Phosphate Pathway, Galactose and Fructose Metabolism 146

Chapter 17. The Pyruvate Dehydrogenase Complex and the Krebs Cycle 152

Chapter 18. Biological Oxidations 164

Chapter 19. Amino Sugars and Glycoconjugates 178

Chapter 20. Lipid Metabolism 188

Chapter 21. Lipid Digestion and Lipoproteins 210

Chapter 22. Cholesterol Metabolism 221

Chapter 23. Nitrogen Metabolism 229

Chapter 24. Nonessential Amino Acid Synthesis 241

Chapter 25. Metabolic Fate of Amino Acids 253

Chapter 26. Biologically Active Nitrogen Compounds 266

Chapter 27. Nucleotide Metabolism 277

VI. METABOLIC INTEGRATION_____ 296

Chapter 28. Interrelationships of the Major Organs 296

VII. BIOLOGICAL INFORMATION PROCESSING_____ 306

Chapter 29. DNA Synthesis 306

Chapter 30. RNA Synthesis 318

Chapter 31. Protein Synthesis 329

Chapter 32. Control of Gene Expression 342

VIII. CELLULAR REGULATION _____ 351

Chapter 33. Growth Factors and Cytokines 351

Chapter 34. Mechanisms of Signal Transduction 358

Chapter 35. Growth Factors and Proto-Oncogenes in Cancer 366

IX. NUTRITION 372

Chapter 36. Vitamins and Minerals 372

X. SPECIAL TOPICS 391

Chapter 37. Blood Coagulation 391

Chapter 38. Recombinant DNA Technologies in Medicine 400

Index 417

Preface

This book is intended to help students of biochemistry, and especially medical biochemistry, to review and prepare for regular course examinations as well as Part I of the National Board of Medical Examiners test (USMLE) following the second year of medical school. The text is organized into a concise review of the major areas of biochemistry and medical biochemistry. It begins by covering the basic elements of these areas and progresses to concise reviews of "intermediary metabolism," including carbohydrate, lipid, amino acid, and nucleotide metabolism. Moreover, significant areas of medical biochemistry have been included, such as kidney function in acid-base balance, blood coagulation, growth factors and receptors and their roles in oncogenesis, signal transduction, and tools of modern molecular medicine. The review of these subjects distinguishes this text from other medical biochemistry board examination reviews. Concise objectives begin each chapter, allowing the reader to comprehend at a glance the scope of covered material. Nearly all chapters contain clinically relevant correlations, providing the medical student reviewer with important information about the direct applications of biochemistry.

Another helpful feature is the review questions section containing several A-type (best answer) and B-type (matching) questions (10–20) which are the two forms now encountered in the Part I National Board Examination (USMLE). In addition, several clinical case-based questions are included in most chapters. Answers for each question are provided. Finally, each chapter offers highly illustrative and succinct figures and diagrams (many of which are further enhanced by the addition of a second color) to assist the user in understanding the covered material.

In closing, the authors would like to acknowledge Carolyn Morgan and Cherri Rutan for their artistic expertise and dedication in the production of numerous figures for this book.

Walter X. Balcavage, PhD
Michael W. King, PhD
May 1995

Part I: Chemical Properties of Biological Systems

Ionic Equilibria and Acid Base Balance

1

Objectives

- To define the following terms: K_{eq}, K_w, pH, pK_a, pI, zwitterion, ampholyte, polyampholyte, Bohr effect, chloride shift.
- To understand the derivation and usage of the Henderson-Hasselbalch equation.
- To define a buffer.
- To understand the role of salts in macromolecular interactions.
- To know the major buffers of blood.
- To define the role of carbonic anhydrase in blood buffering.
- To distinguish between metabolic and respiratory acidosis and alkalosis.
- To define the three major roles of the kidneys in blood buffering; sodium bicarbonate resorption, acid excretion, and ammonia secretion.

Concepts

Water is the universal medium of biological systems. All the unique characteristics of molecules in living matter, all their biochemically and physiologically important characteristics, depend on the ability of biomolecules to interact with water. The major interaction of this type takes place through the formation of **hydrogen bonds (H-bonds)**, which are noncovalent interactions between polar molecules (Chapter 4). However, water plays an equally important role in the maintenance of intracellular and extracellular pH. Therefore, one must consider the function of water in the dissociation of ions from biological molecules. Although essentially a neutral molecule, water will ionize to a small degree. This can be described by a simple equilibrium equation:

$$H_2O \Leftrightarrow H^+ + OH^-$$

1.1

K_{eq}, pK_a, K_w and pH

The **equilibrium constant, K_{eq}**, as written for reaction 1.1, (or for any reaction) can be calculated (where brackets define concentration values):

$$K_{eq} = \frac{[H^+][OH^-]}{[H_2O]}$$

1.2

Since the concentration of H_2O is very high (55.5M) relative to that of the $[H^+]$ and $[OH^-]$, its consideration is generally removed from the equation by multiplying both sides by 55.5M. This yields a new term, K_w:

$$K_w = [H^+][OH^-]$$

1.3

This term is referred to as the **ion product**. In pure water, to which no acids or bases have been added:

$$K_w = 1 \times 10^{-14} \, M^2 \qquad \text{1.4}$$

In conditions of pure water K_w is a constant, therefore:

$$[H^+] = [OH^-] = 1 \times 10^{-7} \, M \qquad \text{1.5}$$

Equation 1.5 indicates that in pure water both the hydrogen ion and hydroxyl ion concentrations are equivalent and, therefore, a solution is said to be neutral when the $[H^+] = 10^{-7}$. In order to avoid exponential terms, the logarithm of the $[H^+]$ is used. Converting equation 1.5 by the use of logarithms produces a scale that reflects the hydrogen ion concentration of any solution. This is termed the **pH**, where:

$$pH = -\log[H^+] \qquad \text{1.6}$$

Acids and bases are classified as proton donors and proton acceptors, respectively. When a proton dissociates from an acid into the surrounding medium, the resultant molecule is termed **the conjugate base of the acid**. This means that the conjugate base of a given acid will carry a net charge that is more negative than the corresponding acid. Conversely, the conjugate acid of a given base will carry a net charge that is more positive than the corresponding base.

As an example, the dissociation of a proton from lactic acid would yield the conjugate base, lactate ion:

$$CH_3CHOHCOOH \Leftrightarrow H^+ + CH_3CHOHCOO^- \qquad \text{1.7}$$

Similarly, dissociation of a proton from pyruvic acid would yield the conjugate base, pyruvate ion:

$$CH_3COCOOH \Leftrightarrow H^+ + CH_3COCOO^- \qquad \text{1.8}$$

Acids (and bases) are defined as strong if they readily ionize in an aqueous medium (ie pK_a values of 3 and below), whereas they are defined as weak if they ionize poorly. In body fluids there exist many compounds that are weak acids and bases, such as the acidic and basic amino acids, nucleotides, lactic acid, pyruvic acid, phospholipids, etc.

Weak acids and bases in solution do not fully dissociate and, therefore, there is an equilibrium between the acid and its conjugate base. Like the ionization of H_2O seen in reaction 1.1, the equilibrium constant, K_{eq} can be calculated for ionizations of weak acids and bases. In the case of weak acid and weak base ionizations K_{eq} is commonly identified as K_a.

In the reaction of a weak acid:

$$HA \Leftrightarrow A^- + H^+ \qquad \text{1.9}$$

the equilibrium constant can be calculated from the following equation:

$$K_a = \frac{[H^+][A^-]}{[HA]} \qquad \text{1.10}$$

The definition of pH from the ion product was obtained by the use of logarithms. Similarly, equation 1.10 can be redefined to describe the term **pK_a**:

$$pK_a = -\log K_a \qquad \text{1.11}$$

In obtaining the negative log of both sides of equation 1.11, which describes the dissociation of a weak acid, we arrive at the following equation:

$$-\log K_a = -\log \left\{ \frac{[H^+][A^-]}{[HA]} \right\} \qquad \text{1.12}$$

Since, as indicated above, $-\log K_a = pK_a$, and taking into account the laws of logarithms:

$$pK_a = -\log[H^+] - \log\left\{\frac{[A^-]}{[HA]}\right\} \qquad \textbf{1.13}$$

Substitution of pH from equation 1.6 yields:

$$pK_a = pH - \log\left\{\frac{[A^-]}{[HA]}\right\} \qquad \textbf{1.14}$$

From equation 1.14 it can be seen that the smaller the value of pK_a the stronger is the acid. Or, to put it the other way, the stronger an acid is, the more readily it will give up H^+ ions; therefore, the value of [HA] in the above equation will be relatively small.

Henderson-Hasselbalch

By rearranging equation 1.14 we arrive at the **Henderson-Hasselbalch** equation:

$$pH = pK_a + \log\left\{\frac{[A^-]}{[HA]}\right\} \qquad \textbf{1.15}$$

The Henderson-Hasselbalch equation is used to determine the pH of a solution of any weak acid, (for which the equilibrium constant is known) at any concentration of the acid, [HA], and its conjugate base [A^-].

Buffering

During the pH titration of a weak acid a point is reached where the concentration of the conjugate base, [A^-], is equal to the concentration of the acid, [HA]. This is the mid-point in a titration, and the Henderson-Hasselbalch equation transforms to:

$$pH = pK_a + \log[1] \qquad \textbf{1.16}$$

The log of 1 = 0, therefore, at the mid-point of a titration of a weak acid:

$$pK_a = pH \qquad \textbf{1.17}$$

In other words, the term pK_a is that pH at which an equivalent distribution of acid and conjugate base (or base and conjugate acid) exists in solution. A typical acid titration is shown in Figure 1–1.

It is apparent from Figure 1–1 that during the titration of a weak acid, near the pK_a, the pH of the solution does not change appreciably even when large equivalents of base are added. This phenome-

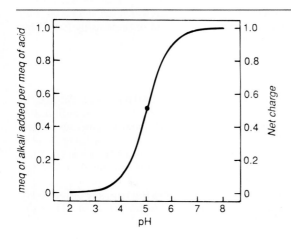

Figure 1–1. Titration curve for an acid of the type HA. The dot (•) indicates the pK, 5.0. (Reproduced with permission, from Murray, RK: *Harper's Biochemistry*, 23rd ed. Appleton & Lange, 1993)

non is known as **buffering** and is defined as the ability of a solution to resist a change in pH when acid or base is added. Weak acids and bases are recognized to be useful buffers in solutions that are 1 pH unit above or below their pK_a.

Electrical Properties

Many substances in nature contain both acidic and basic groups, and often many different types of these groups in the same molecule (eg, proteins). Molecules that contain more than one acidic and basic group are called **ampholytes** (one acidic and one basic group) or **polyampholytes** (many acidic and basic groups).

A term used to describe the ionic state of ampholytes and polyampholytes (eg, proteins) is the **iso-electric pH (or isoelectric point), pI**. The pI is the pH at which the effective net charge on a molecule is zero. This occurs when the number of positively charged groups equals the number of negatively charged groups, in other words the molecule is electrically neutral. A molecule will not migrate in an electric field in a solution where the pH is equal to pI, since it has no net charge. A molecule that exists at its isoelectric pH is in its **zwitterionic** form.

For the case of a simple ampholyte like the amino acid glycine, the pI is easily calculated from the Henderson-Hasselbalch equation and is shown to be the average of the pK_a for the α-COOH group and the pK_a for the α-NH_2 group:

$$pI = pK_{a-COOH} + pK_{a-NH_3^+} \qquad \textbf{1.18}$$

For more complex molecules such as polyampholytes, the pI is approximately the average of the pK_a values that represent the boundaries of the zwitterionic form of the molecule.

The pI value, like that of pK_a, is very informative about the physical nature of a molecule. A molecule with a low pI would contain a predominance of acidic groups, whereas a high pI indicates predominance of basic groups.

Salt Effects

Depending on the pH of a solution, macromolecules such as proteins—which contain many charged groups—will carry substantial net charge, either positive or negative. Cells of the body and blood contain many **polyelectrolytes** (molecules that contain multiple like charges, eg DNA and RNA) and **polyampholytes** that are in close proximity. The close association allows these molecules to interact through opposing charged groups. In blood and in tissue cells are found numerous small charged ions (eg Na^+, Cl^-, Mg^+, Mn^+, K^+), which interact with the larger macroions. This interaction can effectively **shield** the electrostatic charges of like-charged molecules, which in turn allows macroions to become more closely associated than could be predicted from their expected charge repulsion from one another. The net effect of the presence of small ions is to maintain the solubility of macromolecules at pH ranges near their pI. This interaction between solute (proteins, DNA, RNA, etc) and solvent (blood) is termed **solvation or hydration**. The opposite effect of solvation occurs when the salt (small ion) concentration increases to such a level that it interferes with the solvation of proteins by H_2O. This results from the H_2O forming **hydration shells** around the small ions.

Blood Buffering

The pH of blood is maintained in a narrow range around 7.4. Even relatively small changes in this value of blood pH can lead to severe metabolic consequences. Therefore, blood buffering is extremely important in order to maintain homeostasis. Although the blood contains numerous cations (eg, Na^+, K^+, Ca^+ and Mg^+) and anions (eg, Cl^-, PO_4^{3-} and SO_4^{2-}) that can, as a whole, play a role in buffering, the primary buffers in blood are hemoglobin in erythrocytes and bicarbonate ion (HCO_3^-) in the plasma. Buffering by hemoglobin is accomplished by ionization of the imidazole ring of histidines in the protein.

The formation of bicarbonate ion in blood from CO_2 and H_2O allows the transfer of relatively insoluble CO_2 from the tissues to the lungs, where it is expelled. The major source of CO_2 in the tissues comes from the oxidation of ingested carbon compounds.

Carbonic acid is formed from the reaction of dissolved CO_2 with H_2O. The relationship between carbonic acid and bicarbonate ion formation is shown in reactions 1.19 and 1.20.

$$CO_2 + H_2O \Leftrightarrow H_2CO_3 \qquad\qquad \textbf{1.19}$$

Reactions 1.19 and 1.20 occur predominately in the erythrocytes, since nearly all of the CO_2 leaving tissues via the capillary endothelium is taken up by these cells. This reaction is catalyzed by **carbonic anhydrase**. Ionization of carbonic acid then occurs, yielding bicarbonate ion.

$$H_2CO_3 \Leftrightarrow H^+ + HCO_3^- \qquad\qquad \textbf{1.20}$$

Carbonic acid is a relatively strong acid with a pK_a of 3.8. However, carbonic acid is in equilibrium with dissolved CO_2. Therefore, the equilibrium equation for the sum of reactions 1.19 and 1.20 requires a conversion factor, since CO_2 is a dissolved gas. This factor has been shown to be approximately 0.03 times the partial pressure of CO_2 (PCO_2). When this is entered into the Henderson-Hasselbalch equation:

$$pH = 6.1 + \log\left\{\frac{[HCO_3^-]}{(0.03)(PCO_2)}\right\} \qquad\qquad \textbf{1.21}$$

where the apparent pK_a for bicarbonate formation, 6.1, has been introduced into 1.21.

The PCO_2 in the peripheral tissues is approximately 50mm Hg, whereas in the blood entering the peripheral tissues it is approximately 40mm Hg. This difference results in the diffusion of CO_2 from the tissues into the blood in the capillaries of the periphery. When the CO_2 is converted to H_2CO_3 within the erythrocytes and then ionizes, the H^+ ions are buffered by hemoglobin. The production of H^+ ions within erythrocytes and their subsequent buffering by hemoglobin results in a reduced affinity of hemoglobin for oxygen. This leads to a release of O_2 to the peripheral tissues—a phenomenon termed the **Bohr effect** (see Figure 1–2). As CO_2 passes from the tissues to the plasma a minor amount

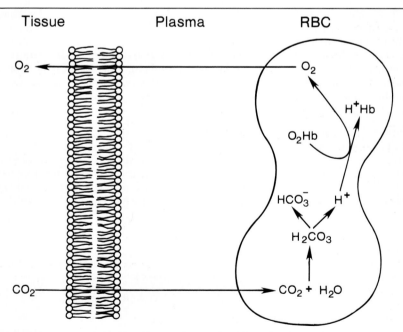

Figure 1–2. Representation of the transport of CO_2 from peripheral tissues to the blood, with the concomitant release of O_2 from the blood to the tissues. (The chloride shift and ionization of CO_2 in the plasma are not shown.) Transport of O_2 from the pulmonary alveoli to the blood, with the resulting exchange of CO_2 to the alveoli, is essentially a reversal of this process. O_2Hb represents oxygenated hemoglobin, H^+Hb represents protonated, deoxygenated hemoglobin.

of carbonic acid takes form and ionizes. The H^+ ions are then buffered predominantly by proteins and phosphate ions in the plasma. As the concentration of bicarbonate ions rises in erythrocytes, an osmotic imbalance occurs. The imbalance is relieved as bicarbonate ion leaves the erythrocytes in exchange for chloride ions from the plasma. This phenomenon is known as the **chloride shift**. Therefore, the majority of the bicarbonate ion formed as CO_2 leaves the peripheral tissues and is transported by the plasma to the lungs.

The partial pressure of O_2 (PO_2) in the pulmonary alveoli is higher than the PO_2 of the entering erythrocytes that contain predominantly deoxygenated hemoglobin. This increased PO_2 leads to oxygenation of hemoglobin and release of H^+ ions from the hemoglobin. The released H^+ ions combine with the bicarbonate ions to form H_2CO_3. Cellular carbonic anhydrase then catalyzes the reverse of reaction 1.19, leading to release of CO_2 from erythrocytes. Owing to the PCO_2 gradient (described above), the CO_2 diffuses from the blood to the alveoli where it is expelled.

The great utility of bicarbonate as a physiological buffer stems from the fact that if excess acid is added to the blood the concentration of bicarbonate ion declines and the level of CO_2 increases. The CO_2 then passes from capillaries in the pulmonary alveoli and is expelled. As a consequence, the H^+ ion concentration drives reaction 1.20 to the left and bicarbonate ion acts as a buffer until all of the hydrogen ion is consumed. Conversely, when excess base is added to the blood, CO_2 is consumed by carbonic acid and replaced by metabolic reactions within the body.

If blood is not adequately buffered, the result may be **metabolic acidosis** or **metabolic alkalosis**. These physiological states can be reached if a metabolic defect results in the inappropriate accumulation or loss of acidic or basic compounds. These compounds may be ingested, or they may accumulate as metabolic by-products such as **acetoacetic acid** and **lactic acid**. Both of these will ionize, thereby increasing the level of H^+ ions that in turn remove bicarbonate ions from the blood and alter blood pH. The predominant defect in acid or base elimination arises when the excretory system of the kidneys is impaired. Alternatively, if the lungs fail to expel accumulated CO_2 adequately and CO_2 accumulates in the body, the result will be **respiratory acidosis**. If a decrease in PCO_2 within the lungs occurs, as during hyperventilation, the result will be **respiratory alkalosis**.

Role of the Kidneys in Acid-Base Balance

The kidneys function to filter the plasma that passes through the nephrons. Filtration of the plasma occurs in the glomerular capillaries of the nephron. These capillaries allow the passage of water and low molecular weight solutes (less than 70 kDa) into the **capsular space**. The filtrate then passes through the **proximal and distal convoluted tubules** where reabsorption of water and many solutes takes place. In the course of **glomerular filtration and tubule reabsorption** the composition of the plasma changes generating the typical composition of urine (Table 1–1). From a biochemical standpoint the kidneys serve important roles in the regulation of **plasma acid-base balance and the elimination of nitrogenous wastes**.

Sodium Bicarbonate Reabsorption

Regulation of plasma acid-base balance is primarily effected within the kidney through control over HCO_3^- reabsorption and secretion of H^+. Secretion of H^+, in excess of its capacity to react with HCO_3^- in the tubular lumenal fluid, requires the presence other buffers (see below). The generation of HCO_3^- and H^+ occurs by dissociation of **carbonic acid** (H_2CO_3), formed in the tubule cells from H_2O and CO_2, through the action of **carbonic anhydrase** (Figure 1–3). Secretion of H^+ into the lumen

Table 1–1. Urinary and plasma concentrations of some physiologically important substances.

Substance	Concentration in		U/P Ratio
	Urine (U)	Plasma (P)	
Glucose (mg/dL)	0	100	0
Na^+ (meq/L)	90	150	0.6
Urea (mg/dL)	900	15	60
Creatinine (mg/dL)	150	1	150

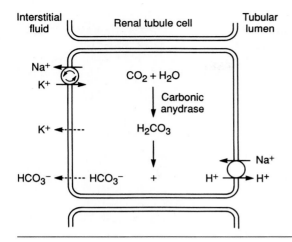

Figure 1–3. Secretion of acid by proximal tubular cells in the kidney. H^+ is transported into the tubular lumen by an antiport in exchange for Na^+. Active transport by the Na^+-K^+ ATPase is indicated by arrows int he circle. Dashed arrows indicate diffusion. (Reproduced with permission, from Ganong, WF: *Review of Medical Physiology*, 16th ed. Appleton & Lange, 1993)

of the tubule is accompanied by an exchange for Na^+ (Figure 1–3). This reabsorption of Na^+ occurs by an antiport mechanism during the exchange for H^+. Reduction in the intracellular concentration of Na^+ occurs by an active transport process involving a Na^+/K^+-ATPase pump which pumps the excess Na^+ into the interstitial fluid (Figure 1–3). The intracellular HCO_3^- then diffuses from the tubule cell into the interstitial fluid (Figure 1–3).

The capacity of the kidney to secrete H^+ is regulated by the maximal H^+ gradient that can form between the tubule and lumen and still allow transport mechanisms to operate. This gradient is determined by the pH of the urine which in humans is near 4.5. The capacity to secrete H^+ would be rapidly reached if it were not for the presence of buffers within the interstitial fluid. The H^+ secreted into the tubular lumen can undergo three different fates depending upon the concentration of the three primary buffers of the interstitial fluid (Figure 1–4). These buffers are HCO_3^-, HPO_4^{2-} and NH_3. Reaction of H^+ with HCO_3^- forms H_2O and CO_2 which diffuse back into the tubule cell. The net result of this process is the regeneration of HCO_3^- within the tubule cell. This process is termed **reabsorption of sodium bicarbonate** (Figure 1–4). The reabsorption of sodium bicarbonate takes place primarily within the proximal convoluted tubules.

Excretion of Acid

As the concentration of HCO_3^- in the tubular lumen drops the pH of the fluid drops due to an increasing concentration of H^+. The pH of the tubular fluid gradually approaches the pK_a for the dibasic/monobasic phosphate buffering system ($pK_a = 6.8$). The excess H^+ reacts with dibasic phosphate (HPO_4^{2-}) forming monobasic phosphate ($H_2PO_4^-$). The $H_2PO_4^-$ so formed is not reabsorbed and its excretion results in the net excretion of H^+ (Figure 1–4). The greatest extent of $H_2PO_4^-$ formation occurs within the distal convoluted tubules and the collecting ducts.

Ammonia Secretion

Buffering of H^+ is also accomplished by reaction with NH_3 to form ammonium ion, NH_4^+ (Figure 1–4). Elimination of NH_4^+ is the major contributory factor in the ability of the body to excrete acid. Because the pK_a of NH_4^+ is 9.3 excretion of acid in this form can be accomplished without lowering the pH of the urine. Additionally important is the fact that excretion of acid in the form of NH_4^+ occurs without depleting Na^+ nor K^+.

Two principal reactions within tubule cells result in the generation of NH_3, conversion of glutamine to glutamate and conversion of glutamate to α-ketoglutarate. These reactions are catalyzed by **glutaminase** and **glutamate dehydrogenase,** respectively (Equations 1.22 and 1.23). Both of these enzymes are abundant in tubule cells. Ammonia is lipid soluble and will diffuse down its concentration gradient out of the tubule cell into the tubular fluid. There it reacts with H^+ to yield NH_4^+ which is excreted in the urine.

$$\text{Glutamine} \rightarrow \text{Glutamate} + NH_4^+ \qquad \textbf{1.22}$$

$$\text{Glutamate} \rightarrow \alpha\text{-Ketoglutarate} + NH_4^+ \qquad \textbf{1.23}$$

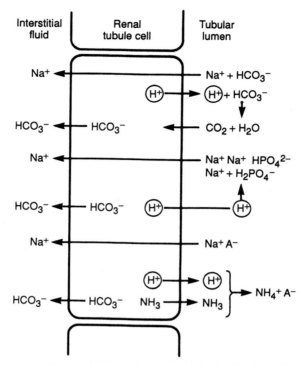

Figure 1–4. Fate of H⁺ secreted into a tulule in exchange for Na⁺. ***Top:*** Reabsorption of filtered bicarbonate. ***Middle:*** Formation of monobasic phosphate. ***Bottom:*** Ammonium formation. Note that in each instance one Na⁺ ion and one HCO₃⁻ ion enter the bloodstream for each H⁺ ion secreted A⁻, anion. (Reproduced with permission, from Ganong, WF: *Review of Medical Physiology*, 16th ed. Appleton & Lange, 1993)

Acidosis and Alkalosis

The kidneys play an important role in the control of acidosis by responding with an increase in the excretion of H^+. When H^+ is excreted as a titratable acid such as $H_2PO_4^-$ or when the anions of strong acids such as acetoacetate are excreted there is a requirement for simultaneous excretion of cations to maintain electrical neutrality. The principal cation excreted is Na^+. As the level of excretable Na^+ is depleted excretion of K^+ increases. In conditions of acidosis the kidney will increase the production of NH_3 from tubular amino acids or amino acids absorbed from the plasma (Figure 1–4). As indicated the NH_3 can diffuse across the tubule cell membrane where it will react with H^+ to form the excretable ammonium ion without a concomitant requirement for cation excretion. This demonstrates that an inability of the kidney to generate NH_3 would rapidly lead to fatal acidosis.

When the kidneys fail to modulate HCO_3^- excretion, metabolic alkalosis will develop. Alkalosis is normally countered quite effectively by the kidney allowing HCO_3^- to freely escape. Alkalosis generally only becomes problematic if the kidneys are restricted in their ability to secrete HCO_3^-. This situation can occur in patients taking diuretics since several of this class of drug cause a reduction in the ability of the kidney to reabsorb an anion (eg Cl^-) concomitant with the reabsorption of Na^+.

Questions

DIRECTIONS (items 1–15): Each numbered item or incomplete statement is followed by answers or by completions of the statement. Select the ONE lettered answer or completion that is BEST in each case.

1. The equilibrium constant for the dissociation of a weak acid of the type:

$$HA \Leftrightarrow H^+ + A^-$$

would be:
   a. $= [A^-][HA]$
   b. $= [HA]/[A^-]$
   c. $= [HA]/[H^+][A^-]$
   d. $= [H^+][A^-]/[HA]$
   e. $= [HA][A^-]/[H^+]$

2. The term "ion product" refers to:
   a. The concentration of H^+ ions in solution.
   b. The product of the concentration of H^+ ions and OH^- ions.
   c. The concentration of OH^- ions in solution.
   d. The negative log of the concentration of OH^- ions in solution.
   e. The negative log of the concentration of H^+ ions in solution.

3. The conjugate base of any given acid is:
   a. The form of the acid exhibiting a net negative charge relative to the acid.
   b. The form of the acid exhibiting a net positive charge relative to the acid.
   c. The form of the acid that will exist in solution at a pH below the pK_a for the acid.
   d. The form of the acid that has accepted H^+ ions from the surrounding medium.
   e. The form of the acid exhibiting electrical neutrality at its pK_a.

4. The definition of a weak acid is:
   a. An acid that has a pK_a in the range of 3.
   b. An acid that exists in equilibrium with its conjugate base at a pH equal to its pK_a.
   c. An acid that readily donates its H^+ ions to the surrounding medium.
   d. An acid that has a pK_a in the range of 10.
   e. An acid that has a pK_a in the range of 11.

5. The term pH is defined as:
   a. The ion product of pure water.
   b. The logarithm of the product of $[H^+]$ and $[OH^-]$.
   c. The negative logarithm of $[HA]$.
   d. The logarithm of $[H^+]$.
   e. The negative logarithm of $[H^+]$.

6. During the titration of a weak acid, the following can be said to be true:
   a. At high pH the acid will be protonated.
   b. At physiological pH the acid will be fully ionized.
   c. It readily releases protons to the surrounding medium.
   d. At a pH equal to the pK_a all of the acid will be protonated.
   e. At a pH equal to the pK_a all of the acid will be ionized.

7. The definition of a good buffer is:
   a. A strong acid with a pK_a near physiological pH.
   b. A weak acid with a pK_a near physiological pH.
   c. A weak base with a pK_a near physiological pH.
   d. An acid that will accept large amounts of H^+ at physiological pH.
   e. An acid that can absorb large OH^- equivalents with little effect on pH.

8. At which point during the titration of a weak acid will the pH change the greatest for the least addition of base?
   a. When exactly half of the acid has been titrated by base.
   b. When the pH reaches the approximate pI of the acid.
   c. When 90% of the acid has been titrated by the base.
   d. When the pH approaches the pK_a of the acid.
   e. When 50% of the acid has been titrated by the base.

9. Which of the following acid/conjugate base pairs would function best as a buffer at physiological pH?
   a. Lactic acid/lactate ion $-pK_a = 3.86$.
   b. Carbonic acid/bicarbonate ion $-pK_a = 6.37$.
   c. Bicarbonate ion/carbonate ion $-pK_a = 10.25$.
   d. Dihydrogen phosphate ion/monohydrogen phosphate ion $-pK_a = 6.86$.
   e. Acetic acid/acetate ion $-pK_a = 4.76$.

10. A 27-year-old male is rushed into the emergency room in a coma and experiencing respiratory depression. Blood pH is determined to be 7.22 and total CO_2 concentration is 26.3 mmol/L. Additional case history indicates the patient is suffering from a narcotic overdose. What is the immediate cause for the acidic nature of the blood?
    a. The narcotic overdose affected the kidneys' ability to adjust to a decreased concentration of dissolved CO_2.
    b. Utilizing the value of total CO_2 concentration the Henderson-Hasselbalch equation allows one to determine that the PCO_2 is elevated. This results in an increase of bicarbonate ion, thereby increasing the H^+ concentration.
    c. The narcotic overdose led to a decreased respiratory rate which led to an increase in dissolved CO_2.
    d. The narcotic overdose led to an increased respiratory rate which led to a decrease in dissolved CO_2.
    e. The narcotic overdose affected the kidneys' ability to adjust to an increased concentration of dissolved CO_2.

11. The Henderson-Hasselbach equation allows one to:
    a. Calculate the pK_a of an acid from the pH of a solution of the acid.
    b. Calculate the pH of a solution of an acid from the pK_a of the acid.
    c. Calculate the molar ratio of an acid and its conjugate base from the pK_a of the acid.
    d. Calculate the molar ratio of an acid and its' conjugate base from the pK_a of the acid and the pH of the solution of the acid.
    e. Calculate the pH of a solution of a base from the pK_a of its conjugate acid.

12. With respect to blood buffering:
    a. The predominant proton donor in plasma is phosphate ion.
    b. The major buffer in erythrocytes is phosphate ion.
    c. Protons from carbonic acid formation in plasma bring about the chloride shift in erythrocytes.
    d. The predominant proton donor in erythrocytes is hemoglobin.
    e. The predominant proton acceptor in erythrocytes is hemoglobin.

13. Carbonic acid has a pK_a of 3.8, yet its ionization plays an important role in the buffering of blood pH. The explanation for this is:
    a. It exists in an equilibrium with its conjugate base, bicarbonate ion.
    b. The pK_a is affected by the concentration of dissolved O_2 in the plasma.
    c. It exists in an equilibrium with dissolved CO_2.
    d. The presence of a high concentration of phosphate ions alters the observed pK_a.
    e. The pK_a is affected by the presence of Cl^- ions within the plasma.

14. As the concentration of CO_2 increases in erythrocytes in the peripheral tissues:
    a. The increased concentration of bicarbonate ion leads to a decrease in PO_2.
    b. The increased concentration of bicarbonate ion leads to an exchange for Cl^- ions in the plasma.
    c. The resultant bicarbonate ion which is generated facilitates a release of O_2 from hemoglobin.
    d. The increased concentration of bicarbonate ion leads to an increase in PO_2.
    e. None of the above is correct.

15. The Bohr effect is:
    a. The effect of an increase in H^+ ion concentration in erythrocytes.
    b. The effect of an increase in H^+ ion concentration in the plasma.
    c. The effect of an increase in CO_2 concentration in peripheral tissues.

   d. The result of an reduced affinity of hemoglobin for CO_2.
   e. The result of an increased affinity of hemoglobin for O_2.

DIRECTIONS (items 16–23): Match the lettered descriptive phrase with the descriptions below. Some of the lettered answers may be used more than once or not at all.
   a. Excreted in exchange for resorbed K^+.
   b. Necessary for resorption of bicarbonate.
   c. Result of diuretic repression of anion resorption.
   d. Result of inhibition of carbonic anhydrase.
   e. Antiport transporter.
   f. Active transporter.
   g. Excreted without concomitant loss of K^+.
   h. Excreted in exchange for resorbed Na^+.
   i. Required for excretion of H^+, without depletion of tubular Na^+.

16. Carbonic anhdrase.

17. Bicarbonate.

18. Ammonia.

19. Metabolic acidosis.

20. Metabolic alkalosis.

21. Monobasic phosphate.

22. Glutamate dehydrogenase.

23. Na^+/K^+ ATPase.

Answers

1. d	**9.** d	**17.** h
2. b	**10.** c	**18.** g
3. a	**11.** d	**19.** d
4. c	**12.** e	**20.** c
5. e	**13.** c	**21.** h
6. b	**14.** b	**22.** i
7. e	**15.** a	**23.** f
8. c	**16.** b	

2 Thermodynamic Principals Applied to the Life Process

Objectives

- To review thermodynamic principals as they apply to the life process.
- To characterize the thermodynamic nature of living systems.
- To define the 1st and 2nd laws of thermodynamics as they apply to living systems.
- To define free energy and its importance to biochemical processes.
- To define the relationship between free energy, entropy, enthalpy, and internal energy of biological reactions.
- To define the relationship between free energy, internal energy, and entropy for any reaction or process at constant temperature and pressure.
- To define the relationship between the value of ΔG and the spontaneity of a reaction.
- To define the terms exergonic and endergonic as they apply to biological reactions.
- To list the conditions that define the standard state for biological reactions.
- To define the relationship between standard free energy, the concentration of reactants and products and the free energy available from a reaction system.
- To compare the standard free energy of metabolic reactions with the actual free energy available under *in vivo* conditions.
- To derive the relationship between the equilibrium constant and the standard free energy of a reaction.

Concepts

All chemical reactions and life processes are ultimately limited by the laws of thermodynamics. Humans expend the greatest proportion of their available energy simply to maintain a thermodynamic steady state. In order to understand the life process, we must concern ourselves with the rules of thermodynamics. This chapter reviews thermodynamic principals as they apply to the human body.

Classical thermodynamics usually recognizes two basic kinds of systems: open and closed. The human body is an open thermodynamic system—that is, one characterized by a flow of both mass (food and excreta in animals) and energy between the system and its environment. In contrast, a closed thermodynamic system (such as a bomb calorimeter) exchanges only energy with its environment. This chapter reviews the first and second laws of thermodynamics, which govern energy interconversions in the living organisms.

The first law of thermodynamics states that the total energy of a system plus that of its environment remains constant. This law declares that energy is neither created nor destroyed, although energy *can* be converted between a system and its surroundings. The human body, as a subsystem of the universe uses energy interconversions to maintain the normal living state (homeostasis). As open thermodynamic systems, we consume and use the energy stored in complex foods to maintain life processes. At the same time, we do mechanical work, we radiate heat energy, and we excrete simpler chemical products such as carbon dioxide, water, feces, and urea. Work energy and radiative energy represent exchanges of energy with our environment. The ingestion of food and excretion of metabolic products represent exchanges of mass.

The process of consuming complex substances from our environment and excreting simpler breakdown products is also a reflection of the second law of thermodynamics. This states that a system and its surroundings always proceed to a state of maximum disorder or maximum entropy (entropy and disorder are synonymous in thermodynamics). In the absence of the transfer of mass (food) from our surroundings into the human body we soon starve. In this case, the system proceeds to its state of maximum entropy with death being one point on the path to maximum entropy.

In the conversion of complex foods such as glucose ($C_6(H_2O)_6$) to simpler products such as CO_2 and H_2O, energy conversions occur. It is these energy changes that are available to perform the chemical and physical work that keeps us alive; the energy associated with them is known as free energy. The entropy change associated with any reaction is qualitatively reflected by a change in the ordered spatial relationship of atoms as reactants transform into products. In our example it is clear that the atoms of glucose ($C_6(H_2O)_6$) are much more highly ordered than the product atoms in CO_2 and H_2O.

In the same chemical reaction, we find another kind of energy transformation that changes with reaction state. This is known as **enthalpy (H)** or heat energy.

RELATIONSHIPS BETWEEN ENTROPY, ENTHALPY, AND FREE ENERGY

The relationship between the different forms of energy in a system, or in a reaction, going from one state to another is given by equation 2.1, known as the **Gibbs equation**:

$$\Delta G = \Delta H - T\Delta S \qquad\qquad\qquad \textbf{2.1}$$

The Gibbs equation applies to all reactions and processes. A general example is the equilibrium reaction, 2.2:

$$A + B \Leftrightarrow C + D \qquad\qquad\qquad \textbf{2.2}$$

In the Gibbs equation T is the absolute temperature of the system, and it is clear that the magnitude of the entropy effect is dependent on temperature ($T\Delta S$). However, enthalpy, or heat content of the system, is more closely tied to temperature. If the temperature of the system is constant, as in the human body, molecular motions and collisions remain the same from state to state and enthalpy changes are often negligible. At constant temperature the principal energy component of the enthalpy term is the energy change associated with chemical bonds. The energy is known as **Internal Energy.** The relationship between internal energy and enthalpy, at constant temperature, is shown in equation 2.3.

$$\Delta H = \Delta E \qquad\qquad\qquad \textbf{2.3}$$

In equation 2.3, ΔE is known as the internal energy; for a chemical reaction, this can be viewed as the sum of all of the energy stored in the chemical bonds and atomic orbitals of the reactants and products.

Combining equations 2.1 and 2.3 yields:

$$\Delta G = \Delta E - T\Delta S \qquad\qquad\qquad \textbf{2.4}$$

Equation 2.4 states that as a consequence of a reaction or process going from one constant temperature state to another the available useful energy (ΔG) equals the difference between the changes in internal energy (ΔE) and the changes in organization (ΔS) of the atoms involved in the reaction.

The sign ($+$ or $-$) and magnitude of each thermodynamic term in the Gibbs' equation is important in determining whether the reaction or process described by the equation will proceed spontaneously. If $T\Delta S$ is positive and greater than ΔE, then ΔG will be negative and reactions, such as 2.2, will proceed spontaneously to the right as written. Reactions with a negative ΔG are called exergonic.

Reactions having a positive ΔG are called endergonic. These require the input of energy in some form, in order for the reaction to proceed in the direction written; by contrast, exergonic reactions proceed spontaneously with a decrease in free energy in the system. In biochemical systems the free energy decrease of exergonic reactions is usually associated with a corresponding increase in entropy, although internal energy changes can be important. Table 2–1 summarizes the relationships discussed above.

As an example of these thermodynamic relationships, we can consider oxidation of one mole of glucose in water (reaction 2.5) proceeding

$$C_6(H_2O)_6 + 6O_2 \qquad\qquad\qquad \textbf{2.5}$$

to the equilibrium state under standard conditions according to reaction 2.6.

Table 2–1. Basic thermodynamic principles of exergonic and endergonic reactions.

Exergonic Reactions	Endergonic Reactions
ΔG is negative	ΔG is positive
K_{eq} is > 1.	K_{eq} is < 1.
Spontaneous as written	Spontaneous in the reverse direction

$$C_6(H_2O)_6 + 6O_2 \Leftrightarrow 6CO_26 + 6H_2O \qquad \Delta G^\circ + -686 \text{ kcal/mole} \qquad 2.6$$

Standard free energy changes of systems operating under classical chemical conditions are denoted by the symbol ΔG°. However, biochemists have defined slightly different standard conditions to reflect those under which the life process exits. Standard biological conditions are defined as 760 mm Hg (1 atmosphere), pH 7, 298° K, 55.5 molar water and 1 molar concentration of all other reactants and products. The symbols for energy changes that occur in these condition are given a prime mark to signify that they refer to reactions taking place under biological standard conditions. For example, the notation of standard free energy change for reaction 2.6 is ΔG°. The standard conditions for the reaction described by equation 2.6 are 1 M glucose, 55.5 M water, 1 atmosphere oxygen, and carbon dioxide, pH 7.0, 298K. Thus, the biochemical standard free energy (ΔG°) is that available as the reaction proceeds from the biological standard state to the equilibrium state under biological standard conditions.

ΔG° for reaction 2.6 is approximately $-686,000$ calories per mole. The units of free energies are either calories per mole (cal/mol) or Joules per mole (J/mol). Since both terms are currently in common use, it is important to recall that 1 calorie is equal to 4.184 Joules. The value of -686 kcal/mol ($-2,870$ kJ/mol) for glucose oxidation is a large negative standard free energy change indicating that the reaction will proceed vigorously in the forward direction, which is also the direction in which it proceeds in a living organism.

As a consequence of the above considerations it follows that energy available from a reaction, with an initial state different from the standard state, will not be equal to that specified as ΔG°. When the initial reactant concentrations are other than 1 Molar, the free energy of a reaction is given by:

$$\Delta G' = \Delta G^\circ + 2.303 \text{ RT log [products]/[reactants]} \qquad 2.7$$

In this expression R is the gas constant, 1.987 calories per mole per degree Kelvin, or 8.134 Joules per mole per degree Kelvin. An important observation related to equation 2.7 is that when the reaction is at its standard state the ratio of reactants to products is 1, the log of 1 is zero and $\Delta G = \Delta G^\circ$.

Equation 2.7 reflects, in mathematical terms, the fact that the actual energy available from a reaction depends on its standard free energy plus an energy contribution determined by the prevailing concentration of reactants and products. Clearly, in the human body the concentrations of reactants and products for metabolic reactions are almost never that of the standard state. Therefore, *in vivo* ΔG° is almost never representative of the actual energy available from a metabolic reaction.

Reactions that are at equilibrium cannot proceed spontaneously to any new reaction state, and thus they cannot provide any useful work. Thermodynamically, it can be said that the cause of death is that the reactions responsible for maintaining life have come to equilibrium. A more precise way of expressing these ideas is to note that **at equilibrium $\Delta G' = 0$.** As a consequence of the latter relationship, equation 2.7 can be algebraically modified to yield equation 2.8.

$$\Delta G^\circ = -2.303 \text{ RT log [[products]/[reactants]]} \qquad 2.8$$

Since, at equilibrium, the ratio of [product]/[reactant] is equal to the equilibrium constant (K_{eq}), the latter equation is often written as:

$$\Delta G^\circ = -2.303 \text{ RT log}K_{eq} \qquad 2.9$$

From equation 2.9 it is apparent that at 298° K ΔG° is equal to K_{eq} multiplied by a collection of constants. In equation 2.10 we evaluate all the constants that affect K_{eq} and simplify equation 2.10 as shown in 2.11.

$$\Delta G^\circ = -2.303 \times 8.134 [\text{Joules/(degrees)(Mol)}] \times 298 \text{ degrees} \times \log K_{eq} \qquad \textbf{2.10}$$

$$\Delta G^\circ = -5582 [\text{Joules/Mol}] \times \log K_{eq} \qquad \textbf{2.11}$$

Thus at 25° the standard free energy of a biological reaction is simply -5.58 kJ/Mol multiplied by the log of the equilibrium constant. The equivalent value for body temperature, 310°K, is 5.80 kJ/Mol.

Questions

DIRECTIONS (items 1–4): Each numbered item or incomplete statement is followed by answers or by completions of the statement. Select the ONE lettered answer or completion that is BEST in each case.

1. Open thermodynamic systems are:
    a. Systems that exchange only energy with their environment.
    b. Systems that exchange energy and mass with their environment.
    c. Characteristic of living organisms.
    d. Two of the above are correct.
    e. None of the above are correct.

2. As a thermodynamic system, an individual human:
    a. Obeys the first and second laws of thermodynamics.
    b. Transfers energy with the environment only in the form of mass.
    c. Decreases the entropy of the universe.
    d. Obeys only the second law of thermodynamics.
    e. Two of the above are correct.

3. Examples of processes that decrease the entropy of the universe include:
    a. Synthesis of glucose from carbon dioxide and water.
    b. Formation of DNA from precursor nucleotides.
    c. Photosynthesis.
    d. All of the above are correct.
    e. None of the above are correct.

DIRECTIONS (items 4–7): Match descriptions with the correct term from the lettered list. Each term may be used once, more than once, or not at all.
    a. Free Energy.
    b. Entropy.
    c. Enthalpy.
    d. Second Law of Thermodynamics.
    e. None of the above.

4. Synonymous with disorder in the universe.

5. A thermodynamic variable that is usually considered to be unchanging when describing the reactions taking place in humans.

6. A reflection of the human processes involved in consuming complex molecules and excreting simpler breakdown products.

7. The energy of metabolism that is available to perform chemical synthesis, maintain ion imbalances across membranes and carry out other work processes.

DIRECTIONS (items 8–11): Each numbered item or incomplete statement is followed by answers or by completions of the statement. Select the ONE lettered answer or completion that is BEST in each case.

8. From the data in the box below choose the best answer.

DATA:
Related to the reaction:

$$\text{Glucose} + \text{ATP} \Leftrightarrow \text{Glucose} - 6 - \text{phosphate} + \text{ADP}$$

The following is known:

Glucose − 6 − phosphate + H$_2$O ⇔ Phosphate + Glucose	**$\Delta G° = -3.3$ kcal/mole**
ATP + water ⇔ ADP + Inorganic Phosphate	**$\Delta G° = -7.3$ kcal/mole**

---

    a. The equilibrium constant is a number less than 1.
    b. The equilibrium constant is about 820.
    c. The equilibrium constant is about 8.5.
    d. The equilibrium constant is larger than that for the hydrolysis of ATP.
    e. Glucose will not be significantly phosphorylated.

9. The standard free energy change for a reaction A going to B at 25° is:

$$\text{A} \Leftrightarrow \text{B} \qquad \Delta G° = +1400 \text{ cal/mole}$$

What is the value closest to the equilibrium ratio of A to B?
    a. 1 to 10.
    b. 1 to 100.
    c. 10 to 1.
    d. 100 to 1.
    e. 1 to 1.

10. The value of the standard free energy change for a reaction at 27° is $-6,900$ J/mole. The value of K_{eq} is.
    a. 10^5.
    b. $10^{2.5}$.
    c. 10^{-5}.
    d. $10^{-2.5}$.
    e. $10^{3.8}$.

11. Metabolic energy available to perform biosynthesis, maintain ion balances across membranes and perform other work processes.
    a. Entropy.
    b. Enthalpy.
    c. Free energy.
    d. Standard free energy.
    e. None of the above.

DIRECTIONS (items 12–15): Match the following descriptions with the best response from the lettered list. Each term may be used once, more than once, or not at all.
    a. About 0.239 calories.
    b. Free energy is negative.
    c. Endergonic reaction.
    d. ΔS.
    e. None of the above.

12. ΔG is positive.

13. Exergonic reaction.

14. A Joule.

15. Enthalpy change.

DIRECTIONS (items 16–19): Choose the most appropriate response from among the lettered choices below.

    a. $A + B \rightarrow C + D$ $\Delta G = 2{,}400$ calories/mole.

    b. $Z + Y \rightarrow W + X$ $\Delta G = -2{,}400$ calories/mole.

    c. glucose $+ O_2 \rightarrow CO_2 + H_2O$ $\Delta G = -686$ kcal/mole.

    d. $CO_2 + H_2O \rightarrow$ glucose $+ O_2$ $\Delta G = 686$ kcal/mole.

    e. glucose $+ O_2 \rightarrow CO_2 + H_2O$ $\Delta G = 0.$

16. The most exothermic reaction.

17. The most endothermic reaction.

18. The largest equilibrium constant.

19. At equilibrium.

Answers

1. d	**8.** b	**15.** e
2. a	**9.** c	**16.** c
3. e	**10.** a	**17.** d
4. b	**11.** c	**18.** c
5. e	**12.** c	**19.** e
6. d	**13.** b	
7. a	**14.** e	

Part II: Amino Acids and Proteins, Structure and Function

3

Amino Acids

Objectives

- To describe the structure of the amino acids found in proteins.
- To define the functional consequences of the physical character of several key amino acid R-groups.
- To illustrate the role of amino acid analysis in the diagnosis of certain disease states.
- To define the nature of the bond linking amino acids in proteins.

Concepts

Amino acids constitute the building blocks of peptides and proteins. All peptides and proteins are polymers of the α-amino acids, of which 20 different types are found in mammalian proteins. The types of amino acids, as well as their spatial orientations within proteins, control the overall three-dimensional structures of proteins, and consequently their functions.

Several other amino acids and amino acid derivatives are found in the body, either free or in combined states (ie, not associated with peptides or proteins). These non-protein amino acids perform specialized functions, for example, the aspartate derivative N-methyl-D-aspartate, **(NMDA)** is a neurotransmitter. Several of the amino acids found in proteins also serve functions distinct from the formation of peptides and proteins, such as tyrosine in the formation of thyroid hormones, glutamine, aspartate, and glycine in the formation of purine nucleotides, glutamine in the formation of pyrimidine nucleotides, glycine in the formation of porphyrins and glutamate acting as a neurotransmitter.

CHEMICAL NATURE OF AMINO ACIDS

The α-amino acids in peptides and proteins (excluding proline) consist of an α-carboxylic acid (α-**COOH**) and an α-amino (α-**NH$_2$**) functional group, both attached to a tetrahedral carbon atom known as the α-carbon. Distinct R-groups (Table 3–1), that distinguish one amino acid from another, also are attached to the α-carbon (except in the case of glycine, where the R-group is hydrogen). The fourth substitution on the tetrahedral α-carbon of amino acids is hydrogen.

Amino Acid Classifications

Each of the 20 α-amino acids found in proteins can be distinguished by the R-group substitution on the α-carbon atom. There are two broad classes of amino acids, those in which the R-group is hydrophobic and those in which it is hydrophilic (Table 3–2). Hydrophobicity and hydrophilicity refer to the tendency of molecules to avoid or to prefer, respectively, an aqueous environment. The hydrophobic amino acids do not ionize nor participate in the formation of H-bonds, whereas the hydrophilic amino acids readily ionize and are often involved in the formation of H-bonds with other amino acids or with surrounding water molecules.

Table 3–1. L-α-Amino acids present in proteins.*

Name	Symbol	Structural Formula
With Aliphatic Side Chains		
Glycine	Gly [G]	$H - CH - COO^-$ $\quad\;\; {}_+NH_3$
Alanine	Ala [A]	$CH_3 - CH - COO^-$ $\qquad\;\; {}_+NH_3$
Valine	Val [V]	H_3C $\quad\;\; CH - CH - COO^-$ $H_3C \qquad\;\; {}_+NH_2$
Leucine	Leu [L]	H_3C $\quad\;\; CH - CH_2 - CH - COO^-$ $H_3C \qquad\qquad\; {}_+NH_3$
Isoleucine	Ile [I]	CH_3 CH_2 $CH - CH - COO^-$ $CH_3 \quad {}_+NH_3$
With Side Chains Containing Hydroxylic (OH) Groups		
Serine	Ser [S]	$CH_2 - CH - COO^-$ $OH \quad\;\; {}_+NH_3$
Threonine	Thr [T]	$CH_3 - CH - CH - COO^-$ $\qquad\; OH \quad\;\; {}_+NH_3$
Tyrosine	Tyr [Y]	See below.
With Side Chains Containing Sulfur Atoms		
Cysteine†	Cys [C]	$CH_2 - CH - COO^-$ $SH \quad\;\; {}_+NH_3$
Methionine	Met [M]	$CH_2 - CH_2 - CH - COO^-$ $S - CH_3 \qquad {}_+NH_3$
With Side Chains Containing Acidic Groups or Their Amides		
Aspartic acid	Asp [D]	$^-OOC - CH_2 - CH - COO^-$ $\qquad\qquad\quad {}_+NH_3$
Asparagine	Asn [N]	$H_2N - C - CH_2 - CH - COO^-$ $\qquad\;\; \overset{\|}{O} \qquad\quad {}_+NH_3$
Glutamic acid	Glu [E]	$^-OOC - CH_2 - CH_2 - CH - COO^-$ $\qquad\qquad\qquad\quad {}_+NH_3$
Glutamine	Gln [Q]	$H_2N - C - CH_2 - CH_2 - CH - COO^-$ $\qquad\;\; \overset{\|}{O} \qquad\qquad\quad {}_+NH_3$

Table 3–1. L-α-Amino acids present in proteins (cont'd).*

Name	Symbol	Structural Formula
With Side Chains Containing Basic Groups		
Arginine	Arg [R]	$H-N-CH_2-CH_2-CH_2-CH-COO^-$ with $C=NH_2$, $+NH_2$ and $+NH_3$
Lysine	Lys [K]	$CH_2-CH_2-CH_2-CH_2-CH-COO^-$ with $+NH_3$ and $+NH_3$
Histidine	His [H]	imidazole ring $-CH_2-CH-COO^-$ with $+NH_3$
Containing Aromatic Rings		
Histidine	His [H]	See above.
Phenylalanine	Phe [F]	benzene ring $-CH_2-CH-COO^-$ with $+NH_3$
Tyrosine	Tyr [Y]	$HO-$ benzene ring $-CH_2-CH-COO^-$ with $+NH_3$
Tryptophan	Trp [W]	indole ring $-CH_2-CH-COO^-$ with $+NH_3$
Amino Acids		
Proline	Pro [P]	pyrrolidine ring with $+N$, H_2, COO^-

*Except for hydroxylysine (Hyl) and hydroxyproline (Hyp), which are incorporated into polypeptide linkages as lysine and proline and subsequently hydroxylated, specific transfer RNA molecules exist for all the amino acids listed in Table 3-1. Their incorporation into proteins is thus under direct genetic control.
†Cystine consists of 2 cysteine residues linked by a disulfide bond.

$$^-OOC-CH(NH_3^+)-CH_2-S-S-CH_2-CH(NH_3^+)-COO^-$$

Acid-base Properties

The α-COOH and α-NH$_2$ groups in amino acids are capable of ionizing (as are the acidic and basic R-groups of the amino acids). As a result of their ionizability the following ionic equilibrium reactions may be written:

$$R\text{-}COOH \leftrightarrow R\text{-}COO^- + H^+ \tag{3.1}$$

$$R\text{-}NH_3^+ \leftrightarrow R\text{-}NH_2 + H^+ \tag{3.2}$$

Table 3–2. Classification of L-α-amino acids based upon hydrophibicity and hydrophilicity.

Hydrophobic	Hydrophilic
Alanine	Arginine
Glycine	Asparagine
Isoleucine	Aspartic acid
Leucine	Cysteine
Methionine	Glutamic acid
Phenylalanine	Glutamine
Proline	Histidine
Tryptophan	Lysine
Valine	Serine
	Threonine
	Tyrosine

The equilibrium reactions, as written, demonstrate that amino acids contain at least two weakly acidic groups. However, the carboxyl group is a far stronger acid than the amino group. At physiological pH (around 7.4) the carboxyl group will be unprotonated and the amino group will be protonated. An amino acid with no ionizable R-group would be electrically neutral at this pH. As described in Chapter 1, this species is termed a **zwitterion** (see Figure 3–1).

As is true for most organic acids, the acidic strength of the carboxyl, amino and ionizable R-groups in amino acids can be defined by the association constant, K_a, or more commonly the negative logarithm of K_a, the pK_a. The **net charge** (the algebraic sum of all the charged groups present) of any amino acid, peptide or protein, will depend upon the pH of the surrounding aqueous environment. As the pH of a solution of an amino acid or protein changes, so too does the net charge. This phenomenon can be observed during the titration of any amino acid or protein (see Figure 3–2). When the net charge of an amino acid or protein is zero, the pH will be equivalent to the **pI** (see Chapter 1).

Functional Significance of R-Groups

In solution, the amino acid R-groups dictate the structural orientation and the function of peptides and proteins. Hydrophobic amino acids are generally located in the interior of proteins, shielded from direct contact with water. Conversely, hydrophilic amino acids are generally found on the exterior of proteins as well as in the active centers of enzymatically active proteins. Indeed, the very nature of certain amino acid R-groups (Table 3–3) is what allows enzyme reactions to occur.

The imidazole ring of histidine allows it to act as either a proton donor or a proton acceptor at physiological pH. Hence, it is frequently found in the reactive center of enzymes. Equally important is the ability of histidines in hemoglobin to buffer the H^+ ions from carbonic acid ionization in red blood cells. It is this property of hemoglobin that allows it to exchange O_2 and CO_2 at the tissues or lungs, respectively (see Chapters 1 and 8).

The primary alcohol of serine and threonine and the thiol (-SH) of cysteine allow these amino acids to act as nucleophiles during enzymatic catalysis. Additionally, the thiol of cysteine is able to form a **sulfhydryl bond** with other cysteines:

$$\textbf{Cysteine-SH + HS-Cysteine} \rightarrow \textbf{Cysteine-S-S-Cysteine} \qquad \textbf{3.3}$$

This simple disulfide is identified as **cystine**. The formation of disulfide bonds between cysteines that are present within proteins is important for the formation of active structural domains in a large number of proteins. Disulfide bonding between cysteines in different polypeptide chains of oligomeric proteins plays a crucial role in ordering the structure of complex proteins, such as the insulin receptor (see Chapter 37).

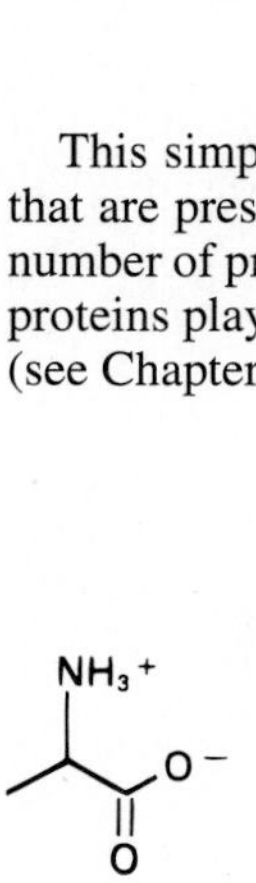

Figure 3–1. Isoelectric or "zwitterionic" structure of alanine. Although charged, the zwitterion bears no *net* charge and hence does not migrate in a direct current electric fluid. (Reproduced with permission, from Murray, RK: *Harper's Biochemistry*, 23rd ed. Appleton & Lange, 1993.)

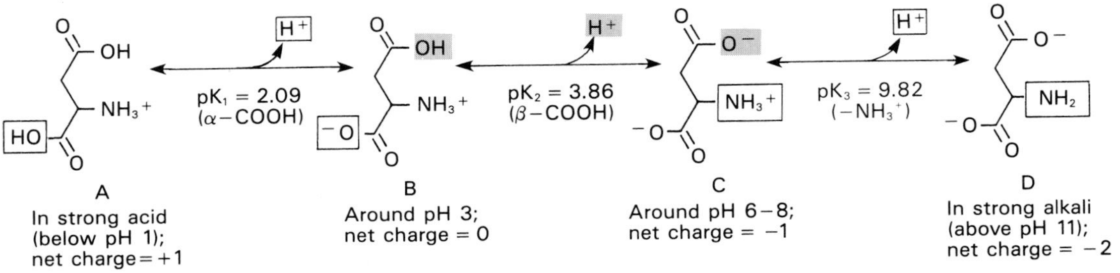

Figure 3–2. Protonic equilibria for aspartic acid. (Reproduced with permission, from Murray, RK: *Harper's Biochemistry*, 23rd ed. Appleton & Lange, 1993.)

Optical Properties

A tetrahedral carbon atom with 4 different substituents is said to be structurally asymmetric or **chiral**. The one amino acid not exhibiting chirality is glycine, since its "R-group" is a hydrogen atom. Chirality describes the so-called handedness of a molecule—its ability to rotate the plane of polarized light either to the right (**dextrorotatory, signified by D**) or to the left (**levorotatory, signified by L**). All the amino acids in proteins exhibit the same absolute steric configuration as **L-glyceraldehyde**. Therefore, they are all L-α-amino acids. D-amino acids are never found in proteins, although they exist in nature; for instance, they are often found in polypetide antibiotics.

The aromatic R-groups in amino acids absorb ultraviolet light with an absorbance maximum in the range of 280nm. The ability of proteins to absorb ultraviolet light is due predominantly to the presence of the amino acid tryptophan, which strongly absorbs ultraviolet light at 280nm (Figure 3–3).

Table 3–3. Characteristic pK'_a Values for the Common Acid Groups in Proteins

Where acid group is found	Acid form		Base form	Approximate pK_a range for group
NH$_2$ terminal residue in peptides, lysine	R—NH$_3^+$ Amino	$\rightleftharpoons$	R—NH$_2$ + H$^+$ Amine	7.6–10.6
COOH terminal residue in peptides, glutamate, aspartate	R—COOH Carboxylic acid	$\rightleftharpoons$	R—COO$^-$ + H$^+$ Carboxylate	3.0–5.5
Arginine	R—NH—C$\overset{+}{=\!\!=\!\!=}$NH$_2$ | NH$_2$ Guanidinium	$\rightleftharpoons$	R—NH—C$=$NH + H$^+$ | NH$_2$ Guanidino	11.5–12.5
Cysteine	R—SH Thiol	$\rightleftharpoons$	R—S$^-$ + H$^+$ Thiolate	8.0–9.0
Histidine	R—C$=$CH HN $^+$NH C H Imidazolium	$\rightleftharpoons$	R—C$=$CH HN N + H$^+$ C H Imidazole	6.0–7.0
Tyrosine	R ⬡ OH Phenol	$\rightleftharpoons$	R ⬡ O$^-$ + H$^+$ Phenolate	9.5–10.5

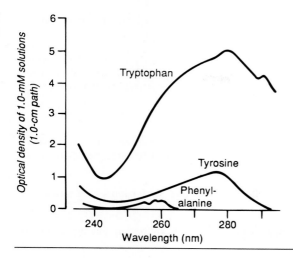

Figure 3–3. The ultraviolet absorption spectra of tryto-phan, tyrosine, and phenylalanine. (Reproduced with permission, from Murray, RK: *Harper's Biochemistry*, 23rd ed. Appleton & Lange, 1993.)

The Peptide Bond

The condensation reaction between an α-amino group of one amino acid and the α-carboxyl group of another leads to the formation of an amide bond, the peptide bond (Figure 3–4). Peptides are small, consisting of just a few amino acids. The simplest form, a **dipeptide**, contains a single peptide bond. A number of hormones and neurotransmitters are peptides, as are several antibiotics and antitumor agents. Proteins, as polypeptides, exist in a great variety of lengths.

The presence of the carbonyl group in the peptide bond allows for stabilization of the electron resonance, such that the peptide bond exhibits rigidity not unlike the typical carbon-carbon double bond. The peptide bond is, therefore, said to have **partial double-bond character**.

CLINICAL SIGNIFICANCE

Aminoacidurias

Numerous clinical disorders are associated with a rise in amino acid levels in the plasma or urine. These conditions are called aminoacidurias.

A. Phenylketonuria (PKU): Phenylketonuria is manifested by the appearance of high levels of phenylalanine, phenylpyruvate, and phenyllactate in the plasma and urine. This condition results from a defect in the gene that encodes **phenylalanine hydroxylase**, the enzyme responsible for catalyzing the conversion of phenylalanine to tyrosine. The prevalence of this condition has led to standard blood testing for PKU at birth; if left undiagnosed, PKU leads to severe mental retardation. When diagnosed, PKU can be controlled simply by diet.

B. Hartnup Disease: Hartnup disease is the result of a defect in the epithelial cell transport of monoamino monocarboxylic acids, leading to their accumulation in the urine. The physical symptoms are related to a deficiency in tryptophan and include a pellagra-like rash (pellagra results from a deficiency in vitamin B_3 which can be derived from tryptophan; see Chapter 36) and cerebellar ataxia (due to increased levels of indole that result from bacterial degradation of unabsorbed tryptophan).

C. Fanconi's Syndrome: Fanconi's syndrome is associated with hypophosphatemia and the elevated excretion of amino acids and glucose. This syndrome is the result of abnormal resorption of glucose, amino acids, and phosphate by the kidneys. Chemotherapeutic regimens involving the use of **ifos-**

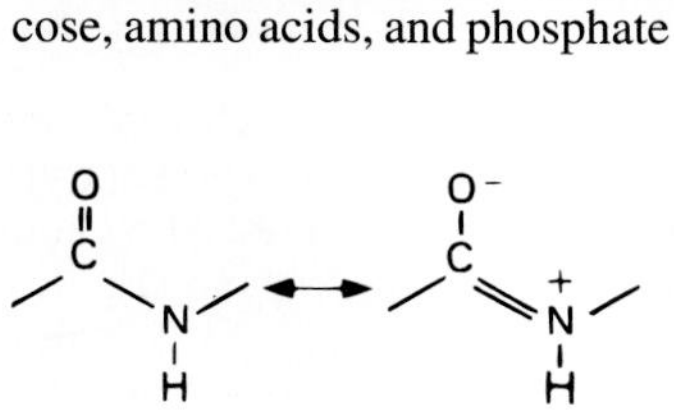

Figure 3–4. Resonance stabilization of the peptide bond confers partial double-bond character, and hence rigidity, on the C—N bond. (Reproduced with permission, from Murray, RK: *Harper's Biochemistry*, 23rd ed. Appleton & Lange, 1993.)

famide (eg, in treatment of pediatric malignant mesenchymal tumors of the head and neck) can lead to the development of Fanconi's syndrome because of its effects on the kidneys.

D. Cystinuria: Cystinuria is manifested by elevated excretion of the amino acids cystine, lysine, arginine, and urea cycle derived ornithine. The condition results from a genetically transmitted defect in membrane transport of these amino acids. An additional symptom of cystinuria is the formation of kidney stones that contain cystine.

Tyrosinemias and Other Disorders

A. Tyrosinemias: Excretion of elevated levels of tyrosine and its related metabolites can indicate tyrosinemias, which result from genetic defects in enzymes involved in tyrosine metabolism (see Chapter 25).

1. Tyrosinemia type I (hepatorenal tyrosinemia) results in renal tubular dysfunction, liver failure, and rickets.

2. Tyrosinemia type II (oculocutaneous tyrosinemia) results in eye and skin lesions and mental retardation.

B. Alcaptonuria: Alcaptonuria is the result of a deficiency in **homogentisate oxidase** (involved in tyrosine metabolism, Chapter 25) and leads to excretion of the tyrosine metabolite, homogentisic acid, in the urine. Upon exposure to air the urine turns an intense dark color (brown), owing to oxidation of the homogentisate.

C. Maple Syrup Urine Disease: Defects in the enzyme **branched chain ketoacid oxidase**, which is involved in the metabolism of leucine, valine, and isoleucine, results in maple syrup urine disease. This disease is characterized by a "maple syrup" odor in the urine and can cause severe mental retardation.

Questions

DIRECTIONS (items 1–8): Each numbered item or incomplete statement is followed by answers or by completions of the statement. Select the ONE lettered answer or completion that is BEST in each case.

1. The ability of proteins to absorb ultraviolet light is a property of the:
    a. Aromatic amino acids.
    b. Aliphatic amino acids.
    c. Hydroxulated amino acids.
    d. Acidic amino acids.
    e. Basic amino acids.

2. Which of the following are characteristics of a peptide bond?
    a. The —C—N— bond exhibits double-bond character.
    b. The R-groups of the peptide bonded amino acids exhibit free rotation, whereas the peptide bond itself exhibits restricted rotation.
    c. The —C=O and —N—H bonds lie in a plane parallel to each other.
    d. The —C=O and —N—H groups of the bond form H-bonds with each other and contribute to the rigidity of the peptide bond.
    e. Two of the above are correct.

3. A 2-year-old had been treated for an embryonal sarcoma with multiagent chemotherapy, including ifosfamide. Following therapy the child exhibited elevated urinary excretion of tryptophan and presented with phosphopenic rickets. The rickets symptoms were ameliorated by administration of phosphorous and vitamin D_3. The most likely diagnosis for these symptoms is:
    a. Phenylketonuria.
    b. Fanconi's syndrome.

    c. Hepatorenal tyrosinemia.
    d. Alcaptonuria.
    e. Cystinuria.

4. Which of the following amino acids would exist in their zwitterionic states nearest physiological pH?
    a. G.
    b. H.
    c. D.
    d. E.
    e. R.

5. Which one of the following amino acids would exhibit the greatest negative charge at physiological pH?
    a. A.
    b. G.
    c. H.
    d. R.
    e. E.

6. A 6-month-old child presents with intermittent dystonic posture of the legs and eczematous dermatitis without ataxia. Her cranial computed tomography, metrizamide myelogram, EEG, and posterior tibial somatosensory evoked potentials are normal. Her spinal fluid hydroxyindoleacetic acid concentration is below normal. Oral tryptophan loading results in a two-fold rise in the hydroxyindolacetic acid concentration of the cerebral spinal fluid. The most likely cause of her unexplained dystonia is:
    a. Fanconi's syndrome.
    b. Phenylketonuria.
    c. Hartnup disease.
    d. Maple syrup urine disease.
    e. Cystinuria.

7. Which one of the following amino acids would exhibit the greatest positive charge at physiological pH?
    a. R.
    b. H.
    c. K.
    d. D.
    e. E.

8. With regard to the chirality of amino acids it can be said that:
    a. V, L, and I all have multiple chiral carbon atoms.
    b. A, S, and T each have one chiral carbon.
    c. E, K, and R each contain two chiral carbons.
    d. G is without chirality.

DIRECTIONS (items 9–15): Match the following lettered amino acids with the properties described. Some of the amino acids may be used more than once or not at all:

    a. A.
    b. Q.
    c. C.
    d. H.
    e. Y.

9. Possesses an R-group that is capable of forming covalent bonds within and between peptide chains.

10. Is in the same amino acid classification as G.

11. Possesses a phenolic R-group.

12. Possesses an R-group capable of forming H-bonds.

13. Would carry a net charge of $+1$ at physiological pH.

14. Is a physiologically significant proton acceptor and/or donor.

15. Would carry a net charge of 0 at physiological pH.

Answers

1. a	6. c	11. e
2. e; both a and c are correct	7. b	12. b
3. b	8. d	13. d
4. a	9. c	14. d
5. e	10. a	15. b

4

Protein Structure

Objectives

- Define the four basic structural characteristics of polypeptides; primary, secondary, tertiary, and quaternary structure.
- Define the meaning of super-secondary structure.
- Describe the common techniques for analysis of primary structure.
- Define the α-helix.
- Define β-sheets and the difference between parallel and anti-parallel sheets.
- Describe the four major forces that control tertiary structure.
- Develop an appreciation for the clinical significance of proper protein structure.
- Describe several common techniques for the analysis of polypeptides.

Concepts

In healthy nonobese human beings, proteins constitute the most abundant type of biomolecule. Proteins serve numerous functions within cells. They function as enzymes, the chemical catalysts of the cell (Chapter 5); they also play a role in processes of signal transduction as hormones, hormone receptors on the surface of cells or, intracellularly, and growth factors and their receptors (Chapters 36 and 37). They function to transport biologically important molecules, such as O_2 and CO_2, metal ions, sugars and lipids. Additionally, they function in highly specialized processes: muscle contraction (Chapter 40), biological defense via the immune system (Chapter 8), photoreception (Chapter 39) and nerve signal transmission (Chapter 41).

In order to appreciate fully the complexities of the functions that proteins perform, one must have an understanding of their structure. It is the very nature of protein structure, in all its diversity, that imparts functional specificity to these biomolecules.

PRIMARY STRUCTURE

Peptides and proteins are unbranched linear arrays of amino acids. The primary structure of peptides and proteins refers to the linear number and order of the amino acids present. The convention for designating the order of amino acids is to consider the end that bears the residue with the free α-amino group as the N-terminal end. This end is written to the left and is amino acid residue number 1. The end, that bears the residue containing a free α-carboxyl group is the C-terminal end and has the highest numerical designation.

According to this convention, a peptide containing five amino acids in the following order—glutamic acid, histidine, arginine, leucine, and alanine—would appear in single letter code as:

EHRLA

Analysis of Primary Structure

The initial analysis of primary structure may consist of determining the amino acid composition of the protein. However, with the advent of automated peptide sequencing techniques, determining the amino acid composition is not always necessary. When needed, this can be accomplished by treating the protein with a strong acid (eg, HCl) at 110°C for 24 hours. This treatment cleaves all the peptide bonds, and the released amino acids can then be characterized by chromatographic analysis and compared to known standards. The use of a strong acid has its drawbacks, however: Q and N are deamidated to their acid forms (glutamic acid, E and D, respectively), and W is essentially degraded.

Before sequencing peptides, it is necessary to eliminate any disulfide bonds within and between them. Several different chemical reactions permit the separation of peptide strands and eliminate protein conformations that are dependent upon disulfide bonds. The most common treatments use either **β-mercaptoethanol** or **dithiothreitol**. Both these compounds reduce disulfide bonds like those involving cysteine residues (Figure 4–1). To prevent the disulfide bonds from re-forming, the peptides are treated with **iodoacetic acid**, which alkylates the free sulfhydryls.

N-Terminal Determination

The three major means of sequencing peptides and proteins from the N-terminus are the **Sanger, dansyl chloride and Edman techniques**.

A.   Sanger: N-terminal sequencing by this method uses the compound 2,4-dinitrofluorobenzene (DNFB), which reacts with the N-terminal residue under alkaline conditions. This reagent allows only the N-terminal residue to be determined. The resultant dinitrophenol (DNP)-derivatized amino acid can be hydrolyzed and will be labeled with a dinitrobenzene group that imparts a yellow color to the amino acid. Identification of the DNP-derivatized N-terminal amino acid is accomplished by electrophoresis and comparison with DNP-derivatized amino acid standards.

B.   Dansyl Chloride: Like DNFB, dansyl chloride reacts with the N-terminal residue under alkaline conditions. Similar to the Sanger's reagent, this technique allows only the N-terminal residue to be determined. Analysis of the modified amino acid is carried out similarly to the Sanger method, except that the dansylated amino acid is detected by fluorescence; this imparts a higher sensitivity than that of the Sanger method.

C.   Edman Degradation: The utility of this technique is that it allows the sequencing of additional amino acids, from the N-terminus inward. With this method it is possible to obtain the entire sequence

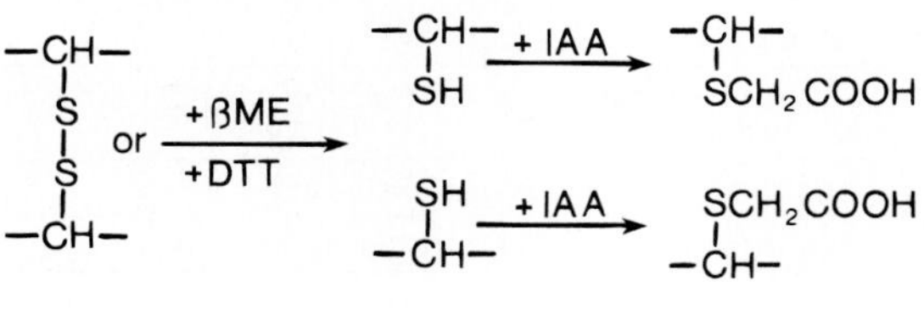

Figure 4–1. Reduction of cystine disulfide bond using either β-mercaptoethanol (β-ME) or dithiothreitol (DTT) followed by carboxymethylation blockage using iodoacetic acid (IAA).

of peptides (with an upper limit of 40–50 amino acids). This method uses phenylisothiocyanate to react with the N-terminal residue under alkaline conditions. The phenylthiocarbamyl derivatized amino acid is hydrolyzed in anhydrous acid. Hydrolysis results in a rearrangement of the released N-terminal residue to a phenylthiohydantoin derivative. As in the Sanger and dansyl chloride methods, the N-terminal residue is tagged with an identifiable marker. However, the added advantage of the Edman technique is that the remainder of the peptide is left intact. The entire sequence of reactions can be repeated as many times as necessary to obtain the sequences of the peptide. This method has now been automated to allow rapid and efficient sequencing of even extremely small quantities of peptide.

C-Terminal Determination

No reliable chemical techniques exist for sequencing the C-terminal amino acid of peptides. However, a certain class of **exopeptidases (protein digesting enzymes)**, has been shown to cleave peptides at the C-terminal residue (Table 4–1). The free amino acid can then be analyzed chromatographically and compared to standard amino acids. This class of exopeptidases is known as the **carboxypeptidases**.

The most reliable chemical technique for identification of C-terminal residues is **hydrazinolysis**. A peptide is treated with **hydrazine, H_2N-NH_2** at high temperature (90°C) for an extended length of time (20–100 hrs). This treatment cleaves all the peptide bonds, yielding amino-acyl hydrazides. All the amino acids are modified except the C-terminal residue; thus the C-terminal amino acid can be identified chromatographically in comparison to amino acid standards. Owing to the high percentage of hydrazine-induced side reactions, this technique is used only on carboxypeptidase-resistant peptides.

Protease Digestion

Because of the limitations of the Edman degradation technique, peptides longer than around 50 residues cannot be sequenced completely. Obtaining peptides of this length from proteins of greater length is facilitated by the use of enzymes, **endopeptidases**, that cleave at specific sites within the primary sequence of proteins (Table 4–2). The resultant smaller peptides can be chromatographically separated and subjected to Edman degradation sequencing.

Chemical Digestion

The most commonly utilized chemical reagent that cleaves peptide bonds by recognition of specific amino acid residues is **cyanogen bromide (CNBr)**. This reagent causes specific cleavage at the C-terminal side of M residues. The number of peptide fragments that result from CNBr cleavage is equivalent to one more than the number of M residues in a protein.

SECONDARY STRUCTURE

The ordered array of amino acids in a protein confers regular conformational forms upon that protein. These conformations constitute the **secondary structures** of a protein. The partial double-bond character of the peptide bond, as well as the R-groups of the amino acids, restricts the possible conformations of a polypeptide chain. Within a single protein, different regions of the polypeptide chain may assume different conformations determined by the primary sequence of the amino acids.

The α-Helix

The α-helix is a secondary structural motif encountered in most proteins of the globular class; it predominates in most filamentous proteins, such as myosin, collagen and α-keratin. The formation of the α-helix is spontaneous and is stabilized by H-bonding between amide nitrogens and carbonyl carbons

Table 4–1. Specificities of several exopeptidases.

Enzyme	Source	Specificity
Carboxypeptidase A	Bovine pancreas	Will not cleave when C-terminal residue = R, K or P, or if P resides next to terminal residue
Carboxypeptidase B	Bovine pancreas	Will not cleave when C-terminal residue = R or K or when P resides next to terminal residue
Carboxypeptidase C	Citrus leaves	Removes all free C-terminal residues, pH optimum = 3.5
Carboxypeptidase Y	Yeast	Removes all free C-terminal residues, slowly at G residues

Table 4–2. Specificities of several endoproteases.

Enzyme	Source	Specificity	Additional Points
Trypsin	Bovine pancreas	Peptide bond C–terminal to R, K, but not if next to P	Highly specific for positively charged residues
Chymotrypsin	Bovine pancreas	Peptide bond C–terminal to F, Y, W, but not if next to P	Prefers bulky hydrophobic residues, cleaves slowly at N, H, M, L
Elastase	Bovine pancreas	Peptide bond C–terminal to A, G, S, V, but not if next to P	
Thermolysin	*Bacillus thermoproteolyticus*	Peptide bond N–terminal to I, M, F, W, Y, V, but not if next to P	Prefers small neutral residues, can cleave at A, D, H, T
Pepsin	Bovine gastric mucosa	Peptide bond N–terminal to L, F, W, Y, but when next to P	Exhibits little specificity, requires low pH
Endopeptidase V8	*Staphylococcus aureus*	Peptide bond C–terminal to E	

of peptide bonds spaced approximately 4 residues apart (Figure 4–2). There are 3.6 amino acid residues per 360° turn of an α-helix. The formation of the α-helix places the R-groups of the amino acids on the exterior surface of the helix and perpendicular to it (Figure 4–3).

Owing to steric constraints of the R-groups, not all amino acids favor the formation of the α-helix. Amino acids such as A, D, E, I, L, and M favor the formation of α-helices, whereas G and P favor disruption of the helix. This is particularly true for P, since it is a pyrrolidine-based imino (HN=) acid whose structure significantly restricts movement about the peptide bond and thereby interferes with extension of the helix. This interference has important consequences: it introduces additional folding of the polypeptide backbone to allow the formation of globular proteins.

β-SHEETS

Whereas an α-helix is composed of a single linear array of helically disposed amino acids, β-sheets are composed of 2 or more different stretches of at least 5–10 amino acids arranged side by side. The folding and alignment of stretches of the polypeptide backbone alongside one another to form β-sheets is stabilized by H-bonding between amide nitrogens of one strand and carbonyl oxygens of the adjacent strand (Figure 4–4). This contrasts with the intrachain H-bonding that stabilizes the α-helix. Because of the positioning of the α-carbons of the peptide bond, which alternates above and below the plane of the sheet, the β-sheets are said to be pleated.

β-sheets are either **parallel or antiparallel**. In parallel sheets, adjacent peptide chains proceed in the same direction (ie, the direction of N-terminal to C-terminal ends is the same), whereas in antiparallel sheets the adjacent chains proceed in opposite directions (see Figure 4–4).

SUPER-SECONDARY STRUCTURE

Some proteins contain an ordered organization of secondary structures that form distinct functional domains or structural motifs. Examples include the helix-turn-helix domain of bacterial proteins that regulate transcription and the leucine zipper, helix-loop-helix, and zinc finger domains of eukaryotic transcriptional regulators. These domains are termed **super-secondary structures**.

TERTIARY STRUCTURE

Tertiary structure refers to the complete three-dimensional structure of the polypeptide units of a given protein. Included in this description is the spatial relationship of different secondary and super-secondary structures to one another within a polypeptide chain and the ways in which secondary structures themselves fold into the three-dimensional form of the protein. Secondary structures of proteins often constitute **distinct domains that themselves have tertiary structure**. Therefore, the tertiary structure of the entire protein describes the relationship of different domains to one another within a protein.

A   B   C

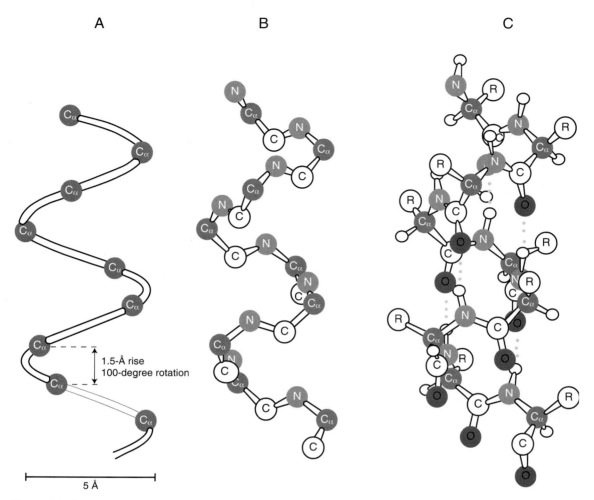

1.5-Å rise
100-degree rotation

5 Å

Figure 4–2. Models of a right-handed α-helix: (A) only the α-carbon atoms are shown on a helical thread; (B) only the backbone nitrogen (N), α-carbon (C$_a$) carbon (C) atoms are shown; (C) entire helix. H-bonds are denoted by dots. (Reproduced, with permission from Stryer, L: *Biochemistry* W.H. Freeman, 1988.)

In general proteins fold into two broad classes of structure, termed, **globular proteins,** and **fibrous proteins.** Globular proteins are compactly folded and coiled, whereas fibrous proteins are more filamentous or elongated. The majority of proteins in the body are of the globular class. Specialized proteins such as the collagens of connective tissue and the myosins of muscle fibers are fibrous proteins.

The interactions of different domains within and between proteins are governed by several forces. These forces include **hydrogen bonding, hydrophobic interactions, electrostatic interactions** and **van der Waals forces.**

Forces Controlling Structure

A. Hydrogen Bonding: Polypeptides contain numerous proton donors and acceptors both in their backbone and in the R-groups of the amino acids. The environment in which proteins are found also contains the ample H-bond donors and acceptors of the water molecule. H-bonding, therefore, occurs not only within and between polypeptide chains but with the surrounding aqueous medium.

B. Hydrophobic Forces: Proteins are composed of amino acids that contain either hydrophilic or hydrophobic R-groups. The nature of the interaction of the different R-groups with the aqueous environment plays a major role in shaping protein structure. The spontaneous folded state of globular

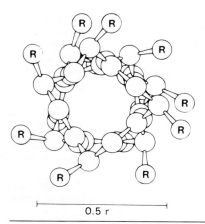

0.5 r

Figure 4–3. View down the axis of an α-helix. The side chains ® are on the outside of the helix. The van der Waals radii of the atoms are larger than shown here; hence, there is almost no free space inside the helix. (Slightly modified and reproduced, with permission from Stryer, L: *Biochemistry*, 2nd ed. W.H. Freeman and Co, 1981.)

proteins is a reflection of a balance between the opposing energetics of H-bonding between hydrophilic R-groups and the aqueous environment, and repulsion from the aqueous environment by the hydrophobic R-groups. The hydrophobicity of certain amino acid R-groups tends to drive them away from the aqueous environment at the exterior of proteins and into the interior. This driving force restricts the available conformations into which a protein may fold.

C. Electrostatic Forces: Electrostatic forces consist mainly of three types; **charge-charge, charge-dipole and dipole-dipole**. Typical charge-charge interactions that favor protein folding are those between oppositely charged R-groups such as K or R and D or E. A substantial component of the energy involved in protein folding comes from charge-dipole interactions: the interaction of ionized R-groups of amino acids with the dipole of the water molecule. The slight dipole moment that exists in the polar R-groups of amino acid also influences their interaction with water. It is, therefore, understandable that the majority of the amino acids found on the exterior surfaces of proteins contain charged or polar R-groups.

D. van der Waals Forces: There are both **attractive and repulsive van der Waals forces** that control protein folding. Attractive van der Waals forces involve the interactions among induced dipoles that arise from fluctuations in the charge densities occurring between adjacent uncharged non-bonded atoms. Repulsive van der Waals forces involve the interactions that take place when uncharged non-

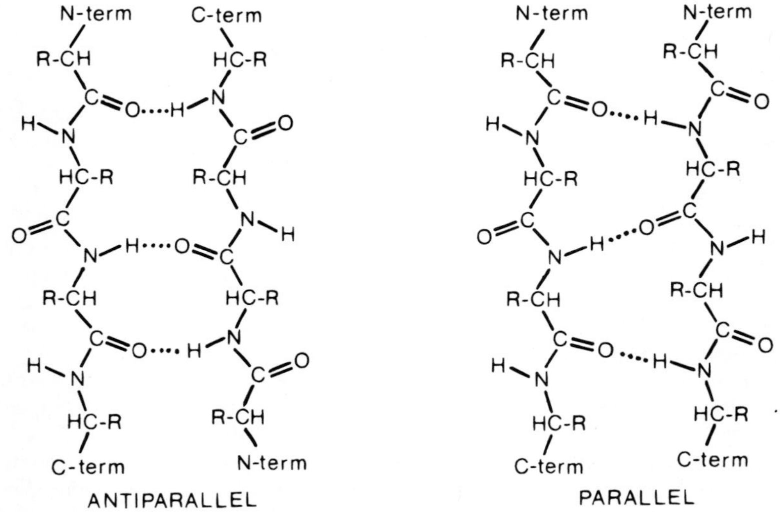

Figure 4–4. Orientation of peptide bonds and R-groups in the antiparallel and parallel forms of β-sheets.

bonded atoms come very close together but do not induce dipoles. These forces are the result of the electron-electron repulsion that occurs as two clouds of electrons begin to overlap.

Although van der Waals forces are extremely weak, (relative to other forces governing conformation), the huge number of such interactions that occur in large protein molecules nevertheless makes them significant to the folding of proteins.

QUATERNARY STRUCTURE

Many proteins contain two or more different polypeptide chains, held in association by the same noncovalent forces that stabilize the tertiary structure of proteins. Proteins with multiple polypeptide chains are termed **oligomeric** proteins. The structure formed by monomer-monomer interaction in an oligomeric protein is known as **quaternary structure**.

Oligomeric proteins can be composed of multiple identical polypeptide chains or multiple distinct polypeptide chains. Proteins with identical subunits are known as **homo-oligomers**. Proteins containing several distinct polypeptide chains are known as **hetero-oligomers**.

Hemoglobin, the oxygen-carrying protein of the blood, contains two α and two β subunits arranged with a quaternary structure in the form $\alpha_2\beta_2$. Hemoglobin is therefore a hetero-oligomeric protein (refer to Chapter 8).

COMPLEX PROTEIN STRUCTURES

Proteins also are found to be covalently conjugated with carbohydrates. These modifications occur during and after the synthesis (translation) of proteins (Chapter 31) and are therefore termed cotranslational modifications. These forms of modification impart specialized functions to the resultant proteins. Proteins covalently associated with carbohydrates are **glycoproteins** (Chapter 19). Glycoproteins are of two classes, N-linked and O-linked, referring to the site of covalent attachment of the sugar moieties. N-linked sugars are attached to the amide nitrogen of the R-group of asparagine; O-linked sugars are attached to the hydroxyl groups of either serine or threonine and occasionally to the 5-hydroxyl group of the modified amino acid, hydroxylysine.

Several extremely important glycoproteins are found on the surface of erythrocytes. Variability in the composition of the carbohydrate portions of the glycoproteins and glycolipids of erythrocytes is the feature that determines blood group specificities. There are at least 100 blood group determinants, most of them are due to carbohydrate differences. The most common blood groups, A, B, and O, are specified by the activity of specific gene products whose functions are to incorporate distinct sugar groups onto erythrocyte membrane lipids (See Chapter 10).

Structural complexes of proteins associated with lipids through noncovalent interactions are termed **lipoproteins** (Chapter 21). The distinct roles of lipoproteins will be reviewed later. Their major function in the body is to aid in the storage and transport of lipid and cholesterol.

CLINICAL SIGNIFICANCE

The substitution of a hydrophobic amino acid (V) for an acidic amino acid (E) in the β-chain of hemoglobin (HbS) results in **sickle cell anemia**. This change of a single amino acid alters the structure of hemoglobin molecules in such a way that the deoxygenated proteins polymerize and precipitate within the erythrocyte, leading to their characteristic sickle shape.

Collagens are the most abundant proteins in the body (see Chapter 8). Alterations in collagen structure arising from abnormal genes or abnormal processing of collagen proteins results in numerous diseases, including Larsen syndrome, scurvy, osteogenesis imperfecta and Ehlers-Danlos syndrome.

Ehlers-Danlos syndrome is actually the name associated with at least ten distinct disorders that are biochemically and clinically distinct yet all manifest structural weakness in connective tissue as a result of defective collagen structure.

Osteogenesis imperfecta also encompasses more than one disorder. At least four biochemically and clinically distinguishable maladies have been identified as osteogenesis imperfecta, all of which are characterized by multiple fractures and resultant bone deformities.

Marfan's syndrome manifests itself as a disorder of the connective tissue and was originally believed to be the result of abnormal collagens. However, recent evidence has shown that Marfan's syndrome results from mutations in the extracellular protein, **fibrillin**, which is an integral constituent of the non-collagenous microfibrils of the extracellular matrix.

Several forms of **familial hypercholesterolemia** are the result of genetic defects in the gene encoding the receptor for low-density lipoprotein (LDL) (see Chapter 21). These defects result in the synthesis of abnormal LDL receptors that are incapable of binding to LDLs, or that bind LDLs but the receptor/LDL complexes are not properly internalized and degraded. The outcome is an elevation in serum cholesterol levels and increased propensity toward the development of atherosclerosis.

A number of proteins can contribute to cellular transformation and carcinogenesis when their basic structure is disrupted by mutations in their genes. These genes are termed **proto-oncogenes** (see Chapter 38). For some of these proteins, all that is required to convert them to the **oncogenic** form is a single amino acid substitution. The cellular gene, **c-ras**, is observed to sustain single amino acid substitutions at positions 12 or 61 with high frequency in colon carcinomas. Mutations in c-ras are the most frequently observed genetic alterations in colon cancer.

Analysis of Proteins

A. Size Exclusion Chromatography: This chromatographic technique is based upon the use of a porous gel in the form of insoluble beads placed into a column. As a solution of proteins is passed through the column, small proteins can penetrate into the pores of the beads and, therefore, are retarded in their rate of travel through the column. The larger a protein is the less likely it will enter the pores. Different beads with different pore sizes can be used, depending upon the protein sizes that are of interest.

B. Ion Exchange Chromatography: Each individual protein exhibits a distinct overall net charge at a given pH. Some proteins will be negatively charged and some will be positively charged at the same pH. This property of proteins is the basis for ion exchange chromatography, which uses fine resins that are either negatively (**cation exchanger**) or positively (**anion exchanger**) charged. When a solution of proteins is passed through the column, proteins that bear the opposite charge from that of the resin are bound and retained in the column. The bound proteins are then eluted by passing a solution of ions bearing a charge opposite to that of the column. By utilizing a gradient of increasing ionic strength, proteins with increasing affinity for the resin are progressively eluted.

C. Affinity Chromatography: Proteins have high affinity for their substrates, co-factors, prosthetic groups, receptors or antibodies raised against them. This affinity can be exploited in the purification of proteins. A column of beads bearing the high-affinity compound can be prepared and a solution of protein passed through the column. The bound proteins are then eluted by passing a solution of unbound soluble high-affinity compound through the column.

D. High Performance Liquid Chromatography (HPLC):   In column chromatography, the smaller and more tightly packed a resin is, the greater the separation capability of the column. In gravity flow columns the limiting factors are how tightly a column can be packed and how long it takes to pass the solution of proteins through the column. HPLC uses very tightly packed, fine-diameter resins for increased resolution and overcomes the flow limitations by pumping the solution of proteins through the column under high pressure. Like standard column chromatography, HPLC columns can be used for size exclusion or charge separation. An additional separation technique commonly used with HPLC is to use hydrophobic resins to retard the movement of nonpolar proteins. The proteins are then eluted from the column with a gradient of increasing concentration of an organic solvent. This latter form of HPLC is termed **reversed-phase HPLC**.

E. Electrophoresis: Proteins also can be characterized according to size and charge by separation in an electric current (electrophoresis) within semi-solid sieving gels made from polymerized and cross-linked acrylamide. The most commonly used technique is termed **SDS polyacrylamide gel electrophoresis (SDS-PAGE)**. The gel is a thin slab of acrylamide polymerized between two glass plates. This technique uses a negatively charged detergent (sodium dodecyl sulfate SDS) to denature and solubilize the proteins. SDS-denatured proteins have a uniform negative charge, so that all proteins will migrate through the gel in the electric field solely on the basis of size. The larger the protein, the more slowly it will move through the matrix of the polyacrylamide. After electrophoresis, the migration distance of unknown proteins is assessed by various staining or radiographic detection techniques and compared with known standards.

Polyacrylamide gel electrophoresis also can be used to determine the isoelectric charge of proteins (pI), in a technique termed **isoelectric focusing**. This procedure uses a thin tube of polyacrylamide made in the presence of a mixture of small positively and negatively charged molecules termed **ampholytes**. The ampholytes have a range of pIs that establish a pH gradient along the gel when current is applied. Proteins will, therefore, cease migration in the gel when they reach the point where the ampholytes have established a pH equal to the proteins' pI.

F. Centrifugation: Proteins will sediment through a solution in a centrifugal field dependent upon their mass. Analytical centrifugation measures the rate at which various proteins sediment; the solution most commonly used for this purpose is a linear gradient of sucrose (generally 5–20%). Proteins are layered atop the gradient in an ultracentrifuge tube and subjected to centrifugal fields in excess of $10^5 \times g$. The sizes of unknown proteins can then be determined by comparing their migration distance in the gradient with those of known standard proteins.

Questions

DIRECTIONS (items 1–15): Each numbered item or incomplete statement is followed by answers or by completions of the statement. Select the ONE lettered answer or completion that is BEST in each case.

1. Protein tertiary structure is controlled by which of the following factors:
   a. The partial double-bond character of each peptide bond.
   b. The ionic character of amino acid R-groups.
   c. The hydrophobic effect of polar amino acids in the protein.
   d. The ability of amino acid R-groups to form H-bonds.
   e. All of the above.

2. The alpha helix:
   a. Is stabilized by the presence of R-group H-bonds.
   b. Is the result of the H-bonding between adjacent peptide bond atoms.
   c. Has 3.6 amino acid residues per 360° turn.
   d. Is stabilized by the presence of G.
   e. Predominantly begins with a P.

3. Treatment of the peptide GEDTY with phenylisothiocyanate followed by anhydrous acid hydrolysis would yield:
   a. GEDT and PTH-Y.
   b. PTH-G and EDTY.
   c. PTH-GEDTY.
   d. G and EDTY.
   e. GEDT and Y.

4. Quaternary structure is defined as:
   a. The ordered organization of secondary structures within a protein.
   b. The overall ordered array of amino acids within a protein.
   c. The structure obtained through interactions between different proteins.
   d. The structure obtained through interactions of tertiary structures within proteins.
   e. None of the above.

5. Which of the following is the most important determinant of tertiary structure?
   a. H-bonding.
   b. Hydrophobic interactions.
   c. van der Waals interactions.
   d. Disulfide bonds.
   e. Electrostatic interactions.

6. A male infant, delivered at 38 weeks' gestation, is presented with severe bowing of long bones, blue sclera and craniotabes at birth. Radiographs show severe generalized osteoporosis, broad and crumpled long bones, beading ribs and poorly mineralized skull. Histological examination of the long bones revealsed the trabecula of the calcified cartilage with an abnormally thin layer of osteoid, and the bony trabeculae are thin and basophilic. These symptoms are the result of the synthesis of a defective collagen and are associated with which disease?

    a. Marfan's syndrome.
    b. Osteogenesis imperfecta.
    c. Ehlers-Danlos syndrome.
    d. Scurvy.
    e. Larsen syndrome.

7. The β-sheet structure is stabilized by:
    a. The presence of R-group electrostatic interactions.
    b. The presence of R-group H-bonding.
    c. The presence of interchain H-bonding between peptide bond atoms.
    d. The presence of intrachain H-bonding between peptide bond atoms.
    e. All of the above.

8. A 58-year-old-male suffers from severe angina pectoris. Upon complete examination and testing he is found to have dilated coronary arteries incidental with adult polycystic kidney disease. The angina pectoris of the patient is explained by left ventricular hypertrophy and coronary heart disease. Multiple liver cysts, mitral valve prolapse, and coronary aneurysms are also found in this patient. These conditions are associated with which disease?
    a. Marfan's disease.
    b. Osteogenesis imperfecta.
    c. Ehlers-Danlos syndrome.
    d. Scurvy.
    e. Larsen syndrome.

9. The definition of protein secondary structure is:
    a. The restricted conformations that regions of a protein assumes as a consequence of the various R-groups in the amino acids.
    b. The structural motifs of a protein that form at physiological pH.
    c. The nature of conformational domains relative to one another within a protein.
    d. Two of the above statements are correct.
    e. None of the above properly define secondary structure.

10. Treatment of the peptide HISTKENE with trypsin would yield:
    a. H and ISTKENE.
    b. HISTK and ENE.
    c. HIST and KENE.
    d. HISTKEN and E.
    e. HISTKE and NE.

11. An alteration in the primary sequence of a protein will affect the overall tertiary structure if:
    a. A hydrophobic amino acid is substituted for a hydrophilic amino acid.
    b. An aliphatic amino acid is substituted for a polar amino acid at the N-terminus of the protein.
    c. Proline is substituted for any other amino acid.
    d. Proline is substituted for any other amino acid within an α-helix.
    e. An aliphatic amino acid is substituted for a polar amino acid at the C-terminus of the protein.

12. Which of the following amino acids are most likely to be encountered within an α-helix?
    a. D, E, and K.
    b. L, G, and F.
    c. R, V, and P.
    d. M, A, and W.

13. Hydrophobic interactions can best be characterized as:
    a. Interactions between opposite charges.
    b. Interactions between water molecules ions in solution.
    c. Interactions between polar groups and water.
    d. Interactions between nonpolar and polar groups.

14. Electrostatic forces are defined as:
    a. The interactions that can occur between the R-groups of closely spaced aliphatic amino acids.
    b. The interactions that can occur between the R-groups of closely spaced acidic and basic amino acids.
    c. The interactions that can occur between the R-groups of closely spaced aromatic amino acids.
    d. The interactions that can occur between the R-groups of closely spaced aromatic and acidic amino acids.
    e. The interactions that can occur between the R-groups of closely spaced aromatic and aliphatic amino acids.

15. The specific ordered organization of certain secondary structures:
    a. Contributes to the formation of distinct domains within proteins.
    b. Leads to the formation of super-secondary structures.
    c. Is the result of interactions between different region of secondary structure.
    d. Contributes to the overall organization of domains within proteins.
    e. All of the above are correct.

Answers

1. e	**6.** b	**11.** d
2. c	**7.** c	**12.** a
3. b	**8.** a	**13.** d
4. c	**9.** a	**14.** b
5. b	**10.** b	**15.** e

5 Introduction to Enzymes

Objectives

- To define the macromolecular composition and function of enzymes.
- To define the I.U.B. enzyme classification system.
- To characterize the reactions carried out by the six I.U.B. enzymes classes.
- To outline enzyme classification systems based on structural complexity.
- To define the terms apoenzyme, holoenzyme, coenzyme, and prosthetic group.
- To outline the relationship between vitamins and coenzymes.
- To define the functional role of coenzymes.
- To outline the relationship between enzymic activity and the types of molecules that a specific enzyme can attack.
- To define the relationship between subunit composition and enzymic activity of isozymes.
- To outline the significance of enzymes to clinical practice.
- To define the relationship between the rate of an enzyme reaction and the concentration of the enzyme in the reaction system.
- To define a unit of enzymic activity.
- To outline a typical scheme for enzyme purification.
- To outline the relationship between enzyme activity and enzyme purity.
- To define the terms specific activity and turnover number.
- To define the role of restriction endonucleases in analysis of the nucleotide sequences of genes.

Concepts

Enzymes are biological catalysts responsible for supporting almost all of the chemical reactions that maintain animal homeostasis. Because of their role in maintaining life processes, the assay and pharmacological regulation of enzymes have become key elements in clinical diagnosis and therapeutics. The macromolecular components of almost all enzymes are composed of protein, except for a class of RNA modifying catalysts known as **ribozymes**. Ribozymes are molecules of ribonucleic acid that catalyze reactions on the phosphodiester bond of other RNAs.

Enzymes are found in all tissues and fluids of the body. Intracellular enzymes catalyze the reactions of metabolic pathways. Plasma membrane enzymes regulate catalysis within cells in response to extracellular signals, and enzymes of the circulatory system are responsible for regulating the clotting of blood. Almost every significant life process is dependent on enzyme activity.

ENZYME CLASSIFICATION

Traditionally, enzymes were simply assigned names by the investigator who discovered the enzyme. As knowledge expanded, systems of enzyme classification became more comprehensive and complex. Currently enzymes are grouped into **six functional classes** by the **International Union of Biochemists (I.U.B.)**. In this classification system, each enzyme is assigned a four-digit number, with each digit separated from the next by a period. The first number defines the enzyme's functional class; the remaining numbers refer to other characteristics of the enzymic reaction. These rules give each enzyme a unique number. The I.U.B. system also specifies a textual name for each enzyme. The enzyme's name is comprised of the names of the substrate(s), the product(s), and the enzyme's functional class. For example, the approved nomenclature for the enzyme commonly known as alcohol dehydrogenase is:

$$1.1.1.1. \quad \text{or} \quad \text{alcohol:NAD}^+ \text{ oxidoreductase}$$

Because many enzymes, such as alcohol dehydrogenase, are widely known in the scientific community by their common names, the change to I.U.B.-approved nomenclature has been slow. In everyday usage, most enzymes are still called by their common name.

The number, name, and type of reaction catalyzed by the six I.U.B. classes of enzymes are shown in Table 5–1.

Enzyme Classification by Structural Complexity

Enzymes are also classified on the basis of their composition. Enzymes composed wholly of protein are known as **simple enzymes** in contrast to **complex enzymes**, which are composed of protein plus a relatively small organic molecule. Complex enzymes are also known as **holoenzymes**. In this terminology the protein component is known as the **apoenzyme**, while the non-protein component is known as the **coenzyme** or **prosthetic group**. ("Prosthetic group" describes a complex in which the small organic molecule is bound to the apoenzyme by covalent bonds; when the binding between the

Table 5–1. Biochemical properties of the six major classifications of enzymes.

Number	Enzyme Classification	Biochemical Property
1.	Oxidoreductases	Act on many chemical groupings to add or remove hydrogen atoms
2.	Transferases	Transfer functional groups between donor and acceptor molecules. Kinases are specialized transferases that regulate metabolism by transferring phosphate from ATP to other molecules
3.	Hydrolases	Add water across a bond, hydrolyzing it
4.	Lyases	Add water, ammonia or carbon dioxide across double bonds, or remove these elements to produce double bonds
5.	Isomerases	Carry out many kinds of isomerization: L to D isomerizations, mutase reactions (shifts of chemical groups) and others
6.	Ligases	Catalyze reactions in which two chemical groups are joined (or "ligated") with the use of energy from ATP

apoenzyme and non-protein components is non-covalent, the small organic molecule is called a "coenzyme.") Many prosthetic groups and coenzymes are water-soluble derivatives of vitamins. It should be noted that the main clinical symptoms of dietary vitamin insufficiency generally arise from the malfunction of enzymes, which lack sufficient cofactors derived from vitamins to maintain homeostasis. The principal vitamin-derived coenzymes or prosthetic groups are shown in Table 5–2.

The non-protein component of an enzyme may be as simple as a metal ion or as complex as a small non-protein organic molecule. Enzymes that require a metal in their composition are known as **metalloenzymes** if they bind and retain their metal atom(s) under all conditions, that is with very high affinity. Those which have a lower affinity for metal ion, but still require the metal ion for activity, are known as metal-activated enzymes.

The Functional Role of Coenzymes

The functional role of coenzymes is to act as transporters of chemical groups from one reactant to another. Examples of this function appear in the following reactions, where E represents the apoenzyme of each coenzyme shown:

$$\text{Ethanol} + \text{E} - \text{NAD} \Leftrightarrow \text{E-NADH} + \text{Acetaldehyde} + \text{H}^+ \qquad 5.1$$

$$\text{Amino Acid} + \text{E-PyrPhos} \Leftrightarrow \text{E} - \text{PyrPhos-NH}_2 + \text{Keto Acid} \qquad 5.2$$

$$\text{Succinic Acid} + \text{E} - \text{FAD} \Leftrightarrow \text{Fumaric Acid} + \text{E} - \text{FADH}_2 \qquad 5.3$$

The chemical groups carried can be as simple as the hydride ion ($\text{H}^+ + 2e\text{-}$) carried by NAD or the mole of hydrogen carried by FAD; or they can be even more complex than the amine ($-\text{NH}_2$) carried by pyridoxal phosphate in reaction 5.2.

Since coenzymes in the above examples are chemically changed as a consequence of enzyme action, it is often useful to consider coenzymes to be a special class of substrates, or **second substrates,** which are common to many different holoenzymes. In all cases, the coenzymes donate the carried chemical grouping to an acceptor molecule and are thus regenerated to their original form. This regeneration of coenzyme and holoenzyme fulfills the definition of an enzyme as a chemical catalyst, since (unlike the usual substrates, which are used up during the course of a reaction) coenzymes are generally regenerated.

Enzymic Activity and Type of Substrate

Although enzymes are highly specific for the kind of reaction they catalyze, the same is not always true of substrates they attack. For example, while succinic dehydrogenase (SDH) always catalyzes an oxidation reduction reaction and its substrate is invariably succinic acid, alcohol dehydrogenase (AHD) always catalyzes oxidation reduction reactions but attacks a number of different alcohols, ranging from methanol to butanol. Generally, enzymes having broad substrate specificity are most active against one particular substrate. In the case of ADH, ethanol is the preferred substrate.

Enzymes also are generally specific for a particular steric configuration (optical isomer) of a substrate. Enzymes that attack D sugars will not attack the corresponding L isomer. Enzymes that act on L amino acids will not employ the corresponding D optical isomer as a substrate. The enzymes known as **racemases** provide a striking exception to these generalities; in fact, the role of racemases

Vitamin Source	Coenzyme/Prosthetic Group
Niacin (B3)	Nicotinamide adenine dinucleotide (NAD +, NADP)
Pyridoxine (B6)	Pyridoxal phosphate (PP)
Riboflavin (B2)	Flavin nucleotides (FAD, FMN)
Pantothenic acid (B5)	Coenzyme A (CoA)
Folic acid	Tetrahydrofolate
Cobalamine (B12)	Cobalamine (B12)
Thiamine (B1)	Thiamine pyrophosphate (TPP)
Biotin	Biotin

Table 5–2. Vitamins as coenzymes.

is to convert D isomers to L isomers and vice versa. Thus racemases attack both D and L forms of their substrate.

As enzymes have a more or less broad range of substrate specificity, it follows that a given substrate may be acted on by a number of different enzymes, each of which uses the same substrate(s) and produces the same product(s). The individual members of a set of enzymes sharing such characteristics are known as **isozymes**. These are the products of genes that vary only slightly; often, various isozymes of a group are expressed in different tissues of the body. The best-studied set of isozymes is the **lactate dehydrogenase (LDH)** system. LDH is a tetrameric enzyme composed of all possible arrangements of two different protein subunits; the subunits are known as H (for heart) and M (for skeletal muscle). The isozymes of LDH are:

$$
\begin{array}{c}
\textbf{H H H H} \\
\textbf{H H H M} \\
\textbf{H H M M} \\
\textbf{H M M M} \\
\textbf{M M M M}
\end{array}
$$

The all-H isozyme is characteristic of that from heart tissue, and the all-M isozyme is typically found in skeletal muscle and liver. These isozymes all catalyze the same chemical reaction, but they exhibit differing degrees of efficiency.

CLINICAL SIGNIFICANCE

Aside from their unsurpassed importance in maintaining homeostasis, enzymes are essential to the diagnosis and treatment of almost all diseases. Since the vast majority of enzymes are normally found within cells, damage to tissues often means that intracellular enzymes are released from the damaged tissue and appear in the general circulation. The diagnosis of cardiac infarct, liver cirrhosis, and other disease states routinely includes measuring the circulating plasma concentration of enzymes specific to these tissues. A differential diagnosis of tissue damage—for example cardiac versus skeletal muscle—is aided by identifying the quantity and form of circulating LDH.

Enzyme Concentration and Rates of Reaction

Except when limited by the quantity of substrate, the rate of an enzyme-catalyzed reaction is directly proportional to the concentration of the enzyme. In clinical settings, this known relationship is useful for diagnostic purposes. Enzyme activity is generally reported as units of activity per milliliter of body fluid. The conventional units of measure employed (μU, nU, or pU) refer to the number of moles of substrate converted per minute per milliliter of body fluid.

In addition to their diagnostic utility, enzymes often present opportunities to combat disease states. As an example, the debilitating symptoms of rheumatoid arthritis (RA) are initiated in part by the action of the enzyme cyclooxygenase, which initiates the conversion of arachidonic acid to prostaglandins and thromboxanes. Salicylates and their derivatives (for instance, aspirin) inhibit cyclooxygenase, providing remarkable symptomatic treatment of RA and other cyclooxygenase-dependent disorders. In another clinical setting, an understanding of enzyme function can be applied directly to human behavior, as when lithium is used to treat manic depression. It is known that lithium inhibits an intracellular enzyme, a phosphatase, involved in converting inositol trisphosphate (IP_3) to inositol and inorganic phosphate. Since manic-depressive disorder is characterized by an unusually rapid turnover of IP_3, the inhibitory action of lithium on a specific phosphatase has the power to modify behavior significantly in affected individuals.

Enzyme Purification

To develop clinically acceptable diagnostic tests and therapies based on a specific enzyme, it is essential to understand the characteristic behavior of that enzyme in its pure state. Enzyme purification and characterization has therefore assumed a major role in clinical biochemistry. Modern purification techniques include tissue homogenization, centrifugation of the homogenate, salt fractionation ((NH_4)$_2SO_4$), and some combination of batch or column chromatography steps (see Chapter 4). Group separations based on ionic charge are made on ion exchange chromatographic materials. Group separations based on molecular size are made on size exclusion chromatographic materials. The most powerful purification method is affinity chromatography, in which the chromatographic material is custom designed to select one specific enzyme out of all those enzymes in a crude homogenate. With affinity chromatography, large quantities of pure enzyme can be isolated rapidly, with maximum efficiency.

Enzyme activity is monitored at each step in a purification scheme. Generally the activity at each step is expressed as the **specific activity**, whose units are **moles of substrate converted per milligram of protein**. At each succeeding purification step the quantity of contaminating protein decreases, while the amount of enzyme remains constant; consequently, specific activity increases at each step in the purification process. At the end of the purification process comes an assay of the enzyme's molecular homogeneity. Analytical ultracentrifugation is a sensitive method for analyzing the purity and determining the molecular weight of enzymes in a preparation. More recently, polyacrylamide gel electrophoresis (PAGE) has come into favor; another technique that is increasingly popular for assessing enzyme purity is high pressure liquid chromatography (HPLC) (see Chapter 4). When an enzyme is purified to homogeneity and its molecular weight characterized by an analytical technique such as PAGE, it is possible to determine its **turnover number (TN)**, the number of moles of substrate converted to product in one minute per mole of catalytic site. Only after carefully characterizing the activity of a purified enzyme is it possible to begin using that enzyme for clinical diagnosis or as a point of therapeutic attack.

Restriction Endonucleases in Analysis of the Nucleotide Sequences of Genes

One of the most significant events in helping to define the basis of genetic diseases has been the purification and characterization of the class of bacterially derived hydrolases known as **restriction endonucleases**. These enzymes cleave DNA molecules at specific nucleotide sequences known as **restriction sites** (see Chapter 43). Because of their great specificity, restriction endonucleases are widely employed to prepare specific small segments of genes, reliably and reproducibly. The nucleotide sequence of each small gene segment can easily be determined and the sequence of all the small segments summed to yield the unique nucleotide sequence of a gene. As a consequence of this kind of analysis, the basis of many genetic diseases is currently being catalogued.

Questions

DIRECTIONS (items 1–2): Each numbered item or incomplete statement is followed by an answer or by completions of the statement. Select the ONE lettered answer or completion that is BEST in each case.

1. Biological reactions are accelerated by:
   a. Carbohydrate molecules.
   b. Lipid molecules.
   c. DNA molecules.
   d. RNA molecules.

2. The I.U.B. system of enzyme classification includes:
   a. A numbering system in which six digits are assigned to each enzyme.
   b. A system of nomenclature in which the name assigned to an enzyme includes the substrate(s), cofactor(s), and one of six class names.
   c. A one- or two-word name describing the enzyme, with the suffix "ase."
   d. A numbering system in which four digits are assigned to each enzyme.
   e. Two of the above.

DIRECTIONS (items 3–5): Match the best response from the lettered list below to each statement. Each lettered response may be used once, more than once, or not at all.

   a. Holoenzyme.
   b. Isozyme.
   c. Lyase.
   d. Transferase.
   e. Ligase.

3. Enzymes which join two chemical groups with the use of ATP to supply the driving force for the reaction.

4. The combination of an apoenzyme and a coenzyme.

5. Enzymes that catalyze the same reaction and are gene products that vary from one tissue type to another.

DIRECTIONS (items 6–14): Each numbered item or incomplete statement is followed by an answer or by completions of the statement. Select the ONE lettered answer or completion that is BEST in each case.

6. Enzyme class names assigned by the I.U.B.
   a. Dehydrogenases.
   b. Oxidases.
   c. Kinases.
   d. Lyases.

7. Select the non-protein organic molecule(s) that form(s) part of a complex protein:
   a. Holoenzyme.
   b. Apoenzyme.
   c. Coenzyme.
   d. Isozyme.

8. Select the non-protein organic molecule(s) that form(s) part of a complex protein:
   a. Holoenzyme.
   b. Apoenzyme.
   c. Coenzyme.
   d. Prosthetic group.
   e. Two of the above.

9. In humans, the rate of an enzyme-catalyzed reaction is directly proportional to enzyme concentration when:
   a. The enzyme is completely saturated with coenzyme, required metal, or prosthetic group.
   b. Additional substrate or other reactants will not yield a further increase in rate.
   c. The temperature is 37° and the pressure is 1 atmosphere.
   d. The enzyme is in its native intracellular, interstitial, or plasma environment.

10. Isoenzymes:
    a. Are different gene products that catalyze identical reactions.
    b. Catalyze reactions independently of cofactors, coenzymes, prosthetic groups, or metal ions.
    c. Are clinically useful for diagnosing specific tissue pathology.
    d. Are composed of subunits that are structural isomers.
    e. Two of the above are correct.

11. Turnover Number (TN) is:
    a. The number of moles of substrate converted to product.
    b. The number of moles of substrate converted to product per unit time.
    c. The number of moles of substrate converted to product per unit time per mole of enzyme.
    d. The number of moles of substrate converted to product per unit time per weight unit of enzyme.
    e. The number of moles of substrate converted to product per unit time per mole of catalytic site.

12. Coenzyme components of human enzymes arise as:
    a. Intermediates in human metabolic pathways.
    b. Unmodified dietary components that form a complex with proteins.
    c. Modified dietary components that form a complex with with proteins.
    d. End products of special human metabolic pathways.
    e. Two of the above are correct.

13. Used in an unmodified form as coenzymes or cofactors:
   a. Vitamin B1.
   b. Vitamin B12.
   c. Riboflavin.
   d. Biotin.
   e. Two of the above are correct.

14. Dietary constituents that are modified and used as coenzymes or cofactors are:
   a. Vitamin B6.
   b. Vitamin B12.
   c. Biotin.
   d. Riboflavin.
   e. Two of the above are correct.

Answers

1. d	**6.** d	**11.** e
2. e	**7.** c	**12.** e
3. e	**8.** e	**13.** e
4. a	**9.** b	**14.** e
5. b	**10.** e	

6

Kinetics

Objectives

- To review the way that substrates and Michaelis-Menten type enzymes interact.
- To explore the way that drugs inhibit enzyme activity.
- To define the role of reactant concentration and rate constants on the rate of a reaction.
- To show the relationship between rate constants, reactant concentration, and the equilibrium constant.
- To define the meaning of certain rate equations.
- To define the rates of forward and reverse reactions that constitute the equilibrium state of a chemical reaction.
- To define the meaning of reaction order.
- To determine the reaction order for a chemical reaction.
- To define the biological role of enzymes.
- To define the thermodynamic basis for rate enhancement by enzymes.
- To compare the effect of increasing temperature on reaction rate for uncatalyzed and enzyme catalyzed reactions.

Concepts

In all organisms and in all cells, homeostasis depends on an interconnected series of chemical reactions. Most of these reactions are catalyzed, and their rates of reaction controlled, by enzymes which normally function in a relatively constant milieu. However, many conditions in humans, such as abnormal temperature, pH, or salt concentration, can result in pathologic states that result directly from the effects these variables have on chemical reaction rates. Consequently, it is important to understand how environmental factors influence chemical reactions and how pharmacological intervention can be employed to restore homeostasis.

Chemical Reactions and Reaction Rates

According to the conventions of biochemistry, **the rate of a chemical reaction is described by the number of molecules of reactant(s) that are converted into product(s) in a specified time period**. Reaction rate is always dependent on the **concentration** of the chemicals involved in the process and on **rate constants** that are characteristic of the reaction. For example, the reaction in which A is converted to B is written as follows:

$$A \rightarrow B \qquad\qquad \textbf{6.1}$$

The rate of this reaction is expressed algebraically as either a decrease in the concentration of reactant A:

$$-[A] = k[B] \qquad\qquad \textbf{6.2}$$

or an increase in the concentration of product B:

$$[B] = k[A] \qquad\qquad \textbf{6.3}$$

In equation 6.2 the negative sign signifies a decrease in concentration of A as the reaction progresses, brackets define concentration in molarity and the "k" is known as a rate constant. Rate constants are simply proportionality constants that provide a quantitative connection between chemical concentrations and reaction rates. Each chemical reaction has characteristic values for its rate constants (k); these in turn directly relate to the **equilibrium constant** for that reaction. Thus, reaction 6.1 can be rewritten (below) as an equilibrium expression in order to show the relationship between reaction rates, rate constants, and the equilibrium constant for this simple case:

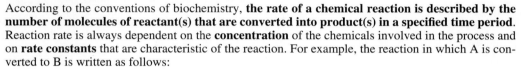

$$A \underset{k_{-1}}{\overset{k_1}{\Leftrightarrow}} B \qquad\qquad \textbf{6.4}$$

Rate Constants, Reactant Concentration, and Equilibrium Constant

At equilibrium the rate (v) of the forward reaction (A → B) is—by definition—equal to that of the reverse or back reaction (B → A), a relationship which is algebraically symbolized as:

$$v_{forward} = v_{reverse} \qquad\qquad \textbf{6.5}$$

where, for the forward reaction:

$$v_{forward} = k_1[A] \qquad\qquad \textbf{6.6}$$

and for the reverse reaction:

$$v_{reverse} = k_{-1}[B] \qquad\qquad \textbf{6.7}$$

In equations 6.6 and 6.7, k_1 and k_{-1} represent rate constants for the forward and reverse reactions, respectively. The negative subscript k_{-1} refers only to a reverse reaction, not to an actual negative value for the constant. To put the relationships represented by equations 6.6 and 6.7 into words, we state that the rate of the forward reaction ($v_{forward}$) is equal to the product of the forward rate constant (k_1) and the molar concentration of A. The rate of the reverse reaction is equal to the product of the reverse rate constant (k_{-1}) and the molar concentration of B.

The Equilibrium State of a Chemical Reaction

At equilibrium, the rate of the forward reaction is equal to the rate of the reverse reaction. This is expressed by:

$$k_1[A] = k_{-1}[B] \qquad 6.8$$

Rearranging equation 6.8 shows the relationship between reactant concentrations, rate constants, and the equilibrium constant:

$$\frac{[B]}{[A]} = \frac{k_1}{k_{-1}} = K_{eq} \qquad 6.9$$

From equation 6.9 it is clear that the equilibrium constant for a chemical reaction is not only equal to the equilibrium ratio of product and reactant concentrations, but is *also* equal to the ratio of the characteristic rate constants of the reaction.

Chemical Reaction Order Classifications

Reaction order refers to the number of molecules involved in forming a reaction complex that is competent to proceed to product(s). Empirically, order is easily determined by summing the exponents of each concentration term in the rate equation for a reaction. Reaction 6.1 is an example of a first-order reaction. It is characterized by the conversion of one molecule of A to one molecule of B with no influence from any other reactant or solvent. The exponent on the substrate concentration in the rate equation for this reaction (Equation 6.2) is 1. The best known examples of first-order reactions include radioactive decay and some isomerization reactions. An example of a second-order chemical reaction might be the formation of ATP by the condensation of ADP and orthophosphate:

$$ADP + H_2PO_4 \overset{k_1}{\underset{k_{-1}}{\rightleftharpoons}} ATP + H_2O \qquad 6.10$$

For the latter reaction the forward reaction rate is given by:

$$v_{forward} = k_1[ADP][H_2PO_4] \qquad 6.11$$

In this second-order reaction two different molecules, ADP and phosphoric acid, combine; the sum of exponents on the concentration terms of equation 6.11 is 2. However, the reactants in second- and higher-order reactions need not be different chemical species. They may be the same:

$$R + R \overset{k_1}{\underset{k_{-1}}{\rightleftharpoons}} Product_1 + Product_2 \qquad 6.12$$

One such reaction is that in which two molecules of acetyl coenzyme A (acetylCoA) interact to produce one molecule of coenzyme A (CoA) and one molecule of butyrylCoA:

$$AcetylCoA + AcetylCoA \rightarrow CoA + ButyrylCoA \qquad 6.13$$

The expression for the forward reaction rate is:

$$v_{forward} = k_1[AcetylCoA]^2 \qquad 6.14$$

Again the sum of the exponents on the concentration term of the rate equation (6.14) is 2. Table 6–1 summarizes the relationship between reaction order and number of reactants.

Enzymes as Biological Catalysts

In cells and organisms most reactions are catalyzed by enzymes, which are regenerated during the course of a reaction. These **biological catalysts** are physiologically important because they speed up the rates of reactions that would otherwise be too slow to support life. Enzymes increase reaction rates—sometimes by as much as one million fold, but more typically by about one thousand fold. Catalysts speed up the forward and reverse reactions proportionately so that, although the *magnitude* of the rate constants of the forward and reverse reactions is increased, the *ratio* of the rate constants

Table 6–1. Summary of the relationship between reaction order and number of reactants.

Order	Definition	Examples
Zero	Reaction rate **appears** independent of chemicals being varied	Enzyme reactions at high reactant concentrations
1st	One molecule of reactant per reaction	[A] or [B]
2nd	Two molecules reactant per reaction	[A] [B] or [A]² or [B]²
3rd	Three molecules reactant per reaction	[A], [B], [C], .[A]³ ....

remains the same in the presence or absence of enzyme. Since the equilibrium constant is equal to a ratio of rate constants, it is apparent that enzymes and other catalysts have no effect on the equilibrium constant of the reactions they catalyze.

Enzymes increase reaction rates by decreasing the amount of energy required to form a complex of reactants that is competent to produce reaction products. This complex is known as the activated state or **transition state complex** for the reaction, and its formation is exemplified by the following expression:

$$\textbf{reactant}_1 + \textbf{reactant}_2 \Leftrightarrow \textbf{transition state} \Leftrightarrow \textbf{product(s)} \qquad \textbf{6.15}$$

Enzymes and other catalysts accelerate reactions by lowering the energy of the transition state. The left panel of Figure 6–1 illustrates the relative free-energy relationships of molecules in the initial state (reactants), the transition state, and the final state (products) in an uncatalyzed reaction. The right panel illustrates the corresponding relationships for the same, isothermal, catalyzed reaction. In this example, the amount of energy required to achieve the transition state is lowered; consequently, at any instant a greater proportion of the molecules in the population can achieve the transition state The result is that the reaction rate is increased.

Increasing the reaction temperature accelerates reactions by imparting more energy to each molecule of reactant. Thus at higher temperatures a greater number of productive (high energy), reactant collisions will occur in a given time period and the reaction rate will be increased. However, at 55° and above many proteins become denatured, and therefore raising the temperature of enzyme-catalyzed reactions much above physiological levels usually leads to enzyme denaturation and decreased reaction rates. Two noteworthy exceptions to this, however, are the enzymes papain, used to tenderize meat in many households, and Taq DNA polymerase, which catalyzes DNA synthesis in the **polymerase chain reaction (PCR)**. Both retain their native state and catalytic properties at temperatures approaching 100°.

Michaelis-Menten Analysis of Enzyme Kinetics

In typical enzyme-catalyzed reactions, reactant and product concentrations are usually hundreds or thousands of times greater than the enzyme concentration. Consequently, each enzyme molecule catalyzes the conversion to product of many reactant molecules. In biochemical reactions, reactants are commonly known as **substrates**. The catalytic event that converts substrate to product involves the formation of a transition state, and it occurs most easily at a specific binding site on the enzyme. This site, called the **catalytic site** of the enzyme, has been evolutionarily structured to provide specific, high-

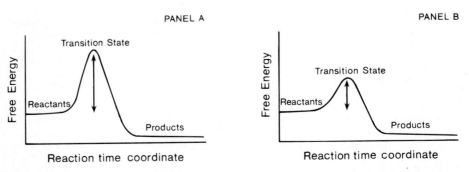

Figure 6–1. Illustrates the energy changes that take place during the conversion of reactants to products. **A**. Uncatalyzed reaction. **B**.Catalyzed reaction.

affinity binding of substrate(s) and to provide an environment that favors the catalytic events. The complex that forms when substrate(s) and enzyme combine is called the **enzyme substrate (ES) complex**. Reaction products arise when the ES complex breaks down releasing free enzyme. These events are usually depicted as follows:

$$\text{Substrate(s)} + \text{Enz} \underset{k_{-1}}{\overset{k_1}{\Longleftrightarrow}} \text{ES} - \text{Complex} \xrightarrow{k_2} \text{Product(s)} + \text{Enz} \qquad \textbf{6.16}$$

Between the binding of substrate to enzyme, and the reappearance of free enzyme and product, a series of complex events must take place. At a minimum an ES complex must be formed; this complex must pass to the transition state (ES*); and the transition state complex must advance to an enzyme product complex (EP). The latter is finally competent to dissociate to product and free enzyme. The series of events can be shown thus:

$$\textbf{E} + \textbf{S} \Longleftrightarrow \textbf{ES} \Longleftrightarrow \textbf{ES*} \Longleftrightarrow \textbf{EP} \Longleftrightarrow \textbf{E} + \textbf{P} \qquad \textbf{6.17}$$

The kinetics of simple reactions like 6.16, were first characterized by biochemists Michaelis and Menten. The concepts underlying their analysis of enzyme kinetics continue to provide the cornerstone for understanding metabolism today, and for the development and clinical use of drugs aimed at selectively altering reaction rates and interfering with the progress of disease states. The **Michaelis-Menten equation**:

$$v_1 = \frac{v_{max}[S]}{K_M + [S]} \qquad \textbf{6.18}$$

is a quantitative description of the relationship between the rate of an enzyme-catalyzed reaction (v_1), the concentration of substrate [S] and two constants, V_{max} and K_M (which are set by the particular equation). The symbols used in the Michaelis-Menten expression are defined in Table 6–2.

The Michaelis-Menten equation (equation 6.18) can be used as shown below to demonstrate that **at the substrate concentration that produces exactly half of the maximum reaction rate, i.e., ½ V_{max}, the substrate concentration is numerically equal to K_M**. This fact provides a simple yet powerful bioanalytical tool that has been used to characterize both normal and altered enzymes, such as those that produce the symptoms of genetic diseases. Rearranging equation 6.18 leads to:

$$\frac{v_1}{V_{max}[S]} = \frac{1}{K_M + [S]} \qquad \textbf{6.19}$$

And inverting and rearranging again yields:

$$\frac{V_{max}[S]}{v_1} - [S] = K_M \qquad \textbf{6.20}$$

Table 6–2. Definition of the symbols used in the Michaelis-Menten expression.

Symbol	Description	Definition
v_1	Reaction rate (velocity)	Moles of product appearing or substrate disappearing per unit time. Usually given as moles/liter/sec
V_{max}	Maximum reaction rate	Maximum change in product or substrate concentration at a given enzyme concentration
[S]	Substrate concentration	Concentration of substrate at the time v_1 is measured. Note that [S] changes during the course of the reaction
K_M	Michaelis-Menten constant	A thermodynamic constant for each enzyme catalyzed reaction. Like K_{eq}. It is a ratio of the individual velocity constants of the reaction $$K_M = \frac{k_{-1} + k_2}{k_1}$$

Simplifying equation 6.20 results in:

$$[S]\left(\frac{V_{max}}{V_1 - 1}\right) = K_M \qquad\qquad 6.21$$

From equation 6.21 it should be apparent that when the substrate concentration is that required to support half the maximum rate of reaction, the observed rate, v_1, will, be equal to V_{max} divided by 2; in other words, $v_1 = (V_{Max}/2)$. At this substrate concentration V_{Max}/v_1 will be exactly equal to 2, with the result that

$$[S](1) = K_M \qquad\qquad 6.22$$

The latter is an algebraic statement of the fact that, for enzymes of the Michaelis-Menten type, when the observed reaction rate is half of the maximum possible reaction rate, the substrate concentration is numerically equal to the Michaelis-Menten constant. In this derivation, the units of K_M are those used to specify the concentration of S, usually Molarity.

The Michaelis-Menten equation has the same form as the equation for a rectangular hyperbola; graphical analysis of reaction rate (v) versus substrate concentration [S] produces a hyperbolic rate plot (Figure 6–2).

The key features of Figure 6–2 are marked by points A, B, and C. At high substrate concentrations the rate represented by point C is almost equal to V_{max}, and the difference in rate at nearby concentrations of substrate is almost negligible. We will see below that if the Michaelis-Menten plot is extrapolated to infinitely high substrate concentrations, the extrapolated rate is equal to V_{max}. When the reaction rate becomes independent of substrate concentration, or nearly so, the rate is said to be zero order. (Note that the reaction is zero order only with respect to this substrate. If the reaction has two substrates, it may or may not be zero order with respect to the second substrate). The very small differences in reaction velocity at substrate concentrations around point C reflect the fact that at these concentrations almost all of the enzyme molecules are bound to substrate ([ES] = [E]$_{total}$) and the rate is virtually independent of substrate, hence zero order. At lower substrate concentrations, such as at points A and B, the lower reaction velocities indicate that at any moment only a portion of the enzyme molecules are bound to the substrate. In fact, at the substrate concentration denoted by point B, exactly half the enzyme molecules are in an ES complex at any instant and the rate is exactly one half of V_{max}. At substrate concentrations near point A the rate appears to be directly proportional to substrate concentration, and the reaction rate at points around A is said to be first order.

The assumptions listed in Table 6–3 define the limits of applicability of the Michaelis-Menten equation.

Analysis of Enzyme Reactions Using Linear Kinetic Plots

To avoid dealing with curvilinear plots like that illustrated in Figure 6–2, biochemists Lineweaver and Burk introduced an analysis of enzyme kinetics based on the following rearrangement of the Michaelis-Menten equation:

$$\frac{1}{v} = \frac{K_M}{V_{Max}}\frac{1}{[S]} + \frac{1}{V_{Max}} \qquad\qquad 6.23$$

Plots of 1/v versus 1/[S] yield straight lines having a slope of K_M/V_{Max} and an intercept on the y-axis at $1/V_{Max}$, as illustrated in Figure 6–3.

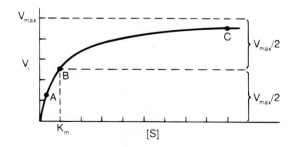

Figure 6–2. Effects of substrate concentration on the velocity of an enzyme-catalyzed reaction.

Table 6–3. Limits of applicability of the Michaelis-Menten equation.

	Michaelis-Menten Assumptions
1.	[S] >> [E], When reaction rates are determined, the concentration of substrate is much greater than that of the enzyme
2.	Reaction velocities, v, are measured only at the beginning of the reaction when the concentration of product is infinitely small and the back reaction (conversion of product to substrate) can be ignored
3.	k_{-1} >> k_2. Under this condition the rate-limiting step, or slowest step in the overall reaction, is the conversion of ES complex to free enzyme and product. Corollaries of this assumption are: a. $v = k_2[ES]$ b. $[ES] \cong [E_{total}]$ at high substrate concentrations c. $V_{Max} = k_2[E_{total}]$ d. K_M is equal to the dissociation constant of the ES complex to free enzyme and substrate, i.e., $$K_M = K_d = \frac{[E][S]}{[ES]} = \frac{k_{-1}}{k_1} = K_{eq.} \text{ for ES dissociation}$$

An alternative linear transformation of the Michaelis-Menten equation is the Eadie-Hofstee transformation:

$$\frac{v}{[S]} = -v\frac{1}{K_M} + \frac{V_{Max}}{K_M} \qquad\qquad 6.24$$

and when v/[S] is plotted on the x-axis versus v on the y-axis, the result is a linear plot with a slope $-1/K_m$ and the value V_{max}/K_M as intercept on the y-axis.

CHARACTERIZATION OF ENZYME INHIBITORS

Since most clinical drug therapy is based on inhibiting the activity of enzymes, analysis of enzyme reactions using the tools described above has been fundamental to the modern design of pharmaceuticals. Well-known examples of such therapy include the use of methotrexate in cancer chemotherapy to semi-selectively inhibit DNA synthesis of malignant cells, the use of aspirin to inhibit the synthesis of prostaglandins which are at least partly responsible for the aches and pains of arthritis, and the use of sulfa drugs to inhibit the folic acid synthesis that is essential for the metabolism and growth of disease-causing bacteria. In addition, many poisons—such as cyanide, carbon monoxide and polychlorinated biphenols (PCBs)—produce their life-threatening effects by means of enzyme inhibition.

Enzyme inhibitors fall into two broad classes: those causing irreversible inactivation of enzymes and those whose inhibitory effects can be reversed. Inhibitors of the first class usually cause an inactivating, covalent modification of enzyme structure. Cyanide is a classic example of an irreversible enzyme inhibitor: by covalently binding mitochondrial cytochrome oxidase, it inhibits all the reactions associated with electron transport. **The kinetic effect of irreversible inhibitors is to decrease the concentration of active enzyme, thus decreasing the maximum possible concentration of ES complex.** Since the limiting enzyme reaction rate is often $k_2[ES]$, it is clear that under these circumstances the reduction of enzyme concentration will lead to decreased reaction rates. Note that when enzymes in cells are only partially inhibited by irreversible inhibitors, the remaining unmodified enzyme molecules are not distinguishable from those in untreated cells; in particular, they have the same turnover number and the same K_M. **Turnover number**, related to V_{max}, is defined as the maximum number of moles of substrate that can be converted to product per mole of catalytic site per second. Irreversible inhibitors are usually considered to be poisons and are generally unsuitable for therapeutic purposes.

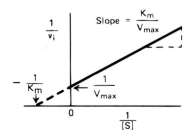

Figure 6–3. Double reciprocal or Lineweaver-Burk plot of 1/v, versus 1/[S] used to evaluate K_{max} and V_{max}

Competitive, Noncompetitive, and Uncompetitive Inhibitors

Reversible inhibitors can be divided into two main categories—competitive and noncompetitive—with a third category, uncompetitive inhibitors, rarely encountered. The hallmark of all the reversible inhibitors is that when the inhibitor concentration drops, enzyme activity is regenerated. Usually these inhibitors bind to enzymes by non-covalent forces and the inhibitor maintains a reversible equilibrium with the enzyme, as illustrated in reaction 6.25 (I signifies "inhibitor").

$$E + I \underset{k_{reverse}}{\overset{k_{forward}}{\rightleftharpoons}} E - I - Complex \qquad\qquad 6.25$$

The equilibrium constant for the dissociation of enzyme inhibitor complexes (the reverse of reaction 6.25) is known as K_I:

$$K_I = \frac{[E][I]}{[E - I - Complex]} \qquad\qquad 6.26$$

The importance of K_I is that in reversible reactions where substrate, inhibitor and enzyme interact, the normal K_M and or V_{max} for substrate enzyme interaction appear to be altered. These changes are a consequence of the influence of K_I on the overall rate equation for the reaction. The effects of K_I are best observed in Lineweaver-Burk plots.

Interactions of Substrate and Inhibitor

Probably the best known reversible inhibitors are **competitive inhibitors**, which always bind at the catalytic or active site of the enzyme. Most drugs that alter enzyme activity are of this type. Competitive inhibitors are especially attractive as clinical modulators of enzyme activity because they offer two routes for the reversal of enzyme inhibition, while other reversible inhibitors offer only one. First, as with all kinds of reversible inhibitors, a decreasing concentration of the inhibitor reverses the equilibrium shown in reaction 6.25, regenerating active free enzyme. Second, since substrate and competitive inhibitors both bind at the same site, they compete with one another for binding, as illustrated in reaction 6.27.

$$Enz + I + S \underset{k_{reverse}}{\overset{k_{forward}}{\rightleftharpoons}} Enz - I - Complex + Enz - S - Complex \qquad\qquad 6.27$$

Raising the concentration of substrate (S), while holding the concentration of inhibitor constant, provides the second route for reversal of competitive inhibition. The greater the proportion of substrate, the greater the proportion of enzyme present in competent ES complexes. The characteristic features of reversible enzyme inhibitors are summarized in Table 6–4 and illustrated in Figure 6–4.

Basis for the Kinetic Effect of Reversible Inhibitors

As noted earlier, high concentrations of substrate can displace virtually all competitive inhibitor bound to active sites. Thus, it is apparent that V_{max} should be unchanged by competitive inhibitors. This characteristic of competitive inhibitors is reflected in the identical vertical-axis intercepts of Lineweaver-Burk plots, with and without inhibitor (Figure 6–4, panel B). Since attaining V_{max} requires appreciably

Table 6–4. Characteristics of reversible inhibitors.

Inhibitor Type	Binding Site on Enzyme	Kinetic Effect
Competitive	Specifically at the catalytic site, where it competes with substrate for binding in a dynamic equilibrium-like process. Inhibition is reversible by substrate	V_{max} is unchanged K_M, as defined by [S] required for 1/2 maximal activity, is increased
Noncompetitive	Binds E or ES complex other than at the catalytic site. Substrate binding unaltered, but ESI complex cannot form products. Inhibition cannot be reversed by substrate	K_M appears unaltered V_{max} is decreased proportionately to inhibitor concentration
Uncompetitive	Binds only to ES complexes at locations other than the catalytic site. Substrate binding modifies enzyme structure, making inhibitor-binding site available. Inhibition cannot be reversed by substrate	Apparent V_{max} is decreased K_M, as defined by [S] required for 1/2 maximal activity, is decreased

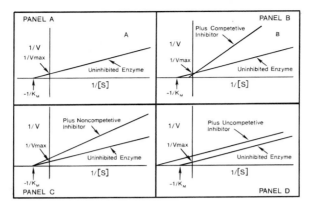

Figure 6–4. Effects of reversible inhibitors on Lineweaver-Burk plots.

higher substrate concentrations in the presence of competitive inhibitor, K_M (the substrate concentration at half maximal velocity) is also higher, as demonstrated by the differing negative intercepts on the horizontal axis in panel B.

Analogously, panel C illustrates that noncompetitive inhibitors appear to have no effect on the intercept of the negative abscissa implying that **noncompetitive inhibitors have no effect on K_M of the enzymes they inhibit.** This observation can be explained by recalling two facts: first, $K_M = (k_{-1} + k_2/k_1)$ (see Table 6–2), and second, k_{-1} is usually assumed to be so much larger than k_2 that $K_M \sim k_{-1}/k_1$. Since this class of inhibitor does not interfere in the equilibration of enzyme, substrate, and ES complexes, the K_MS of Michaelis-Menten type enzymes are not expected to be affected by noncompetitive inhibitors, as demonstrated by the abscissa intercepts in panel C. However, because complexes that contain inhibitor (ESI) are incapable of progressing to reaction products, **the effect of a noncompetitive inhibitor is to reduce the concentration of ES complexes that can advance to product.** Since $V_{max} = k_2 [E_{total}]$ (Table 6–3), and the concentration of competent E_{total} is diminished by the amount of ESI formed, noncompetitive inhibitors are expected to decrease V_{max}, as illustrated by the ordinate intercepts in panel C.

A corresponding analysis of uncompetitive inhibition leads to the expectation that these inhibitors should change the apparent values of K_M as well as V_{max}. Changing both constants leads to double reciprocal plots, in which intercepts on the vertical and horizontal axis are proportionately changed; this leads to the production of parallel lines in inhibited and uninhibited reactions.

Questions

DIRECTIONS (items 1–11): Each numbered item or incomplete statement is followed by answers or by completions of the statement. Select the ONE lettered answer or completion that is BEST in each case.

1. The rate of enzymatic reactions taking place in human cells is normally set by the following factors:
   a. Concentration of substrates and products.
   b. The concentration of substrates, products and the rate constants of individual reactions.
   c. Reactant concentrations, rate constants, and temperature.
   d. Reactant concentrations, rate constants, temperature, and enzyme concentration.

2. The expression:

$$A \underset{k_{-1}}{\overset{k_1}{\rightleftarrows}} B$$

    a. Is a second-order reaction.

    b. Has a rate for the forward reaction of k_{-1} [A].

    c. Is a first-order reaction and has a rate for the forward reaction of k_{-1} [A].

    d. Has an equilibrium constant, K_{eq}, that is equal to k_1/k_{-1}.

3. For the following expression the forward reaction is said to be second-order because:

$$ADP + H_2PO_4 \overset{k_1}{\underset{k_{-1}}{\rightleftharpoons}} ATP + H_2O$$

    a. The reaction has two rate constants.

    b. It has two molecules of product per reaction.

    c. The sum of exponents on the concentration terms in the rate equations is 2.

4. The mechanism by which enzymes exert their catalytic effect:

    a. Raising the ratio of products to reactants, [products]/[reactants], that are found in the equilibrium state.

    b. Decreasing the free energy of the transition state for the reaction.

    c. Possessing specific substrate binding sites for the reactants.

5. In the Michaelis-Menten description of enzyme-substrate interactions, where the rate constants are k_1, k_{-1}, and k_2, K_M is:

    a. Equal to the concentration of substrate when the substrate concentration supports a reaction velocity equal to $\frac{1}{2} V_{max}$.

    b. Equal to: $k_{-1} + k_2/k_1$.

    c. Generally taken to be equal to the equilibrium constant for the binding of substrate to enzyme $E + S \rightleftharpoons ES$.

    d. Equal to k_2, [ES].

    e. Two of the above are correct.

6. The value of K_M is graphically obtained from Michaelis-Menten plots:

    a. As the concentration of substrate that supports V_{max}.

    b. As one-half the substrate concentration that supports V_{max}.

    c. As the substrate concentration that supports one half of V_{max}.

7. Where the rate constants are k_1, k_{-1}, and k_2, enzyme reactions are said to be zero order:

    a. When the reaction rate approximates V_{max} and is taking place at a substrate concentration at which slightly higher or lower concentrations have no appreciable effect on the rate.

    b. At substrate concentrations where $v_{forward} = k_2$ [E]$_{total}$.

    c. When k_{-1}[ES] = k_2[ES].

    d. Two of the above are correct.

8. In Lineweaver-Burk plots:

    a. K_M is numerically equal to the intercept on the negative abscissa.

    b. V_{max} is numerically equal to the intercept on the ordinate.

    c. 1/v is plotted versus 1/[S].

    d. K_M is equal to the slope of the plotted line.

9. Reversible enzyme inhibitors have the following properties:

    a. They form enzyme-substrate-inhibitor complexes (ESI complexes).

    b. Their effects can be reversed by reducing the concentration of inhibitor in the reaction.

    c. They bind to the active site of the enzyme.
    d. The principal mode of enzyme-inhibitor interaction is covalent in nature.

10. In the presence of a competitive inhibitor, enzyme reactions have the following characteristics:
    a. The formation of enzyme-inhibitor complexes (EI) can be reversed by substrate.
    b. $v_{forward} = k_2 [ES]$.
    c. The inhibitor is always structurally analogous to the substrate.
    d. Two of the above are correct.

11. The main reason for treating severe carbon monoxide poisoning by placing the affected individuals in an oxygen tent is:
    a. They have difficulty breathing.
    b. Carbon monoxide poisons oxygen transport across lung alveoli.
    c. Carbon monoxide is a competitive inhibitor of oxygen binding proteins.
    d. Carbon monoxide combines with oxygen to form carbon dioxide ($CO + Oxygen \rightarrow CO_2$ and this production of nontoxic carbon dioxide reduces the concentration of poison.

DIRECTIONS (item 12–15): Answered each question with reference to the accompanying figure, which represents a graphical description of enzyme catalyzed reactions: In each panel, the results are shown for reactions in the presence and absence of an enzyme inhibitor.

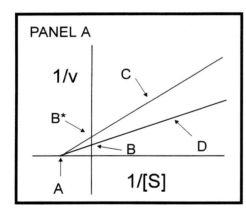
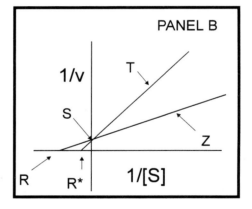

12. Panel A illustrates:
    a. The effect of K_I on V_{max} of an enzyme.
    b. Results obtained for the effect of a typical noncompetitive inhibitor on an enzyme reaction.
    c. The V_{max} of the uninhibited enzyme as point B*.
    d. Two of the above are correct.

13. Panels A and B represent:
    a. Eadie-Hofstee plots.
    b. Lineweaver-Burk plots.
    c. Michaelis-Menten plots.

14. Panel B illustrates:
    a. The effect of K_I on the V_{max} of an enzymatic reaction.
    b. The fact that competitive inhibitors do not change the V_{max} of an enzyme.
    c. Reciprocal rates versus reciprocal substrate concentration for an uninhibited enzyme as line Z.
    d. Two of the above.

15. In a comparison of Panel A and Panel B:
    a. Panel B illustrates the effects of a noncompetitive inhibitor while Panel A illustrates effects of a competitive inhibitor.
    b. Points B and B* are related to K_M while R and R* are related to V_{max}.
    c. Lines Z and D represent the uninhibited reactions, while C and T represent the inhibited condition.
    d. Panels A and B illustrate effects of reversible enzyme inhibitors.
    e. Two of the above are correct.

Answers

1. d	6. c	11. c
2. d	7. d	12. d
3. c	8. c	13. b
4. b	9. b	14. d
5. e	10. d	15. e

Mechanisms and Regulation of Enzyme Activity

7

Objectives

- To review the mechanisms of enzyme mediated catalysis, including strain catalysis, acid-base catalysis, covalent catalysis, and catalysis by proximity and orientation effects.
- To review the physiologically important mechanisms by which the catalytic activity of enzymes is regulated, including covalent modification, zymogen activation and allosteric interactions.
- To outline the induced fit mechanism of enzyme substrate interaction.
- To illustrate how strain mechanisms and other mechanisms of catalysis combine to catalyze a reaction.
- To outline how an enzyme's acid groups and basic groups can participate in catalyzing covalent bond modification.
- To review covalent catalysis using digestive and blood clotting proteases as models for the process.
- To review how substrate or enzyme concentration can regulate enzyme catalyzed reactions.
- To define the process of allosteric regulation of enzyme activity.
- To define the role of effectors in regulation of allosteric enzymes.
- To define the role of allosteric enzymes in metabolic pathways.
- To identify the effectors that regulate enzyme activity.
- To define the regulation of allosteric enzymes made up of regulatory and catalytic subunits.
- To review the mechanism of regulating protein catalytic activity by reversible covalent modification.

Concepts

The mechanisms of enzyme catalysis that have developed over hundreds of millions of years of evolution are identical to those contrived by organic chemists in the course of the last century. Understanding these mechanisms, therefore, is central to appreciating the unique role of enzymes in biology.

Intermediary metabolism—the breakdown and transformation of food into building materials for the body—comprises a series of linked reaction pathways that are regulated by enzymes. The mechanisms by which enzymes carry out this regulation are what underlie normal metabolism. Moreover, at certain "branch points" in the pathways of metabolism, regulated enzymes can often serve as targets for treatment of disease. Such target enzymes tend to be exploited for therapeutic use through the special mechanisms that regulate their activity.

Enzyme-Substrate Interaction

The favored model of enzyme substrate interaction is known as the **induced fit model**. This model proposes that the initial interaction between enzyme and substrate is relatively weak, but that these weak interactions rapidly induce conformational changes in the enzyme that strengthen binding and bring catalytic sites close to substrate bonds to be altered. After binding takes place, one or more mechanisms of catalysis generates transition-state complexes and reaction products. The possible mechanisms of catalysis are four in number: catalysis by bond strain; catalysis by proximity and orientation; acid-base catalysis; and catalysis by formation of covalent enzyme substrate intermediates.

A.   Catalysis by Bond Strain: In this form of catalysis, the induced structural rearrangements that take place with the binding of substrate and enzyme ultimately produce strained substrate bonds, which more easily attain the transition state. Figure 7–1 depicts the alignment of a model hexasaccharide in a binding site on the surface of the enzyme, lysozyme. After initial alignment, stronger hydrophobic bonds are formed between sugar residues and amino acid R groups because of induced protein conformational changes (arrows). The new conformation often forces substrate atoms and bulky catalytic groups, such as aspartate and glutamate, into conformations that strain existing substrate bonds.

B.   Catalysis by Proximity and Orientation: Enzyme-substrate interactions orient reactive groups and bring them into proximity with one another. In addition to inducing strain, groups such as aspartate are frequently chemically reactive as well, and their proximity and orientation toward the substrate thus favors their participation in catalysis.

C.   Catalysis Involving Proton Donors (Acids) and Acceptors (Bases):   Other mechanisms also contribute significantly to the completion of catalytic events initiated by a strain mechanism. Illustrated in Figure 7–2, for example, is the use of glutamate as a general acid catalyst (proton donor).

D.   Covalent Catalysis: In catalysis that takes place by covalent mechanisms, the substrate is oriented to active sites on the enzymes in such a way that a covalent intermediate forms between the

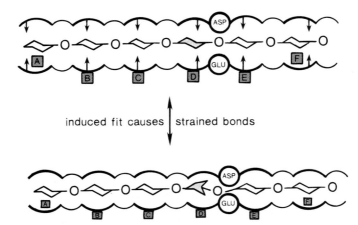

induced fit causes | strained bonds

Figure 7–1. Model demonstrating how a conformational change in an enzyme (termed induced fit) can lead to a strain in the bonds of the substrate resulting in increased catalysis.

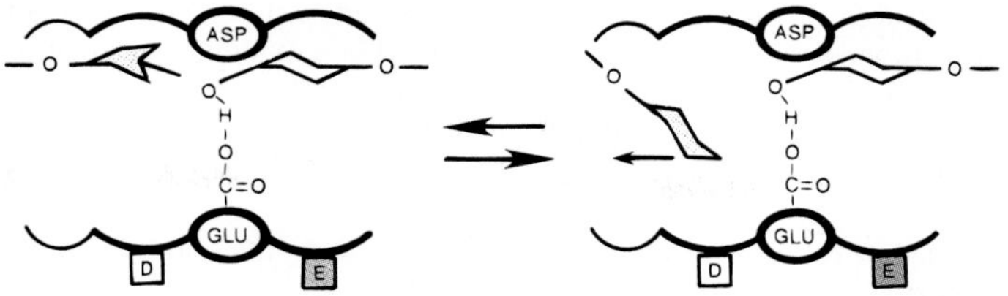

| Concerted H bond formation and covalent bond lysis | Strain on D ring relieved and D ring trisaccharide leaves |

Figure 7–2. Model depicting the R-group of an acidic amino acid, in the enzyme active-site, performing the role of a proton-donor during catalysis.

enzyme or coenzyme and the substrate. One of the best-known examples of this mechanism is that involving proteolysis by **serine proteases**, which include both digestive enzymes (trypsin, chymotrypsin, and elastase) and several enzymes of the blood-clotting cascade. These proteases contain an active site serine whose R group hydroxyl forms a covalent bond with a carbonyl carbon of a peptide bond, thereby causing hydrolysis of the peptide bond.

MECHANISMS FOR REGULATING ENZYME ACTIVITY

The concentration of substrates and enzymes can regulate individual biochemical reactions and whole metabolic pathways.

While it is clear that enzymes are responsible for the catalysis of almost all biochemical reactions, it is important to also recognize that rarely, if ever, do enzymatic reactions proceed in isolation. The most common scenario is that enzymes catalyze individual steps of multi-step metabolic pathways, as is the case with glycolysis, gluconeogenesis or the synthesis of fatty acids. As a consequence of these lock-step sequences of reactions, any given enzyme is dependent on the activity of preceding reaction steps for its substrate. In sequence 7.1, the rate of reaction catalyzed by enzyme C is dependent on the production of fructose-6-P by enzyme B. Clearly, if enzymes A or B are missing or inhibited from action, the rate of the final reaction will be zero, since its necessary substrate will not have been produced.

$$\text{glucose} \underset{A}{\Leftrightarrow} \text{glucose-6-P} \underset{B}{\Leftrightarrow} \text{fructose-6-P} \underset{C}{\Leftrightarrow} \text{fructose-1,6-bisP} \qquad 7.1$$

This sequence demonstrates that enzyme activity can be regulated by substrate concentration. Additionally, the concentration of a particular enzyme, such as A, B, or C in reaction 7.1, also regulates the rate of each individual reaction and thus the throughput of the entire metabolic pathway. For example, given non-limiting amounts of substrate, the rate of production of fructose-1,6-bis-phosphate in 7.1 will be directly proportional to the concentration of enzymes A, B, and C. In humans, substrate concentration is dependent on food supply and is not usually a physiologically important mechanism for the routine regulation of enzyme activity. Enzyme concentration, by contrast, is continually modulated in response to physiological needs. Three principal mechanisms are known to regulate the concentration of active enzyme in tissues:

1. **Regulation of gene expression** controls the quantity and rate of enzyme synthesis.
2. **Proteolytic enzyme activity** determines the rate of enzyme degradation.
3. **Covalent modification of** preexisting pools of inactive proenzymes produces active enzymes.

Enzyme synthesis and proteolytic degradation are comparatively slow mechanisms for regulating enzyme concentration, with response times of hours, days or even weeks. Proenzyme activation is a more rapid method of increasing enzyme activity but, as a regulatory mechanism, it has the disadvantage of not being a reversible process. Proenzymes are generally synthesized in abundance, stored in secretory granules and covalently activated upon release from their storage sites. Examples of important proenzymes include pepsinogen, trypsinogen and chymotrypsinogen, which give rise to the proteolytic digestive enzymes. Likewise, many of the proteins involved in the cascade of chemical reactions responsible for blood clotting are synthesized as proenzymes. Other important proteins, such as peptide hormones and collagen, are also derived by covalent modification of precursors.

Another mechanism of regulating enzyme activity is to sequester enzymes in compartments where access to their substrates is limited. For example, the proteolysis of cell proteins and glycolipids by enzymes responsible for their degradation is controlled by sequestering these enzymes within the lysosome.

In contrast to regulatory mechanisms that alter enzyme concentration, there is an important group of regulatory mechanisms that do not affect enzyme concentration, are reversible and rapid in action, and actually carry out most of the moment-to-moment physiological regulation of enzyme activity. These mechanisms include allosteric regulation, regulation by reversible covalent modification and regulation by control proteins such as calmodulin.

Allosteric Enzyme Regulation

In addition to simple enzymes that interact only with substrates and inhibitors, there is a class of enzymes that bind small, physiologically important molecules and modulate activity in ways other than those described above. These are known as **allosteric enzymes**; the small regulatory molecules to which they bind are known as **effectors**. Allosteric effectors bring about catalytic modification by binding to the enzyme at distinct **allosteric sites**, well removed from the catalytic site, and causing conformational changes that are transmitted through the bulk of the protein to the catalytically active site(s). Figure 7–3 illustrates these events for a monomeric enzyme.

The hallmark of effectors is that when they bind to enzymes, they alter the catalytic properties of an enzyme's active site. Those that increase catalytic activity are known as **positive effectors**. In path (a) of Figure 7–3, positive effector (+E) is shown opening the active site and allowing substrate (S) ready access to its binding site. Effectors that reduce or inhibit catalytic activity are **negative effectors**, designated in Figure 7–3 as (−E). In path (b), −E is depicted as closing the active site, denying entry to the substrate. Note that even in the absence of effectors, allosteric enzymes usually have some base level of activity, as illustrated in path (c).

Most allosteric enzymes are **oligomeric** (consisting of multiple subunits); generally they are located at or near branch points in metabolic pathways, where they are influential in directing substrates along one or another of the available metabolic paths. The effectors that modulate the activity of these allosteric enzymes are of two types. Those activating and inhibiting effectors that bind at allosteric sites (depicted as +E and −E in Figure 7–4) are called **heterotropic effectors**. (Thus there exist both positive and negative heterotropic effectors.) These effectors can assume a vast diversity of chemical forms, ranging from simple inorganic molecules to complex nucleotides such as cyclic adenosine monophosphate (cAMP). Their single defining feature is that they are not identical to the substrate.

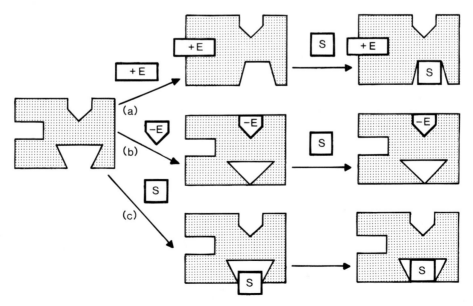

Figure 7–3. Model of allosteric effector interaction with a monomeric enzyme.

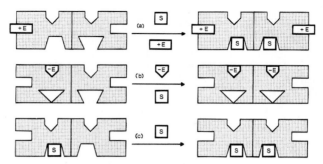

Figure 7–4. Model of allosteric effector interaction with a oligomeric enzyme.

In many cases the substrate itself induces distant allosteric effects when it binds to the catalytic site. Substrates acting as effectors are said to be **homotropic effectors**. When the substrate is the effector, it can act as such, either by binding to the substrate-binding site, or to an allosteric effector site. When the substrate binds to the catalytic site it transmits an activity-modulating effect to other subunits of the molecule. Often used as the model of a homotropic effector is hemoglobin, although it is not a branch-point enzyme and thus does not fit the definition on all counts.

There are two ways that enzymatic activity can be altered by effectors: the V_{max} can be increased or decreased, or the K_M can be raised or lowered. Enzymes whose K_M is altered by effectors are said to be "K-type" enzymes and the effector a "K-type" effector. If V_{max} is altered, the enzyme and effector are said to be "V-type." Many allosteric enzymes respond to multiple effectors with V-type and K-type behavior. Here again, hemoglobin is often used as a model to study allosteric interactions, although it is not strictly an enzyme. The allosteric properties of hemoglobin will be reviewed in Chapter 8.

Characteristically, long metabolic sequences regulated by allosteric enzymes use end products of the pathway as negative effectors, or feedback inhibitors, on the allosteric enzyme. In this way substrates are conserved, and the wasteful accumulation of end product is prevented.

Enzymes With Separate Regulatory and Catalytic Subunits

In the preceding discussion we assumed that allosteric sites and catalytic sites were homogeneously present on every subunit of an allosteric enzyme. While this is often the case, there is another class of allosteric enzymes that are comprised of separate catalytic and regulatory subunits. The archetype of this class of enzymes is cAMP-dependent protein kinase (**PKA**), whose mechanism of regulation is illustrated in Figure 7–5. The enzyme is tetrameric ($\alpha_2\beta_2$), containing two catalytic subunits and two regulatory subunits, and enzymatically inactive. When intracellular cAMP levels rise, one molecule of cAMP binds to each regulatory subunit, causing the tetramer to dissociate into one regulatory dimer and two catalytic monomers. In the dissociated form, the catalytic subunits are fully active; they catalyze the phosphorylation of a number of other enzymes, such as those involved in regulating glycogen metabolism. The regulatory subunits have no catalytic activity.

The ubiquitously distributed calcium-binding protein known as **calmodulin** serves much the same role as the regulatory subunit of PKA. However, while PKA regulatory subunits specifically bind to PKA catalytic subunits, calmodulin appears to be a regulatory component of many different enzymes. It thus provides a coordinated physiological response by many enzymes to fluxes in calcium concentration.

Covalent Modification by Phosphorylation and Prenylation

This mechanism of regulation involves the covalent addition of a small regulatory molecule, which leads to profound changes in the catalytic properties of the enzyme. The best examples, again, come from studies on the regulation of glycogen metabolism. Catalytically active PKA (see above) catalyzes the phosphorylation of glycogen synthase and glycogen phosphorylase kinase, with the result that glycogen degradation is stimulated while glycogen synthesis is coordinately inhibited (see Chapter 15). These covalent phosphorylations can be reversed by a separate sub-subclass of enzymes known as **phosphatases**. Recent research has indicated that the aberrant phosphorylation of growth factor and hormone receptors, as well as of proteins that regulate cell division, often leads to unregulated cell growth or cancer (see Chapter 38). The usual sites for phosphate addition to proteins are the serine, threonine, and tyrosine R group hydroxyl residues.

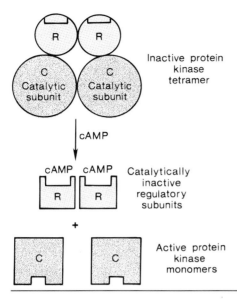

Figure 7–5. cAMP activation of the tetrameric enzyme, PKA. Binding of cAMP to the regulatory subunits (R) leads to their dissociation from the complex and activation of the catalytic subunits.

An increasing number of covalent modifications are now being found to lead to altered enzyme activity. The addition of hydrophobic fatty acids or polyisoprene residues (obtained from the metabolic pathway of cholesterol) to specific residues on enzymes (a modification termed prenylation) is often necessary to anchor an enzyme in a membrane where it can express its biological function.

Questions

DIRECTIONS (items 1–5): Each numbered item or incomplete statement is followed by answers or by completions of statement. Select the ONE lettered answer or completion that is BEST in each case.

1. Prior to the formation of a transition state complex, the interaction of an enzyme with its substrate is often accompanied by:
   a. Covalent modification of the enzyme.
   b. Non-covalent rearrangement of enzyme conformation.
   c. Non-covalent and covalent modification of the substrate.

2. Mechanisms of enzyme catalysis include:
   a. Strain catalysis.
   b. Acid-base catalysis and strain catalysis.
   c. Covalent catalysis, acid-base catalysis, and strain catalysis.
   d. Proximity and orientation effects, covalent catalysis, acid-base catalysis, and strain catalysis.

3. Intracellular catalysis by enzymes:
   a. Shunts substrates along one or another path by differentially changing the equilibrium concentration of substrates and products for the various pathways.
   b. Often involves locating reactive groups closer to one another than they might otherwise be.
   c. Is often accompanied by covalent modification of enzyme peptide bonds.
   d. Shunts substrates along one or another path by selectively enhancing the reaction in the forward direction and simultaneously reducing the reaction in the reverse direction.

4. Processes not generally associated with enzyme catalysis:
   a. Change in K_{eq} to a value more favorable for maintaining homeostasis.
   b. Protonation of the substrate.

    c. Covalent bond formation between enzyme R groups and substrate.
    d. Covalent bond formation between substrates and peptide-bonded carboxyl carbons.
    e. Two of the above are correct.

5. Serine proteases are:
    a. Trypsin, chymotrypsin, and thrombin.
    b. Examples of enzymes that form covalent bonds with their substrate during catalysis.
    c. Enzymes that have the R group "S" at the active site.
    d. Two of the above are correct.

DIRECTIONS (items 6–8): Match each term with the best definition from the lettered list below.

    a. Structural rearrangements of protein R groups at sites other than the substrate binding site of enzymes.
    b. Covalent bond formation between two substrates to produce one product.
    c. Protonation of substrate by protons derived from weak acid or base R groups at the active site of enzymes.
    d. Thrombin.
    e. None of the above.

6. Covalent catalysis is exemplified by:

7. Acid-base catalysis is exemplified by:

8. Strain catalysis is exemplified by:

9. The concentration of catalytically competent enzyme in an organism, can be regulated by:
    a. Covalent activation of otherwise inactive protein.
    b. Altered regulation of gene transcription.
    c. Altered protease activity.
    d. Two of the above are correct.
    e. A, B, and C are correct.

10. Proenzymes are catalytically important because:
    a. Their activation and inactivation are rapid, providing a readily reversible mechanism by which enzyme activity can be regulated.
    b. Their activation generally involves reactions on their peptide bonds producing active catalyst from inactive precursor pools.
    c. Inactive pools of precursors are activated by covalent modification of R groups at or near the active site.
    d. Compared to other catalytic molecules, their concentration is preferentially regulated by transcriptional mechanisms.

11. Processes by which the activity of allosteric enzymes is reversibly regulated include:
    a. Association/dissociation of regulatory proteins.
    b. Covalent modification of the enzyme's peptide bonds.
    c. Association/dissociation of small molecular weight regulatory ligands.
    d. Binding non-substrate inhibitors at the catalytic site.
    e. Two of the above are correct.

12. Regulatory ligands of allosteric enzymes have the following properties:
    a. They all bind at sites distant from the substrate binding site.
    b. They all bind at the substrate binding site.
    c. They are often related to the substrate by being metabolically modified products.

13. Heterotropic effectors can:
    a. Increase enzyme activity.

    b. Decrease enzyme activity.

    c. Cause dissociation or association of oligomeric proteins and increase or decrease enzyme activity.

    d. Cause dissociation or association of oligomeric proteins and increase or decrease enzyme activity by binding at the catalytic site.

    e. None of the above are correct.

14. For a given enzyme, homotropic effectors are similar to heterotropic effectors in the following way(s):

    a. They can both be called ligands, with each having its own ligand-binding site distant from the catalytic site.

    b. For a given allosteric enzyme, they have the same effect on enzyme catalysis.

    c. Compared to proteins they are generally small molecular weight molecules.

    d. They are downstream products of metabolic pathways regulated by the enzyme for which they are effectors.

    e. Two of the above are correct.

15. Allosteric enzymes can have the following properties:

    a. Contain homotropic and heterotropic binding sites.

    b. Often are oligomeric and contain homotropic and heterotropic binding sites.

    c. Often are oligomeric and contain homotropic binding sites on one class of monomer and heterotropic binding sites on another class of monomer.

    d. Are found at metabolic branch points where they regulate the metabolic fate of substrates.

    e. All of the above are correct.

Answers

1. b	**6.** d	**11.** e
2. d	**7.** c	**12.** c
3. b	**8.** d	**13.** c
4. e	**9.** e	**14.** c
5. e	**10.** b	**15.** e

8   Integration of Protein Structure and Function

Objectives

- To review the relationship between protein structure and function, how similar structures are used for common functions and how seemingly minor modifications of protein structure induce significant changes in protein function.
- To define the relationship between amino acid composition and the molecular structure of beta pleated sheets in the silk fibroin model.

- To compare the amino acid composition and tertiary structure of collagen with that of a typical globular protein.
- To describe the quaternary structure of tropocollagen.
- To define the role of collagen in the human body.
- To compare the polyproline helix of a collagen polypeptide with that of an alpha helix.
- To identify the significance of the fact that every third amino acid in a mature collagen strand is glycine.
- To identify the subunit composition, secondary and tertiary structure, and prosthetic group of hemoglobin and myoglobin.
- To identify the oxygen binding site in hemeproteins.
- To identify the commonly occurring hemoglobins and their subunit composition.
- To compare the oxygen binding properties of myoglobin and hemoglobin and identify the basis of the differences.
- To identify the role of hemoglobin in transport of CO_2 from peripheral tissues to lungs.
- To identify the relationship between the T and R state of hemoglobin and its oxygen binding properties.
- To define the oligomeric structure of IgG and the inter polypeptide bonds that maintain the quaternary structure of IgG's.
- To discuss the nature of the domains in IgGs, including number, distribution, secondary structure and similarity.

Concepts

This chapter reviews the relationship between structure and biological function for the selected group of proteins listed in Table 8–1. The two general functions of proteins considered are their role as relatively static structural components of body tissues and their more dynamic participation in the body as biological catalysts, transport molecules, and active constituents of the immune system.

Generally, proteins that comprise structural elements of biological tissues are long and narrow resulting in their being described as fibrous in tertiary structure. The secondary structure of fibrous proteins is generally that of an extended α-helix or a β-structure with extensive participation of hydrophobic amino acids whose exposed R-groups make these proteins relatively insoluble in biological fluids. In tissues structural proteins are characteristically stacked together to form three dimensional aggregates such as tendons and basement membranes.

Proteins with catalytic and similar dynamic functions typically have globular tertiary structures formed from complex mixtures of β-structures and α helices that are folded to produce their characteristic three dimensional configuration. Many catalytic proteins consist of more than one globular polypeptide chain with the interaction between these subunits being responsible for their characteristic biological activity.

A common theme in many catalytic proteins is the presence of repeating homologous domains of secondary structure that contribute to the final protein product. Antibody structure, specifically that of immunoglobulin G (IgG), provides a clear example of proteins with repeating structural domains.

Silk Fibroin a Model of Antiparallel Beta Pleated Sheet

In silk fibroin (molecules that comprise silk thread), glycine, alanine, and serine comprise about 85% of the amino acids, with glycine occurring at every other residue in the polypeptide chains. The secondary

Table 8–1. Structure/function relationships of a selected group of proteins.

Protein	Structure/Function Relationship
Collagen	Major structural protein of the human body
Silk fibroin	Antiparallel β pleated sheets
Myoglobin	A globular protein
Hemoglobin	An oligomer (4×) of myoglobin-like molecules
Immunoglobulin G	Multiple conserved domain structures

structure of fibroin is antiparallel β-pleated sheet with the individual polypeptides existing in an almost fully extended or stretched form. In fibroin β-sheets, illustrated in Figure 8–1, every other R-group is arranged on the same side of the sheet, so that in fibroin all of the glycine residues are on one side of a sheet and R-groups of the other amino acids, mainly alanine, are on the other side of the sheet. Laterally, peptide strands in the same sheet are structurally stabilized by extensive hydrogen bonding between adjacent α amino hydrogen atoms and oxygen atoms of carboxyl residues in peptide bonds.

Note that one surface of each β-sheet contains all of the alanine R-groups and the other surface of the same sheet contains the glycine R-group which is a hydrogen atom. In the layering of β-sheet on β-sheet to form a three dimensional array the Ala residues of one sheet nestle between the Ala residues of an adjacent sheet. The zipper-like interdigitation of the Ala residues keeps these two sheets of beta structure from sliding over each other. The exposed upper and lower surfaces of such a stacked pair of sheets is comprised mainly of the hydrogen atoms from glycine and these surfaces provide optimal hydrophobic interaction for stabilizing stacked pairs of β-sheets. Coupled with the fact that the peptide chains of β-sheets are fully extended the resulting silk threads have very little stretch and are excellent for forming material like parachutes and specialized suture material.

COLLAGEN: THE MAJOR PROTEIN IN THE HUMAN BODY

Collagen, a trimeric, fibrous glycoprotein is the dominant insoluble material of animal connective tissue. It is the most abundant body protein, accounting for about 30% of total protein, or about 6% of the total body weight. There are at least 12 different forms of collagen molecules, designated I-XII. Collagen molecules are classified by their amino acid content, their relative abundance of carbohydrate and their subunit composition.

The amino acid composition of collagen is remarkable in that ⅓ of the amino acids are glycine with, proline plus hydroxyproline (Hyp) making up another ⅓ of the amino acids. **Hydroxyproline is unique to the collagens.** A number of other unusual amino acids are found in the collagens (Figure 8–2). The occurrence of glycine at every third residue in the collagen polypeptide chain generates a 3–amino acid repeat, **Gly-X-Y**, where the amino acid of the X position is predominantly proline and that of the Y is predominantly Hyp. In combination, hydroxyproline and proline determine the secondary structure of collagen polypeptides which is known as a proline helix since its helical pitch is similar to that formed by synthetic polypeptides comprised entirely of proline. Hydroxylysine provides the principal attachment sites for carbohydrate residues. Allysine is formed extracellularly by removal of the ε amino of lysine and oxidation of the ε carbon atom to an aldehyde. Allysine provides an important site of intermolecular collagen crosslinks that stabilize the final collagen structure.

The posttranslational reactions that hydroxylate procollagen require vitamin C (ascorbic acid), O_2 and α-ketoglutaric acid. Consequently, individuals with vitamin C deficiency (scurvy) make collagen lacking the derived amino acids and these molecules are incapable of forming normal collagen fibrils. Ex-

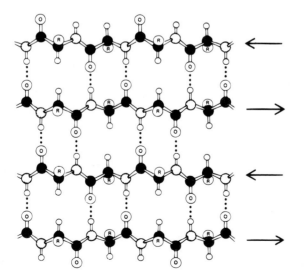

Figure 8–1. The structure of antiparallel β pleated sheets in silk fibroin is characterized by extensive lateral hydrogen bonding between peptide strands in the same sheet and extensive hydrophobic and steric interactions between the R groups of individual β sheets that are stacked on top of each other. (Reproduced with permission, from Murray, RK: *Harpers Biochemistry*, 23rd ed. Appleton & Lange, 1993.)

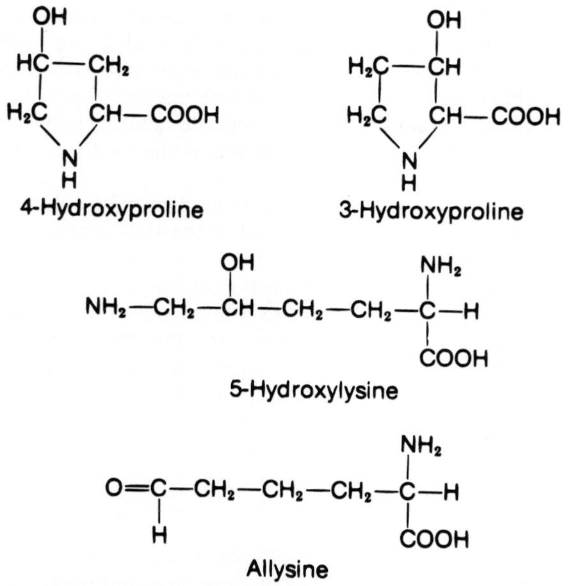

OH
|
HC—CH₂
| |
H₂C   CH—COOH
\   /
N
|
H
4-Hydroxyproline

OH
|
H₂C—CH
| |
H₂C   CH—COOH
\   /
N
|
H
3-Hydroxyproline

NH₂—CH₂—$\overset{\overset{\text{OH}}{|}}{\text{CH}}$—CH₂—CH₂—$\overset{\overset{\text{NH}_2}{|}}{\underset{\underset{\text{COOH}}{|}}{\text{C}}}$—H

5-Hydroxylysine

O=$\overset{}{\underset{\underset{\text{H}}{|}}{\text{C}}}$—CH₂—CH₂—CH₂—$\overset{\overset{\text{NH}_2}{|}}{\underset{\underset{\text{COOH}}{|}}{\text{C}}}$—H

Allysine

Figure 8–2. Structure of unusual amino acids found in relatively high abundance in mature collagen molecules. All of these amino acids are formed posttranslationally and are required for formation of normal mature tissue collagen. (Reproduced, with permission from Devlin, T: *Textbook of Biochemistry with Clinical Correlations*, 3rd ed. Wiley-Liss, Inc. 1992.)

tracellular conversion of lysyl residues to allysine involves the pyridoxal phosphate requiring enzyme **lysyl oxidase**. Thus, vitamin B_6 deficiency also leads to formation of abnormal collagen.

Mature collagen fibrils are composed of numerous long narrow hetero or homotrimers stacked adjacent to each other forming long cylindrical fiber bundles. A mature collagen trimer is known as tropocollagen. As illustrated in Figure 8–3, each polypeptide in a tropocollagen molecule is known as an α chain. As noted above α chains posses characteristic helical secondary structure (proline helix) determined by their unusually high content of proline and hydroxyproline. Three α chains are helically wound around each other to form a tropocollagen triple helix. The proline helix of the collagen α chains has 3 amino acids per full turn and consequently is not an α-helix, since an α-helix is characterized by having 3.6 amino acid residues per turn of the helix.

Since one third of the amino acids in collagen are glycine, and since there are 3 amino acids per helical turn of a collagen α chain, every third amino acid in the polypeptide is a glycine. The result is that

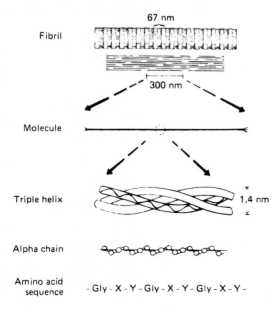

Figure 8–3. Molecular features of collagen structure from primary sequence up to the fibril. (Slightly modified and reproduced with permission, from Eyre DR: Collagen: Molecular diversity in the bodys protein scaffold. Science 1980;207:1315. Copyright © 1980 by the American Association for the Advancement of Science.)

one edge or surface of each polypeptide helix has a glycine R-group (a hydrogen atom) arrayed linearly along the strand from one end to the other. The minimal bulk of this R-group along a distinct linear region of the polypeptide helix permits three helical collagen polypeptides to approach each other very closely leading to extensive hydrophobic interactions that greatly stabilize the tropocollagen molecule. Collagen fibrils are stabilized by extensive intermolecular hydrogen bonding between hydrogen and oxygen atoms, of nearby peptide bonds that are oriented perpendicular to the axis of the triple helix.

The mature tropocollagen molecule is derived from a precursor known as procollagen. Procollagen polypeptides have long regions on the C- and N- terminal ends that do not contain the usual amino acid triplets characteristic of mature tropocollagen α chains. Consequently these regions would pack very badly into a tropocollagen molecule and they are excised during post translational processing.

When tropocollagen molecules assemble into connective tissue sheaths outside cells a final processing step takes place. Allysine residues react with other allysyl R-groups, or with adjacent lysine ε-amino groups in the same, and in adjacent tropocollagen molecules to form covalently cross linked structures of great stability. These cross linking reactions continue for the life of the collagen so that the number of collagen cross links in tissues increases with the age of an individual.

MYOGLOBIN AND HEMOGLOBIN

Myoglobin and hemoglobin are hemeproteins whose physiological importance is principally related to their ability to bind molecular oxygen. Myoglobin is a monomeric heme protein found mainly in muscle tissue where it serves as an intracellular storage site for oxygen. During periods of oxygen deprivation oxymyoglobin releases its bound oxygen which is then used for metabolic purposes. Hemoglobin is an $\alpha_2\beta_2$ tetrameric hemeprotein found in erythrocytes where it is responsible for binding oxygen in the lung and transporting the bound oxygen throughout the body where it is used in aerobic metabolic pathways.

The tertiary structure of myoglobin is that of a typical water soluble globular protein. Its secondary structure is unusual in that it contains a very high proportion (75%) of α helices. A myoglobin polypeptide is comprised of 8 separate right handed α helices, designated A through H, that are connected by short non helical regions as illustrated in Figure 8–4. Amino acid R groups packed into the interior of the molecule are predominantly hydrophobic in character while those exposed on the surface of the molecule are generally hydrophilic, thus making the molecule relatively water soluble.

Each myoglobin molecule contains one heme prosthetic group inserted into a hydrophobic cleft in the protein (Figure 8–5). The heme residue is formed from 4 substituted pyrrole rings connected by methylene bridges to form a substituted cyclic tetrapyrrole. Each heme residue contains one central coordinately bound iron atom that is normally in the +2, or ferrous, oxidation state. The oxygen carried by hemeproteins is bound directly to the ferrous iron atom of the heme prosthetic group. Oxida-

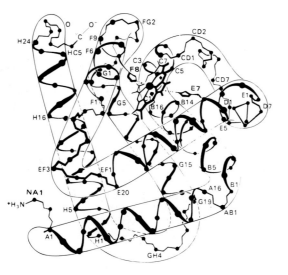

Figure 8–4. A model of myoglobin at low resolution. Only the α-carbon atoms are shown. The α-helical regions are named A through H. (Based on Dickerson RE in: *The Proteins*, 2nd ed. Vol. 2. Neurath H [editor]. Academic Press. 1964. Reproduced with permission.)

Proximal His (F8)

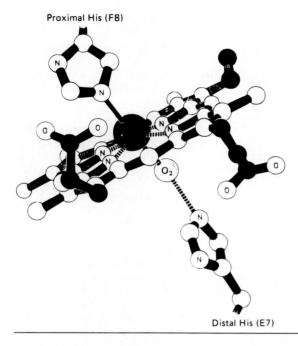

Distal His (E7)

Figure 8–5. Addition of oxygen to heme iron in onygenation. Shown also are the imidazole side chains of the 2 important histidine residues of globin that attach to the heme iron. (Reproduced with permission, from Harper HA et al: *Physiologische Chemie*, Springer-Verlag, 1975.)

tion of the iron to the $+3$, ferric, oxidation state renders the molecule incapable of normal oxygen binding. Hydrophobic interactions between the tetrapyrrole ring and hydrophobic amino acid R groups on the interior of the cleft in the protein strongly stabilize the heme protein conjugate. In addition a nitrogen atom from a histidine R group located above the plane of the heme ring is coordinated with the iron atom further stabilizing the interaction between the heme and the protein. In oxymyoglobin the remaining bonding site on the iron atom (the 6th coordinate position) is occupied by the oxygen, whose binding is stabilized by a second histidine residue (Figure 8–5).

Carbon monoxide also binds coordinately to heme iron atoms in a manner similar to that of oxygen, but the binding of carbon monoxide to heme is much stronger than that of oxygen. The preferential binding of carbon monoxide to heme iron is largely responsible for the asphyxiation that results from carbon monoxide poisoning.

Hemoglobin

Hemoglobins are tetrameric molecules consisting of two each of two different polypeptides. Each subunit of a hemoglobin tetramer has a heme prosthetic group identical to that described for myoglobin. The common peptide subunits are designated α, β, γ, and δ which are arranged into the most commonly occurring functional hemoglobins shown in Table 8–2.

Although the secondary and tertiary structure of various hemoglobin subunits are similar, reflecting extensive homology in amino acid composition, the variations in amino acid composition that do exist impart marked differences in hemoglobins oxygen carrying properties. In addition, the quaternary structure of hemoglobin leads to physiologically important allosteric interactions between the subunits, a property lacking in monomeric myoglobin which is otherwise very similar to the α subunit of he-

Table 8–2. Classification of the most commonly occurring functional hemoglobins.

Hemoglobin Name	Abbreviation	Composition
Normal adult hemoglobin	HbA	$\alpha_2\beta_2$
Fetal hemoglobin	HbF	$\alpha_2\gamma_2$
Sickle cell hemoglobin	HbS	α_2S_2
Minor adult hemoglobin	HbA$_2$	$\alpha_2\delta_2$

moglobin. In this section we review how interactions between small ligands (oxygen and 2,3-bisphosphoglycerate) and large proteins can significantly alter their physiological properties and how small covalent (carbamate formation) and non covalent (proton dissociation) changes in proteins profoundly alter their functional properties.

Comparison of the oxygen binding properties of myoglobin and hemoglobin (Figure 8–6) illustrate the allosteric properties of hemoglobin that results from its quaternary structure and differentiate hemoglobin's oxygen binding properties from that of myoglobin.

Hemoglobin's sigmoidal oxygen binding curve is typical of allosteric proteins in which the substrate, in this case oxygen, is a positive homotropic effector. When oxygen binds to the first subunit of deoxyhemoglobin it increases the affinity of the remaining subunits for oxygen. As additional oxygen is bound to the second and third subunits oxygen binding is further, incrementally, strengthened, so that at the oxygen tension in lung alveoli, hemoglobin is fully saturated with oxygen. As oxyhemoglobin circulates to deoxygenated tissue, oxygen is incrementally unloaded and the affinity of hemoglobin for oxygen is reduced. Thus at the lowest oxygen tensions found in very active tissues the binding affinity of hemoglobin for oxygen is very low allowing maximal delivery of oxygen to the tissue. As illustrated in Figure 8–6 the oxygen binding curve for myoglobin is hyperbolic in character indicating the absence of allosteric interactions in this process.

The allosteric oxygen binding properties of hemoglobin arise directly from the interaction of oxygen with the iron atom of the heme prosthetic groups and the resultant effects of these interactions on the quaternary structure of the protein. When oxygen binds to an iron atom of deoxyhemoglobin it pulls the iron atom into the plane of the heme. Since the iron is also bound to histidine F8, this residue is also pulled toward the plane of the heme ring. The conformational change at histidine F8 is transmitted throughout the peptide backbone resulting in a significant change in tertiary structure of the entire subunit. Conformational changes at the subunit surface lead to a new set of binding interactions between adjacent subunits. The latter changes include disruption of salt bridges and formation of new hydrogen bonds and new hydrophobic interactions, all of which contribute to the new quaternary structure.

The latter changes in subunit interaction are transmitted, from the surface, to the heme binding pocket of a second deoxy subunit and result in easier access of oxygen to the iron atom of the second heme and thus a greater affinity of the hemoglobin molecule for a second oxygen molecule. The tertiary configuration of low affinity, deoxygenated hemoglobin (Hb) is known as the taut (T) state. Conversely, the quaternary structure of the fully oxygenated high affinity form of hemoglobin (HbO$_2$) is known as the relaxed (R) state.

In addition to transporting oxygen from lungs to peripheral tissues, hemoglobin molecules play an important role in transporting CO$_2$ in the opposite direction. As illustrated in equation 8.1, N-terminal amino groups of the T form of hemoglobin are available for reaction with CO$_2$ and about 15% of the CO$_2$ formed in tissues is carried to the lung covalently bound to the N-terminal nitrogens as carbamate.

$$CO_2 + (Hb - NH_3) \Leftrightarrow 2H^+ + (Hb - NH - COO^-) \qquad \textbf{8.1}$$
$$\text{T form} \qquad\qquad\qquad \text{T form Carbamate}$$

In the lung the high oxygen partial pressure favors formation of the R form resulting in reversal of the carbamate with release and exhalation of the CO$_2$.

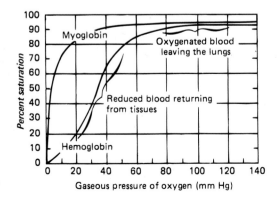

Figure 8–6. Oxygen binding curves for myoglobin and hemoglobin. Oxygen tension in arteries is about 100 mm Hg. At this concentration hemoglobin and myoglobin are fully saturated with oxygen. In venous blood the oxygen tension is generally between 20 and 40 mm Hg resulting in a significantly reduced oxygen content for hemoglobin, but only limited dissociation of oxygen from myoglobin. In very active muscle the oxygen tension can fall below 5 mm Hg at which point the reserve supply of oxygen bound to myoglobin is almost fully released and made available for metabolic purposes. (Modified, with permission, from Stanbury JB, Wyngaarden JB, Fredrickson DS [editors]: *Ther Metabolic Basis of Inherited Disease*, 4th ed. McGraw-Hill, 1978.)

During the conversion from T form to R form several R groups on the surface of hemoglobin subunits also become available to dissociate protons as shown in equation 8.2.

$$4O_2 + Hb \text{ (T form)} \Leftrightarrow nH^+ + Hb(O_2)_4 \text{ (R form)} \qquad \textbf{8.2}$$

Equations 8.1 and 8.2 illustrate that the conformation of hemoglobin and its oxygen binding are sensitive to hydrogen ion concentration. These effects of hydrogen ion concentration are responsible for the well known **Bohr effect** in which increases in hydrogen ion concentration decrease the amount of oxygen bound by hemoglobin at any oxygen concentration (partial pressure).

Tissue CO_2 is also carried to the lungs as the dissolved gas and as bicarbonate formed spontaneously and by the erythrocyte enzyme carbonic anhydrase as illustrated in equation 8.3.

$$CO_2 + H_2O \xrightarrow[\text{Anhydrase}]{\text{Carbonic}} H_2CO_3 \xrightarrow{\text{spontaneous}} H^+ + HCO_3^- \qquad \textbf{8.3}$$

Finally, the compound **2,3-bisphosphoglycerate (BPG)**, derived from the glycolytic intermediate 1,3-bisphosphoglycerate, is a potent allosteric effector on the oxygen binding properties of hemoglobin. In the deoxygenated T conformer, a cavity capable of binding BPG forms in the center of the molecule. BPG can occupy this cavity stabilizing the T state. Conversely, when BPG is not available, or not bound in the central cavity, Hb can be converted to HbO_2 more readily. Thus, like increased hydrogen ion concentration, increased BPG concentration favors conversion of R form Hb to T form Hb and decreases the amount of oxygen bound by Hb at any oxygen concentration. Hemoglobin molecules differing in subunit composition are known to have different BPG binding properties with correspondingly different allosteric responses to BPG. For example, HbF binds BPG much less avidly than HbA with the result that HbF in fetuses of pregnant women binds oxygen with greater affinity than the mothers HbA, thus giving the fetus preferential access to oxygen carried by the mothers circulatory system.

ANTIBODIES

This review of antibody structure is directed at antibodies as models of proteins with multiple, well defined protein domains, and while antibodies provide the body with a major defense mechanism against a host of toxic compounds and microbes we will not focus our attention on antibody function. An antibody is a protein formed by the immune system in response to foreign substances. The foreign materials are called antigens and the specific chemical group on an antigen, against which an antibody is formed, is known as an **epitope**.

The most common immunoglobulin (Ig) is immunoglobulin G (IgG), it has a well defined quaternary structure and a mass of about 150 kd. IgG is an $\alpha_2\beta_2$ tetramer held together by interchain disulfide bonds and extensive regions of non covalent bonding. The molecule is illustrated schematically in Figure 8–7. It is comprised of 2 heavy chains (H) and two light chains (L). Each L chain is bonded to an H chain by one S-S bond and the H chains are bonded to each other by at least one S-S bond. The protease, papain, specifically cleaves IgGs to produce the S-S bonded carboxyl terminus of the H chains and two fragments called F_{ab} (for "fragment antibody binding") each composed of the amino terminal half of a H chain and a complete light chain.

A number of features of a typical IgG molecule (Figure 8–7) are worthy of noting:

1. The molecule contains 12 distinct domains that are composed of a bilayer of antiparallel β-pleated sheets.
2. Four of the domains contain variable amino acid sequences, but very similar secondary structure, they are designated VH and VK and comprise the antigen-binding sites of the molecule.
3. Eight regions contain repeat amino acid sequences and are designated C. The domain structure is virtually identical for these regions.
4. The amino acid secondary structure connecting all of these domains is said to be unstructured, or random coil in nature.

Thus the structure of IgG illustrates the use of a common protein structural motif used multiple times in the same molecule to achieve different functional results.

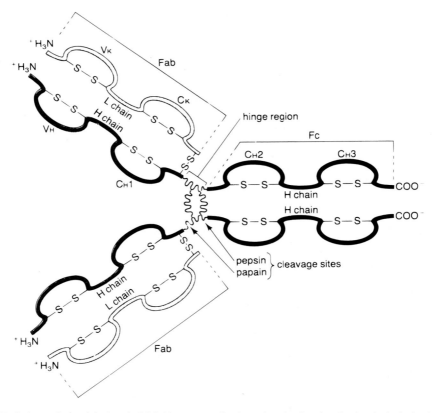

Figure 8–7. Schematic model of an IgG1 (κ) human antibody molecule showing the basic 4-chain structure and domains (V$_H$,C$_H$1, etc). Sites of enzymatic cleavage by pepsin and papain are shown. (Reproduced with permission, from Stites, DP: *Basic & Clinical Immunology*, 8th ed. Appleton & Lange, 1994.)

Questions

DIRECTIONS (items 1–14): Each numbered item or incomplete statement is followed by answers or by completions of the statement. Select the ONE lettered answer or completion that is BEST in each case.

1. The following general relationships exist between protein structures and their biological function:
    a. Proteins that are principally structural are globular in tertiary structure and are composed of numerous regions of β-sheet structure.
    b. Non structural proteins with dynamic activity, such as enzymes are globular in tertiary structure and are composed of a wide variety of secondary structure.
    c. Fibrous structural proteins rarely contain helical secondary structures.
    d. Water soluble globular proteins contain very small proportions of hydrophobic amino acids.

2. In silk threads quaternary structure is a consequence of:
    a. β-sheets with alanine residues on one side interacting with the glycine R-group containing side of contiguous β-sheets.
    b. Alanine residues on one side of a β-sheet interacting, in a zipper-like fashion, with alanine residues on one side of an adjacent sheet. and pairs of such sheets interacting glycine side with glycine side.
    c. Hydrogen bonding between peptide bond atoms in laterally adjacent β-strands within the same β-sheet.

3. Collagen molecules can be described as follows:
   a. They are glycoproteins all of which contain the same proportion of glycosy residues.
   b. All collagen molecules are homotrimers.
   c. Collagen molecules contain extensive regions of β-secondary structure.
   d. Collagen polypeptides are principally α-helical in secondary structure.
   e. None of the above are correct.

4. The principal amino acids of collagen are:
   a. Serine, alanine, and glycine.
   b. Glycine alanine, and proline.
   c. Proline, glycine, and hydroxyproline.
   d. Hydroxyproline, serine, and glycine.

5. Post translational modifications to collagen include the following:
   a. Formation of allysine, an aldehyde, from the δ carbon of lysine.
   b. Removal of N-terminal and C-terminal peptide stretches that contain amino acids that are uncharacteristic of the mature molecule.
   c. Formation of hydroxyproline by a vitamin C requiring process.
   d. All of the above.

6. The allosteric nature of myoglobin is responsible for:
   a. Its oxygen binding properties.
   b. Its interaction with creatine phosphate in energy and oxygen deficient cells.
   c. Its interaction with hemoglobin in oxygen depleted cells.
   d. Intermolecular (between myoglobin molecules) heme heme interactions.
   e. None of the above.

7. Myoglobin can be considered a typical globular protein because:
   a. Its secondary structure is more than 60% α helix.
   b. Its hydrophilic amino acid R-groups are arranged on its spherical surface and its hydrophobic amino acids are in the interior of the sphere.
   c. It contains a small organic prosthetic group.
   d. It has the same distribution of secondary peptide structures as found in the majority of globular proteins.
   e. None of the above are correct.

8. The prosthetic group of hemoglobin and myoglobin contains:
   a. Cobalt.
   b. Four furan rings connected by methylene bridges.
   c. Two propionic acid substituents on a fully conjugated nitrogen/carbon ring system.
   d. 4 vinyl residues that are responsible for its covalent interaction with the apoprotein.

9. The following statements regarding the central metal ion in the prosthetic group of hemoglobin and myoglobin are correct:
   a. When oxygen is bound to the molecule the metal ion undergoes an obligatory change in oxidation state from.
   b. The normal oxidation state of oxyhemoglobin or oxymyoglobin is +3.
   c. It is the binding site for oxygen and carbon monoxide.
   d. Coordinate covalent bonds between the metal and 5 separate carbon atoms are responsible for stabilizing the metal in the prosthetic group.

10. All hemoglobins that are formed from normal (not diseased) human gene products have the following subunit structure:
    a. $\alpha_2\gamma_2$.
    b. $\alpha_2\beta_2$.
    c. $\alpha_2\delta_2$.
    d. α_2S_2.
    e. None of the above.

11. The Bohr effect is:
    a. The name applied to the allosteric effect that oxygen binding has on hemoglobin.
    b. The name of the effect caused by the activity of carbonic anhydrase.
    c. The name of the effect caused by the influence of hydrogen ion concentration on oxygen binding.
    d. The name applied to the chemical reaction that produces carbamates on the N-terminal nitrogen of hemoglobin molecules.

12. Lower pH and higher concentrations of 2,3-bisphosphoglycerate (BPG) have the following effect on hemoglobin:
    a. Lower pH decreases the affinity of hemoglobin for oxygen and higher BPG also decreases the affinity of hemoglobin for oxygen.
    b. Lower pH decreases the affinity of hemoglobin for oxygen while higher BPG increases the affinity of hemoglobin for oxygen.
    c. Lower pH increases the affinity of hemoglobin for oxygen while higher BPG decreases the affinity of hemoglobin for oxygen.
    d. Lower pH increases the affinity of hemoglobin for oxygen and higher BPG also increases the affinity of hemoglobin for oxygen.

13. The T and the R states of hemoglobin can be characterized as follows:
    a. T and R refer to states of hemoglobin secondary structure that are principally dependent on heterotropic effectors and are independent of oxygen binding.
    b. The T state is favored by increased oxygen binding while the R state is favored by oxygen dissociation.
    c. The T state is favored by decreased oxygen binding while the R state is favored by increased oxygen binding.
    d. T and R refer to states of hemoglobin tertiary structure that are principally dependent on heterotropic effectors and are independent of oxygen binding.

Answers

1. b	**6.** e	**11.** c
2. b	**7.** b	**12.** a
3. e	**8.** c	**13.** c
4. c	**9.** c	
5. d	**10.** e	

Part III: Structure of Carbohydrates, Lipids, Nucleic Acids

Carbohydrates

9

Objectives

- To describe the basis for the common nomenclature for the conformation of carbohydrates.
- To define the differences between aldoses and ketoses.
- To describe the structure of glycogen.
- To describe the structure of starch including the differences between amylose and amylopectin.
- To describe the basic composition of glycosaminoglycans (GAGs).
- To describe the major localizations of hyaluronates, chondroitin sulfates, heparan sulfates, heparins, dermatan sulfates, and keratan sulfates.
- To describe the basic composition of proteoglycans.
- To develop an appreciation for the clinical significance of GAGs in lysosomal storage diseases.

Concepts

Carbohydrates are carbon compounds that contain large quantities of hydroxyl groups. The simplest carbohydrates also contain either an aldehyde moiety (these are termed **polyhydroxyaldehydes**) or a ketone moiety (**polyhydroxyketones**) All carbohydrates can be classified as **monosaccharides, oligosaccharides** or **polysaccharides**. Anywhere from two to ten monosaccharide units, linked by glycosidic bonds, make up an oligosaccharide. Polysaccharides are much larger, containing hundreds of monosaccharide units. The presence of the hydroxyl groups allows carbohydrates to interact with the aqueous environment and to participate in hydrogen bonding, both within and between chains. Derivatives of the carbohydrates can contain nitrogens, phosphates and sulfur compounds. Carbohydrates also can combine with lipid to form **glycolipids** (Chapter 10), or with protein to form **glycoproteins** (Chapter 19).

Carbohydrate Nomenclature

The predominant carbohydrates encountered in the body are structurally related to the aldotriose **glyceraldehyde** and to the ketotriose **dihydroxyacetone**. All carbohydrates contain at least one asymmetrical (chiral) carbon and are, therefore, optically active. In addition, carbohydrates can exist in either of two conformations, as determined by the orientation of the hydroxyl group about the asymmetric carbon farthest from the carbonyl. With a few exceptions, those carbohydrates that are of physiological significance exist in the D-**conformation**. The mirror-image conformations, called **enantiomers**, are in the L-conformation.

Monosaccharides

The monosaccharides commonly found in humans are classified according to the number of carbons they contain in their backbone structures (Table 9–1). The major monosaccharides contain four to six carbon atoms (Figures 9–1 and 9–2).

The aldehyde and ketone moieties of the carbohydrates with five and six carbons will spontaneously react with alcohol groups present in neighboring carbons to produce intramolecular **hemiacetals** or **hemiketals**, respectively. This results in the formation of five- or six-membered rings. Because the

Table 9–1. Carbohydrate classification.

Number of Carbons	Category Name	Examples
3	Triose	Glyceraldehyde, dihydroxyacetone
4	Tetrose	Erythrose
5	Pentose	Ribose, ribulose, xylulose
6	Hexose	Glucose, galactose, mannose, fructose
7	Heptose	Sedoheptulose
9	Nonose	Neuraminic acid

five-membered ring structure resembles the organic molecule **furan**, derivatives with this structure are termed **furanoses**. Those with six-membered rings resemble the organic molecule **pyran** and are termed **pyranoses**.

Such ring structures can be depicted by either **Fischer-** or **Haworth-** style diagrams (Figure 9–3). The numbering of the carbons in carbohydrates proceeds from the carbonyl carbon, for aldoses, or the carbon nearest the carbonyl, for ketoses.

The rings can open and re-close, allowing rotation to occur about the carbon bearing the reactive carbonyl yielding two distinct configurations (α and β) of the hemiacetals and hemiketals (Figure 9–3). The carbon about which this rotation occurs is the **anomeric carbon**, and the two forms are termed **anomers**. Carbohydrates can change spontaneously between the α and β configurations—a process known as **mutarotation**. When drawn in the Fischer projection, the α configuration places the hydroxyl attached to the anomeric carbon to the right, toward the ring. When drawn in the Haworth projection, the α configuration places the hydroxyl downward (see Figure 9–3).

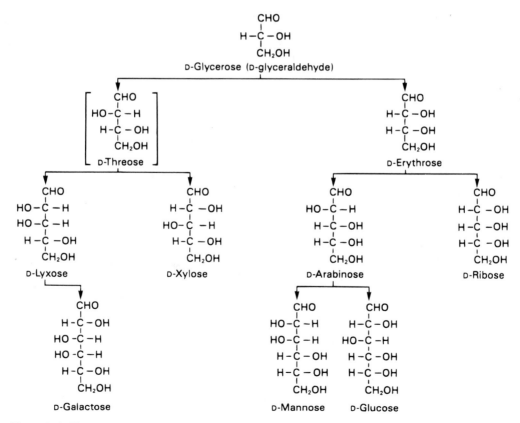

Figure 9–1. The structural relations of the aldoses, D series. D-Threose is not of physiologic significance. The series is built up by the theoretical addition of a CH₂O unit to the −CHO group of the sugar. (Reproduced with permission, from Murray, RK: *Harpers Biochemistry*, 23rd ed. Appleton & Lange, 1993.)

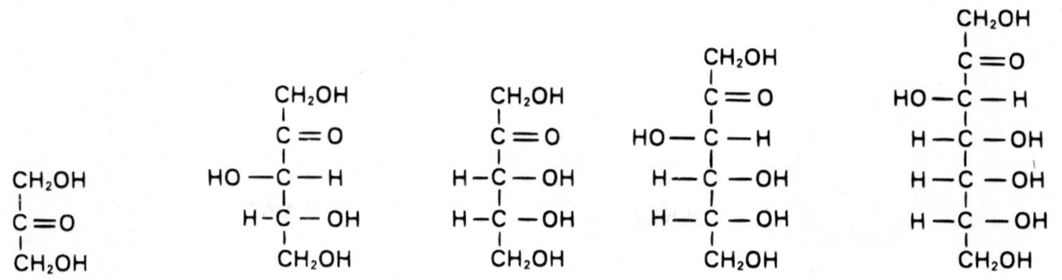

Dihydroxyacetone | D-Xylulose | D-Ribulose | D-Fructose | D-Sedoheptulose

Figure 9–2. Examples of Ketoses of physiologic significance. (Reproduced with permission, from Murray, RK: *Harpers Biochemistry*, 23rd ed. Appleton & Lange, 1993.)

The spatial relationships of the atoms of the furanose and pyranose ring structures are more correctly described by the two conformations identified as the **chair form** and the **boat form** (see Figure 9–3). The chair form is the more stable of the two. Constituents of the ring that project above or below the plane of the ring are **axial**, and those that project parallel to the plane are **equatorial**. In the chair conformation, the orientation of the hydroxyl group about the anomeric carbon of α-D-glucose is axial and equatorial in β-D-glucose.

Disaccharides

Covalent bonds between the anomeric hydroxyl of a cyclic sugar and the hydroxyl of a second sugar (or another alcohol containing compound) are termed **glycosidic bonds**, and the resultant molecules are **glycosides**. The linkage of two monosaccharides to form disaccharides involves a glycosidic bond. Several physiologically important disaccharides are sucrose, lactose and maltose, whose structures are shown in Figure 9–4.

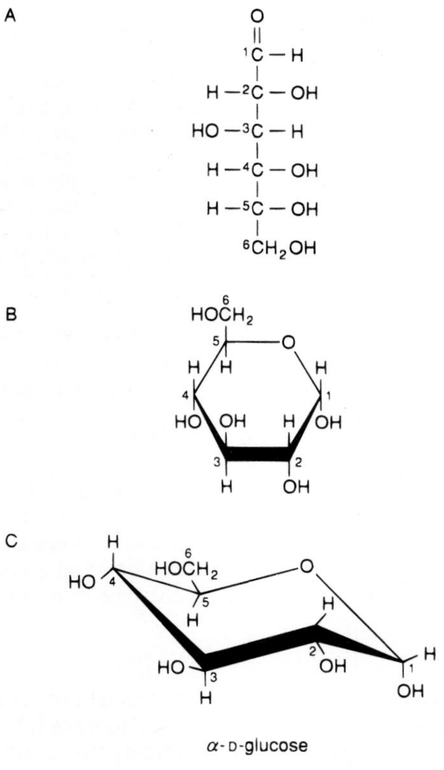

α-D-glucose

Figure 9–3. D-Glucose. ***A:*** straight chain Fischer form. ***B:*** α-D-glucose; Hawthorne projection. ***C:*** α-D-Glucose; chair form. (Reproduced with permission, from Murray, RK: *Harpers Biochemistry*, 23rd ed. Appleton & Lange, 1993.)

Maltose

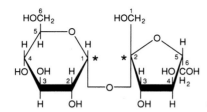

O-α-D-Glucopyranosyl-(1→4)-α-D-glucopyranose

Sucrose

O-α-D-Glucopyranosyl-(1→2)-β-D-fructofuranoside

Lactose

O-β-D-Galactopyranosyl-(1→4)-β-D-glucopyranose

Figure 9–4. Structures of important disaccharides. The −α and β refer to the configuration at the anomeric carbon atom (*). When the anomeric carbon of the second residue takes part in the formation of the glycosidic bond, as in sucrose, the residue becomes a glycoside known as a furanoside or pyranoside. As the disaccharide no longer has a free anomeric carbon with a free potential aldehyde or ketone group, it no longer exhibits reducing properties. (Reproduced with permission, from Murray, RK: *Harpers Biochemistry*, 23rd ed. Appleton & Lange, 1993.)

Sucrose, prevalent in sugar cane and sugar beets, is composed of glucose and fructose through an $\alpha(1,2)\beta$ glycosidic bond. **Lactose** is found exclusively in the milk of mammals and consists of galactose and glucose in a $\beta(1,4)$ glycosidic bond **Maltose,** the major degradation product of starch, is composed of 2 glucose monomers in an $\alpha(1,4)$ glycosidic bond.

Polysaccharides

Most of the carbohydrates found in nature occur in the form of high molecular weight polymers called **polysaccharides**. The monomeric building blocks used to generate polysaccharides can be varied (Figure 9–5); in all cases, however, the predominant monosaccharide found in polysaccharides is D-glucose. When polysaccharides are composed of a single monosaccharide building block, they are termed **homopolysaccharides**. Polysaccharides composed of more than one type of monosaccharide are **heteropolysaccharides**.

Glycogen

Glycogen is the major form of stored carbohydrate in animals. This crucial molecule is a homopolymer of glucose in $\alpha(1,4)$ linkage; it is also highly branched, with $\alpha(1,6)$ branch linkages occurring every 8–10 residues (Figure 9–6). Glycogen is a very compact structure that results from the

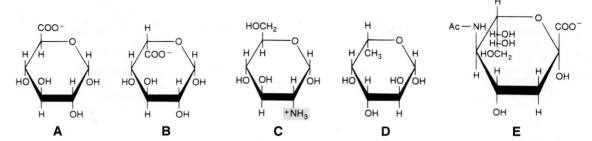

Figure 9–5. A: α-D-Glucuronate (left) and β-L-iduronate (right). **B:** Glucosamine (2-amino-D-galactopyranose. Both glucosamine and galactosamine occur as N-acetyl derivatives in more complex carbohydrates, eg, glycoproteins. **C:** β-L-Fucose (6-deoxy-β-L-galactose). **D:** Stucture of N-acetylneuraminic acid, a sialic acid (Ac = $CH_3 - CO-$). (Reproduced with permission, from Murray, RK: *Harpers Biochemistry*, 23rd ed. Appleton & Lange, 1993.)

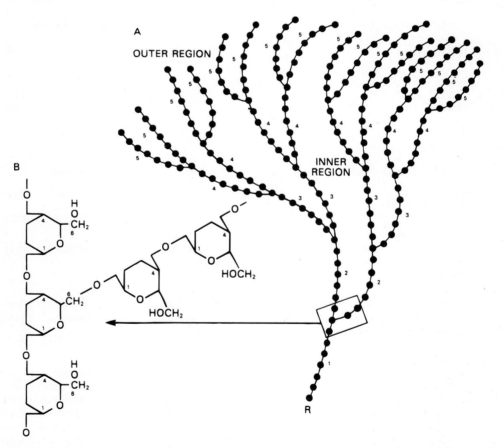

Figure 9–6. The glycogen molecule **A:** General structure. **B:** Enlargement of structure at a branch point. The numbers in **A** refer to equivalent stages in the growth of the macromolecule. R, primary glucose residue, which is the only glucose residue in the structure shown that contains a free reducing group on C_1. The branching is ore variable than shown, the ratio of 1 → 4 to 1 → 6 bonds being from 10 to 18. (Reproduced with permission, from Murray, RK: Harpers Biochemistry, 23rd ed. Appleton & Lange, 1993.)

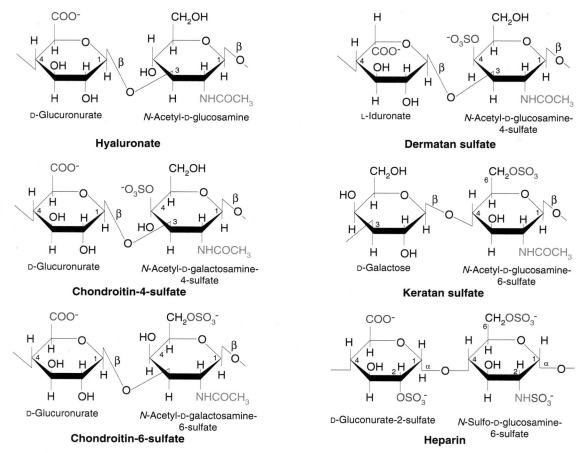

Figure 9–7. The disaccharide repeating units of the common glycosaminoglycans. (Reproduced, with permission from Voet, D. and Voet, JG: *Biochemistry*, 1st ed. John Wiley and Sons, Inc. 1990.)

coiling of the polymer chains. This compactness allows large amounts of carbon energy to be stored in a small volume, with little effect on cellular osmolarity.

Starch

Starch is the major form of stored carbohydrate in plant cells. Its structure is identical to glycogen, except for a much lower degree of branching (about every 20–30 residues). Unbranched starch is called **amylose**; branched starch is called **amylopectin**.

Glycosaminoglycans

The most abundant heteropolysaccharides in the body are the **glycosaminoglycans (GAGs)**. These molecules are long unbranched polysaccharides containing a repeating disaccharide unit. The disaccharide units contain either of two modified sugars—[*N*-acetylgalactosamine (GalNAc)] or [*N*-acetylglucosamine (GlcNAc)]—and a uronic acid such as **glucuronate** or **iduronate** (see Figure 9–5). GAGs are highly negatively charged molecules, with an extended conformation that imparts high viscosity to the solution. GAGs are located primarily on the surface of cells or in the extracellular matrix (ECM). Along with the high viscosity of GAGs comes low compressibility, which makes these molecules ideal for a lubricating fluid in the joints. At the same time, their rigidity provides structural integrity to cells and provides passageways between cells, allowing for cell migration. The specific GAGs of physiological significance are **hyaluronic acid, dermatan sulfate, chondroitin sulfate, heparin, heparan sulfate, and keratan sulfate** (Figure 9–7).

Hyaluronic acid is unique among the GAGs in that it does not contain any sulfate and is not found covalently attached to proteins as a proteoglycan. It is, however, a component of non-covalently formed complexes with proteoglycans in the ECM. Hyaluronic acid polymers are very large (with molecular weights of 10^5-10^7) and can displace a large volume of water. This property makes them excellent lubricators and shock absorbers.

Proteoglycans

The majority of GAGs in the body are linked to core proteins, forming **proteoglycans** (also called mucopolysaccharides). The GAGs extend perpendicularly from the core in a brush-like structure. The linkage of GAGs to the protein core involves a specific trisaccharide composed of two galactose residues and a xylulose residue (**GAG-GalGalXyl-O-CH₂-protein**). The trisaccharide linker is coupled to the protein core through an O-glycosidic bond to a S residue in the protein (Figure 9–8). Some forms of keratin sulfates are linked to the protein core through an N-asparaginyl bond. The protein cores of proteoglycans are rich in S and T residues, which allows multiple GAG attachments.

CLINICAL SIGNIFICANCE

There are at least 15 forms of inherited disease that are related to deficiencies in the enzymes involved in glycogen metabolism. These disorders are known collectively as the **glycogen storage diseases** (see Chapter 15).

Proteoglycans and GAGs perform numerous vital functions within the body, some of which still remain to be studied. One well-defined function of the GAG heparin is its role in preventing coagulation of the blood. As indicated in Table 9–2, heparin is abundant in the granules of the mast cells that line blood vessels. The release of heparin from these granules, in response to injury, and its subsequent entry into the serum leads to an inhibition of blood clotting, in the following manner. Free heparin complexes with and activates antithrombin III, which in turn inhibits all the serine proteases of the coagulation cascade (see Chapter 42). This phenomenon has been clinically exploited in the use of heparin injection for anti-coagulation therapies.

Several genetically inherited diseases, for example the **lysosomal storage diseases**, result from defects in the lysosomal enzymes responsible for the metabolism of complex membrane-associated GAGs. These specific diseases, termed **mucopolysaccharidoses (MPS)** lead to an accumulation of GAGs

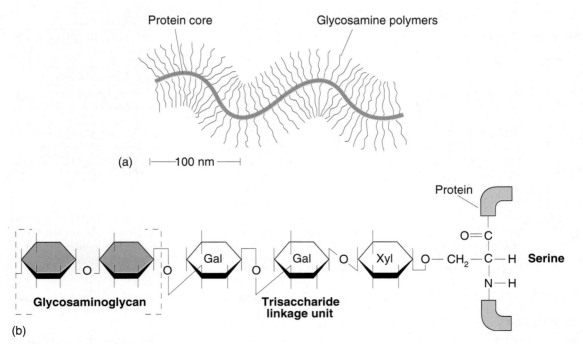

Figure 9–8. (a) Structure of a typical proteoglycan. (b) The attachment site between a serine in the core protein and the glycosaminoglycan. (Reproduced, with permission from Zubay, G: *Biochemistry*, 3d ed. Wm. C. Brown Communications, Inc. 1993.)

Table 9–2. Characteristics of GAGs (glycosaminoglycans).

GAG	Localization; Comments
Hyaluronate	Synovial fluid, vitreous humor, ECM of loose connective tissue; large polymers, shock absorbing
Chondroitin sulfate	Cartilage, bone, heart valves; most abundant GAG
Heparan sulfate	Basement membranes, components of cell surfaces; contains higher acetylated glucosamine than heparin
Heparin	Component of intracellular granules of mast cells lining the arteries of the lungs, liver and skin; more sulfated than heparan sulfates
Dermatan sulfate	Skin, blood vessels, heart valves
Keratan sulfate	Cornea, bone, cartilage aggregated with chondroitin sulfates

Table 9–3. Several diseases of mucopolysaccharide metabolism.

Type: Syndrome	Enzyme Defect	Symptoms
I-H: Hurler's	α-L-iduronidase	Corneal clouding; dwarfism; mental retardation; early mortality
I-S: Scheie's	α-L-iduronidase	Corneal clouding; aortic valve disease; joint stiffening; normal intelligence and life span
I-H/S: Huler/Scheie	α-L-iduronidase	Similar to both IH and IS
II: Hunter's	L-iduronate-2-sulfatase	Mild and severe forms, X-linked, also a possible autosomal form, facial and physical deformities; mental retardation
III-A: Sanfilippo(A)	Heparan N-sulfatase	Skin, brain, lungs, heart and skeletal muscle affected in all four types of MPSIII
III-B: Sanfilippo(B)	N-acetyl-α-D-glucosaminidase	Congestive heart failure; progressive mental retardation
III-C: Sanfilippo(C)	N-acetylCoA:α-glucosamine-N-acetyltransferase	Coarse facial features; organomegaly
III-D: Sanfilippo(D)	N-acetyl-α-glucosamine-6-sulfatase	Moderate physical deformities; progressive mental retardation
IV-A: Morquio's(A)	N-acetylgalactosamine-6-sulfatase	Corneal clouding, thin enamel, aortic valve disease, skeletal abnormalities
IV-B: Morquio's(B)	β-galactosidase	Mild skeletal abnormalities, normal enamel, hypoplastic odontoid, corneal clouding
VI: Maroteaux-Lamy	N-acetylgalactosamine-4-sulfatase	Three distinct forms from mild to severe, aortic valve disease, normal intellect, corneal clouding, coarse facial features
VII: Sly	β-glucuronidase	Hepatosplenomegaly, dystosis multiplex

within cells. There are at least 14 known types of lysosomal storage diseases that affect GAG catabolism (Table 9.3); some of the more commonly encountered examples are **Hurler's syndrome, Hunter's syndrome, Sanfilippo syndrome, Maroteaux-Lamy syndrome, and Morquio's syndrome.** All of these disorders, excepting Hunters syndrome, are inherited in an autosomal recessive manner.

Questions

DIRECTIONS (items 1–15): Each numbered item or incomplete statement is followed by answers or by completions of the statement. Select the ONE lettered answer or completion that is BEST in each case.

1. A 2-year-old female from a large consanguineous Turkish family was presented with coarse facial features, organomegaly, radiological signs of dystosis multiplex and elevated urinary heparan sulfate secretion. An assay for N-acetylCoA:α-glucosamine-N-acetyltransferase activity in fibroblast from the patient showed significantly lower activity than that of control fibroblasts taken from a normal sibling. These symptoms and clinical observations suggest the child is suffering from:
   a. Sanfilippo syndrome Type B.
   b. Sanfilippo syndrome Type C.
   c. Hurlers syndrome.
   d. Sly syndrome.
   e. Hunters syndrome.

2. A proteoglycan with no N- or O-glycosidic linkages is:
   a. Heparan Sulfate.
   b. Dermatan Sulfate.
   c. Keratan Sulfate.
   d. Hyaluronic acid.
   e. Chondroitin Sulfate.

3. A 30-month-old child exhibiting disproportionate short-trunk dwarfism underwent histological and electron-microscopic examination of the long bone growth plate. These results indicated marked irregularities in chondrocyte orientation within the growth plate, disruption of the normal columnar architecture, and vacuolization and enlargement of the cellular border. The childs physical stature and the ultrastructural analysis of bone development indicate the child is suffering from:
   a. Hurler's syndrome.
   b. Sanfilippo syndrome Type A.
   c. Hunter's syndrome.
   d. Morquio's syndrome Type B.
   e. Maroteaux-Lamy syndrome.

4. Which of the following are aldohexoses?
   a. Erythrose and ribose.
   b. Erythrose and glucose.
   c. Ribulose and xylulose.
   d. Fructose and xylulose.
   e. Glucose and mannose.

5. The anomeric carbon(s) of carbohydrates:
   a. Is the carbon bearing the reactive carbonyl.
   b. Is the carbon next to the reactive carbonyl.
   c. Is the carbon furthest from the reactive carbonyl.
   d. Are all the carbons bearing hydroxyl groups.
   e. Is the carbon with which the reactive carbonyl interacts to form a ring structure.

6. Mucopolysaccharidoses (MPS) are inherited disorders resulting from:
   a. The synthesis of polysaccharides with altered structures that results in abnormal cell shape and functioning.
   b. The synthesis of GAGs with abnormal structures that results in abnormal cell shape and functioning.
   c. Elevated biosynthesis and accumulation of GAGs.
   d. Defective lysosomal enzymes involved in the degradation of GAGs.
   e. Defective enzymes involved in the synthesis of GAGs, leading to reduced levels of these molecules.

7. Glycosides are:
   a. Molecules resulting from the covalent linkage of two alcohols.
   b. Sugars that exist in the linear conformation.

   c. Molecules resulting from the covalent linkage of a sugar and an acid such as phosphate.

   d. Molecules resulting from the covalent linkage of a sugar and an alcohol such as a second sugar.

   e. Sugars that exist in the hemiacetal or hemiketal form.

8. A 12-month-old female is presented with severe developmental delay, macrocephaly, dysmorphic facies, hypotonia, hepatosplenomegaly and Alder-Reilly bodies in peripheral blood leukocytes. The activity of iduronate sulfatase in the plasma is significantly below normal. These symptoms are indicative of:

   a. Hurler's syndrome.

   b. Sanfilippo syndrome Type B.

   c. Hunter's syndrome.

   d. Morquio's syndrome Type B.

   e. Maroteaux-Lamy syndrome.

9. The prevalent GAG found in the composition of basement membranes and associated with cell surfaces is:

   a. Keratan sulfate.

   b. Chondroitin sulfate.

   c. Heparan sulfate.

   d. Dermatan sulfate.

   e. Heparan.

10. The prevalent disaccharide composed of fructose and glucose is:

   a. Maltose.

   b. Sucrose.

   c. Lactose.

   d. Cellobiose.

   e. Isomaltose.

11. A 6-year-old female is presented with corneal clouding, coarse facial features, increased urinary GAG excretion, white cell metachromasia and decreased N-acetylgalactosamine-4-sulfatase (sometimes referred to as arylsulfatase B) activity in peripheral blood leukocytes. These symptoms are indicative of which disease?

   a. Scheies syndrome.

   b. Sanfilippo syndrome Type C.

   c. Hunter's syndrome.

   d. Sly syndrome.

   e. Maroteaux-Lamy syndrome.

12. Carbohydrates that are enantiomers have which of the following features?

   a. They differ by the orientation of one hydroxyl group.

   b. One would exist in the boat conformation, the other in the chair conformation.

   c. They are mirror images of each other.

   d. They differ by the orientation of the hydroxyl about the anomeric carbon.

   e. One would be a pyranose, the other a furanose.

13. Glycogen is the major cellular store of carbohydrate. This molecule is composed of:

   a. Glucose in $\alpha(1,6)$ linkage and $\alpha(1,4)$ branches.

   b. Glucose in $\alpha(1,4)$ linkage and $\alpha(1,6)$ branches.

   c. Glucose in $\beta(1,4)$ linkage and $\alpha(1,6)$ branches.

   d. Glucose in $\alpha(1,4)$ linkage and $\beta(1,6)$ branches.

   e. Glucose in $\beta(1,4)$ linkage and $\beta(1,6)$ branches.

14. A typical proteoglycan aggregate may contain:

   a. Hyaluronate non-covalently linked to a protein core.

   b. Protein non-covalently linked to a chondroitin sulfate core.

   c. Keratan sulfates non-covalently linked to a protein core.

   d. Hyaluronate covalently linked to a protein core.

   e. Heparan sulfates covalently linked to keratan sulfates.

15. Keratan sulfates contain:
    a. Iduronate and GlcNAc.
    b. Glucose and GalNAc.
    c. Glucuronate and GlcNAc.
    d. Galactose and GlcNAc.
    e. Glucuronate and GalNAc.

Answers

1. b	**6.** d	**11.** e
2. d	**7.** d	**12.** c
3. a	**8.** c	**13.** b
4. e	**9.** c	**14.** a
5. a	**10.** b	**15.** d

Lipids

10

Objectives

- To describe the structure of a fatty acid.
- To define the physiologically relevant fatty acids.
- To describe the structure of triacylglycerols and their role in the body.
- To describe the basic structure of the phospholipids.
- To describe the difference between phosphatidylcholines, phosphatidylethanolamines, phosphatidylserines, phosphatidylglycerols, cardiolipins, and phosphatidylinositols.
- To describe the basic structure of sphingolipids.
- To describe the structure of sphingomyelin.
- To describe the structure of glycosphingolipids.
- To describe several clinical significances of the phospholipids.
- To develop an appreciation for the clinical significance of glycoshingolipids in lysosomal storage diseases.
- To describe the structure of cholesterol.

Concepts

Biological molecules that are insoluble in aqueous solutions and soluble in organic solvents are classified as **lipids**. The lipids of physiological importance for humans have four major functions:

1. They serve as structural components of biological membranes;
2. They provide energy reserves, predominantly in the form of triacylglycerols;
3. Both lipids and lipid derivatives serve as vitamins and hormones;
4. Lipophilic bile acids aid in lipid solubilization.

Fatty Acids

Fatty acids are long-chain hydrocarbon molecules containing a carboxylic acid moiety at one end (Table 10–1). The numbering of carbons in fatty acids begins with the carbon of the carboxylate group. At physiological pH, the carboxyl group is readily ionized, rendering a negative charge onto fatty acids in bodily fluids.

Fatty acids that contain no carbon-carbon double bonds are termed **saturated fatty acids**; those that contain double bonds are **unsaturated fatty acids**. The numeric designations used for fatty acids come from the number of carbon atoms, followed by the number of sites of unsaturation (eg, palmitic acid is a 16-carbon fatty acid with no unsaturation and is designated by 16:0). The site of unsaturation in a fatty acid is indicated by the symbol $\triangle$ and the number of the first carbon of the double bond (eg, palmitoleic acid is a 16-carbon fatty acid with one site of unsaturation between carbons 9 and 10 and is designated by $16:1^{\triangle}$.

Saturated fatty acids of less than eight carbon atoms are liquid at physiological temperature, whereas those containing more than ten are solid. The presence of double bonds in fatty acids significantly lowers the melting point relative to a saturated fatty acid.

The majority of body fatty acids are acquired in the diet and subsequently utilized for energy production or stored for future use (Figure 10–1). However, the lipid biosynthetic capacity of the body (fatty acid synthase and other modifying enzymes) can supply the body with all the various fatty acid structures needed. The two key exceptions to this are the highly unsaturated fatty acids known as **linoleic acid** and **linolenic acid**, containing unsaturation sites beyond carbons 9 and 10. These two fatty acids cannot be synthesized from precursors in the body, and are thus considered the **essential fatty acids**; essential in the sense that they must be provided in the diet. Since plants are indeed capable of synthesizing linoleic and linolenic acid, humans can acquire these fats by consuming a variety of plants, or else by eating the meat of animals that have consumed these plant fats.

Fatty acids fill two major roles in the body: as the **components of more complex membrane lipids** (see Chapter 12) and as the **major components of stored fat**, in the form of triacylglycerols (Figure 10–2).

Triacylglycerols

Fatty acids are stored for future use as esters of glycerol. Most are stored in the form of triacylglycerols (Figure 10–2), with all three hydroxyls of glycerol esterified to a fatty acid. Triacylglycerols are stored predominantly within the cells of adipose tissue. Mono- and diacylglycerols exist only as temporary metabolic byproducts of triacylglycerol metabolism and are found in very small amounts.

Phospholipids

The major lipids of biological membranes are the **phospholipids, glycosphingolipids** and **cholesterol**. The phospholipids comprise the phosphoglycerides and sphingomyelins (see below).

Phosphoglycerides are composed of **a polar head group and two nonpolar long chain aliphatic tails**. The structural core of these molecules is glycerol-3-phosphate, in which the phosphate moiety constitutes the polar head group and fatty acids esterified to carbons 1 and 2 constitute the non-polar tail. The basic form of this class of molecule is **phosphatidic acid** (Figure 10–3).

Table 10–1. Physiologically relevant fatty acids.

Numerical Symbol	Common Name	Structure	Comments
14:0	Myristic acid	$CH_3(CH_2)_{12}COOH$	Often found attached to the N-term of plasma membrane-associated cytoplasmic proteins
16:0	Palmitic acid	$CH_3(CH_2)_{14}COOH$	End product of mammalian fatty acid synthesis
$16:1^{\triangle 9}$	Palmitoleic acid	$CH_3(CH_2)_5C=C(CH_2)_7COOH$	
18:0	Stearic acid	$CH_3(CH_2)_{16}COOH$	
$18:1^{\triangle 9}$	Oleic acid	$CH_3(CH_2)_7C=C(CH_2)_7COOH$	
$18:2^{\triangle 9,12}$	Linoleic acid	$CH_3(CH_2)_4C=CCH_2C=C(CH_2)_7COOH$	Essential fatty acid
$18:3^{\triangle 9,12,15}$	Linolenic acid	$CH_3CH_2C=CCH_2C=CCH_2C=C(CH_2)_7COOH$	Essential fatty acid
$20:4^{\triangle 5,8,11,14}$	Arachidonic acid	$CH_3(CH_2)_3(CH_2C=C)_4(CH_2)_3COOH$	Precursor for eicosanoid synthesis

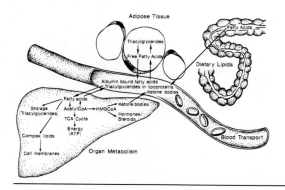

Figure 10–1. The metabolic interrelationships between dietary intake, plasma transport and organ utilization/storage of fatty acids.

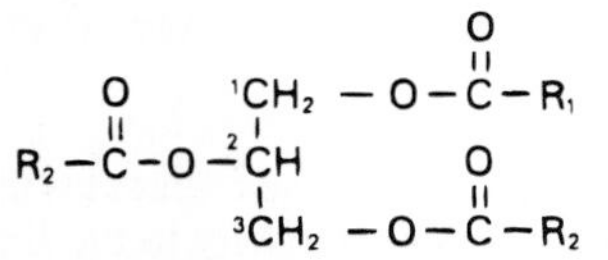

Figure 10–2. Triacylglycerol. (Reproduced with permission, from Murray, RK: *Harpers Biochemistry*, 23rd ed. Appleton & Lange, 1993.)

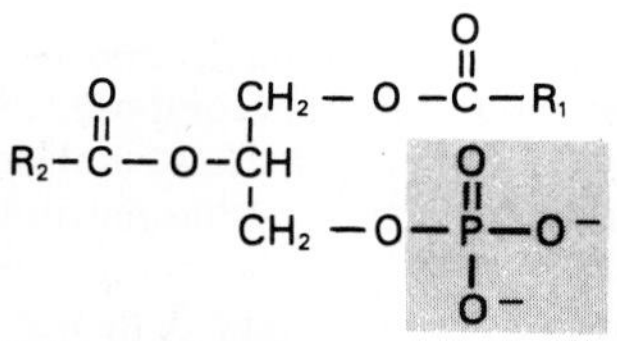

Figure 10–3. Phosphatidic acid. (Reproduced with permission, from Murray, RK: *Harpers Biochemistry*, 23rd ed. Appleton & Lange, 1993.)

The fatty acid on carbon 1 is almost always saturated. If an unsaturated fatty acid is present in a phosphoglyceride, it is almost always found on carbon 2. The phosphate group is a bridge to additional functional groups, predominantly nitrogenous bases. Phospholipids lacking one of the fatty acids at carbon 1 or carbon 2 are termed **lysophospholipids**. The vast majority of lysophospholipids lack a fatty acid at carbon 2; these compounds make up only a minor proportion of the total phospholipid in the cell.

Abundant Phosphoglycerides Found in Mammalian Membranes

A. Phosphatidylcholines (Figure 10–4): This class of phospholipids is also called the **lecithins**. At physiological pH, phosphatidylcholines are neutral zwitterions. They contain primarily palmitic or stearic acid at carbon 1 and primarily oleic, linoleic or linolenic acid at carbon 2. The lecithin di-palmitoyllecithin, is a component of **lung or pulmonary surfactant**. It contains palmitate at both carbon 1 and 2 of glycerol and is the major (80%) phospholipid found in the extracellular lipid layer lining the pulmonary alveoli.

B. Phosphatidylethanolamines (Figure 10–5): These molecules are neutral zwitterions at physiological pH. They contain primarily palmitic or stearic acid on carbon 1 and a long chain unsaturated fatty acid (eg, 18:2, 20:4, and 22:6) on carbon 2.

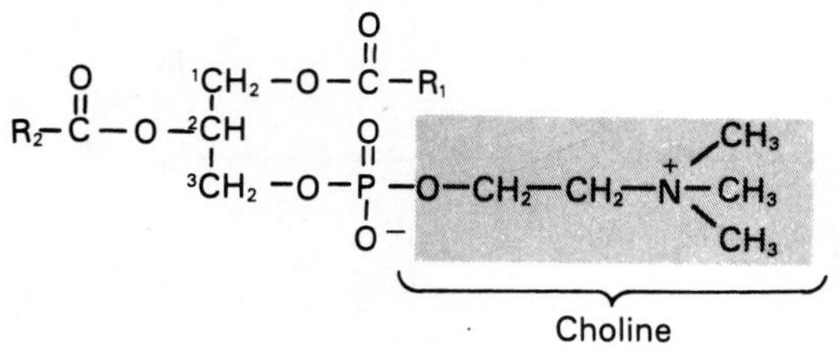

Choline

Figure 10–4. 3-Phosphatidylcholine. (Reproduced with permission, from Murray, RK: *Harpers Biochemistry*, 23rd ed. Appleton & Lange, 1993.)

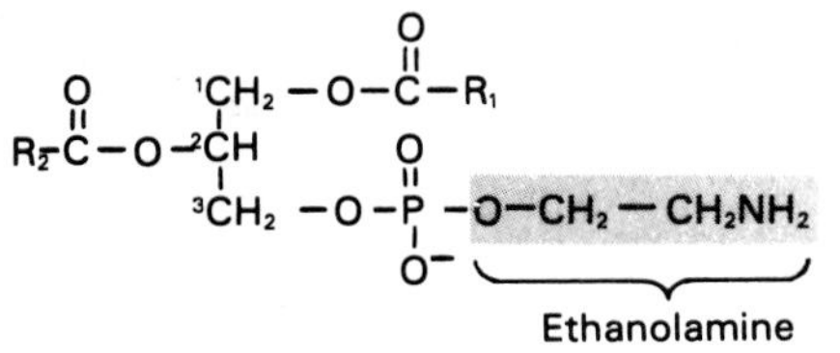

Figure 10–5. 3-Phosphatidylethanolamine. (Reproduced with permission, from Murray, RK: *Harpers Biochemistry*, 23rd ed. Appleton & Lange, 1993.)

C. Phosphatidylserines (Figure 10–6): Phosphatidylserines will carry a net charge of -1 at physiological pH and are composed of fatty acids similar to the phosphatidylethanolamines.

D. Phosphatidylglycerols (Figure 10–7): Phosphatidylglycerols exhibit a net charge of -1 at physiological pH. These molecules are found in high concentration in mitochondrial membranes and as components of **pulmonary surfactant**. Phosphatidylglycerol also is a precursor for the synthesis of cardiolipin.

E. Diphosphatidylglycerols (cardiolipins, Figure 10–7): These molecules are very acidic, exhibiting a net charge of -2 at physiological pH. They are found primarily in the inner mitochondrial membrane and also as components of **pulmonary surfactant**.

F. Phosphatidylinositols (Figure 10–8): These molecules contain almost exclusively stearic acid at carbon 1 and arachidonic acid at carbon 2. Phosphatidylinositols composed exclusively of non-phosphorylated inositol exhibit a net charge of -1 at physiological pH. These molecules exist in membranes with various levels of phosphate esterified to the hydroxyls of the inositol. Molecules with phosphorylated inositol are termed **polyphosphoinositides**. The polyphosphoinositides are important intracellular transducers of signals emanating from the plasma membrane (Chapter 38).

An additional class of phospholipids containing an alkyl ether or an alkenyl ether substitution at carbon 1 is the **plasmalogens** (Figure 10–9).

Ethanolamine plasmalogen is prevalent in myelin. Choline plasmalogen is abundant in cardiac tissue. **Platelet activating factor (PAF)** is a physiologically significant alkyl ether choline plasmalogen (Figure 10–10). It functions as a mediator of hypersensitivity, acute inflammatory reactions and anaphylactic shock. PAF is synthesized in response to the formation of antigen-IgE complexes on the sur-

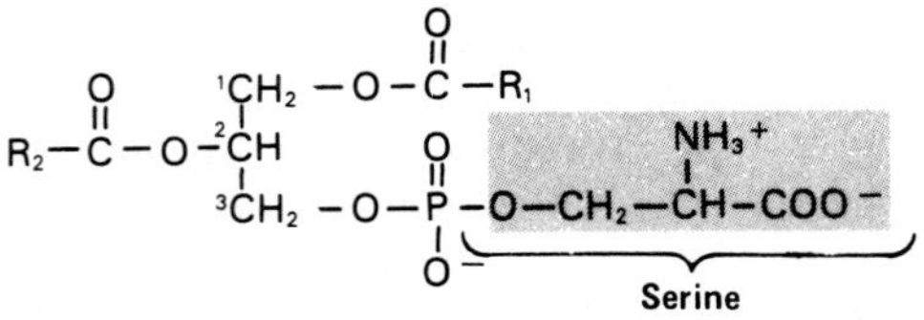

Figure 10–6. 3-Phosphatidylserine. (Reproduced with permission, from Murray, RK: *Harpers Biochemistry*, 23rd ed. Appleton & Lange, 1993.)

Phosphatidylglycerol

Diphosphatidylglycerol (cardiolipin)

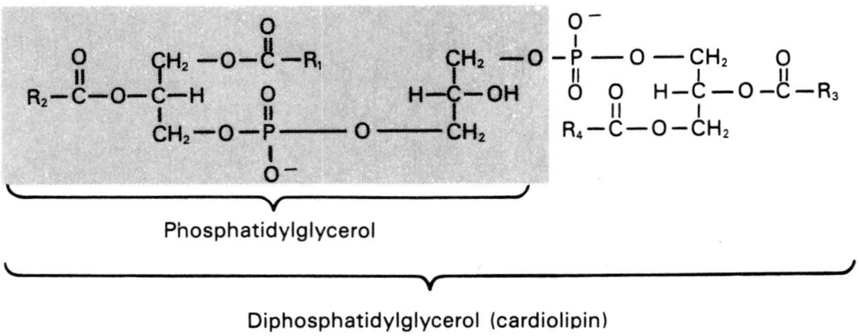

Figure 10–7. Cardiolipin (diphosphatidylglycerol). (Reproduced with permission, from Murray, RK: *Harpers Biochemistry*, 23rd ed. Appleton & Lange, 1993.)

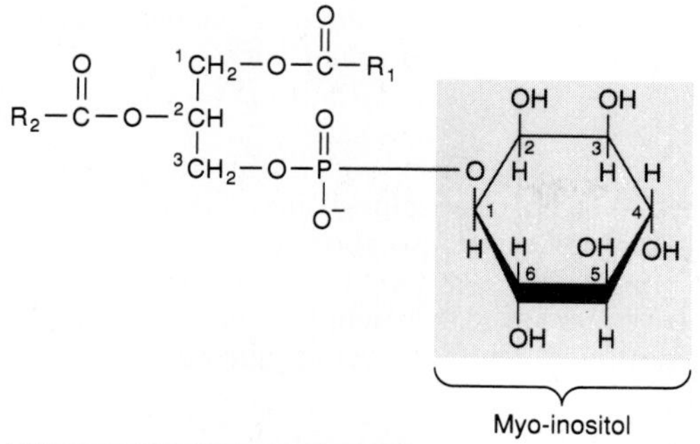

Figure 10–8. 3-Phosphatidylinositol. (Reproduced with permission, from Murray, RK: *Harpers Biochemistry*, 23rd ed. Appleton & Lange, 1993.)

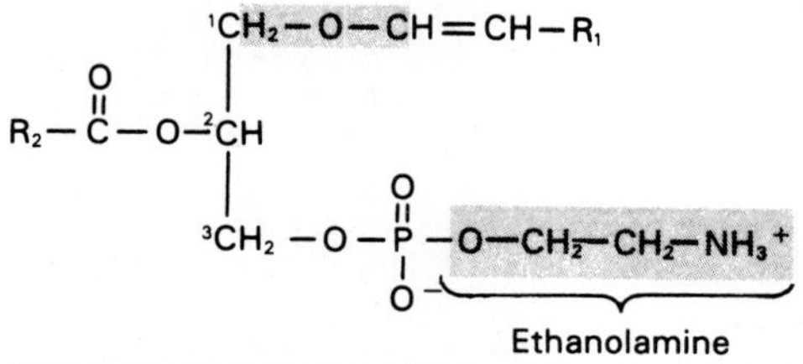

Figure 10–9. Plasmalogen (phosphatidal ethanolamine.) (Reproduced with permission, from Murray, RK: *Harpers Biochemistry*, 23rd ed. Appleton & Lange, 1993.)

faces of basophils, neutrophils, eosinophils, macrophages, and monocytes. The synthesis and release of PAF from cells leads to platelet aggregation and the release of serotonin from platelets. PAF also produces responses in liver, heart, smooth muscle, and uterine and lung tissues.

Sphingolipids

The sphingolipids, like the phospholipids, are composed of a polar head group and two nonpolar tails. The core of sphingolipids is the long-chain amino alcohol, **sphingosine** (Figure 10–11). Amino acylation, with a long chain fatty acid, at carbon 2 of sphingosine yields a **ceramide** (Figure 10–11). The esterification of carbon 1 of a ceramide with phosphocholine yields a **sphingomyelin** (Figure 10–11).

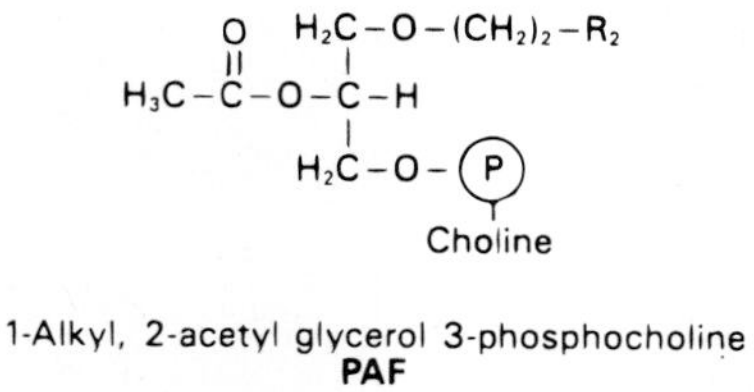

1-Alkyl, 2-acetyl glycerol 3-phosphocholine
PAF

Figure 10–10. 1,alkyl, 2-acetyl glycerol 3-phosphocholine (platelet activating factor, PAF). (Reproduced with permission, from Murray, RK: *Harpers Biochemistry*, 23rd ed. Appleton & Lange, 1993.)

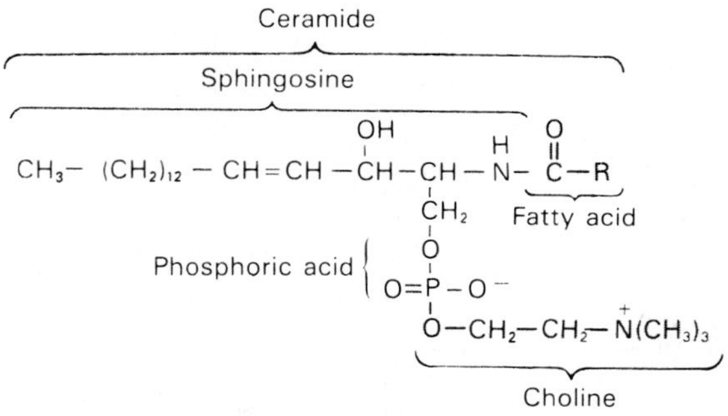

Figure 10–11. A sphingomyelin. (Reproduced with permission, from Murray, RK: *Harpers Biochemistry*, 23rd ed. Appleton & Lange, 1993.)

Sphingomyelins are important structural lipid components of nerve cell membranes. The predominant sphingomyelins contain palmitic or stearic acid N-acylated at carbon 2 of sphingosine.

Glycosphingolipids

Glycosphingolipids, or glycolipids, are composed of a ceramide backbone with a wide variety of carbohydrate groups (mono- or oligosaccharides) attached to carbon 1 of sphingosine. The three principal classes of glycosphingolipids are the cerebrosides, globosides, and gangliosides.

A. Cerebrosides: Cerebrosides have a single sugar group linked to ceramide (Figure 10–12). The most common of these is galactose (galactocerebrosides), with a minor level of glucose (glucocere-

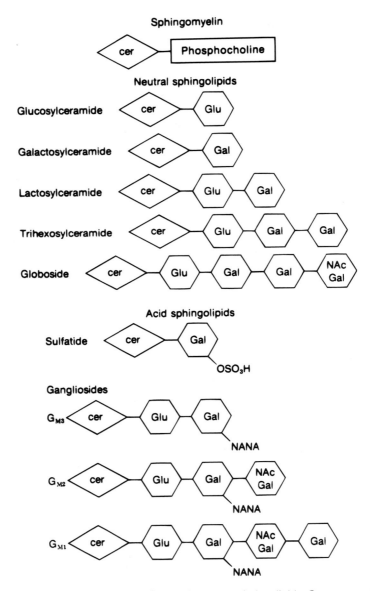

Figure 10–12. Diagrammatic representation of several common shpingolipids. Cer = ceramide, Glu = glucose, Gal = galactose, NAcGal = N-acetylgalactosamine, NANA = N-acetylneuraminic acid (sialic acid). (Reproduced with permission, from Devlin, T: *Textbook of Biochemistry with Clinical Correlations*, 3rd ed. Wiley-Liss, Inc. 1992.)

brosides). Galactocerebrosides are found predominantly in neuronal cell membranes. By contrast, glucocerebrosides are not normally found in membranes, especially neuronal membranes; instead, they represent intermediates in the synthesis or degradation of more complex glycosphingolipids.

B. Globosides: Globosides have multiple sugar groups linked to ceramide (Figure 10–12).

C. Gangliosides: Gangliosides are the most complex of the glycosphingolipids (Figure 10–12). Like globosides, they contain multiple sugar groups linked to ceramide. The distinguishing feature of gangliosides is the presence of **N-acetylneuraminic acid** (or NANA, one of a family of C_9 sugars termed sialic acids, see Figure 9–5).

The specific names for gangliosides are a key to their structure. The letter G refers to "ganglioside," and the subscripts M, D, T, and Q indicate that the molecule contains a mono-, di-, tri- and quatra(tetra)-sialic acid. The numerical subscripts 1, 2 and 3 refer to the carbohydrate sequence that is attached to ceramide: 1 stands for GalGalNAcGalGlc-ceramide, 2 for GalNAcGalGlc-ceramide and 3 for GalGlc-ceramide.

Cholesterol

Cholesterol is a major structural lipid of membranes, lending increased "stiffness" to the lipid bilayer. Cholesterol also is the precursor of the steroid hormones and the bile acids. This important compound is composed of the steroid nucleus, consisting of four fused rings, and an alkyl side chain (Figure 10–13). The presence of a hydroxyl attached to the A ring of the nucleus imparts a polar head group to cholesterol, allowing it to orient within the lipid bilayer. The specifics of cholesterol metabolism and utilization will be discussed in Chapter 22.

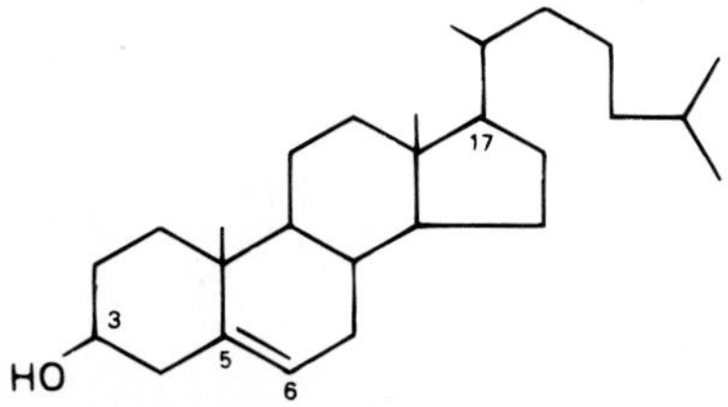

Figure 10–13. Cholesterol. (Reproduced with permission, from Murray, RK: *Harpers Biochemistry*, 23rd ed. Appleton & Lange, 1993.)

CLINICAL SIGNIFICANCE

A significant cause of death in premature infants and, on occasion, in full term infants is **respiratory distress syndrome (RDS)** or hyaline membrane disease. This condition is caused by an insufficient amount of pulmonary surfactant. Under normal conditions, the surfactant is synthesized by type II endothelial cells and is secreted into the alveolar spaces to prevent atelectasis following expiration during breathing. Surfactant is comprised primarily of dipalmitoyllecithin; additional lipid components include phosphatidylglycerol and phosphatidylinositol along with proteins of 18 and 36 kDa (termed **surfactant proteins**). During the third trimester the fetal lung synthesizes primarily sphingomyelin, and type II endothelial cells convert the majority of their stored glycogen to fatty acids and then to dipalmitoyllecithin. Fetal lung maturity can be determined by measuring the ratio of lecithin to sphingomyelin (L/S ratio) in the amniotic fluid. An L/S ratio less than 2.0 indicates a potential risk of RDS. The risk is nearly 75–80% when the L/S ratio is 1.5.

The sphingolipids are involved in various recognition functions at the surface of cells. As with the complex glycoproteins, an understanding of all of the functions of the glycolipids is far from complete. Glycosphingolipids on the surface of erythrocytes comprise the blood group antigens, A, B, and O (Figure 10–14). The ganglioside, G_{M1} present on the surface of intestinal epithelial cells, is the site of attachment of cholera toxin, the protein secreted by *Vibrio cholerae*.

Several inborn diseases of complex glycosphingolipid metabolism have been identified. Like the mucopolysaccharidoses, these diseases are the result of abnormal lysosomal enzymes involved in the metabolism of glycosphingolipids; they are therefore termed **lysosomal or sphingolipid storage diseases** (Table 10–2).

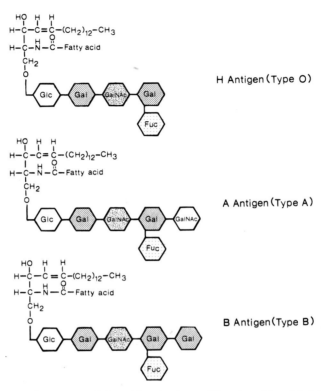

H Antigen (Type O)

A Antigen (Type A)

B Antigen (Type B)

Figure 10–14. Diagramatice representation of the structures of the glycoshingolipids that comprise the three major blood group antigens, A, B, and O. Glc = glucose, Gal = galactose, Fuc = fucose, GalNAc = N-acetyl-galactoseamine.

Table 10–2. Disorders associated with abnormal sphingolipid metabolism.

Disorder	Enzyme Deficiency	Accumulating Substance	Symptoms
Tay-Sachs disease	Hexosaminidase A	G_{M2} ganglioside	Mental retardation, blindness, early mortality
Gaucher's disease	Glucocerebrosidase	Glucocerebroside	Hepatosplenomegaly, mental retardation in infantile form, long bone degeneration
Fabry's disease	α-Galactosidase A	Globotriaosylceramide or Ceramide trihexoside (CTH)	Kidney failure, skin rashes
Niemann-Pick disease	Sphingomyelinase	Sphingomyelin	Mental retardation, hepatosplenomegaly
Krabbe's disease; globoid leukodystrophy	Galactocerebrosidase	Galactocerebroside	Mental retardation, myelin deficiency
Sandhoff-Jatzkewitz disease	Hexosaminidase A and B	Globoside or G_{M2} ganglioside	Same symptoms as Tay-Sachs, progresses more rapidly
G_{M1} gangliosidosis	G_{M1} ganglioside: β-galactosidase	G_{M1} ganglioside	Mental retardation, skeletal abnormalities, hepatomegaly
Sulfatide lipodosis; metachromatic leukodystrophy	Arylsulfatase A	Sulfatide	Mental retardation, metachromasia of nerves
Fucosidosis	α-L-Fucosidase	Pentahexosylfucoglycolipid	Cerebral degeneration, thickened skin, muscle spasticity
Farber's lipogranulomatosis	Acid ceramidase	Ceramide	Hepatosplenomegaly, painful swollen joints

Questions

DIRECTIONS (items 1–15): Each numbered item or incomplete statement is followed by answers or by completions of the statement. Select the ONE lettered answer or completion that is BEST in each case.

1. Linoleic acid is an essential fatty acid because:
   a. Humans cannot synthesize fatty acids longer than 16 carbon atoms.
   b. It contains a $\Delta 9$ site of unsaturation and humans are incapable of introducing this unsaturation into fatty acids.
   c. Humans cannot introduce more than one site of unsaturation into fatty acids.
   d. It contains sites of unsaturation beyond carbon atom #9.
   e. When synthesized *de novo* in animals, linoleic acid contains only one site of unsaturation instead of two.

2. Which of the following fatty acids would exhibit the lowest melting temperature?
   a. Linolenic.
   b. Linoleic.
   c. Oleic.
   d. Stearic.
   e. Palmitoleic.

3. The major lipids of biological membranes are:
   a. Triacylglycerols and cholesterol.
   b. Triacylglycerols and phospholipids.
   c. Triacylglycerols, phospholipids, and sphingolipids.
   d. Phospholipids and cholesterol.
   e. Phospholipids and glycosphingolipids.

4. An adult male suffered from stable angina pectoris for 15 years, during which time there was progressive heart failure and repeated pulmonary thromboembolism. Upon his death at age 63, autopsy disclosed enormous cardiomyopathy (1100 g), cardiac storage of globotriaosylceramide (11 mg lipid/g wet weight) and restricted cardiocytes. Other tissues were free of glycolipid storage. The lipid storage disease indicated by these results would most likely be:
   a. Fabry's disease.
   b. Gaucher's disease.
   c. Niemann-Pick disease.
   d. Tay-Sachs disease.
   e. Krabbe's disease.

5. Lysophospholipids are composed of:
   a. Ceramide and two fatty acids.
   b. Sphingosine and two fatty acids.
   c. Phosphatidic acid and one fatty acid.
   d. Phosphatidic acid and two fatty acids.
   e. Sphingosine and one fatty acid.

6. Physical characteristics of the lecithins include:
   a. Inositol, phosphatidic acid, and a net charge of -1 at physiological pH.
   b. Choline, phosphatidic acid and a net charge of 0 at physiological pH.
   c. Ethanolamine, phosphatidic acid, and a net charge of 0 at physiological pH.
   d. Choline, phosphatidic acid, and a net charge of -1 at physiological pH.
   e. Ethanolamine, phosphatidic acid, and a net charge of -1 at physiological pH.

7. A 42-year-old woman exhibiting the following symptoms is suffering from which disease? Marked hepatomegaly, ascites, variceal bleeding, chronic bilateral pulmonary infiltrates, multiple stigmata of chronic liver disease, hepatic encephalopathy, and only 5% of normal sphingomyelinase activity in peripheral leukocytes:
   a. Fabry's disease.
   b. Gaucher's disease.
   c. Niemann-Pick disease.
   d. Tay-Sachs disease.
   e. Krabbe's disease.

8. The ganglioside G_{M2} contains which of the following?
   a. GalGalNAcGalGlc-ceramide.
   b. GalGlc-ceramide.
   c. GalNAcGalGlc-ceramide.
   d. GalGalNAcGalGlc-ceramide.
   e. GalGalNAcGlcGlc-ceramide.

9. Components of pulmonary surfactant include:
   a. Phosphatidylcholine and phosphatidylinositol.
   b. Phosphatidylethanolamine and cholesterol.
   c. Phosphatidylcholine and phosphatidylethanolamine.
   d. Phosphatidylserine and phosphatidylcholine.
   e. Phosphatidylserine and phosphatidylethanolamine.

10. Structural characteristics of the sphingomyelins include:
    a. Ceramide and sphingosine.
    b. Sphingosine, ethanolamine, and phosphatidic acid.
    c. Ceramide and phosphorylcholine.
    d. Sphingosine and phosphorylserine.
    e. Ceramide, inositol, and phosphatidic acid.

11. Which disease would explain the following symptoms presented in a 42-year-old male? Hepatomegaly, jaundice, refractory ascites, and renal insufficiency, with peripheral leukocytes exhibiting only 20% of normal glucocerebrosidase activity:
    a. Fabry's disease.
    b. Gaucher's disease.
    c. Niemann-Pick disease.
    d. Tay-Sachs disease.
    e. Krabbe's disease.

12. In the case described in question 11, which of the following therapies would be expected to be most beneficial?
    a. Administration of a galactosidase analogue.
    b. Liver transplant.
    c. Kidney transplant.
    d. Administration of a hexosaminidase analogue.
    e. Administration of a sphingomyelinase analogue.

13. The glycosphingolipids identified as gangliosides:
    a. Contain a single unit of glucose in addition to a sialic acid.
    b. Contain sialic acid in addition to several other monosaccharide units, one of which is always galactose.
    c. Contain a mole of galactose and glucose as their carbohydrate component.
    d. Contain sialic acid as the sole carbohydrate.
    e. Contain sphingosine and one mole of galactose as their sole carbohydrate.

14. Gangliosides and cerebrosides differ in that:
    a. Gangliosides contain phosphate and cerebrosides do not.
    b. Cerebrosides contain phosphate and gangliosides do not.
    c. Gangliosides are phospholipids and cerebrosides are not.
    d. Cerebrosides contain a more complex oligosaccharide modification than do gangliosides.
    e. Gangliosides contain a more complex oligosaccharide modification than do cerebrosides.

15. Lipid storage diseases result from:
    a. Abnormal or missing enzymes involved in the synthesis of complex glycolipids.
    b. Abnormal or missing enzymes involved in the synthesis of fatty acids.
    c. Abnormal or missing enzymes involved in the catabolism of fatty acids.
    d. Defective production of lysosomes that are involved in lipid catabolism.
    e. Abnormal or missing enzymes involved in the catabolism of complex glycolipids.

Answers

1. d		**6.** b		**11.** b	
2. a		**7.** c		**12.** b	
3. d		**8.** c		**13.** b	
4. a		**9.** a		**14.** e	
5. d		**10.** c		**15.** e	

Nucleic Acids

11

Objectives

- To describe and define the roles of the purine and pyrimidine nucleotides.
- To describe the structure of the nucleotides.
- To illustrate the clinical significance of modified nucleotides.
- To define the formation and structure of the bonds in oligonucleotides.
- To define the double helical structure of DNA.
- To demonstrate the fluidity of the DNA helix.
- To review several DNA analytical techniques relevant to biomedical sciences.

Concepts

As a class, the nucleotides are among the most important metabolites of the cell. Nucleotides are found primarily as the monomeric units comprising the major nucleic acids, RNA, and DNA. However, they also are required within the cell for numerous other important functions, including:

1. Serving as energy stores for future use in phosphate transfer reactions. These reactions are predominantly carried out by ATP.
2. Forming a portion of several important coenzymes such as NAD^+, $NADP^+$, FAD, and coenzyme A.
3. Serving as mediators of numerous important cellular processes such as "second messengers" in signal transduction events. The predominant second messenger is cyclic-AMP (cAMP), a cyclic derivative of AMP formed from ATP.
4. Controlling numerous enzymatic reactions through allosteric effects on enzyme activity.
5. Serving as activated intermediates in numerous biosynthetic reactions. These may include S-adenosylmethionine, which participates in methyl transfer reactions, as well as the many sugar-coupled nucleotides involved in glycogen and glycoprotein synthesis.

Structure and Nomenclatures of the Nucleosides and Nucleotides

A. Base Conformations: The nucleotides found in cells are derivatives of two heterocyclic, highly basic compounds: purine and pyrimidine (Figure 11–1).

It is the chemical basicity of the nucleotides that has given them the common term "bases" as they are associated with the nucleotides of DNA and RNA. Five major bases are found in cells: the derivatives of purine are called **adenine** and **guanine**, and the derivatives of pyrimidine are called **thymine, cytosine** and **uracil** (Figure 11–2). The common abbreviations used for these five bases are **A, G, T, C,** and **U,** respectively.

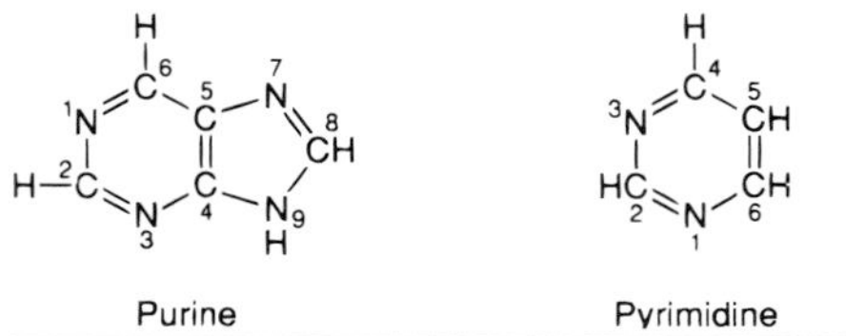

Purine

Pyrimidine

Figure 11–1. Purine and pyrimidine. The positions of the atoms are numbered according to the international system. (Reproduced with permission, from Murray, RK: *Harper's Biochemistry*, 23rd ed. Appleton & Lange, 1993.)

Adenine (A)
(6-aminopurine)

Guanine (G)
(2-amino-6-oxypurine)

Cytosine (C)
(2-oxy-4-aminopyrimidine)

Thymine (T)
(2,4-dioxy-5-methylpyrimidine)

Uracil (U)
(2,4-dioxypyrimidine)

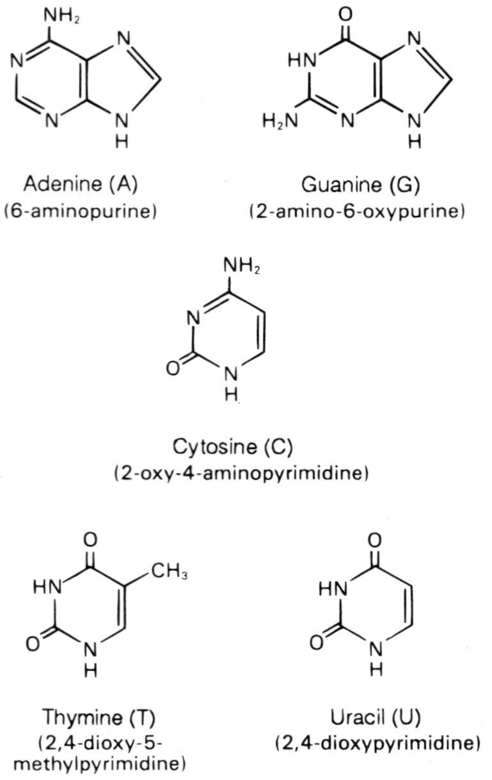

Figure 11–2. The major purine (top) and pyrimidine (bottom) bases of nucleic acids. Adenine, guanine, and cytosine occur in both DNA and RNA. Thymine occurs only in DNA and uracil only in RNA. (Reproduced with permission, from Murray, RK: *Harper's Biochemistry*, 23rd ed. Appleton & Lange, 1993.)

B.   Optical Properties of the Bases: The purines and pyrimidines contain several conjugated double bonds (Figure 11–2). It is this orientation of bonds that allows the bases to absorb ultraviolet (uv) light. The absorbance maximum of the different bases depends upon the pH, however. At physiological pH the bases of the nucleotides in DNA and RNA absorb maximally near 260 nm.

C.   Sugar Conformations: The purine and pyrimidine bases in cells are linked to carbohydrate and in this form are termed **nucleosides**. The covalent linkages form with **D-ribose** or **2′-deoxy-D-ribose** and consist of a β-N-glycosidic bond between the anomeric carbon of the ribose and the N^9 of a purine or N^1 of a pyrimidine (Figure 11–3).

The orientation of the bases about the β-N-glycosidic bond is sterically constrained into two conformations, *syn* and *anti* (Figure 11–4). Most bases in DNA and RNA exist in the *anti* conformation, which allows them to interact in DNA and to form stable hydrogen bonds. The ability of the bases to form these bonds, in turn, produces the Watson-Crick structure of DNA in solution (see below).

Nucleosides are found in the cell primarily in their phosphorylated form, termed **nucleotides**. The most common site for the phosphorylation of nucleotides found in cells is the hydroxyl group attached

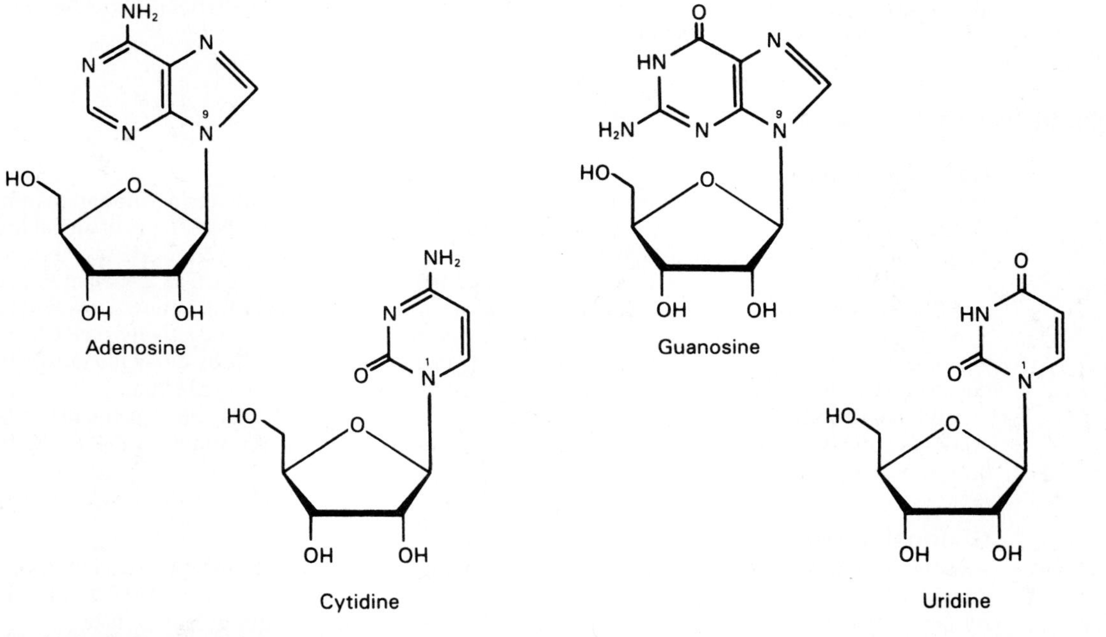

Figure 11–3. Structures of ribonucleosides. Adenosine and guanosine are shown in the most common *syn* conformation. (Reproduced with permission, from Murray, RK: *Harper's Biochemistry*, 23rd ed. Appleton & Lange, 1993.)

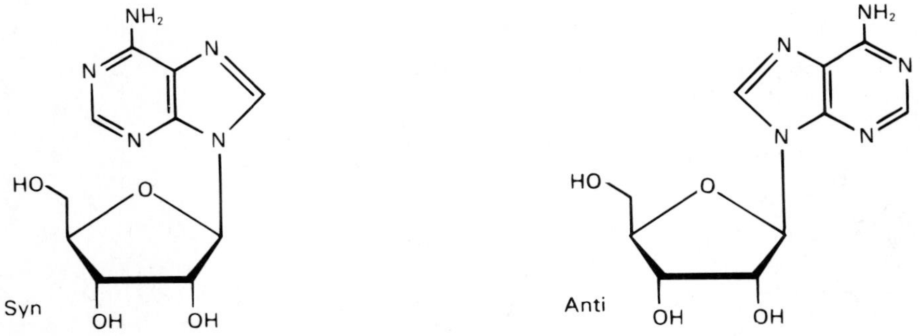

Figure 11–4. The *syn* (left) and *anti* (right) conformations of adenosine. (Reproduced with permission, from Murray, RK: *Harper's Biochemistry*, 23rd ed. Appleton & Lange, 1993.)

to the 5'-carbon of the ribose (The carbon atoms of the ribose present in nucleotides are designated with a prime (') mark to distinguish their numbering from that of the bases. Nucleotides can exist in mono-, di-, or tri-phosphorylated forms.

Nucleotides are given distinct abbreviations to allow easy identification of their structure and state of phosphorylation. The monophosphorylated form of adenosine (adenosine-5'-monophosphate) is written as **AMP**. The di- and tri-phosphorylated forms are written as **ADP** and **ATP**, respectively (Figure 11–5).

The use of these abbreviations assumes that the nucleotide is in the 5'-phosphorylated form. The di- and tri-phosphates of nucleotides are linked by **acid anhydride bonds.** Such bonds have a high ΔG°' for hydrolysis, which imparts a high potential to transfer the phosphates to other molecules. It is this property of the nucleotides that results in their involvement in group transfer reactions in the cell.

The nucleotides found in DNA are unique from those of RNA. First the ribose exists in the 2'-deoxy form, and the abbreviations of the nucleotides contain a "d" designation. The monophosphorylated form of adenosine found in DNA (deoxyadenosine-5'-monophosphate) is written as **dAMP**. Second the nucleotide uridine is never found in DNA; by contrast, thymine is almost exclusively found there. Thymine is found occasionally in tRNAs but not rRNAs nor mRNAs. Several less common bases are also found in DNA and RNA. The primary modified base in DNA is **5-methylcytosine**, while a variety of modified bases appear in the tRNAs (see Chapter 30).

CLINICAL SIGNIFICANCE

Adenosine Derivatives

The most common adenosine derivative is the cyclic form, **3'-5'-cyclic-adenosine-monophosphate**, **cAMP** (Figure 11–6). Activation of adenylate cyclase is an early event in the process of signal transduction that occurs upon ligand-mediated cell-surface receptor activation. Cyclic AMP is a very powerful "second messenger" involved in several signal transduction processes. The principal effect of cAMP is activation of cAMP-dependent protein kinase (PKA). Activated PKA in turn phosphorylates a number of proteins, thereby, affecting their activity either positively or negatively (see Chapters 14,15, and 22). Cyclic AMP is also involved in the regulation of ion channels by direct interaction with the channel proteins, as in the activation of olfactory receptors in the nasal epithelium.

S-adenosylmethionine (SAM or AdoMet) (Figure 11–7) is a form of "activated" methionine which serves as a methyl donor in methylation reactions and as a source of propylamine in the synthesis of polyamines (Chapter 26).

Guanosine Derivatives

A cyclic form of GMP (3'-5'-cyclic-guanosine-monophosphate, or **cGMP**) also is found in cells as a second messenger molecule. In many cases its role is to antagonize the effects of cAMP. Cyclic GMP formation occurs in response to receptor-mediated signals that activate **guanylate cyclase.**

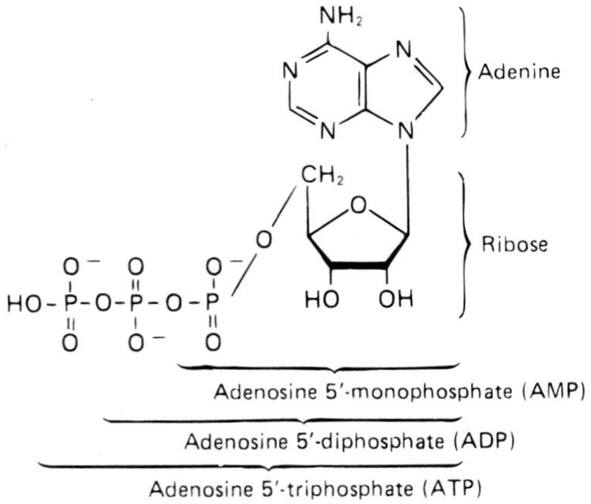

Adenosine 5'-monophosphate (AMP)

Adenosine 5'-diphosphate (ADP)

Adenosine 5'-triphosphate (ATP)

Figure 11–5. ATP, its diphosphate, and its monophosphate. (Reproduced with permission, from Murray, RK: *Harper's Biochemistry*, 23rd ed. Appleton & Lange, 1993.)

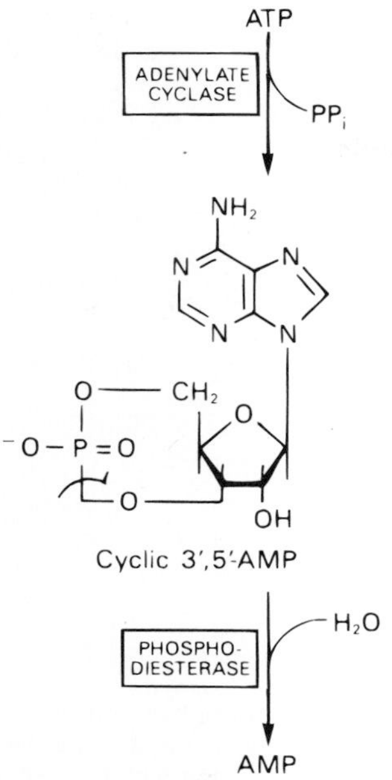

Figure 11–6. Formation of cAMP from ATP by adenylate cyclase and hydrolysis of cAMP by cAMP phosphodiesterase. (Reproduced with permission, from Murray, RK: *Harper's Biochemistry*, 23rd ed. Appleton & Lange, 1993.)

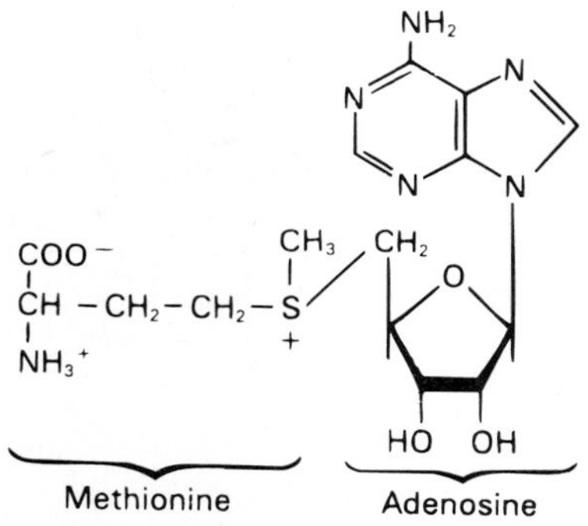

Figure 11–7. S-Adenosylmethionine. (Reproduced with permission, from Murray, RK: *Harper's Biochemistry*, 23rd ed. Appleton & Lange, 1993.)

The most important cGMP-coupled signal transduction cascade is that of photoreception (see Chapter 39). However, in this case stimulation of photoreceptors by light activates a cGMP-specific phosphodiesterase that hydrolyzes cGMP to GMP. This lowers the effective concentration of cGMP bound to gated ion channels, resulting in their closure and a concomitant hyperpolarization of the cell.

Synthetic Nucleotide Analogs

Many nucleotide analogues are chemically synthesized and used for their therapeutic potential. The nucleotide analogues can be utilized to inhibit specific enzymatic activities. A large family of analogues are used as anti-tumor agents, for instance, because they interfere with the synthesis of DNA and thereby preferentially kill rapidly dividing cells such as tumor cells. Some of the nucleotide analogues commonly used in chemotherapy are **6-mercaptopurine, 5-fluorouracil, 5-iodo-2′-deoxyuridine and 6-thioguanine** (Figure 11–8). Each of these compounds disrupts the normal replication process by interfering with the formation of correct Watson-Crick base-pairing.

Nucleotide analogues also have been targeted for use as antiviral agents. Several analogues are used to interfere with the replication of HIV, such as **AZT** (azidothymidine) and **ddI** (dideoxyinosine) (Figure 11–9).

Several purine analogs are used to treat **gout**. The most common is **allopurinol**, which resembles **hypoxanthine** (Figure 11–10). Allopurinol inhibits the activity of **xanthine oxidase**, an enzyme involved in *de novo* purine biosynthesis (Chapter 27). Additionally, several nucleotide analogues are

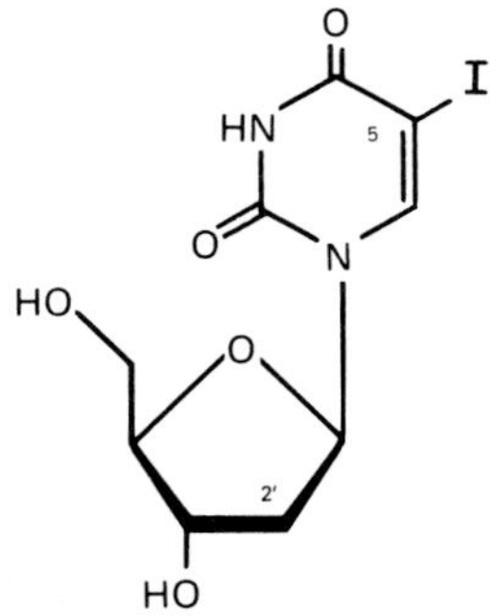

5-Iodo-2'-deoxyuridine

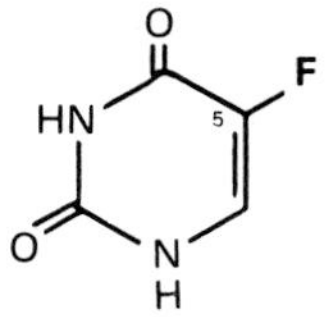

5-Fluorouracil

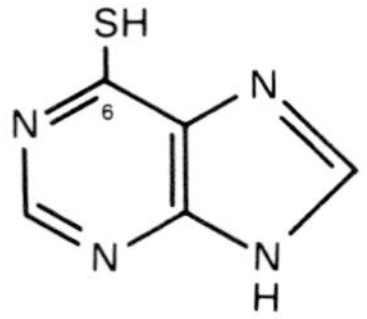

6-Mercaptopurine

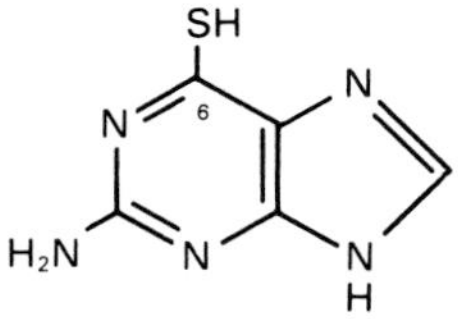

6-Thioguanine

Figure 11–8. Synthetic pyrimidine and purine analogs. (Reproduced with permission, from Murray, RK: *Harper's Biochemistry*, 23rd ed. Appleton & Lange, 1993.)

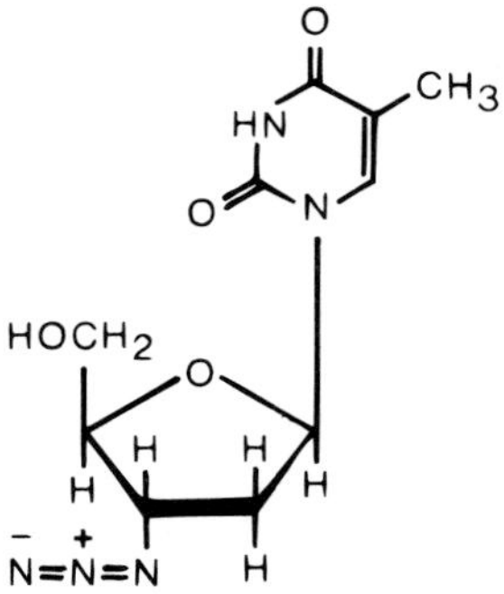

3'-Azido-2',3'-dideoxythymidine(AZT) 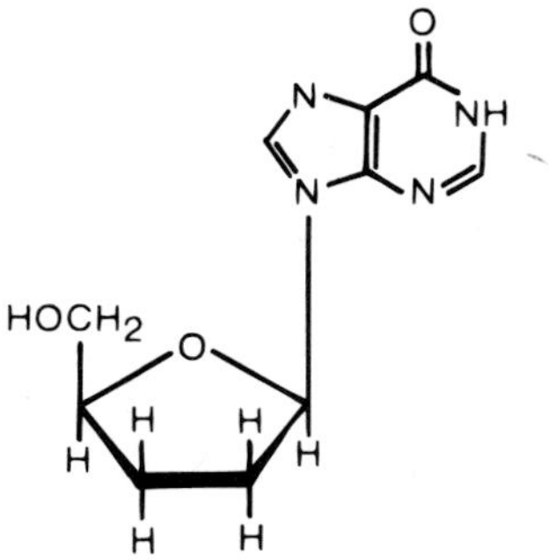   2',3'-Dideoxyinosine (DDI)

Figure 11–9. Structure of the two most common nucleotide analogs used in the treatment of AIDS, AZ and ddl.

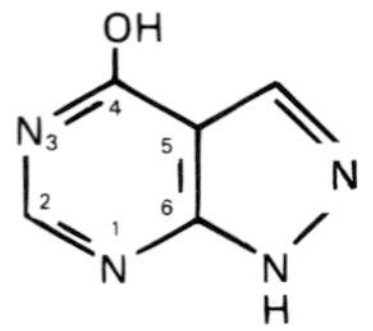

Allopurinol
(lactim)

Figure 11–10. 4-hydroxypyrazolopyrimidine (allopurinol). (Partially reproduced with permission, from Murray, RK: *Harper's Biochemistry*, 23rd ed. Appleton & Lange, 1993.)

used after organ transplantation in order to suppress the immune system and reduce the likelihood of transplant rejection by the host.

Oligonucleotides

Polynucleotides are formed by the condensation of two or more nucleotides. The condensation most commonly occurs between the alcohol of a 5'-phosphate of one nucleotide and the 3'-hydroxyl of a second, with the elimination of H_2O, forming a **phosphodiester** bond (Figure 11–11). This is the bond

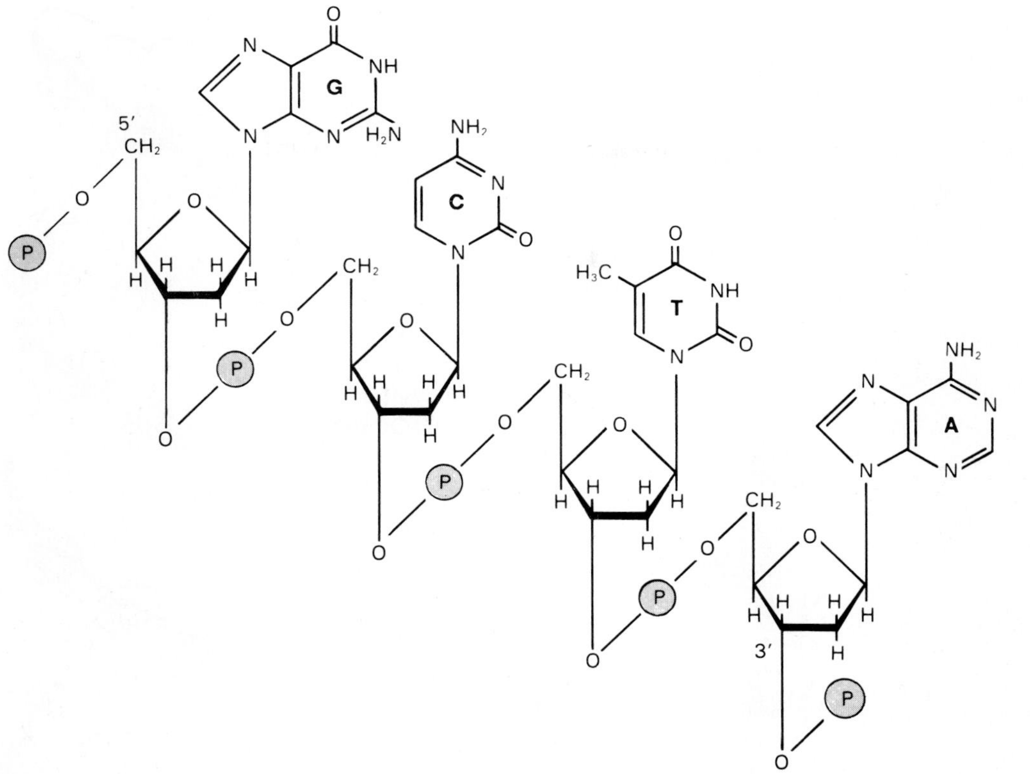

Figure 11–11. A segment of one strand of a DNA molecule in which the purine and pyrimidine bases adenine (A), thymine (T), cytosine (C), and guanine (G) are held together by a phosphodiester backbone between 2'-deoxyribosyl moieties attached to the nucleobases by an N-glycosidic bond. Note that the backbone has polarity (ie, a direction). (Reproduced with permission, from Murray, RK: *Harper's Biochemistry*, 23rd ed. Appleton & Lange, 1993.)

formed by condensation of an alcohol with an alcohol (remember that the bonds between the di- and tri-phosphates of nucleotides are acid anhydrides, not esters).

The formation of phosphodiester bonds in DNA and RNA exhibits directionality. The primary structure of DNA and RNA (the linear arrangement of the nucleotides) proceeds in the $5' \rightarrow 3'$ direction. The common representation of the primary structure of DNA or RNA molecules is to write the nucleotide sequences from left to right synonymous with the $5' \rightarrow 3'$ direction. The $5'$ "end" of a molecule of DNA or RNA is almost always $5'$-phosphorylated and the $3'$ "end" is a free $3'$-hydroxyl. The common way to write polynucleotide sequences is:

5'-pGpApTpC-3'

Double Helical Structure of DNA

Utilizing x-ray diffraction data, obtained from crystals of DNA, James Watson and Francis Crick proposed a model for the structure of DNA. This model (subsequently verified by additional data) predicted that DNA would exist as a helix of two complementary antiparallel strands, wound around each other in a rightward direction and stabilized by H-bonding between bases in adjacent strands. The antiparallel nature of the helix stems from the orientation of the individual strands. From any fixed position in the helix, one strand is oriented in the $5' \rightarrow 3'$ direction and the other in the $3' \rightarrow 5'$ direction (Figure 11–12).

On its exterior surface, the double helix of DNA contains two deep grooves between the ribose-phosphate chains. These grooves are of unequal size and are termed the **major** and **minor grooves**. The difference in their size is due to the asymmetry of the deoxyribose rings and the structurally distinct nature of the upper surface of a base-pair relative to the bottom surface. These grooves allow the DNA to serve as a template for sequence-specific interactions with many types of proteins.

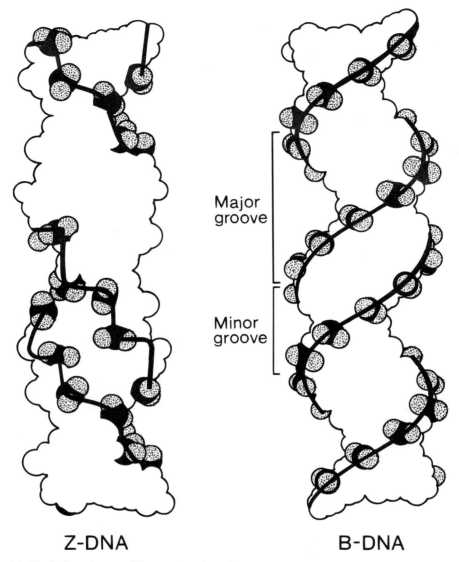

Z-DNA   B-DNA

Figure 11–12. Outline of space-filling models of the Z- and B-forms of DNA helices. The backbone of both helices is highlighted.

The double helix of DNA has been shown to exist in several different forms (Table 11–1), depending upon sequence content and ionic conditions of crystal preparation. The B-form of DNA (Figure 11–12) prevails under physiological conditions of low ionic strength and a high degree of hydration. Regions of the helix that are rich in pCpG dinucleotides can exist in a novel left-handed helical conformation termed Z-DNA (Figure 11–12). This conformation results from a 180° change in the orientation of the bases relative to that of the more common A- and B-DNA.

In the Watson-Crick model, the bases are in the interior of the helix, aligned at a nearly 90° angle to the axis of the helix. Purine bases form hydrogen bonds with pyrimidines, in the crucial phenomenon of **base-pairing**. Experimental determination has shown that, in any given molecule of DNA, the concentration of adenine (A) is equal to thymine (T) and the concentration of cytidine (C) is equal to guanine (G). This means that A will only base-pair with T, and C with G. According to this pattern, known as **Watson-Crick base-pairing**, the base pairs composed of G and C contain three H-bonds, whereas those of A and T contain two H-bonds (Figure 11–13). This makes G-C base pairs more stable than A-T base pairs.

Table 11–1. Parameters of major DNA helices.

Parameters	A Form	B Form	Z-Form
Direction of helical rotation	Right	Right	Left
Residues per turn of helix	11	10	12 base pairs
Rotation of helix per residue	33°	36°	−30°
Base tilt relative to helix axis	20°	6°	7°
Major groove	narrow and deep	wide and deep	Flat
Minor groove	wide and shallow	narrow and deep	narrow and deep
Orientation of N-glycosidic Bond	Anti	Anti	Anti for Pyrimidines, Syn for Purines
Comments		most prevalent within cells	occurs in stretches of alternating purine-pyrimidine base pairs

Thermal Properties of the Helix

As cells divide it is a necessity that the DNA be copied (replicated), in such a way that each daughter cell acquires the same amount of genetic material. In order for this process to proceed the two strands of the helix must first be separated, in a process termed **denaturation**. This process can also be carried out *in vitro*. If a solution of DNA is subjected to high temperature, the H-bonds between bases become unstable, and the strands of the helix separate in a process of **thermal denaturation**.

The base composition of DNA varies widely from molecule to molecule and even within different regions of the same molecule. Regions of the duplex that have predominantly A-T base-pairs will be less thermally stable than those rich in G-C base pairs. In the process of thermal denaturation, a point is reached at which 50% of the DNA molecule exists as single strands. This point is the **melting temperature, T_M** and is characteristic of the base composition of that DNA molecule. The T_M depends upon several factors in addition to the base composition. These include the chemical nature of the solvent and the identities and concentrations of ions in the solution.

When thermally melted DNA is cooled, the complementary strands will again re-form the correct base pairs, in a process termed **annealing** or **hybridization**. The rate of annealing is dependent upon the nucleotide sequence of the two annealing strands of DNA.

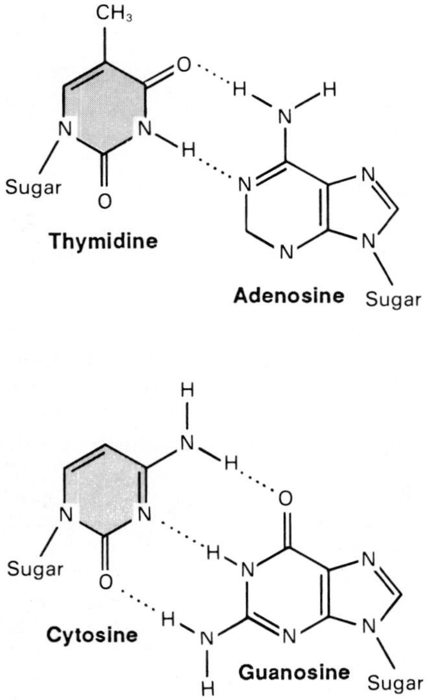

Figure 11–13. Base pairing between deoxyadenosine and thymidine involves the formation of 2 hydrogen bonds. Three such bonds form between deoxycytidine and deoxyguanosine. The broken lines represent hydrogen bonds. In DNA, the sugar moiety is 2 deoxyribose. (Reproduced with permission, from Murray, RK: *Harper's Biochemistry*, 23rd ed. Appleton & Lange, 1993.)

Analysis of DNA Structure

A. Chromatography: Several of the chromatographic techniques available for the characterization of proteins (Chapter 4) can also be applied to the characterization of DNA. The most commonly used technique is HPLC (high performance liquid chromatography). Affinity chromatographic techniques also can be employed. One common affinity matrix is **hydroxyapatite** (a form of calcium phosphate), which binds double-stranded DNA with a higher affinity than single-stranded DNA.

B. Electrophoresis: This procedure can serve the same function with regard to DNA molecules as it does for the analysis of proteins: ie, the characterization of size (Chapter 4). However, since DNA molecules have much higher molecular weights than proteins, the molecular sieve used in electrophoresis of DNA must be different as well. The material of choice is **agarose**, a carbohydrate polymer purified from a salt water algae. It is a copolymer of mannose and galactose that when melted and re-cooled forms a gel with pore sizes dependent upon the concentration of agarose. To the phosphate backbone of DNA is highly negatively charged, therefore DNA will migrate in an electric field. The size of DNA fragments can then be determined by comparing their migration in the gel to known size standards. Extremely large molecules of DNA (in excess of 10^6 base pairs) are effectively separated in agarose gels using **pulsed-field gel electrophoresis (PFGE)**. This technique employs two or more electrodes, placed orthogonally with respect to the gel, that receive short alternating pulses of current. PFGE allows whole chromosomes and large portions of chromosomes to be analyzed.

Questions

DIRECTIONS (items 1–8): Each numbered item or incomplete statement is followed by answers or by completions of the statement. Select the ONE lettered answer or completion that is BEST in each case.

1. The nucleotides found in DNA and RNA absorb ultraviolet light because of:
    a. The presence of the hydroxyls of ribose.
    b. The presence of the phosphodiester bonds.
    c. The resonance stabilization of the N-glycosidic bond.
    d. The aromatic nature of the purine bases.
    e. The aromatic nature of ribose.

2. The structure of the DNA helix can best be described as:
    a. Containing two strands in an antiparallel alignment.
    b. Containing equal molar ratios of adenines and guanines.
    c. Being stabilized by H-bonding between purines on opposite strands.
    d. Containing two strands in a parallel alignment.
    e. Being stabilized by H-bonding between pyrimidines within the same strand.

3. Hydroxyapatite is a useful resin for the analysis of DNA because:
    a. Large molecules of DNA pass more slowly through the resin than do small molecules of DNA.
    b. Double-stranded DNA binds with much higher affinity than single-stranded DNA.
    c. Small molecules of DNA pass more slowly through the resin than do large molecules of DNA.
    d. DNA with a higher molar ratio of G-C base-pairs is retained by the resin more effectively than DNA with a low molar ratio of G-C base-pairs.
    e. Single-stranded DNA binds with much higher affinity than double-stranded DNA.

4. The stability of the DNA helix is predominantly dependent upon:
    a. The concentration of G-C base-pairs.
    b. The configuration of the bases relative to the ribose sugars.
    c. The concentration of ions in solution.
    d. The concentration of A-T base pairs.
    e. The conformation of the ribose sugars.

5. Nucleotides are derived from nucleosides by:
   a. The addition of phosphate to the 3'-OH of the ribose.
   b. The addition of phosphate to the 5'-OH of the ribose.
   c. The addition of ribose to a purine or pyrimidine base.
   d. The removal of phosphate to the 3'-phosphate of the ribose.
   e. The removal of phosphate from the 5'-phosphate of the ribose.

6. The utility of synthetic nucleotide as therapeutics stems from the fact that:
   a. These nucleotides bind to and inhibit DNA polymerase activity.
   b. Modified nucleotides H-bond to the DNA helix, thereby disrupting its overall structure and interfering with DNA synthesis.
   c. Their inability to form correct H-bonds leads to abortive DNA synthesis.
   d. They are degraded to toxic fluoro- and aza-compounds within cells.
   e. They disrupt normal H-bonding within DNA helices following their incorporation during DNA synthesis, resulting in a high rate of mutation of the DNA.

7. The cyclic derivatives of AMP and GMP have been shown to be important for:
   a. Anti-cancer therapy.
   b. Immunosuppression therapy.
   c. Regulation of ribonucleotide reductase activity.
   d. Anti-viral therapy.
   e. Movement of signals from outside the cell to inside the cell.

8. All of the following are characteristic features of DNA, **except**:
   a. The B-form predominates under conditions of low ionic strength and low hydration.
   b. The Z-form appears in stretches of alternating pyrimidines and purines.
   c. The B-form is a right-handed helix containing 10 nucleotides per 360 turn.
   d. The A-form is a right-handed helix containing 11 nucleotides per 360 turn.
   e. All of the above correctly define the structure of DNA.

DIRECTIONS (items 9–15): Match the following diagrams, lettered A–G, to the descriptions in Questions 9–15.

A.

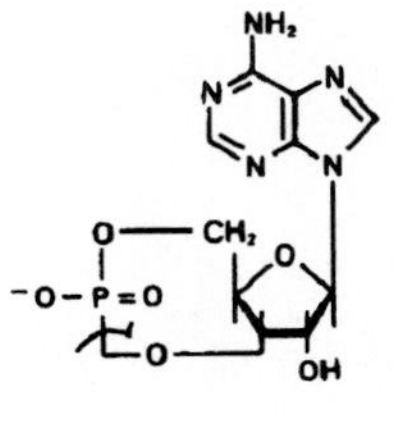

B.

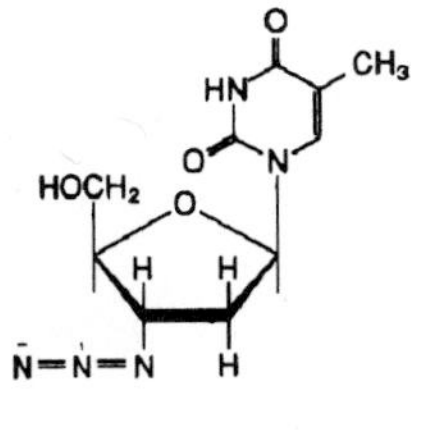

C.

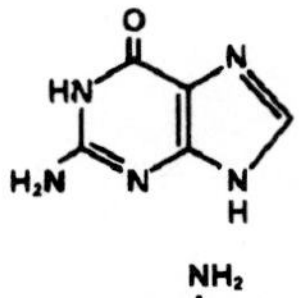

D.

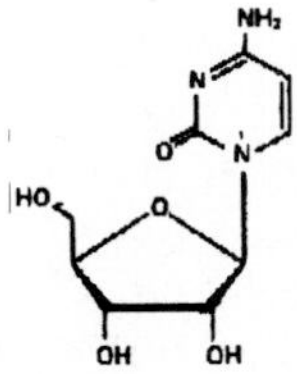

E.

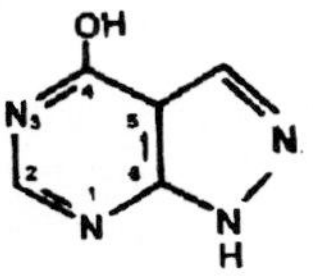

F.

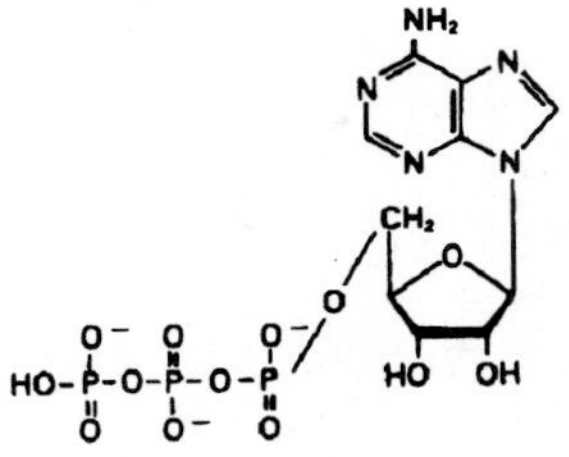

G.

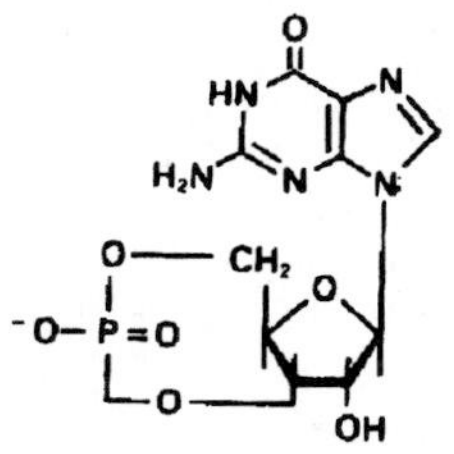

9. Nucleotide analogue used in the treatment of gout.

10. Pyrimidine-derived base.

11. Nucleotide analogue used in the treatment of viral infection.

12. Purine-derived base.

13. Second messenger involved in vision.

14. Purine-derived base involved in energy transfer reactions.

15. Second messenger involved in odor detection.

ANSWERS

1. d	**6.** c	**11.** b
2. a	**7.** e	**12.** c
3. b	**8.** a	**13.** g
4. a	**9.** e	**14.** f
5. b	**10.** d	**15.** a

Part IV: Biological Membranes

Structure and Function of Biological Membrane

12

Objectives

- To review the nature of the lipids and proteins that form the structural elements of biological membranes and their role in the biological activity of membranes.
- To identify the general features of the fluid mosaic model of membrane structure.
- To identify the classes of lipids found in biological membranes, their key structural features, and the most prominent members of each class.
- To identify the process of lipid self-assembly and discriminate between micelles and liposomes.
- To identify how hydrocarbon chain length, degree of unsaturation, and temperature impact the fluidity of a membrane.
- To identify the range of protein content of biological membranes and the role of proteins in the membranes.
- To identify the orientation of domains and the way that integral membrane proteins interact with other proteins and with the lipid bilayer.
- To identify the basis for the constant asymmetry found in the distribution of lipids and proteins of the inner and outer monolayers of a given biological membrane.
- To identify the difference between simple diffusion and facilitated diffusion.
- To identify the difference between facilitated diffusion and active transport.
- To identify the similarities and differences between symports and antiports.

Concepts

The role of biological membranes is to form compartments capable of segregating specific ions, nutrients, and macromolecules, thus producing biological units with specialized functions based on their specific complement of chemicals. Without biological membranes, life as we know it could not exist. In humans, the compartments can range in volume from subcellular organelles, such as lysosome, Golgi apparatus, and mitochondria, to the vascular compartment separated from the remainder of the body by the plasma membrane of endothelial cells. This chapter will review the components of biological membranes and acquaint the reader with their great structural and functional diversity in the human body.

MEMBRANE STRUCTURE

The basic structure of biological membranes is summarized by Singer and Nicholson's "fluid mosaic model." This depicts biological membrane as a lipid bilayer with a characteristic complement of associated proteins; it is the proteins that are largely responsible for the functional differences among various membrane systems. Membranes associated with extensive biological activity (eg, inner mitochondrial membranes) contain up to 80% protein, while less active membranes (such as those of the myelin sheath) contain as little as 20% protein. Although all membranes include a hydrophobic lipid barrier separating aqueous compartments, their lipid composition varies significantly from one membrane to another, and even from one lipid monolayer to another within the same membrane.

Lipid Composition of Membranes

There are three principal kinds of lipids found in most membranes: phosphoglycerides, sphingolipids, and cholesterol. (Structures of all the important members of these classes were presented in Chapter 10 and may be reviewed at this time.) The structural role of phospholipids and sphingolipids is based on their amphipathic nature. The head group, which consists of a charged phosphate residue and a nitrogenous base (or sugar), is polar in character, with the fatty acid tails providing the hydrophobic portion. The range of lipid composition in membranes and the net molecular charge of the principal phospholipids is summarized in Table 12–1.

Because of their amphipathic character, phospholipids and sphingolipids tend to assemble themselves into compact forms at very low lipid concentrations. In artificial systems, when the lipid concentration exceeds the **critical micellar concentration** (which is often less than 10^{-9} M), lipids assemble into micelles and bilayers; structures in which the hydrophobic "tails" (fatty acids) are oriented away from the aqueous environment and the polar "heads" toward the aqueous solution (Figure 12–1). A similar self-assembly appears to occur in the formation of biological membranes, although the process is not as well understood. These lipid bilayers comprise the fundamental component of Nicholson and Singer's fluid mosaic model (Figure 12–2).

Fluid Properties of Membranes

It is important to recognize that although most of the fatty acid esterified to membrane lipid is either palmitic or stearic acid (saturated, C-16, and C-18), the fluid properties of membranes are directly related to their content of C-16 to C-20 unsaturated fatty acids. Lipids containing only saturated fatty acids assemble into very stable, densely packed bilayers. The greater the proportion of saturated fatty acids, the greater the intermolecular van der Waals forces and the more crystal-like is the membrane. Almost all biologically derived unsaturated fatty acids are arranged in *cis* configuration across the double bond, and they are almost always found on carbon 2 of the glycerol backbone of the phospholipids. The *cis* configuration introduces a kink in the fatty acid chain, disrupting the packing of adjacent lipid molecules, and decreasing van der Waals bonding and the tight packing of the membrane. The generalizations summarized in Table 12–2 are based on the latter point.

Cholesterol, found in many plasma membranes, is located near the polar head groups, with the net effect being a decrease in membrane fluidity. However, at the same time, cholesterol increases the mobility of the fatty acyl hydrocarbons in the core of the membrane. Some membranes, such as those of mitochondria, contain little or no cholesterol.

From these observations, it should be clear that the fluid properties of any membrane depend on a complex interplay between temperature, length of the fatty acyl chains of lipids, degree of lipid unsaturation, and the presence of other compounds, such as cholesterol.

Table 12–1. Principal lipids found in biological membranes. The percentage composition is that of the individual lipids, calculated on the basis of total membrane lipid content.

Principal Membrane Lipids	Net Charge on Head Group	Percentage Composition (range)
Phosphoglycerides	0 to −2	50–90%
Phosphatidylcholine	0	40–60%
Phosphatidylethanolamine	0	20–30%
Phosphatidylserine	−1	5–15%
Cardiolipin	−2	0–20%
Phosphatidylinositol	−1	5–10%
Sphingomyelin	0	5–20%
Cholesterol	0	0–10%

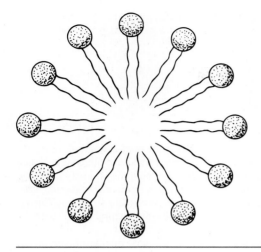

Figure 12–1. Diagrammatic cross section of a micelle. The polar head groups are bathed in water, whereas the hydrophobic hydrocarbon tails are surrounded by other hydrocarbons and thereby protected from water. Micelles are spherical structures. (Reproduced, with permission, from Murray, RK: *Harper's Biochemistry*, 23rd ed. Appleton & Lange, 1993.)

Membrane Proteins

As noted above, the protein content of membranes is quite variable in composition, however, most membranes contain from approximately 40% to 50% protein. Two notable exceptions are mitochondrial inner membranes, with about 80% protein, and myelin, which contains only about 20% protein. The proteins in a membrane determine its functional properties; they also influence its fluid properties, because of extensive protein-lipid interactions.

Membrane proteins are often classified on the basis of their solubility under various conditions. **Peripheral membrane proteins** (Figure 12–2) are those which are only loosely associated with membranes and thus readily solubilized with mild salt solutions. **Integral membrane proteins** are not soluble in mild aqueous solvents and are generally only solubilized with the aid of detergents.

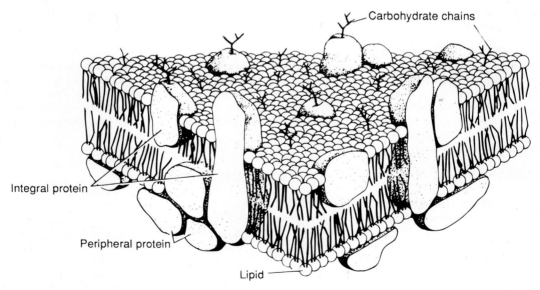

Carbohydrate chains

Integral protein

Peripheral protein

Lipid

Figure 12–2. The fluid mosaic model of membrane structure. The membrane consists of a bimolecular lipid layer with proteins inserted in it or bound to the cytoplasmic surface, integral membrane proteins are firmly embedded in the lipid layers. Some of these proteins completely span the bilayer and are called transmembrane proteins, while others are embedded in either the outer or inner leaflet of the lipid bilayer. Loosely bound to the outer or inner surface of the membrane are the peripheral proteins. Many of the proteins and lipids have externally exposed oligosaccharide chains. (Reproduced, with permission, from Junquiera LC, Caneiro J, Kelley RO: *Basic Histology,* 7th ed. Appleton & Lange, 1992.)

Table 12–2. Relationship of lipid unsaturation to membrane fluidity.

1.	With a constant fatty acyl chain length and constant temperature, the greater the number of double bonds the more fluid the membrane.
2.	With a constant number of double bonds and constant temperature, the longer the acyl chain length the less fluid the membrane.
3.	With any combination of chain length and double bonds, the greater the temperature the more fluid the membrane.

As illustrated in Figure 12–2, integral membrane proteins may penetrate only a single lipid monolayer or, as **transmembrane**, they may extend from one side of the membrane to the other side. Transmembrane proteins generally have distinct domains on each side of the membrane and may have two or more transmembrane domains. The transmembrane domains of integral proteins are most often composed of amino acids having hydrophobic R groups arranged in α-helical secondary structure. The hydrophobic R groups on the periphery of the α-helixes provide strong hydrophobic interactions with the adjacent membrane lipids, stabilizing the protein in the membrane.

Many transmembrane proteins are now known to pass back and forth across the membrane in serpentine fashion, forming as many as 12 separate, α-helical, transmembrane-domains. A red blood cell membrane called "band 3" is responsible for the exchange of Cl^- and bicarbonate across the membrane during the chloride shift as red cells pass from arteries to veins (see Chapter 1), unloading their oxygen, and acquiring bicarbonate. In band 3 protein, the individual transmembrane α-helices are believed to be arranged like rods placed side by side, creating a centrally located protein pore through the membrane.

Membrane Asymmetry

Membranes are asymmetric in both lipid and protein composition. For example, the outer monolayer and the inner monolayer of red blood cell membranes are each composed of about 50% lipid, but whereas the outer monolayer is largely composed of sphingomyelin and phosphatidylcholine, the inner one is largely composed of phosphatidylserine and phosphatidylethanolamine. Similarly, the protein composition of each side of the membrane is different. Most often the C-terminus of transmembrane proteins is found on the cytosolic side of the membrane, whereas the N-terminus is exposed on the extracellular surface. Many membrane lipids and proteins are glycosylated, and all glycosylations are found on the extracellular side of the membrane.

Mobility of Membrane Components

Asymmetry is a permanent feature in the composition of the membrane. Thus, although individual proteins and lipids have a wide range of rotational mobility and lateral translational mobility in the plane of the membrane, they almost never flip from one side of the membrane to the other. The rare occurrence of "flip-flop motion" is due to the thermodynamic difficulty of moving charged groups, usually associated with the exterior surfaces of membranes, through the hydrophobic lipid bilayer that constitutes the core of the membrane.

MEMBRANE PERMEABILITY AND TRANSMEMBRANE SIGNALING

Simple Diffusion and Facilitated Diffusion

The two general processes by which substances move across membranes are **passive transport** and **active transport** (Figure 12–3). Passive transport in turn includes two categories: **simple diffusion** and **facilitated diffusion** involving pores or carrier proteins. While the lipid bilayer restricts the permeability of membranes to hydrophilic and charged substances, gases (O_2, CO_2) and lipophilic substances (steroids and neutral fatty acids) appear to pass through the membrane by simple diffusion—the least complicated transport mechanism. The rate of permeability of these compounds is simply related to their lipid solubility and their concentration difference across the membrane.

Facilitated diffusion through pores or channels (Figure 12–3) and via carrier proteins requires the intervention of special carriers or pore-forming proteins in membranes. In facilitated processes, just as in simple diffusion, substances move down their electrochemical gradient, with the direction of transport depending only on the existing gradient. The principal difference between simple diffusion and the various facilitated processes is that the rate of transport in facilitated processes is limited by the

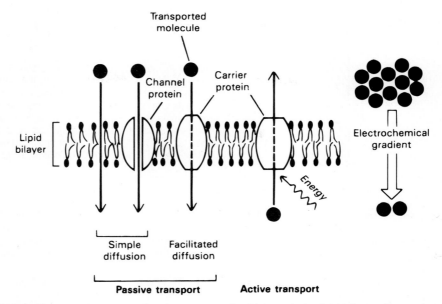

Figure 12–3. Many small uncharged molecules pass freely through the lipid bilayer. Charged molecules, larger uncharged molecules, and some small uncharged molecules are transferred through channels or pores or by specific carrier proteins. Passive transport is always down an electrochemical gradient, toward equilibrium. Active transport is against an electrochemical gradient and requires an input of energy, whereas passive transport does not. (Redrawn and reproduced, with permission, from Alberts B et al: *Molecular Biology of the Cell.* Garland, 1983.)

number of pores or carrier proteins in a membrane; thus, these processes are saturable and can be characterized by a K_M and V_{max} as illustrated in Figure 12–4.

The most complicated of the transport mechanisms is that of **active transport**. Like facilitated processes, active transport requires a special membrane protein and thus exhibits saturation kinetics. However, active transport differs from other facilitated transport processes in that substances move against their electrochemical gradient and because of the unfavorable thermodynamic character of this movement they require a source of energy, most often ATP. The movement of Na^+ and K^+ ions via the ubiquitous Na^+/K^+-ATPase pump (Figure 12–5) is an example of an active transport process.

Symports and Antiports

Membrane transport processes are also often characterized as being **uniports, symports, or antiports**, all of which are illustrated in Figure 12–6. Uniports move a single substance, such as glucose, across a membrane, one molecule at a time. Symports are synergistic in their transport, moving two different substances at the same time and in the same direction across a membrane. Many amino acid and glu-

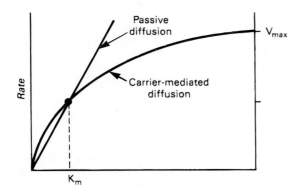

Figure 12–4. A comparison of the kinetics of carrier-mediated (facilitated) diffusion with passive diffusion. The rate of movement in the latter is directly proportionate to solute concentration, whereas the process is saturable when carriers are involved V_{max}, maximal rate. The concentration at half-maximal velocity is equal to the binding constant (K_m) of the carrier for the solute. (Reproduced, with permission, from Murray, RK: *Harper's Biochemistry,* 23rd ed. Appleton & Lange, 1993.)

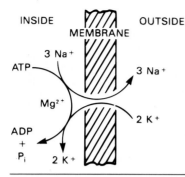

Figure 12–5. Stoichiometry of the Na^+/K^+-ATPase pump. This pump moves 3 Na^+ ions from inside the cell to the outside and brings 2 K^+ ions from the outside to the inside for every molecule of ATP hydrolyzed to ADP by the membrane-associated ATPase. Ouabain and other cardiac glycosides inhibit this pump by acting on the extracellular surface of the membrane. (Reproduced, with permission, from Murray, RK: *Harper's Biochemistry*, 23rd ed. Appleton & Lange, 1993.)

cose transporters are symports, with sodium being the second, or cotransported, substance. In amino acid and glucose symports, the carrier protein has binding sites for Na^+ and for glucose or amino acids on the same side of the membrane. The driving force for the transport through these symports is the Na^+ electrochemical gradient, which is maintained in a favorable orientation by the Na^+/K^+ ATPase pump (Figure 12–5).

Antiports move two different substances, at the same time, in the opposite directions across a membrane. The Na^+/K^+ ATPase shown in Figure 12–5 is a classic example of an antiport, with K^+ moving from outside the membrane to the inside at the same time that Na^+ moves from the interior of the membrane to the outside. Although the detailed mechanisms are not known for any transport process, a popular hypothetical mechanism for the facilitated processes is the "ping-pong" mechanism illustrated in Figure 12–7.

Water Transport

One of the most puzzling problems in the area of membrane transport is that involving the movement of water. Although biological membranes are basically hydrophobic in character, the transport of water across lipid bilayers is among the fastest transport processes that have been measured. Small gaps between molecules in the membrane and special membrane proteins, known as aquapores, appear to provide the pathway for the rapid water movement. Thereafter, the direction of net water flow is simply based on the thermodynamic concentration gradient of water across the membrane.

Ionophores

An interesting group of antibiotics that act on the membrane have been derived from bacteria. These compounds, called ionophores, have the property of transporting small-molecular-weight compounds, predominantly metal ions (Na^+, K^+, Ca^{2+}), across a wide variety of natural and artificial membranes. Most ionophores are quite specific as to the size and charge of substances that they can transport. The process of transport itself is generally passive, with movement of the transported substance down its

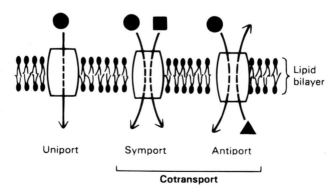

Figure 12–6. Schematic representation of types of transport systmes. Transporters can be classified with regard to the direction of movement and whether one or more unique molecules are moved. (Redrawn and reproduced, with permission, form Alberts B et al: *Molecular Biology of the Cell*. Garland, 1983.)

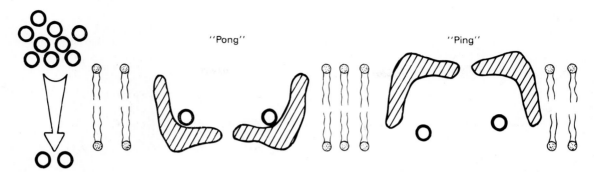

Figure 12–7. The "ping-pong" model of facilitated diffusion. A protein carrier (shaded structure) in the lipid bilayer associates with a solute in high concentration on one side of the membrane. A conformational change ensues ("ping" to "pong"), and the solute is discharged on the side favoring the new equilibrium. The empty carrier then reverts to the original conformation ("ping" to "pong") to complete the cycle. (Reproduced, with permission, from Murray, RK: *Harper's Biochemistry*, 23rd ed. Appleton & Lange, 1993.)

electrochemical concentration gradient. Generally ionophores carry out their antibiotic activity by collapsing membrane potentials or disrupting concentration gradients that are essential for homeostasis.

The ionophores fall into two classes: those which bind or chelate a substrate on one side of a membrane and diffuse through the membrane, releasing the substrate on the opposite side; and those which form permanent pores in the membrane, again allowing the transmembrane movement of substances that have the appropriate size and charge. Some of the more well-known ionophores include valinomycin, which is selective for the transport of K^+; A23187, which is selective for calcium transport; monesin, selective for Na^+ transport; and gramacidin, which has a broad specificity and can transport Na^+, K^+ and several other ions.

Disease States Associated With Abnormal Membrane Function

Since the functional properties of membranes are largely determined by proteins, protein deficiencies here generally lead to disease states characterized by functional abnormalities. Among these are some cases of type II diabetes (non-insulin dependent diabetes, NIDD), which develop when the plasma membrane insulin receptors have a low affinity for plasma insulin. Although these forms of diabetes are generally considered to be due to a protein dysfunction (the ineffective insulin receptors), it should be recognized that the real deficiency in this instance is a failure of the plasma membrane to transmit an insulin signal to the interior of the cell. The plasma membrane also fails to compartmentalize glucose appropriately on the basis of the insulin signal. Moreover, it has been observed that the binding affinity of plasma membrane receptors for their ligands can be significantly altered by the fluidity of the lipid phase of the membrane. Thus, a correlation exists between structural features of membranes (ie, their complement of unsaturated fatty acyl groups) and disease states such as the various forms of NIDD.

Some people suffer from a condition in which glucose uptake from the digestive tract is deficient. In such cases, the Na^+/glucose transporter responsible for transport of glucose into intestinal enterocytes is the defective protein.

Many autoimmune diseases, including rheumatoid arthritis, are caused by the development of antibodies to cell-surface antigenic determinants. While plasma membrane components are not defective in these cases, their complement of antigenic sites permits the membrane to be a target for abnormally produced antibodies and thus to represent the focal point of a disease state.

Questions

DIRECTIONS (items 1–10): Each numbered item or incomplete statement is followed by answers or by completions of the statement. Select the ONE lettered answer or completion that is BEST in each case.

1. According to the fluid mosaic model of membrane structure:
   a. Membranes are lipid bilayers, each associated with the same set of proteins.
   b. Opposing (+ and −) charges on polar lipid head groups are responsible for the attraction between the individual lipid monolayers of a membrane.
   c. Membranes are lipid bilayers, each with a characteristic complement of integral and peripheral proteins.
   d. All membranes have approximately the same proportions of lipids and proteins.

2. The dominant sphingolipids associated with most membranes include:
   a. Sphingomyelins.
   b. Cerebrosides.
   c. Sulfatides.
   d. Gangliosides.

3. Membrane lipids are amphipathic because:
   a. They all contain head groups with a net charge.
   b. They all contain polar head groups.
   c. They all contain head groups with a net charge and hydrophobic tail groups.
   d. They all contain polar head groups and hydrophobic tail groups.

4. Membrane fluidity is increased by:
   a. High proportions of membrane cholesterol.
   b. High proportions of long-chain saturated acyl groups.
   c. Higher proportions of sphingomyelins than phosphatidyl cholines.
   d. Elevated temperatures.

5. Integral membrane proteins:
   a. Are generally arranged with their C terminus on the cytosolic side and their N terminus on the opposite side of a membrane.
   b. Are termed "integral" because they bind to polar head groups, which tightly integrates them into the fabric of the membrane.
   c. Generally are arranged with their N terminus on the cytosolic side and their C terminus on the opposite side of a membrane.
   d. None of the above are appropriate responses.

6. The transmembrane domain of integral membrane proteins is:
   a. Most often composed of hydrophobic amino acids in random coil conformation and with minimal lipid protein interaction.
   b. Most often composed of hydrophilic amino acids in a conformation resulting in minimal lipid protein interaction.
   c. Most often composed of hydrophobic amino acids in a helical conformation.
   d. Composed of hydrophobic amino acids.

7. Glycosyl groups of membranes are:
   a. Associated with lipids.
   b. Associated with proteins.
   c. Localized on the extracellular side of plasma membranes.
   d. Responsible for many antigenic determinants of the membrane.
   e. All of the above.

8. Constituents of biological membranes:
   a. Never exhibit flip-flop motion across the plane of the membrane.
   b. Exhibit flip-flop motion across the plane of the membrane at rates comparable to that of lipid rotational motion.
   c. Exhibit flip-flop motion across the plane of the membrane at rates comparable to that of lateral translational motion.
   d. Rarely exhibit flip-flop motion across the plane of the membrane.

9. The best described cotransport systems involving glucose and sodium are:

    a. Uniports.
    b. Two oppositely oriented uniports.
    c. Two uniports oriented in the same direction.
    d. Antiports.
    e. Symports.

10. Ionophores include:
    a. Pore-forming antibiotics that act by passive transport.
    b. Carrier antibiotics that act by active transport mechanisms.
    c. Pore-forming and carrier antibiotics that act by active transport and passive transport respectively.
    d. Pore-forming and carrier antibiotics that act by passive transport and active transport, respectively.

Answers

1. c	**5.** a	**9.** e
2. a	**6.** c	**10.** a
3. d	**7.** e	
4. d	**8.** d	

Part V: Intermediary Metabolism

13

Introduction to Intermediary Metabolism

Objectives

- To identify the distinguishing characteristics of anabolic and catabolic reactions.
- To identify the features of state 3 metabolism that lead to it being described as amphibolic as well as anapleurotic in character.
- To characterize metabolic reactions sequences known as futile cycles and identify their proposed metabolic function.
- To define, the term basal metabolic activity.
- To compare the energy content of the common food materials determined from oxidation in the body versus oxidation in a bomb calorimeter.
- To identify the basis for the differing specific dynamic action of foods.
- To define the term R.Q. and its importance in characterizing an individual's metabolic state.
- To characterize nature of substances being metabolism in individuals with measured R.Qs. of 1.0, 0.7 and 1.3.
- To define the term homeostasis.
- To identify five mechanisms by which metabolic transport processes are regulated.
- To identify five mechanisms by which enzymic activity is regulated to maintain homeostasis.
- To identify the enzymic steps that are usually regulated to control metabolic pathways.
- To define the terms positive and negative feedback as they relate to control of metabolic pathways.

Concepts

This chapter deals with the general principles that govern the metabolism of cells and organisms and the relationship of long metabolic reaction sequences to the production and use of energy. Subsequent chapters will review the important chemical and regulatory features of individual metabolic reactions, emphasizing the regulatory processes that serve to integrate all elements of metabolism into an efficient biological system.

Anabolic and Catabolic Reactions

Intermediary metabolism is considered to include all of the **anabolic** (synthetic) and **catabolic** (degradative) reactions that take place in a living organism. Metabolism may be considered to be comprised of three stages (see Figure 13–1), with each stage having an anabolic and a catabolic component. Generally, catabolic reactions result in the net production of ATP or other chemical forms of metabolic energy, whereas anabolic reactions result in the net consumption of ATP or of other chemical forms of metabolic energy.

Stage 1 includes the conversion processes that take place between polymeric substances and their monomers (eg, glycogen and glucose interconversions). **Stage 2** includes the processes of conversion between the monomeric products of stage 1 metabolism and Acetyl-CoA, sometimes known as "activated acetate." In **stage 3**, under normal conditions, acetate from stage 2 is catabolized to CO_2 and

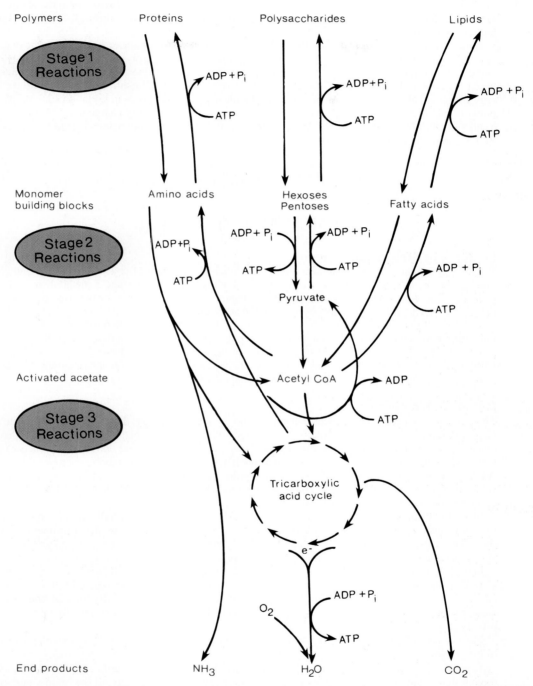

Figure 13–1. Intermediary metabolism is represented by the sum total of catabolic, anabolic, and amphibolic reactions that synthesize and degrade protein, carbohydrates, and lipids. Stage 1 includes all reactions that interconvert the latter compounds with their respective building-block precursors. Stage 2 metabolism involves interconverting the diverse building blocks and products of stage 1 into a common chemical pool, principally acetate or acetylCoA, which is further metabolized in stage 3. The reactions that take place in stage 3 are called amphibolic, since they serve either catabolic or anabolic purposes, depending on the needs of the cell.

H_2O. If there is an excess of energy, compounds of stage 3 are anabolically converted to the building blocks of proteins—polysaccharides and lipids. In extreme cases of energy excess, or in metabolic disease states such as diabetes (in which glucose is badly catabolized), large stores of fat may accumulate in adipose tissue.

Generally, the energy state of the cell determines its net metabolic character: either anabolic or catabolic. Because ATP is the chemical driving force of most metabolic processes that require energy, the energy state of a cell is described by the proportion of the adenine ribonucleotide pool that contains high-energy phosphate (**Energy Charge**), or that is present as ATP (**Phosphorylation Potential**). Energy charge is calculated as the molar ratio of adenine nucleotides, shown in Equation 13.1 and Phosphorylation Potential is calculated as shown in Equation 13.2. The energy charge ranges from zero, in a cell with no ATP and no ADP, to 1.0 in a cell where all the adenine ribonucleotide is present as ATP. In most cells a normal energy charge is in the vicinity of 0.8 to 0.9.

$$\text{Energy Charge} = \frac{[\text{ATP}] + \frac{1}{2}[\text{ADP}]}{[\text{AMP}] + [\text{ADP}] + [\text{ATP}]} \qquad \textbf{13.1}$$

$$\text{Phosphorylation Potential} = \frac{[\text{ATP}]}{[\text{ADP}][\text{Pi}]} \qquad \textbf{13.2}$$

Features of State 3 Metabolism

While stages 1 and 2 most often have clearly identifiable sequences of anabolic and catabolic reactions, generally with separate sets of enzymes for both reaction types, stage 3 metabolism is not so clear-cut. The dominant metabolic pathway of stage 3 is the Krebs cycle, or the tricarboxylic acid (TCA) cycle. This pathway always operates in the same direction, but, depending on the needs of the cell, the effect may be either catabolic or anabolic. Stage 3 reactions are therefore often described as **amphibolic** in nature. When this pathway is in the anabolic mode, intermediates are siphoned off to produce essential cell materials. As intermediates of the cycle become depleted, a series of **anapleurotic** (filling up) reactions are available to restore the normal balance of cycle intermediates. Anapleurotic reactions are discussed in greater detail in Chapter 17.

Metabolic Reactions

There are several reaction sequences that appear simply to waste energy; these are known as **futile cycles**. An example might be the conversion of glucose to acetate and its reconversion to glucose: the net result is that the end product is the same as the starting compound (glucose converted to glucose) and there is no net change in concentration of intermediates. Nevertheless metabolic energy, in this case ATP, is required to drive the interconversions. Futile cycles operate in cells at low levels of activity, where they are thought to keep reaction pathways ready to become highly active in either a catabolic or an anabolic direction in a very short time.

In all organisms, catabolic reaction sequences yield metabolic energy, usually as ATP, but often as reducing equivalents (eg, NADH, NADPH). Anabolic reaction sequences are net consumers of these same high-energy compounds. The catabolic pathway yielding the most energy is glycolysis: the conversion of glucose to pyruvate (aerobic glycolysis) or lactate (anaerobic glycolysis).

Energy consumption among humans can vary greatly, depending on such factors as sex, age, state of health, and ratio of body surface area to mass. As an average, however, an adult of about 70 kg expends about 2 million calories (2,000 kcal) to support basal metabolic activity for a 24-hour period. Basal metabolic activity is measured when an individual is awake and at rest; its clinical values are usually given as kcal or kJ per hour.

Most of the energy required for basal metabolism is used for maintenance processes like neuronal transport, osmotic pumping, pulmonary ventilation, muscle tone, and so on. Only a very small percentage of this energy is available to do useful work. More than 20% is utilized by the brain, even though this organ comprises only about 2% of total body weight. Virtually all the brain's energy is normally derived from the aerobic catabolism of glucose, and only under very unusual circumstances (such as starvation) is the brain capable of utilizing other compounds as a source of energy. Other highly active tissues, including smooth muscle that maintains body tone, and cardiac muscle that constantly pumps blood, derive the bulk of their energy from the oxidation of fatty acids.

For many tissues in the body, the state of activity determines the source of energy. Rapidly contracting skeletal muscle derives its energy from aerobic glycolysis when oxygen is available, from anaerobic glycolysis when oxygen delivery is below that required for ATP production, and from fatty

Table 13–1. The energy value of food categories in kcal/gram.

Food	Oxidized in Body	Oxidized in Calorimeter
Carbohydrate	4.1	4.1
Fat	9.3	9.3
Protein	4.1	5.4
Ethanol	7.1	7.1

acids when at rest. Normally, amino acids are consumed fairly uniformly by all tissues, but if fatty acids and glucose are available, amino acids and the proteins from which they are derived are spared.

Daily energy requirements above the basal level depend on the nature and extent of activity. An individual with a light load of physical work, such as a sedentary student, typically requires an additional 2500 cal per minute; a moderately active individual an extra 5000 cal per minute; and someone engaged in strenuous activities an additional 10,000 cal per minute above the basal level.

The energy value of some common dietary substances are shown in Table 13–1. From this table it is clear that except for the oxidation of proteins (amino acid), metabolic oxidations are as efficient as those carried out in a calorimeter. The rationale behind these figures is that for carbohydrates, fats, and alcohols, the metabolic oxidation products are the same in a cell as in a calorimeter (ie, H_2O and CO_2). In the case of nitrogenous compounds, however, metabolic and nonmetabolic oxidations lead to different nitrogenous products, as indicated in Equations 13.3 and 13.4.

$$\text{Protein} + O_2 \xrightarrow{\text{metabolism}} CO_2 + H_2O + NH_3 \qquad \textbf{13.3}$$

$$\text{Protein} + O_2 \xrightarrow{\text{nonmetabolic reactions}} CO_2 + H_2O + H_2NO_3 \qquad \textbf{13.4}$$

In humans the energy available from the nonmetabolic oxidation of reduced forms of nitrogen (ammonia) is lost, since humans lack the ability to form oxides of nitrogen and consequently excrete most of the reduced nitrogen as urea and ammonia. However, humans and other animals do oxidize a limited amount of nitrogen to the level of an oxide, producing the very important vasodilator, **nitric oxide (NO)**.

The postprandial state is associated with an increase in body temperature, as energy is expended to process food that has been consumed. This extra energy usage, known as the **specific dynamic action** of foods, varies widely, depending on the composition of a meal. The specific dynamic action of the main food groups are presented in Table 13–2 as a percentage of their respective caloric intake.

Respiratory Quotient

The **respiratory quotient (RQ)** is the ratio of CO_2 produced by the body to O_2 consumed (Equation 13.5). The value of RQ is used as an indicator of the class of food (carbohydrate, lipid, protein) being metabolized to produce body energy. In practice, the changes in partial pressure of CO_2 and O_2 in expired air are measured and converted to molar values, with the molar ratio providing the metabolic RQ index used to define the food being metabolized.

Equation 13.6 illustrates the stoichiometry of carbohydrate oxidation. In this example it is apparent that the quantity of oxygen consumed (nO_2) is stoichiometrically equivalent to the quantity of CO_2 produced, with the result that the typical RQ of carbohydrates is 1.0. Equivalent oxidation reactions for

Table 13–2. The specific dynamic action of food categories, expressed as a percentage. The percentages shown are the proportion of the caloric value of ingested food that is required to process that class of food into a form useable by the body. Eg, a 10 g protein meal has a metabolic energy content of about 40 kcal, but 30% of that energy, or 12 kcal, are used just to process the protein leaving only 28 kcal available for other purposes.

Category	Percentage
Carbohydrates	6%
Fats	4%
Proteins	30%

lipid and protein are somewhat more complex, leading to an RQ of 0.7 for lipids (which reflects the greater reduction level of fatty acids), and an average RQ for proteins of about 0.8.

$$\text{R.Q} = \frac{\text{moles } CO_2 \text{ produced}}{\text{moles } O_2 \text{ consumed}} \qquad\qquad 13.5$$

$$C_n(H_2O)_n + nO_2 \longrightarrow nCO_2 + nH_2O \qquad\qquad 13.6$$

The nature of foods being used to provide energy is not necessarily reflective of dietary intake. For example, in unregulated diabetes mellitus, an individual may consume a carbohydrate-rich diet, but as a consequence of the inability to metabolize carbohydrate he or she may have an RQ in the vicinity of 0.7 to 0.8, revealing that the principal energy source is lipid or protein. Likewise, during food deprivation, normal individuals consume and deplete body stores of carbohydrate first. During later stages of starvation the principal energy source is fatty acids, with the consequence that starving individuals exhibit RQs in the vicinity of 0.7. Conversely, in well-nourished individuals, a diet containing high levels of carbohydrates may lead to their conversion and storage as fat. The RQ under these circumstances can exceed a value of 1.0, since there is a net gain of oxygen during the conversion of the oxygen-rich carbohydrates to the highly reduced fatty acids of adipose tissue.

Regulation of Metabolism

Individuals are in a metabolic homeostatic state when all of their intracellular and extracellular reactions and process are proceeding normally to maintain a healthy state. Under such conditions, all the individual reactions and processes in the body are optimally regulated to produce the finely balanced homeostatic state. Among the many mechanisms that combine to produce homeostasis are the regulation of transport to control the entry of metabolites into cells, and the regulation of enzyme activities within cells.

Regulation of Metabolism by Regulating Transport of Biomolecules

With the exception of water, the transmembrane movement of charged biomolecules rarely occurs by simple diffusion. The transport of small molecules into the body's circulatory system is generally mediated by transport proteins localized in mucosal membranes of the gastrointestinal tract. Similar transport systems then regulate the entry of these molecules into cells, where they are used to fuel metabolism. Some transport processes operate against a concentration gradient (a higher concentration inside the cell than outside), and as a consequence the transport process appears to have a positive free energy ($\Delta G^{0\prime}$). Such processes require a source of metabolic energy, such as ATP, to proceed forward and are thus described as **active transport**. Conversely, transport processes with a negative $\Delta G^{0\prime}$, in which metabolites move down their concentration gradient, are described as **passive transport** processes. Active or passive transport can be regulated by the factors described in Table 13–3.

The Committed Step in Metabolic Pathways

In general, metabolic pathways are composed of multiple reaction steps, with only one or a few steps being enzymatically regulated (Table 13–4). The regulated step in a pathway usually occurs at a metabolic branch point, where raw materials are directed toward one or another end product, as illustrated in Figure 13–2. In this example, metabolite M_{in} is a metabolite which can proceed to the products A and B, while E_{A1} and E_{B1} are enzymes at the branch point. An accumulation of excess product from pathway A is shown to produce negative feedback on the first two steps of the A pathway, thus inhibiting additional A accumulation. At the same time, A has positive feedback for enzyme E_{B1} leading

---

Table 13–3. Factors that regulate metabolic transport processes.

1. Transmembrane concentration ratio of the transportable molecules. When substrate concentration regulates transport, the system is said to be **substrate-gated**.

2. Transmembrane electrical potential. When the membrane potential regulates transport, the system is said to be **charge-gated**.

3. The number of available membrane transporter molecules in the membrane.

4. The affinity of the transport system for the transported molecules.

5. In the case of active transport, the available energy supplies.

Table 13–4. Metabolic pathways are regulated by modifying the concentration or activity of key enzymes catalyzing reactions that normally limit flow through the pathway.

Mechanisms that regulate the activity of metabolic pathways by regulating enzyme concentration
1. Regulation of gene expression controls the quantity and rate of enzyme synthesis.
2. Proteolytic enzyme activity determines the rate of enzyme degradation.
3. Covalent modification of preexisting pools of inactive proenzymes produces active enzymes.
Mechanisms that regulate metabolic pathways by regulating the activity of enzymes
1. Allosteric regulation.
2. Covalent modification.

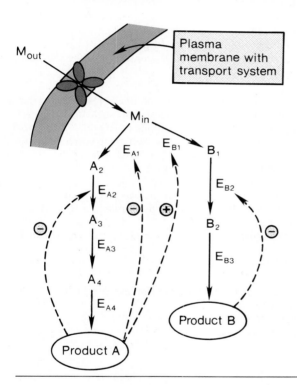

Figure 13–2. Metabolite M is transported to the interior of a cell, where it is metabolized via pathway $A_1 \rightarrow P_A$ or $B_1 \rightarrow P_B$. The products of the two pathways are shown to regulate mass flow to P_B and P_B by allosteric positive and negative feedback loops.

to enhanced utilization of metabolite M via pathway B. Many similar and more complex scenarios are found throughout metabolism.

Questions

DIRECTIONS (items 1–7): Each numbered item or incomplete statement is followed by answers or by completions of the statement. Select the ONE lettered answer or completion that is BEST in each case.

1. With regard to the available metabolic energy content of foods, it can be stated that:
   a. Getting energy from protein consumption is generally more efficient than from carbohydrate consumption.
   b. Getting energy from fat or from carbohydrate consumption is about equally efficient.
   c. In comparing carbohydrate, fat, and protein, the "best" energy source in terms of calories or joules per gram of material is carbohydrate.

   d. The metabolically useful energy available from glucose oxidation is equal to that obtained from glucose combustion in a bomb calorimeter.
   e. None of the above.

2. With regard to respiratory quotient (RQ) it can be said that:
   a. The RQ of fats is less than that of carbohydrates, which is less than that of protein.
   b. The RQ of fat is less than that of protein, which is less than that of carbohydrate.
   c. The RQ of carbohydrates is less than that of protein, which is less than that of fat.
   d. The RQ of carbohydrates is less than that of fats, which is less than that of protein.

3. RQ is a measure of:
   a. The extent of metabolic disease states.
   b. The oxidation capacity of a tissue.
   c. For a specific cell or organism, the ratio of CO_2 produced to O_2 consumed.
   d. For a cell or organism, the ratio of O_2 consumed to the CO_2 produced.
   e. The oxidative capacity of a tissue, assessed from the ratio of O_2 consumed to the CO_2 produced.

4. During clinical tests, apparently normal individuals had their total food intake restricted so that their available carbohydrate stores were depleted. Each individual ingested the standard 100-g carbohydrate (glucose) load, and RQ was determined. Individual A exhibited an RQ of 7.4, while individual B exhibited an RQ of 0.98. The attending physician made the following correct diagnosis of the two cases.
   a. Individual B is metabolically normal, but A has a metabolic disease.
   b. Neither individual exhibits an RQ appropriate to the dietary intake.
   c. There is no significant difference between the metabolic state of the two individuals.
   d. The dominant source of metabolic energy for individual A is lipid, whereas for B the dominant metabolic energy source is carbohydrate.
   e. One cannot determine the metabolic energy source of either individual from the data provided.

5. The energy charge of a cell or tissue is a measure of:
   a. Its adenosine content.
   b. The energy source used to drive adenine ribonucleotide phosphorylation.
   c. Its relative content of phosphorylated adenine nucleotides.
   d. Its relative content of total phosphorylated nucleotides.
   e. The energy state of a cell, ranging from 0 to 10.

6. Stage 3 of cell metabolism can be characterized as follows:
   a. It is a metabolic reaction pathway that uses the same set of enzymes to operate catabolically in the forward direction and anabolically in the reverse direction.
   b. It is principally comprised of the reaction of oxidative phosphorylation.
   c. It is principally comprised of the reactions of glycolysis and gluconeogenesis.
   d. Its reaction pathway(s) are amphibolic in character.
   e. It is only metabolically significant during carbohydrate catabolism.

7. Basal metabolism of an average human adult can be characterized as:
   a. Metabolic reactions requiring expenditure of about 8.4 millions of Joules per day and which support all the normal activities of a 70-kg person.
   b. Metabolic reactions requiring expenditure of 85 kcal per hour to support all the internal maintenance activities of a 70-kg individual.
   c. A measure of metabolism which is equal to 2,000 kcal per hour for a 70-kg person.
   d. A measure of metabolic reactions that varies from one individual to another and is directly proportional to the person's level of physical activity.
   e. The set of metabolic reactions that support all the internal maintenance activities of a 70-kg person.

DIRECTIONS (items 8–13): Match the following descriptive phrases, with the lettered set of responses. Each response may be used once, more than once, or not at all.

    a. 0.7
    b. 1.0
    c. 4.1
    d. 5.4
    e. 7.0
    f. 7.1
    g. 8.3
    h. 10

8. The total caloric energy value of protein in kcal/gram.

9. The metabolic energy value of carbohydrate in kcal/gram.

10. The basal metabolic caloric rate of an average 70-kg adult human in millions of Joules per 24 hours.

11. The RQ for carbohydrate metabolism.

12. The RQ for lipid metabolism.

13. The RQ observed during later stages of starvation.

DIRECTIONS (items 14–15): Each numbered item or incomplete statement is followed by answers or by completions of the statement. Select the ONE lettered answer or completion that is BEST in each case.

14. If a person consumes 10 g of dietary protein, how much would be available to fuel basal metabolism or perform external work?
    a. About 4.1 kcal.
    b. About 5.4 kcal.
    c. About 29 kcal.
    d. About 38 kcal.
    e. About 41 kcal.

15. Which process(es) control(s) metabolic pathways by regulating the activity of individual enzymes?
    a. Regulation of gene expression and/or activation of proenzymes.
    b. Altered proteolysis and covalent enzyme modification.
    c. Regulation of gene expression and allosteric mechanisms.
    d. Covalent enzyme modification and allosteric mechanisms.

Answers

1. d	**6.** d	**11.** b
2. b	**7.** b	**12.** a
3. c	**8.** d	**13.** a
4. d	**9.** c	**14.** c
5. c	**10.** g	**15.** d

14   Glycolysis and Gluconeogenesis

Objectives

- To define the processes by which carbohydrates from ingested carbohydrate polymers gain access to the body.
- To describe the energy yield from glycolysis.
- To identify the reactions of the priming phase of glycolysis.
- To identify the reactions of the energy yielding phase of glycolysis.
- To identify the functional similarities and differences between hexokinases I, II, III, and IV and identify the type that is also known as glucokinase.
- To identify how PFK and fructose-1,6-bisphosphatase might support a futile cycle that could deplete the ATP supply of a cell.
- To define the regulatory function of fructose-2,6-bisphosphate on PFK and on fructose-1,6-bis phosphatase.
- To identify how peptide-hormones regulate cell levels of fructose-2,6-bisphosphate.
- To compare the role of the following allosteric effectors on PFK activity and on F-1-6-bis phosphatase activity: ATP, AMP, Citrate, and F-6-P.
- To define the potential ATP yield of the glyceraldehyde-3-P dehydrogenase reaction.
- To compare the energy yield from glycolytic conversion of 1,3-BP-glycerate to 3-P-glycerate with that obtained from the same conversion via the 2,3-BP-glycerate shunt.
- To compare the activity of fetal and neonatal PK of erythrocytes and identify the physiological significance of the differing activity of the two isozymes.
- To identify the main allosteric regulators of PK and their effect on PK activity.
- To compare the effect of peptide hormones on liver and muscle PK activity and identify the physiological significance of the differences between the two effects.
- To identify the physiological importance of the LDH reaction to erythrocyte glycolysis and to anaerobic muscle glycolysis.
- To identify the enzymic reactions involved in gluconeogenesis which are not involved in glycolysis.
- To identify the subcellular location and the major allosteric effector of pyruvate carboxylase.
- To identify the source of substrate for cytosolic PEPCK and the process by which the substrate appears in the cytosol.
- To identify the importance of glucose-6-phosphatase activity to maintenance of euglycemia.
- To define the Cori cycle and the glucose alanine cycle.

Concepts

Carbohydrates constitute a principal source of dietary calories and a main fuel of the body. This chapter focuses on the conversion of dietary carbohydrate to substances that are assimilated into the body and processed through glycolysis and gluconeogenesis, the catabolic and anabolic arms of stage 2 carbohydrate metabolism.

Digestion of Dietary Carbohydrates

The bulk of dietary carbohydrate enters the body in complex forms, such as disaccharides and the polymers amylose, amylopectin, glycogen, and cellulose. (Humans do not digest cellulose effectively; instead, cellulose largely passes through the digestive tract and is eliminated in its original form.) The first step in the metabolism of digestible carbohydrate is the conversion of the higher polymers to simpler, soluble forms that can be transported across the intestinal wall and delivered to the tissues.

This conversion or breakdown begins right along with food intake. In the mouth, saliva has a slightly acid pH of 6.8 and contains a **lingual amylase** that begins the digestion of carbohydrates, catalyzing conversion of polysaccharides to maltose and higher polymers. The action of lingual amylase is limited to the area of the mouth and the esophagus; it is virtually inactivated by the much stronger acid pH of the stomach. Once the food has arrived in the stomach, acid hydrolysis contributes to its degradation; specific gastric proteases and lipases aid this process for proteins and fats, respectively. The mixture of gastric secretions, saliva, and food, known collectively as **chyme**, moves to the small intestine. There, liver and pancreatic secretions neutralize the acid pH of the chyme by adding a variety of buffers, enzymes, hormones, and detergents.

The main polymeric-carbohydrate digesting enzyme of the small intestine is **α amylase**. This enzyme is secreted by the pancreas and has the same activity as salivary amylase, producing disaccharides and trisaccharides. The latter are converted to monosaccharides by intestinal saccharidases, including maltases that hydrolyze di- and trisaccharides, and the more specific disaccharidases, **sucrase**, lactase, and **trehalase**. The net result is the almost complete conversion of digestible carbohydrate to its constituent monosaccharides.

The resultant glucose and other simple carbohydrates are transported across the intestinal wall to the hepatic portal vein and then to liver parenchymal cells and other tissues. There they are converted to fatty acids, amino acids, and glycogen, or else oxidized to CO_2 and H_2O by the catabolic pathway known as glycolysis.

Intracellular Catabolism of Glucose

The Embden Myerhoff Pathway, better known as **glycolysis**, is the intracellular catabolic pathway that degrades 6-carbon sugars to produce either lactate or pyruvate. Under aerobic conditions, the dominant product in most tissues is pyruvate and the pathway is known as **aerobic glycolysis**. When oxygen is depleted, as for instance during prolonged vigorous exercise, the dominant glycolytic product in many tissues is lactate and the process is known as **anaerobic glycolysis**.

Energy Yield of Glycolysis

Examination of the overall energy economy of aerobic glycolysis reveals, as shown in Equations 14.1 and 14.2, that aerobic glycolysis requires two equivalents of ATP to *prime* the process (Equation 14.1), with the subsequent production of four equivalents of ATP and two equivalents of NADH (Equation 14.2). Thus, conversion of one mole of glucose to two moles of pyruvate is accompanied by the net production of two moles each of ATP and NADH, as shown in Equation 14.3 (the sum of Equations 14.1 and 14.2).

$$\text{Glucose + 2ATP} \rightarrow \text{2 Glyceraldehyde-3-P + 2ADP} \qquad \textbf{14.1}$$

$$\text{Glyceraldehyde-3-P + 4ADP + 2NAD}^+ + \text{2P}_i \rightarrow \text{2Pyruvate + 4ATP + 2NADH + 2H}^+ \qquad \textbf{14.2}$$

$$\text{Glucose + 2ADP + 2NAD}^+ + \text{2P}_i \rightarrow \text{2Pyruvate + 2ATP + 2NADH + 2H}^+ \qquad \textbf{14.3}$$

The NADH generated during glycolysis is used to fuel mitochondrial ATP synthesis via oxidative phosphorylation, producing either two or three equivalents of ATP depending upon the transport mechanism utilized to transport the electrons from cytoplasmic NADH into the mitochondria (see Chapter 18). The net yield from the oxidation of 1 mole of glucose to 2 moles of pyruvate is, therefore, either 6 or 8 moles of ATP. Complete oxidation of the 2 moles of pyruvate, through the TCA cycle (see Chapter 17), yeilds an additional 30 moles of ATP; the total yield, therefore being either 36 or 38 moles of ATP from the complete oxidation of 1 mole of glucose to CO_2 and H_2O.

INDIVIDUAL REACTIONS OF GLYCOLYSIS

The pathway of glycolysis (Figure 14–1) can be seen as consisting of 2 separate phases. The first is the chemical priming phase requiring energy in the form of ATP, and the second is considered the energy-yielding phase. In the first phase, 2 equivalents of ATP are used to convert glucose to fructose-1,6-bisphosphate (F-1,6-BP). In the second phase F-1,6-BP is degraded to pyruvate, with the production of 4 equivalents of ATP and 2 equivalents of NADH. Although this discussion focuses on the conversion of glucose to catabolic products, it is important to recognize that—depending on the energy state of a cell—the flow of intermediates can be in the glycolytic direction or else in the reverse, anabolic direction, in which case the process is known as gluconeogenesis.

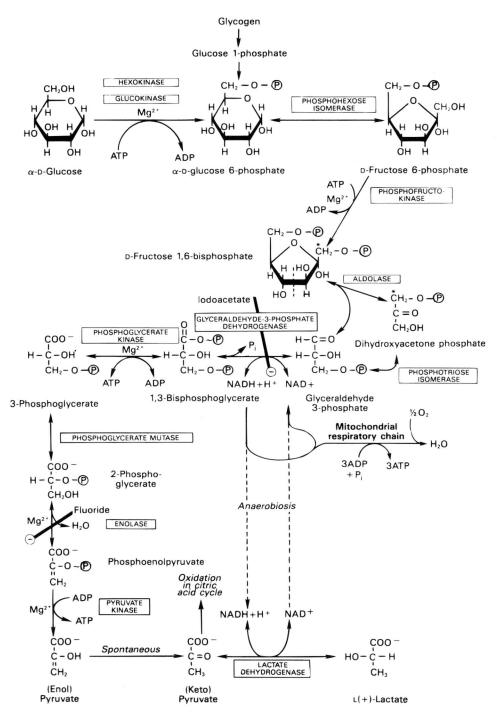

Figure 14–1. The pathway of glycolysis. Ⓟ, $-PO_3^{2-}$; P_i, $HOPO_3^{2-}$; ⊖, inhibition. *Carbon atoms 1–3 of fructose bisphosphate form dihydroxyacetone phosphate, whereas carbons 4–6 form glyceraldehyde 3-phosphate. The term bis-, as in bisphosphate, indicates that the phosphate groups are separated, whereas diphosphate, as in adenosine diphosphate, indicates that they are joined. Notice that NADH from the glyceraldehyde-3-P dehydrogenase step can be used to produce lactate or by mitochondria to produce ATP.

Table 14–1. Characteristics of hexokinases.

Characteristic	Types I, II, III Hexokinases	Glucokinase or Type IV Hexokinase
Sugar specificity	Similar to glucokinase	Similar to hexokinase
Glucose $K_{M(app)}$	0.1 mM	10 mM
Kinetic response to glucose	Hyperbolic, Michaelis-Menten	Sigmoid allosteric type
Allosteric response to G-6P	None	Negative effector
Tissue distribution	All human tissues studied	Liver and pancreas isoforms

The Hexokinase Reaction

The ATP-dependent phosphorylation of glucose to form glucose-6-phosphate (G-6-P) is the first reaction of glycolysis, and is catalyzed by tissue-specific isoenzymes known as hexokinases. The phosphorylation accomplishes two goals: First, the hexokinase reaction converts nonionic glucose into an anion that is trapped in the cell since cells lack transport systems for phosphorylated sugars. Second, the otherwise biologically inert glucose becomes activated into a labile form capable of being further metabolized.

$$\text{Glucose} + \text{ATP} \rightarrow \text{Glucose-6-P} + \text{ADP} \qquad \textbf{14.4}$$

Four mammalian isozymes of hexokinase are known (Types I–IV), with the Type IV isozyme often referred to as **glucokinase**. Glucokinase is the form of the enzyme found in hepatocytes. Contrary to earlier assertions, glucokinase is no more *specific* for glucose than any of the other hexokinases, although it does differ from them in a number of important ways (outlined in Table 14–1).

The high K_M of glucokinase for glucose means that this enzyme is saturated only at very high concentrations of sustrate (Figure 14–2). This feature of hepatic glucokinase allows the liver to "buffer" blood glucose. After meals, when postprandial blood glucose levels are high, liver glucokinase is significantly active, which causes the liver preferentially to trap and to store circulating glucose. When blood glucose falls to very low levels, tissues such as liver and kidney—which contain glucokinases but are not highly dependent on glucose—do not continue to use the meager glucose supplies that remain available. At the same time, tissues such as the brain, which are critically dependent on glucose, continue to scavenge blood glucose using their low K_M hexokinases, and as a consequence their viability is protected. Under various conditions of glucose deficiency, such as long periods between meals, the liver is stimulated to supply the blood with glucose through the pathway of gluconeogenesis. The levels of glucose produced are insufficient to activate glucokinase, allowing the glucose to pass out of hepatocytes and into the blood.

The regulation of hexokinase and glucokinase activities is also different. Hexokinases I, II, and III are allosterically inhibited by product (G-6-P) accumulation, whereas glucokinases are not allosterically effected by G-6-P. The latter further insures liver accumulation of glucose stores during times of glucose excess, while favoring peripheral glucose utilization when glucose is required to supply energy to peripheral tissues.

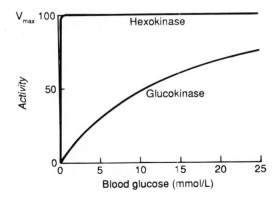

Figure 14–2. Variation in glucose phosphorylating activity of hexokinase and glucokinase with increase in blood glucose concentration. The K_M for glucose of hexokinase is 0.005 mmol/L and of glucokinase is 10 mmol/L.

Phosphohexose Isomerase

$$\text{α-D-glucose 6-phosphate} \Leftrightarrow \text{D-fructose-6-phosphate} \qquad 14.5$$

The second reaction of glycolysis is an isomerization, in which the hemiacetal, glucose-6-P is converted to the hemiketal, fructose-6-P. The enzyme catalyzing this reaction is phosphohexose isomerase (also known as phosphogluco isomerase). The reaction is freely reversible at normal cellular concentrations of the two hexose phosphates and thus the isomerase catalyzes this interconversion during glycolytic carbon flow and during gluconeogenesis.

6-Phosphofructo-1-Kinase (Phosphofructokinase-1, PFK-1)

$$\text{D-fructose 6-phosphate} + \text{ATP} \rightarrow \text{D-fructose 1,6-bisphosphate} + \text{ADP} \qquad 14.6$$

The next reaction of glycolysis (Equation 14.6) involves the utilization of a second ATP to convert fructose-6-P to fructose-1,6-bisphosphate (F-1,6-BP). This reaction is catalyzed by 6-phosphofructo-1-kinase, better known as phosphofructokinase-1 or PFK-1. This reaction is not readily reversible because of its large positive free energy ($\Delta G^{0'} = +5.4$ kcal/mol) in the reverse direction. Nevertheless, as illustrated by Equation 14.7, fructose units readily flow in the reverse (gluconeogenic) direction because of the ubiquitous presence of a hydrolytic enzyme, fructose-1,6-bisphosphatase (F-1,6-BPase).

$$\text{D-fructose 1,6-bisphosphate} + \text{H}_2\text{O} \rightarrow \text{D-fructose 6-phosphate} + \text{P}_\text{I} \qquad 14.7$$

The presence of these two enzymes in the same cell compartment provides an example of a metabolic futile cycle, which if unregulated would rapidly deplete cell energy stores. However, the activity of these two enzymes is so highly regulated that PFK-1 is considered to be the **rate-limiting enzyme of glycolysis** and F-1,6-BPase is considered to be the **rate-limiting enzyme in gluconeogenesis**.

Although there are numerous allosteric feedback regulators of PFK-1 and F-1,6-BPase, one of the most important effectors is fructose-2,6-bisphosphate (F-2,6-BP), which is not an intermediate in glycolysis or in gluconeogenesis. Figure 14–3 illustrates the regulatory role of F-2,6-BP on these two enzymes and the mechanism of hormone-regulated covalent modification that determines the cellular levels of F-2,6-BP. The synthesis of F-2,6-BP is catalyzed by the bifunctional enzyme PFK-2/F-2,6-BPase. In the nonphosphorylated form the enzyme is known as PFK-2 and serves to catalyze the synthesis of F-2,6-BP. The result is that the activity of PFK-1 is greatly stimulated and the activity of F-1,6-BPase is greatly inhibited.

Under these conditions, fructose flow through the two enzymes takes place in the glycolytic direction, with a net production of F-1,6-BP. When the bifunctional enzyme is phosphorylated it no longer exhibits kinase activity, but a new active site hydrolyzes F-2,6-BP to F-6-P and inorganic phosphate. The metabolic result of the phosphorylation of the bifunctional enzyme is that allosteric stimulation of PFK-1 ceases, allosteric inhibition of F-1,6-BPase is eliminated, and net flow of fructose through these two enzymes is gluconeogenic, producing F-6-P and eventually glucose.

The interconversion of the bifunctional enzyme is catalyzed by cAMP-regulated protein kinase (PKA), which in turn is regulated by circulating peptide hormones. When blood glucose levels drop, pancreatic insulin production falls, glucagon secretion is stimulated, and circulating glucagon is highly increased. Hormones such as glucagon bind to plasma membrane receptors on liver cells, activating membrane-localized adenylate cyclase leading to an increase in the conversion of ATP to cAMP. As previously indicated in Chapter 7, cAMP binds to the regulatory subunits of PKA, leading to release and activation of the catalytic subunits (Figure 7–5). PKA phosphorylates numerous enzymes, including the bifunctional PFK-2/F-2,6-BPase. Under these conditions the liver stops consuming glucose and becomes metabolically gluconeogenic, producing glucose to reestablish normoglycemia.

A number of other allosteric regulators of PFK-1 and F-1,6-BPase activity are listed in Table 14–2. It is important to note that the final direction of fructose flow depends on the summed effect of all the individual regulators of PFK and F-1,6-BPase.

Aldolase

$$\text{D-fructose 1,6-bisphosphate} \Leftrightarrow \text{dihydroxyacetone phosphate}$$
$$+ \text{ glyceraldehyde 3-phosphate} \qquad 14.8$$

The aldolase reaction also proceeds readily in the reverse direction, being utilized for both glycolysis and gluconeogenesis.

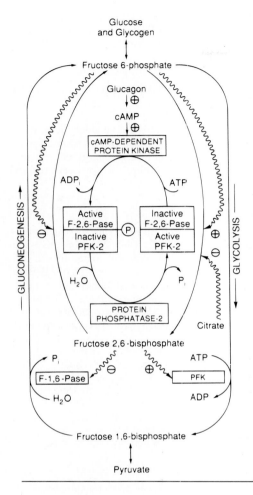

Figure 14–3. Control of glycolysis and gluconeogenesis in the liver by fructose 2,6-bisphosphate and the bifunctional enzyme PFK-2/F-2,6-Pase (6-phosphofructo-2-kinase/fructose 2,6 bisphosphatase). PFK, phosphofructokinase (6-phosphofructo-1-kinase); F-1,6-Pase, fructose-1,6-bisphosphatase. Arrows with wavy shafts indicate allosteric effects.

Triose Phosphate Isomerase

As shown in Equation 14.9, the two products of the aldolase reaction equilibrate readily in a reaction catalyzed by triose phosphate isomerase. Succeeding reactions of glycolysis utilize glyceraldehyde-3-P as a substrate; thus, reaction 14.8 is pulled in the glycolytic direction by mass action principals.

$$\text{Dihydroxyacetone phosphate} \underset{}{\overset{\text{Triose Phosphate Isomerase}}{\rightleftharpoons}} \text{Glyceraldehyde-3-phosphate} \qquad \textbf{14.9}$$

Glyceraldehyde-3-P Dehydrogenase

The second phase of glucose catabolism features the energy-yielding glycolytic reactions that produce ATP and NADH (see Figure 14–1). In the first of these reactions, glyceraldehyde-3-P dehydrogenase

Table 14–2. Allosteric regulators of PFK-1 and F-2,6-BPase.		

Effector	Effect on PFK-1 Activity	Effect on F-1,6-BPase Activity
F-2,6-BP	Positive	Negative
ATP	Negative	Negative
AMP	Positive	Negative
Citrate	Negative	
F-6-P	Positive	

(G3PDH) catalyzes the NAD^+-dependent oxidation of glyceraldehyde 3-P to 1,3-bisphosphoglycerate (1,3-BPG) and NADH (Equation 14.10). As will be discussed below, NADH can potentially give rise to either 2 or 3 ATPs. The G3PDH reaction is reversible, and the same enzyme catalyzes the reverse reaction during gluconeogenesis.

$$\textbf{Glyceraldehyde-3-P} + \textbf{NAD}^+ + \textbf{HPO}_4^{-2} \underset{\text{dehydrogenase}}{\overset{\text{glyceraldehyde-3-P}}{\rightleftharpoons}} \textbf{1,3-BPG} + \textbf{NADH} + \textbf{H}^+ \qquad \textbf{14.10}$$

The key features of this important reaction are diagrammed in Figure 14–4. The aldehyde forms a covalent bond with a cysteine residue of the enzyme; the complex is then oxidized by bound NAD^+ forming a thioester intermediate. Thioesters are energetically equivalent to high-energy phosphates, and the subsequent addition of phosphate across the thioester bond (phosphorolysis) proceeds readily to form 1,3-BPG which is a mixed, phosphoacyl-anhydride at C-1.

Phosphoglycerate Kinase

Next the newly formed, high-energy phosphate of 1,3-BPG is used to form ATP and 3-phosphoglycerate (3-PG) by the enzyme phosphoglycerate kinase, as illustrated in Figure 14–5. Note that this is the only reaction of glycolysis or gluconeogenesis that involves ATP and is reversible under normal cell conditions. Figure 14–5 also illustrates an important reaction of erythrocytes, the formation of 2,3-BPG by the enzyme bisphosphoglycerate mutase. 2,3-BPG is an important regulator of hemoglobin's affinity for oxygen (see Chapter 8). Note that 2,3-bisphosphoglycerate phosphatase degrades 2,3-BPG to 3-phosphoglycerate, a normal intermediate of glycolysis. The 2,3-BPG shunt thus operates with the effective loss of 1 equivalent of ATP per triose passed through the shunt. The process is not reversible under physiological conditions.

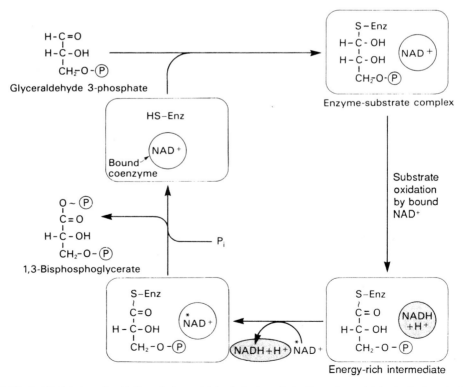

Figure 14–4. Mechanism of oxidation of glyceraldehyde 3-phosphate. Enz, glyceraldehyde-3-phosphate dehydrogenase. The enzyme is inhibited by the −SH poison iodoacetate, which is thus able to inhibit glycolysis. (Reproduced, with permission, from Murray, RK: *Harper's Biochemistry*, 23rd ed. Appleton & Lange, 1993.)

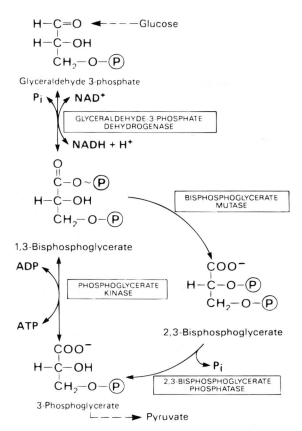

Figure 14–5. 2,3-Bisphosphoglycerate pathway in erythrocytes. (Reproduced, with permission, from Murray, RK: *Harper's Biochemistry*, 23rd ed. Appleton & Lange, 1993.)

Phosphoglycerate Mutase and Enolase

The remaining reactions of glycolysis are aimed at converting the relatively low energy phosphoacyl-ester of 3-PG to a high-energy form and harvesting the phosphate as ATP. The 3-PG is first converted to 2-PG by phosphoglycerate mutase (Equation 14.11) and the 2-PG conversion to phospho-enoylpyruvate (PEP) is catalyzed by enolase (Equation 14.12).

$$3\text{-PG} \Leftrightarrow 2\text{-PG} \qquad\qquad 14.11$$

$$2\text{-PG} \Leftrightarrow \text{PEP} \qquad\qquad 14.12$$

The most important feature of the enolase reaction is that it forms a double bond next to the phospho-ester of PEP (Figure 14-1). The latter makes the PEP phosphate particularly labile, with an exergonic free energy of hydrolysis about 2 times that of an ATP high-energy phosphate (-14.6 kcal/mole or -61.1 kJ/mole).

Pyruvate Kinase

The final reaction of aerobic glycolysis is catalyzed by the highly regulated enzyme pyruvate kinase (PK). In this strongly exergonic reaction, the high-energy phosphate of PEP is conserved as ATP, as illustrated in Equation 14.13.

$$\text{PEP} + \text{ADP} + \text{H}^+ \rightarrow \text{Pyruvate} + \text{ATP} \qquad \delta G^{o\prime} = -7.5\text{kcal/mol} \qquad 14.13$$

The loss of phosphate by PEP leads to the production of pyruvate in an unstable enol form, which spontaneously tautomerizes to the more stable, keto form of pyruvate. This reaction, shown in Figure 14–1, contributes a large proportion of the free energy of hydrolysis of PEP.

A number of PK isozymes have been described. The liver isozyme (L-type), characteristic of a gluconeogenic tissue, is regulated via phosphorylation by PKA, whereas the M-type isozyme found in brain, muscle, and other glucose requiring tissue is unaffected by PKA. As a consequence of these differences, blood glucose levels and associated hormones can regulate the balance of liver gluconeogenesis and glycolysis while muscle metabolism remains unaffected.

In erythrocytes, the fetal PK isozyme has much greater activity than the adult isozyme; as a result, fetal erythrocytes have comparatively low concentrations of glycolytic intermediates. Because of the low steady-state concentration of fetal 1,3-BPG, the 2,3-BPG shunt is greatly reduced in fetal cells and little 2,3-BPG is formed. Since 2,3-BPG is a negative effector of hemoglobin affinity for oxygen, fetal erythrocytes have a higher oxygen affinity than maternal erythrocytes. Therefore, transfer of oxygen from maternal hemoglobin to fetal hemoglobin is favored, assuring the fetal oxygen supply. In the newborn, an erythrocyte isozyme of the M-type with comparatively low PK activity displaces the fetal type, resulting in an accumulation of glycolytic intermediates. The increased 1,3-BPG levels activate the 2,3-BPG shunt, producing 2,3-BPG needed to regulate oxygen binding to hemoglobin.

Genetic diseases of adult erythrocyte PK are known in which the kinase is virtually inactive. The erythrocytes of affected individuals have a greatly reduced capacity to make ATP and thus do not have sufficient ATP to perform activities such as ion pumping and maintaining osmotic balance. These erythrocytes have a short half-life, lyse readily, and are responsible for some cases of hereditary hemolytic anemia.

The liver PK isozyme is regulated by phosphorylation, allosteric effectors, and modulation of gene expression. The major allosteric effectors are fructose-1,6-BP, which stimulates PK activity by decreasing its K_M for PEP, and for the negative effector, ATP. Expression of the liver PK gene is strongly influenced by the quantity of carbohydrate in the diet, with high-carbohydrate diets inducing up to 10-fold increases in PK concentration as compared to low carbohydrate diets. Liver PK is phosphorylated and inhibited by protein kinase A (PKA), and thus it is under hormonal control similar to that described earlier for PFK-2.

Muscle PK (M-type) is not regulated by the same mechanisms as the liver enzyme. Extracellular conditions that lead to the phosphorylation and inhibition of liver PK, such as low blood glucose and high levels of circulating glucagon, do not inhibit the muscle enzyme. The result of this differential regulation is that hormones such as glucagon and epinephrine favor liver gluconeogenesis by inhibiting liver glycolysis, while at the same time, muscle glycolysis can proceed in accord with needs directed by intracellular conditions.

Anaerobic Glycolysis

Under aerobic conditions, pyruvate in most cells is further metabolized via the Kreb's (TCA) cycle. Under anaerobic conditions and in erythrocytes under aerobic conditions, pyruvate is converted to lactate by the enzyme **lactate dehydrogenase** (LDH) (Figure 14–1), and the lactate is transported out of the cell into the circulation. Figure 14–1 also illustrates the relationship between glucose catabolism and the flow of NADH reducing equivalents under anaerobic conditions. It is important to note that unless NAD^+ is continually regenerated, glycolysis becomes NAD^+ limited and ATP becomes unavailable from glycolytic reactions.

GLUCONEOGENESIS

The export of lactate out of a cell under anaerobic conditions potentially represents a large loss of energy. However, when conditions become favorable, circulatory lactate is transported back into cells and oxidized aerobically, or else it is converted to glucose by the anabolic arm of Stage 2 carbohydrate metabolism known as gluconeogenesis. Like glycolysis, gluconeogenesis involves two sets of reactions: First, **endergonic priming reactions that convert pyruvate to PEP**, requiring expenditure of two high-energy phosphates and two new enzymes. Second, **flow of carbon from PEP to glucose-6-phosphate** via reactions and enzymes described earlier in this chapter.

Conversion of Lactate to PEP

Imported cellular lactate is readily converted to pyruvate by the LDH reaction described above. However, the pyruvate kinase reaction is essentially irreversible requiring the conversion from pyruvate to PEP to proceed via the circuitous route illustrated in Figure 14–6. The first step in this conversion is the transport of pyruvate into the mitochondrial matrix where, depending on the energy status of the cell, 1 of 2 possible metabolic fates are followed. If the energy charge is low, pyruvate is oxidized to acetyl-CoA via the pyruvate dehydrogenase (PDH) complex, and the acetate is oxidized to CO_2 and

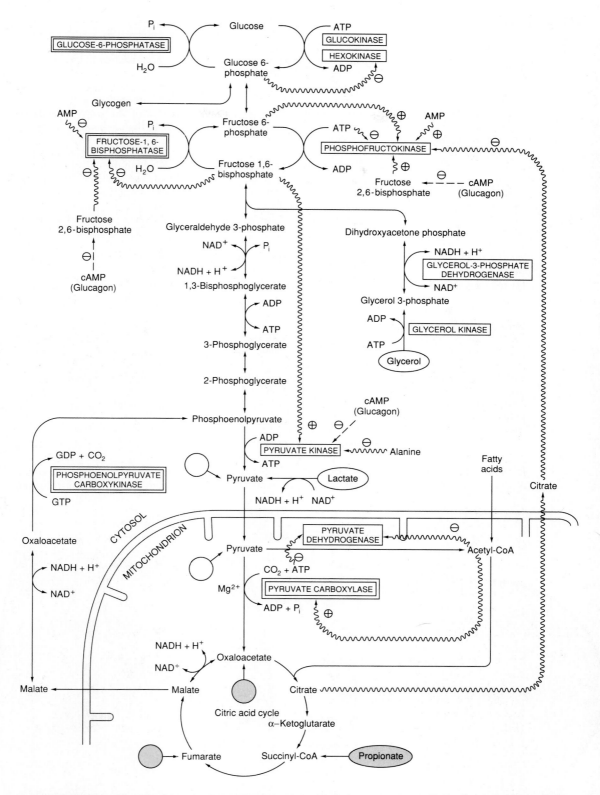

Figure 14–6. Major pathways and regulation of gluconeogenesis and glycolysis in the liver. Entry points of glucogenic amino acids after transamination are indicated by ●→. The key gluconeogenic enzymes are shown thus ▭. The ATP required for gluconeogenesis is supplied by the oxidation of long-chain fatty acids. Propionate is of quantitative importance only in ruminants. ↝, allosteric effects; --→, covalent modification by reversible phosphaorylation. High concentrations of alanine act as a "gluconeogenic signal" by inhibiting glycolysis at the pyruvate kinase step. (Reproduced, with permission, from Murray, RK: *Harper's Biochemistry*, 23rd ed. Appleton & Lange, 1993.)

H_2O by TCA cycle activity. If the energy charge is high, and the CoA pool is already largely in the form of acetyl-CoA, reduced pyruvate carbon will be conserved as fat or glycogen.

The initial step in the pathway of pyruvate to glycogen is the direct conversion of pyruvate to oxaloacetate (OAA) by mitochondrial pyruvate carboxylase. Pyruvate carboxylase activity is absolutely dependent upon acetyl-CoA which is a powerful positive effector of the enzymes' activity. Also required are ATP and CO_2. As with other enzymes that carboxylate their substrates, pyruvate carboxylase utilizes biotin as a cofactor.

Continuing on the gluconeogenic path (Figure 14–6), OAA is either converted to PEP in mitochondria by the enzyme phosphoenolpyruvate carboxykinase (PEPCK) or moved to the cytoplasm by the malate/aspartate transport system and then converted to PEP in the cytosol. The mitochondrial and cytosolic PEPCK isozymes each carry out the reaction shown in Equation 14.14. GTP is required by both enzymes, and the carbon lost in the reaction is the same carbon added by pyruvate carboxylase. The remaining reactions of gluconeogenesis are all cytosolic and involve catalysis by enzymes discussed above and illustrated in Figure 14–6.

$$\text{OAA} + \text{GTP} \rightarrow \text{PEP} + CO_2 + \text{GDP} + P_i \qquad \textbf{14.14}$$

Gluconeogenic Conversion of PEP to Glucose-6-Phosphate
A summary is given in Figure 14–6.

Glucose-6-Phosphatase
In liver, kidney, and intestine the final reaction of gluconeogenesis proceeds 1 step further than in most other tissues. These organs contain the enzyme glucose-6-phosphatase, which catalyzes the production of free glucose (Equation 14.15). Under hypoglycemic conditions the glucose is transported out of the cell, augmenting blood glucose levels. Because of the liver's ability to help maintain euglycemia, the liver is often considered the buffer of blood glucose. Muscle, brain, and erythrocytes, as well as other tissues, lack glucose-6-phosphatase with the result that these tissues cannot supplement blood glucose levels.

$$\text{Glucose-6-P} + H_2O \rightarrow \text{Glucose} + P_i \qquad \textbf{14.15}$$

In addition to the conversion of peripheral tissue lactate to glucose by the liver, peripheral tissue amino acids are a major source of carbon atoms for gluconeogenesis. In particular, muscle tissue protein can

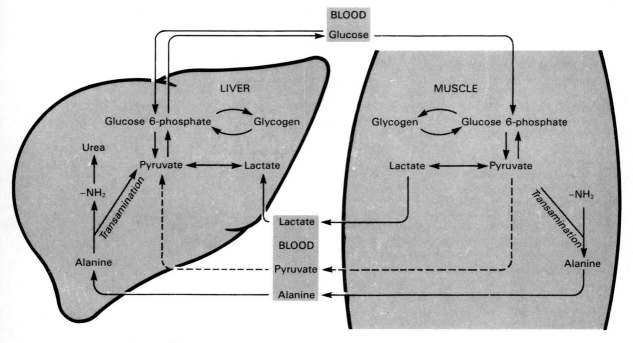

Figure 14–7. The lactic acid (Cori) and glucose-alanine cycle.

be degraded to supply carbon atoms for oxidation and consequent energy production. This process leads to the production of ammonia in muscles which lack the enzymes of the urea cycle (see Chapter 23). By transaminating pyruvate to alanine, the ammonia can be delivered to the liver and the pyruvate carbon atoms used for glucose production. The flow of glucose and lactate between liver and lactate-producing tissues, such as erythrocytes and muscle, is known as the **Cori cycle** and the flow of pyruvate and ammonia from muscle to liver with concommitant glucose synthesis is known as the **glucose-alanine cycle**. These two interconnected pathways are illustrated in Figure 14–7.

Questions:

DIRECTIONS (items 1–10): Each numbered item or incomplete statement is followed by answers or by completions of the statement. Select the ONE lettered answer or completion that is BEST in each case.

1. Which of the following enzymes is NOT found in significant amounts in muscle?
   a. Phosphoglucomutase.
   b. Glycogen synthase.
   c. Enolase.
   d. Glucose-6-phosphatase.

2. Under conditions of low blood glucose:
   a. Muscle glycogen is degraded to glucose to supply the brain with glucose.
   b. Glucokinase becomes the main intracellular "glucose trapping" enzyme.
   c. Liver glycogen is converted to brain glucose.
   d. Liver glycogen is converted to brain glucose in reactions involving glucose-6-phosphatase.
   e. Stored lactate is converted to glucose to supply muscle and brain.

3. Aerobic glycolysis is the conversion of glucose to:
   a. Pyruvate with the net production of two equivalents of ATP and two equivalents of NADH.
   b. PEP with the net production of two equivalents of NADH.
   c. PEP with the net production of two equivalents NADH and two equivalents of ATP.
   d. Lactate with the net production of 2 equivalents of ATP.
   e. Glyceraldehyde-3-phosphate with net production of two equivalents of NADH.

4. 1,3-bis-phosphoglycerate:
   a. Is converted to 3-phosphoglycerate by erythrocytes with a higher efficiency of energy conservation than that of liver.
   b. Is converted to 3-P-Glycerate in liver and red blood cells, with the obligatory production of 1 equivalent of ATP per 1,3-bisphosphoglycerate metabolized.
   c. Is an allosteric effector of Hbg affinity for oxygen.
   d. Is produced only by erythrocytes.
   e. Can be converted to 2-phosphoglycerate without production of ATP.

5. Pyruvate kinase(s) are:
   a. Mammalian isozymes, each of which is tissue-specific and subject to phosphorylation by PKA.
   b. An example of isozymes that have different tissue distributions but the same regulatory mechanism in all tissues.
   c. Positively effected by fructose-2,6-bisphosphate.
   d. Positively effected by high levels of cellular ATP (ie, high energy charge).
   e. Isozymes capable of catalyzing a reaction that is effected by fructose-1,6-bisphosphate.

6. Tissues that dominantly depend on glucose for their source of energy include:
   a. Muscle and brain.
   b. Brain and kidney.
   c. Liver and muscle.

    d. RBCs and brain.

    e. RBCs and kidney.

7. Allosteric regulation of glycolysis includes:

    a. Glucose-6-phosphate on hexokinase.

    b. Fructose-1,6-bisphosphate on glucokinase.

    c. Fructose-6-phosphate on pyruvate kinase.

    d. 2,3-bis phosphoglycerate on glycerate kinase.

    e. Fructose-2,6-bisphosphate on pyruvate kinase.

8. Characteristic products of glucose metabolism include:

    a. CO_2 and H_2O by erythrocytes.

    b. O_2 dependent production of lactate by muscle.

    c. Conversion to lactate by brain.

    d. Aerobic production of lactate by liver.

9. During glycolytic reactions leading to formation of acetyl-CoA, the C-2 and C-5 of glucose are:

    a. Phosphorylated and oxidized to the level of a carboxylic acid.

    b. Phosphorylated and oxidized to CO_2

    c. Phosphorylated, oxidized to the level of a carboxylic acid and covalently linked to a sulfur atom.

    d. Phosphorylated and oxidized to CO_2 which is linked to a sulfur atom.

10. A young child was found to exhibit extremely rapid onset of muscle fatigue, without the appearance of cramps, during periods of high muscle activity. Tissue biopsy revealed that resting levels of NAD/NADH and ATP/ADP were normal, as was the developmental and histological appearance of the tissue. During high-level treadmill testing it was found that serum lactate was unusually low and pyruvate was higher than expected. Continued enzymic analysis of muscle tissue revealed a genetic deficiency in this individual, which was:

    a. Inability to mobilize glycogen and provide glucose for anaerobic glycolysis.

    b. Deficient FPK-1 activity leading to diminished ATP production during exercise.

    c. Deficient pyruvate carboxylase activity leading to accumulation of pyruvate with consequent feedback inhibition on glycolysis during periods of high anaerobic glycolysis.

    d. Deficient lactate dehydrogenase activity leading to inability to regenerate NAD^+ during anaerobic muscle metabolism.

    e. Deficient cAMP dependent protein kinase leading to inability to phosphorylate and down-regulate pyruvate kinase, with attendant production of excess pyruvate.

DIRECTIONS (items 11–16): Match the following lettered responses with the questions below. Each response may be used more than once or not at all.

    a. pyruvate dehydrogenase

    b. PEPCK

    c. pyruvate carboxylase

    d. 1

    e. 2

    f. 3

    g. ATP

    h. acetyl-CoA

    i. CO_2

    j. pyruvate

    k. Citrate

11. The number of high energy phosphates required to convert pyruvate to PEP.

12. A product of the PEPCK reaction.

13. The enzyme involved in converting pyruvate to oxaloacetate.

14. An enzyme localized only in the mitochondrion.

15. The allosteric effector most important in determining the metabolic fate of pyruvate.

16. The number of oxaloacetate molecules required to form a glucose molecule via gluconeo-genesis.

Answers

1. d	**7.** a	**13.** c
2. d	**8.** b	**14.** c
3. a	**9.** c	**15.** h
4. e	**10.** d	**16.** e
5. e	**11.** e	
6. d	**12.** i	

Glycogen Metabolism and Regulation   15

Objectives

- To review the role of glycogen in the energy economy of the cell emphasizing tissue related regulation of catabolic and anabolic pathways.
- To define the form(s) and content of readily mobilizable glucose in the body and its site(s) of storage.
- To identify body tissues that utilize glucose as a principal source of energy.
- To identify the anomeric reducing carbon, the α (1–4) glycosidic linkages and α (1–6) glycosidic linkages of glycogen molecules.
- To identify the role of glycogenin in glycogen metabolism.
- To identify the substrates and products of the glycogenolytic enzyme known as phosphorylase.
- To identify the substrates and products of phosphorylase and the type of linkages broken by the enzyme.
- To identify the enzyme responsible for hydrolyzing α (1–6) bonds of glycogen molecules and the basis for classifying it as a bifuntional enzyme.
- To compare the chemical composition of the monosaccharides produced by phosphorylase and debrancher.
- To compare the enzyme bound intermediates of phosphoglucomutase and phosphoglycerate mutase.
- To define the role of phosphoglucokinase in glycogen metabolism.
- To write the reaction catalyzed by uridylyl transferase showing the fate of all phosphate atoms.
- To identify the substrates and products of the glycogen synthase reaction.
- To identify the chemical nature of glycogenin and its role in glycogen metabolism.
- To discuss the process involved in forming glycogen from the amylose generated by glycogen synthase.
- To identify the energy cost of glycogen synthesis starting from either blood glucose, or from fructose-6-phosphate generated by gluconeogenesis.

Concepts

Previous chapters stressed the importance of glucose to the energy economy of the body, with some elements—such as red blood cells and the brain—almost exclusively dependent on glucose while others are more flexible in their source of energy. Stores of readily available glucose to supply these tissues are found principally in the liver, as glycogen. A second major source of stored glucose is the glycogen of muscle tissue. However, muscle glycogen is not generally available to other tissues, because muscle lacks the enzyme glucose-6-phosphatase. This chapter reviews the pathways and regulation of Stage 1 of carbohydrate metabolism, which is essentially the interconversion of glucose and glycogen.

Normally a sedentary, 70-kg human catabolizes about 160 g of glucose per day. Of this 75%, or 120 g, is consumed by the brain via aerobic pathways. Most of the balance is consumed—by erythrocytes, skeletal muscle, and heart muscle. The body uses glucose from a variety of sources: directly from the diet, from amino acids and lactate via gluconeogenesis, and, to a small extent, from the oxidation of odd-numbered fatty acids. The body's readily available stores of glucose from these sources can be found in two forms: soluble glucose, which exists in all body fluids and amounts to about 20 g; and polymeric glucose, which is stored as glycogen, amounting to about 190 g. Glycogen is considered the principal storage form of glucose and is found mainly in liver and muscle, with kidney and intestines adding minor storage sites. With up to 10% of its weight as glycogen, the liver has the highest specific content of any body tissue. Muscle has a much lower amount of glycogen per unit mass of tissue, but since the total mass of muscle is so much greater than that of liver, total glycogen stored in muscle is about twice that of liver.

When it is measured against daily catabolic requirements, the body's 210 g of available glucose clearly can supply the glucose needs for only a little more than one day. Because of this constraint, total body glucose may vary widely during the periods between meals. Stores of glycogen in the liver are considered the main buffer of blood glucose levels.

GLYCOGEN STRUCTURE

Figure 15–1 illustrates the branched polymeric structure characteristic of glycogen molecules. A key feature of this structure is that there is only one free anomeric or reducing glucose carbon (C-1). The magnified section of the figure illustrates that most of the linkages in the molecule are α (1–4) glycosidic bonds, while the branch points are always α (1–6) linkages. In vivo there are between 10–20 glucose residues between branch points, and it is likely that no reducing ends are available since the initial or reducing glucose of most glycogen molecules is attached to a serine hydroxyl group of a protein known as glycogenin.

Glycogenolysis is the catabolic pathway by which glycogen is degraded to glucose and the anabolic glycogen synthesis pathway is known as **glycogenesis**.

Glycogenolysis
A. Phosphorylase:

Glycogen (n residues)   Glucose 1-phosphate   Glycogen (n-1 residues)   **15.1**

Equation 15.1 illustrates the first reaction of glycogenolysis, which is catalyzed by the highly regulated enzyme known as **phosphorylase**. In its active form, phosphorylase (phosphorylase *a*) catalyzes the phosphorylytic cleavage of α (1–4) glycosydic bonds to produce glucose-1-P (G-1-P).

B.   Debranching Enzyme: Phosphorylase is incapable of phosphorylyzing glucose residues close to the α 1–6 glycosidic branches. Consequently, a second glycogen-degrading enzyme known as **de-**

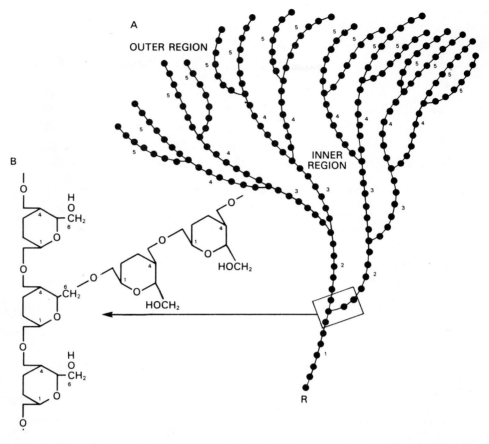

Figure 15–1. The glycogen molecule. **A:** General structure. **B:** Enlargement of structure at a branch point. The numbers in **A** refer to equivalent stages in the growth of the macromolecule. R, primary glucose residue, which is the only glucose residue in the structure shown to contain a free reducing group on C_1. The branching is more variable than shown, the ratio of $1 \rightarrow 4$ to $1 \rightarrow 6$ bonds being from 10 to 18. (Reproduced, with permission, from Murray, RK: *Harper's Biochemistry*, 23rd ed. Appleton & Lange, 1993.)

branching enzyme is involved in metabolizing glucose residues that are inaccessible to phosphorylase. Debranching enzyme is a bifunctional enzyme with two distinct activities. The first, illustrated in Figure 15–2 (top), is as an α [$1 \rightarrow 4$] $\rightarrow \alpha$ [$1 \rightarrow 4$] glucan transferase, in which the outer 3 glucosyl residues from a branch containing 4 glucosyls are transferred to a free C-4 of a second branch. This forms a longer branch, and leaves behind a single glucose residue at the 1–6 branch.

The second role of debrancher, that of a **glycosidase**, is illustrated in greater detail in Figure 15–2 (bottom). The glycosidase activity hydrolyzes the α (1–6) glycosidic linkage remaining at a branch point, releasing the monomer as free glucose rather than as G-1-P.

The result of the combined activity of phosphorylase and debranching enzyme is that most of the glycogen catabolized appears as glucose-1 phosphate, with a small proportion appearing as free glucose. In normal individuals about 1 out of every 12 glucose residues (or 8%) appears as free (unphosphorylated) glucose. Since glycogenolysis can yield some free glucose, it might appear that tissues, like muscle, could help buffer blood glucose. However, since the regulation of muscle glycogenolysis ensures that glycogen is only degraded when muscle energy supplies are needed, glucose in these tissues is almost all phosphorylated to G-6-P, by hexokinase.

C. Phosphoglucomutase: In tissues other than liver, kidney, and intestine, glucose derived from glycogenolysis is almost always destined for catabolism via glycolysis. Therefore most G-1-P is converted to G-6-P by phosphoglucomutase, as illustrated in Equation 15.2.

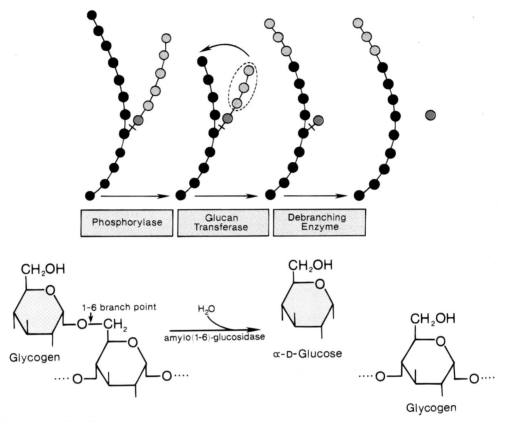

Figure 15–2. Top Glycogen phosphorylase removes glucose residues from branched glycogen until 4 glucose residues from an α-1,6-linkage of a branch point. Glucan transferase transfers 3 of these 4 residues to an adjacent branch forming an α1,4-linkage. **Bottom** The second activity of glucan transferase (amylo(1,6)-glucosidase removes the glucose residue from the α1,6-branch point. The result is the release of a free glucose residue.

15.2

The reaction proceeds readily in either the forward or reverse direction, depending on the needs of the cell. Glucose-1,6-bisphosphate is an intermediate in the reaction, and catalytic amounts of glucose-1,6-bisphosphate are required in order for it to proceed. The required catalytic quantities of glucose-1,6-bisphosphate are supplied by a specific kinase known as phosphoglucokinase, in which G-6-P is the substrate and the phosphate is derived from ATP.

In liver, kidney, and intestine, the fate of the products of glycogenolysis depend on the concentration of blood glucose and the energy needs of the cell. If blood glucose levels are low, glucose-6-phosphatase activity will direct virtually all glucose to the circulatory system. If tissue energy charge is also low, glucose from glycogenolysis will be distributed to the blood and to intracellular, energy yielding pathways in proportions that reflect metabolic needs.

Glycogen Storage Diseases

Since glycogen molecules can become enormously large, an inability to degrade glycogen can cause cells to become pathologically engorged; it can also lead to the functional loss of glycogen as a source of cell energy and as a blood glucose buffer. Although glycogen storage diseases are quite rare, their effects can be most dramatic. A number of these genetic diseases have been linked to mutations in enzymes associated with glycogen processing, as is summarized in Table 15–1.

The debilitating effect of many glycogen storage diseases depends on the severity of the mutation causing the deficiency. In addition, although the glycogen storage diseases are attributed to specific enzyme deficiencies, other events can cause the same characteristic symptoms. For example, Type I disease (von Gierke's disease) is attributed to lack of glucose-6-phosphatase. However, this enzyme is localized on the cisternal surface of the endoplasmic reticulum (ER); in order to gain access to the phosphatase, glucose-6-phosphate must pass through a specific translocase in the ER membrane. Mutation of either the phosphatase or the translocase makes transfer of liver glycogen to the blood a very limited process. Thus, mutation of either gene leads to symptoms associated with von Gierke's disease, which occurs at a rate of about 1 in 200,000 people.

Glycogenesis

A. Phosphoglucomutase: G-6-P, derived from the blood via the hexokinase reaction or from intracellular sources via gluconeogenesis, is considered to be the starting material for glycogenesis. Phosphoglucomutase, which we reviewed earlier, is the first enzyme on the pathway to glycogen, converting G-6-P to G-1-P.

B. Glucose-1-Phosphate: Uridylyltransferase:

Glucose 1-phosphate   UDP-glucose   **15.3**

Table 15–1. Categories of glycogen storage disease.

Type	Name of Disease	Deficiency	Effect of Deficiency
I	von Gierke's disease	Glucose-6-phosphatase activity	Liver, kidney, intestinal cells engorged with glycogen; exercise hypoglycemia
II	Pompe's disease	Lysosomal $1 \to 4$ or $1 \to 6$ glucosidase activity	Often fatal accumulation of lysosomal glycogen deposits
III	Forbe's/Cori's disease	Debranching enzyme	Highly branched glycogen with limited availability of glucose monomers
IV	Andersen's disease	Branching enzyme	Glycogen with few branches. Often fatal during childhood
V	McArdle's syndrome	Muscle phosphorylase	Limited ability to sustain muscle activity; excess muscle glycogen; limited blood lactate after exercise
VI	Her's disease	Liver phosphorylase	Tendency to hypoglycemia and excess accumulation of liver glycogen
VII	Tauri's disease	Phosphofructokinase of muscle and erythrocytes	Limited ability to sustain muscle activity; excess muscle glycogen; limited blood lactate after exercise
VIII		Phosphorylase kinase	Tendency to hypoglycemia and excess accumulation of liver glycogen

Glycogen synthesis is an anabolic process, requiring energy. In fact the next enzymatic step in the pathway provides the energy that is ultimately used to form the α (1 → 4) glycosidic bonds of glycogen. As illustrated in Equation 15.3, the enzyme **glucose-1-phosphate uridylyltransferase** catalyzes the addition of the uridylyl group (UMP) from UTP to the phosphate of glucose-1-phosphate, forming UDP-glucose and inorganic pyrophosphate. Notice that, although the reaction shown in Equation 15.3 is nearly isoenergetic, one of the products is pyrophosphate. The ubiquitously distributed pyrophosphatase activity of cells hydrolyzes the pyrophosphate with a standard free energy of about −7.3 kcal/mole; the net effect is to pull the coupled transferase reaction strongly in the forward direction. It is largely the latter energy that is used to drive the polymerization of glucose into glycogen.

C.   Glycogen Synthase: As shown in Figure 15–3, UDP-glucose is the substrate for the enzyme **glycogen synthase**, which catalyzes the polymerization of UDP-glucose into glycogen. The glycogen synthase reaction proceeds by the addition of an incoming glucose C-1 to an available C-4 of existing **"primer" glycogen**.

Primer glycogen is an absolute requirement for glycogen synthase activity, since the synthase will not form the disaccharide maltose from any combination of glucose and UDP-glucose.

Note that the glycogen synthase reaction is virtually isoenergetic, with no net loss of high-energy bonds in the reaction. As a consequence, the breakdown of glycogen by the phosphorylase reaction proceeds readily, with only inorganic phosphate being required as a phosphate donor. Glycogen synthase, like phosphorylase, is highly regulated by allosteric and phosphorylation mechanisms, but in a reciprocal way (as will be discussed below).

D.   Glycogenin: Until recently, the source of the first glycogen molecule that might act as a primer in glycogen synthesis was unknown. Recently it has been discovered that a protein known as **glycogenin** is located at the core of glycogen molecules. Glycogenin has the unusual property of catalyzing its own glycosylation, attaching C-1 of a UDP-glucose to a tyrosine reside on the enzyme. The attached glucose is believed to serve as the primer required by glycogen synthase.

E.   Branching Enzyme: Glycogen synthase is incapable of forming the 1 → 6 branches characteristic of glycogen; consequently a branching enzyme is involved in forming the tree-like structure of mature glycogen. Branching enzyme, or α (1 → 4) → α (1 → 6) glucosyl transferase, has the function of cleaving an α (1 → 4) linkage of a 7 glycosyl oligosaccharide from a growing glycogen branch and transferring it to C-6 of a glucose at least 4 glucosyl residues from the end branch (Figure 15–4).

With G-6-P as a starting point for glycogen synthesis, the energy balance sheet in Table 15–2 shows that the cost of 1 glucose addition to preexisting glycogen is 1 phosphoanhydride bond.

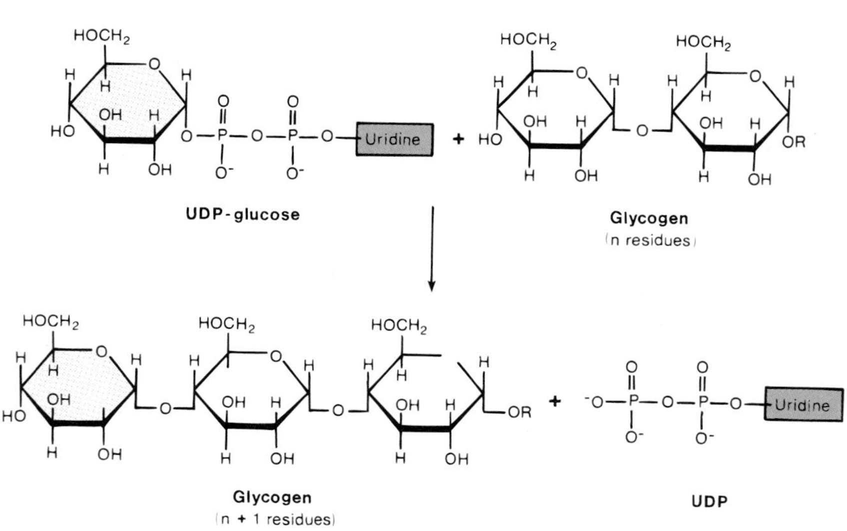

Figure 15–3. Glycogen synthase catalyzes the addition of a glucose residue from UDP-glucose to preformed "primer" glycogen. The products of the reaction are glycogen extended by 1 glucose residue and UDP.

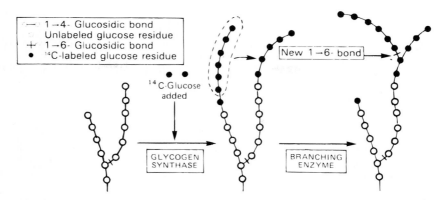

Figure 15–4. The biosynthesis of glycogen. The mechanism of branching as revealed by adding ^{14}C-labeled glucose to the diet in the living animal and examing the liver glycogen at further intervals.

REGULATION OF GLYCOGEN METABOLISM

A convenient way to review the regulation of glycogen metabolism is to consider changes in its metabolism associated with the well-known "fight or flight" response. In this reaction, glycogenolysis increases dramatically throughout the body to produce high levels of intracellular glucose and to maintain high blood glucose levels by mobilizing liver glycogen. Simultaneously, glycogenesis is severely down-regulated.

The trigger of the "fight or flight" response is the appearance of massive quantities of the amino acid—derived hormone **epinephrine** in the circulatory system. The hormone binds to cell surface receptors, which induce the synthesis of the second messenger, cAMP (Figure 15–5) which in turn initiates a cascade of events that culminates in massive glycogenolysis, with a virtually complete shutdown of glycogen synthesis. The following section traces the events that lead from the formation of intracellular cAMP to the stimulation of phosphorylase activity and the down-regulation of glycogen synthase.

REGULATION OF PHOSPHORYLASE ACTIVITY

The up-regulating effects of cAMP on glycogenolysis and down-regulation of glycogenesis are directed at the activation of phosphorylase and the inactivation of glycogen synthase, respectively. However, there are additional steps in the hormone-activated cascade that intervene before these end points are attained.

Active PKA is responsible for phosphorylating and stimulating a second kinase known as **phosphorylase kinase**. The active **"a"** form of phosphorylase kinase is produced by phosphorylation of the less active **"b"** form. The **a** form is responsible for phosphorylating and up-regulating glycogen phosphorylase, see Figure 15–6.

Table 15–2. Reactions of glycogen synthesis and the number of high-energy phosphates expended.

Number	Reactions of Glycogen Synthesis	High-Energy Phosphates Expended
1.	glucose $-6-P$ $\xrightarrow{\text{phosphoglucomutase}}$ glucose $-1-P$	0
2.	glucose $-1-P$ + UTP $\xrightarrow[\text{transferase}]{\text{uridylyl}}$ UDP $-$ glucose + PPi	0
3.	PPi(pyrophosphate) + H_2O $\xrightarrow{\text{pyrophosphatase}}$ 2Pi	1
4.	UDP $-$ glucose + glycogen$_{(n)}$ $\xrightarrow[\text{synthase}]{\text{glycogen}}$ UDP + glycogen$_{(n+1)}$	0
Sum	glucose $-6-P$ + UTP + glycogen$_n$ $\rightarrow$ glycogen$_{n+1}$ + 2Pi + UDP	1

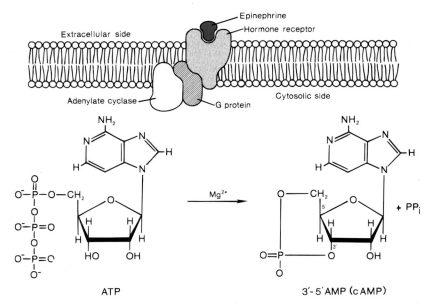

Figure 15–5. Epinephrine binds to plasma membrane receptors, inducing the formation of complexes containing receptor, trimeric G-protein, and adenylate cyclase. The formation of these complexes activates adenylate cyclase, which converts ATP to 3′,5′-cyclic-adenosine monophosphate (cAMP) and pyrophosphate. The formation of cAMP represents a minimal expediter of energy, since the cAMP signal is greatly amplified by a cascade process that stimulates phosphorylase, inhibits glycogen synthase, and culminates in million-fold increases in the rate of cellular ATP production, by up-regulating virtually all regulated enzymes involved in energy (ATP) production.

As with phosphorylase kinase, the mechanism of activation is conversion of an inactive (or less active) **b** form of the enzyme to a phosphorylated, active **a** form. The reversible nature of such an activation event is diagrammed in Figure 15–6, which shows that the end result of cAMP appearance is the covalent activation of phosphorylase **b** to phosphorylase **a**. All the events initiated by cAMP formation are reversed when adenylate cyclase activity ceases and existing cAMP is hydrolyzed to AMP by the ubiquitously distributed enzyme **phosphodiesterase**. Subsequently, phosphoprotein levels fall by the action of specific phosphoprotein phosphatases, which dephosphorylate the **a** forms and regenerate less active **b** forms.

Maximum conversion of phosphorylase kinase **b** and phosphorylase **b** to **a**-forms is augmented by the PKA-mediated inhibition of protein phosphatase activity. PKA phosphorylates a protein known as inhibitor-1, which enables inhibitor-1-P to bind and inactivate the phosphoprotein phosphatase responsible for dephosphorylating and inactivating phosphorylase and phosphorylase kinase. Notice, in Figure 15–6, that the concentration of inhibitor-1-P is itself regulated by phosphoprotein phosphatase. It is a curious turn of events that allows inhibitor-1-P to inhibit the activity of phosphoprotein phosphatase on other proteins, but not on itself.

While the covalent mechanisms described above are dependent on the hormone activation of PKA, there are many other physiological conditions in which glucose stores (glycogen) need to be mobilized or conserved. For example, the liver must mobilize glycogen during prolonged periods between meals, and muscle needs to mobilize glycogen during prolonged periods of moderate exercise. In these circumstances PKA generally remains inactive, but the up-regulation of phosphorylase activity is required to produce glucose for increased cell energy needs. An overlay of allosteric regulation on the enzymes of glycogen metabolism, by the effectors glucose, calcium, AMP and ATP, provide these regulatory effects. When cell energy charge is lowered, as in the moderately exercised muscle, AMP levels rise and AMP acts as a potent positive effector directly on phosphorylase **b**. However, it is important to recognize that the conversion of phosphorylase **b** to phosphorylase **a**, by phosphorylase kinase, maximally stimulates phosphorylase; therefore, AMP has no comparable allosteric effect on phosphorylase **a**.

Increased intracellular calcium levels (Ca^{+2}) also allosterically up-regulate glycogenolysis. Calcium activation takes place at the level of phosphorylase kinase (Figure 15–6) and is made possible by the

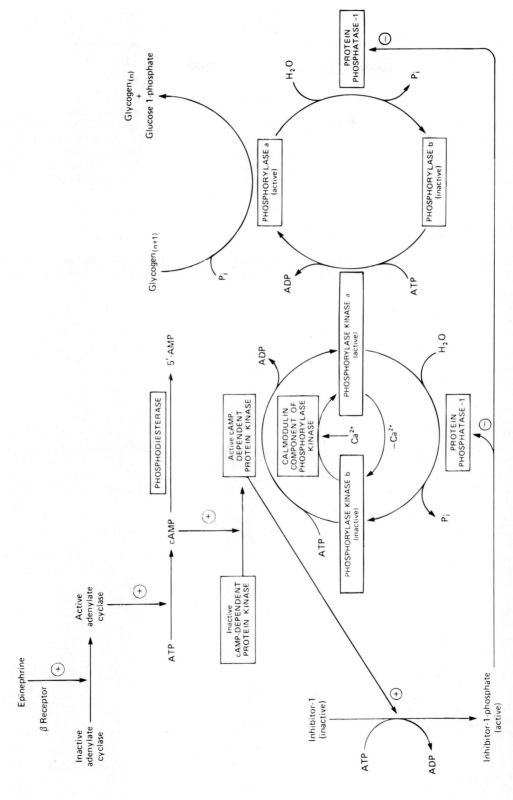

Figure 15–6. Control of phosphorylase in muscle. The sequence of reactions arranged as a cascade allows amplification of the hormonal signal at each step. n = number of glucose residues.

fact that one of the subunits of phosphorylase kinase is the Ca^{+2} binding protein **calmodulin**. Calcium activation of glycogenolysis is particularly important in muscle tissue, where Ca^{+2} discharged from the cisternae of the endoplasmic reticulum binds phosphorylase kinase **b**, raising its activity to about the same level as that of phosphorylase kinase **a**. Calmodulin is a regulatory component of many different enzymes, thus providing a coordinated response of many enzymes to fluxes in intracellular Ca^{+2} levels.

The allosteric down-regulation of **a**-forms of phosphorylase and phosphorylase kinase is of equal physiologic importance. If PKA is hormonally activated in cells with a high energy charge, containing high levels of glucose and G-6-P, there is little need for enhanced glycogenolysis. An example of this state is the physiologic condition encountered in medical students prior to an NBME exam. Under such circumstances the epinephrine stimulation of PKA activity is expected, but enhanced glycogenolysis would be wasteful; therefore, glycogenolysis is not observed to be significantly stimulated. The basis for this regulation is that glucose, or G-6-P, and ATP are potent negative allosteric effectors on phosphorylase **a**.

REGULATION OF GLYCOGEN SYNTHASE (GS)

The mechanisms which regulate the activity of GS are illustrated in Figure 15–7. The key features of the process are: First, the phosphorylation of GS occurs primarily as a result of PKA-mediated acti-

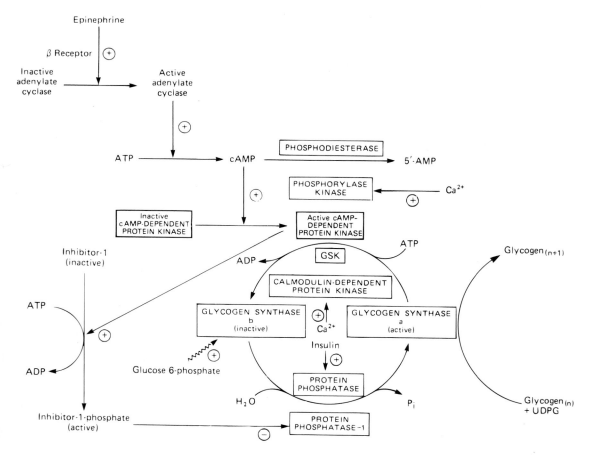

Figure 15–7. Control of glycogen synthase in muscle (n = number of glucose residues). The sequence of reactions arranged in a cascade causes amplification at each step, allowing only nanomol quantites of hormone to cause major changes in glycogen concentration. GSK, glycogen synthase kinase -3, -4, and -5; wavy arrow, allosteric activation. (Reproduced, with permission, from Murray, RK: *Harper's Biochemistry*, 23rd ed. Appleton & Lange, 1993.)

vation of **glycogen synthase kinase** (GSK). Second, the phosphorylation of GS reduces its enzymatic activity; thus, phosphorylated GS is the **b** form. Third, PKA phosphorylation of inhibitor-1 leads to inhibition of the phosphoprotein phosphatase responsible for dephosphorylation of GS **b**, thus maintaining GS in the inactive **b** form. Fourth, allosteric effectors that regulate phosphorylase seem to have a coordinate effect on intracellular GS. Finally, it is important to note that PKA itself can phosphorylate GS and that additional cAMP-independent GS kinases are known, such as **glycogen synthase kinase-3** (GSK-3). GSK-3 has a calmodulin subunit and responds to elevated intracellular Ca^{+2} by phosphorylating a multiplicity of sites on GS leading to its inactivation.

It is important to recognize that all the processes involved in regulating glycogen metabolism have exactly the opposite effect in glycogenolysis and glycogenesis (see Figure 15–8). Thus, glycogen metabolism is tightly controlled in a coordinated way, with the result that events which increase glycogenolysis cause decreased glycogenesis, and vice versa. The comparative regulation of GS and phosphorylase are summarized in Table 15–3.

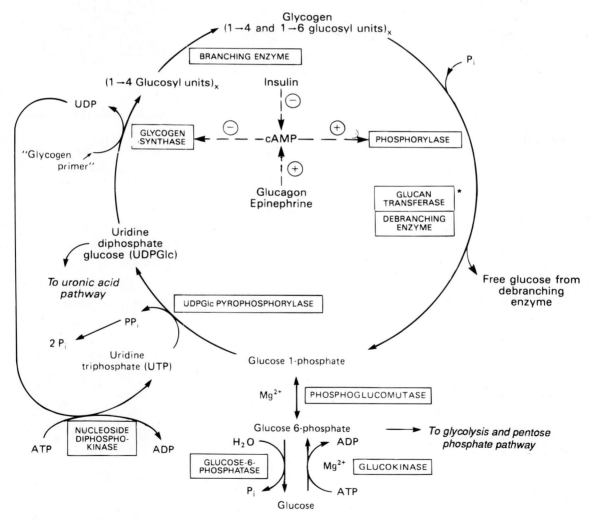

Figure 15–8. Pathway of glycogenesis and of glycogenolysis in the liver. Two high-energy phosphates are used in the incorporation of 1 mol of glucose into glycogen. ⊕, Stimulation; ⊖, inhibition. Insulin decreases the level of cAMP only after it has been raised by glucagon or epinephrine: ie, it antagonizes their action. Glucagon is active in heart muscle but not skeletal muscle. *Glucan transferase and debranching enzyme appear to be 2 separate activities of the same enzyme. (Reproduced, with permission, from Murray, RK: *Harper's Biochemistry*, 23rd ed. Appleton & Lange, 1993.)

Table 15–3. The influence of effectors on Phosphorylase (Pase) and Glycogen Synthase (GS). Vertical arrows indicate increases (up arrow) or decreases (down arrows) in concentrations, energy charge (E.C.) or enzymatic activity. GS = glycogen synthase, Pase = phosphorylase.

Effectors	Effect on Phosphorylase (Pase)		Effect on Glycogen Synthase (GS)	
	Change	Basis of change	Change	Basis of change
⇑cAMP	⇑	Pase **b** ⇒ Pase **a** via ⇑ PKA and ⇑ Pase Kinase	⇓	GS **a** ⇒ GS **b** via ⇑ PKA
⇑cAMP	⇑	Inhibitor-1 ⇒ Inhibitor-1-P causing loss of protein phosphatase, maintaining high Pase **a** levels	⇓	Inhibitor-1 ⇒ Inhibitor-1-P causing loss of protein phosphatase, maintaining high GS **b** levels
⇑ Ca^{+2}	⇑	Binds calmodulin subunit of Pase Kinase with subsequent conversion of Pase **b** ⇒ Pase **a**	⇓	Activates GS-kinase-3 by binding calmodulin subunit, resulting in GS **a** ⇒ GS **b**
⇑E.C.	⇓	ATP is a negative effector on Pase **a**		
⇓E.C.	⇑	AMP is a positive effector on Pase **b**		
⇑ glycogen			⇓	Proportion of GS in **a** form is inversely proportional to cell glycogen content
⇑ G-6-P		⇓ G-6-P is a negative effector on Pase **a**	⇑	G-6-P is a positive effector on GS **b**

Questions

DIRECTIONS (items 1–15): Each numbered item or incomplete statement is followed by answers or by completions of the statement. Select the ONE lettered answer or completion that is BEST in each case.

1. Glycogenolysis can best be characterized as:
   a. The direct production from glycogen of G-1-P.
   b. The direct production from glycogen of G-1-P and unphosphorylated glucose.
   c. The direct production from glycogen of G-1-P and unphosphorylated glucose via a phosphorolytic reaction and hydrolytic reaction, respectively.
   d. The direct production from glycogen of G-1-P unphosphorylated glucose and G-6-P.
   e. The direct production from glycogen of G-1-P and unphosphorylated glucose via hydrolytic reactions.

2. The conversion of blood glucose to glycogen involves:
   a. The net consumption of 1 high-energy phosphate bond.
   b. The net consumption of 2 high-energy phosphate bonds.
   c. The net consumption of 3 high-energy phosphate bonds.
   d. The net consumption of 4 high-energy phosphate bonds.
   e. The net consumption of 5 high-energy phosphate bonds.

3. Increased blood glucagon causes increased blood glucose because:
   a. Liver phosphorylase **a** is formed.
   b. Liver phosphorylase **a** is formed and PFK 1 is down-regulated.
   c. Kidney phosphorylase **a** is formed, and PFK-1 is down-regulated because PFK-2 phosphatase activity is up regulated.
   d. Liver phosphorylase **a** is formed; PFK-1 is down-regulated because PFK-2 phosphatase activity is up-regulated, and glycogen synthase is converted to the **b** form.
   e. Kidney phosphorylase **a** is generated and PFK 1 is down-regulated.

4. The physiological effect of PKA is best described as:
   a. The activation of protein phosphatase inhibitor 1.

    b. The activation of protein phosphatase inhibitor 1 and increased glycogenolysis.

    c. The activation of protein phosphatase inhibitor 1, increased glycogenolysis, and increased liver glycolysis.

    d. Increased glycogenolysis, increased glycolysis, and increased intracellular free glucose.

5. All of the following statements concerning the glucagon receptor are true EXCEPT:
   a. It is found on liver cell membranes.
   b. Involves a G protein in a cAMP synthetic process.
   c. It inactivates the cAMP-dependent phosphorylase system.
   d. It has an extra cellular glucagon-binding domain.
   e. Its activation simulates cAMP production via adenylate cyclase.

6. Which enzyme is effected first by epinephrine binding to epinephrine receptor in a liver cell?
   a. Phosphorylase **b.**
   b. Phosphorylase **a.**
   c. Phosphorylase kinase.
   d. Adenylate cyclase.

7. All of the following statements concerning glycogen synthesis are correct EXCEPT:
   a. It requires UDP-glucose.
   b. Glycogen synthase transfers glucosyl residues from nucleotide sugars to preformed glycogen.
   c. It is stimulated by cAMP.
   d. Glucosyl residues are added to the nonreducing terminal ends of glucose polymers.
   e. A polymeric primer is required for glycogen synthase activity.

8. Under conditions of low blood glucose:
   a. Muscle glucose-6-phosphatase converts G-6-P to glucose to supply the brain with glucose.
   b. Glucokinase becomes the main intracellular "glucose trapping" enzyme.
   c. Liver glycogen can be converted to brain glucose by a pathway involving glucose-6-phosphatase.
   d. Lactate stores are converted to glucose by muscle, to supply brain and other glucose-dependent tissue.

9. Which of the following is NOT characteristic of the enzyme phosphorylase?
   a. Its mechanism of action involves retention of the stereochemical configuration of glucose used in the synthesis of glycogen.
   b. It catalyzes a hydrolytic reaction.
   c. It involves the formation of G-1-P as product.
   d. Its allosteric activation involves AMP.
   e. Its activation is linked to adenylate cyclase activity.

10. Which of the following proteins is *not* a substrate for cAMP dependent protein kinase (PKA)?
    a. Phosphorylase kinase.
    b. Glycogen synthase **a.**
    c. Phosphatase inhibitor-1.
    d. Phosphorylase **b.**
    e. Phosphofructokinase-2.

11. Covalent regulation of enzymes by a phosphorylation mechanism:
    a. Always results in increased activity of the phosphorylated enzyme.
    b. Always results in decreased activity of the phosphorylated enzyme.
    c. Is an irreversible process.
    d. Involves formation of a high-energy bond between phosphate and enzyme.
    e. Involves a phosphoprotein phosphatase.

12. A child referred to pediatric clinic is unable to sustain strenuous activities for a prolonged time. Serum assays performed before, during, and after periods of vigorous exercise reveal that resting glucose is normal and that lactate levels increase with exercise, but only 50% of the maximally expected lactate values are observed. Needle biopsy of muscle and liver indicates no gross pathol-

ogy when examined by light microscopy or electron microscopy. There is no evidence of hemolytic anemia. Tissues stained for glycogen suggested a moderate elevation of glycogen levels in muscle but not in liver. The findings are consistent with all of the following EXCEPT:

a. Type I glycogen storage disease caused by defective glucose-6-phosphatase.
b. Type II glycogen storage disease caused by defective lysosome activity.
c. Type III glycogen storage disease caused by defective debranching enzyme.
d. Type VI glycogen storage disease caused by defective liver phosphorylase.

Answers

1. c	**5.** c	**9.** b
2. b	**6.** d	**10.** d
3. d	**7.** c	**11.** e
4. b	**8.** c	**12.** b

16

Pentose Phosphate Pathway, Galactose and Fructose Metabolism

Objectives

- To review the role of glucose in producing NADPH for reductive biosynthesis and ribose for synthesis of nucleic acids.
- To identify the physiological role of the pentose phosphate pathway.
- To identify the main products of the two stages of the pentose phosphate pathway.
- To identify conditions under which the second stage of the pentose phosphate pathway is particularly active.
- To review the entry of fructose and galactose into glycolysis.
- To identify the products of fructokinase and hexokinase action when fructose is the only available hexose.
- To identify the cause of fructose intolerance.
- To identify the sugar nucleotide involved in galactose metabolism.
- To identify the role of aldose reductase in causing cataracts in people with hypergalactosemia.
- To identify the enzymes that are usually defective in individuals afflicted with hypergalactosemia.

Concepts

Previous chapters stressed the importance of glucose to the energy economy of the cell with its role in ATP production. The current chapter focuses on the importance of glucose in two additional roles: in the production of another high-energy compound, NADPH, which is employed in reductive biosynthetic reactions; and as a carbon donor in the production of 5-carbon sugars that are of fundamental importance in nucleotide metabolism. The metabolism of the next most important hexoses, fructose and galactose, will also be reviewed.

The Pentose Phosphate Pathway

The pentose phosphate, or hexose monophosphate, pathway has three important physiological roles:

1. It is responsible for oxidation of glucose with production of the ribose used in all nucleotide biosynthesis.
2. It is the primary process for generation of the NADPH required for reductive biosynthesis.
3. It interconverts 3-, 4-, 5-, 6-, and 7-carbon sugars as required for other metabolic process or for the metabolism of excess ribose-5-P via glycolysis.

The pentose phosphate pathway can be characterized as having two prominent phases. Initially there is an oxidative phase, in which glucose is converted to ribulose-5-phosphate plus NADPH. Afterward comes a nonoxidative phase, in which 3 equivalents of ribulose-5-phosphate are converted into 1 equivalent of glyceraldehyde-3-phosphate and 2 equivalents of fructose-6-phosphate.

The reactions of the oxidative phase of the pentose phosphate pathway require 3 enzymes, 2 of which are $NADP^+$-specific dehydrogenases (Figure 16–1 and Equation 16.1). The first enzyme in the pathway, **glucose-6-phosphate dehydrogenase**, catalyzes the rate-limiting reaction of the pathway. The flow of carbon atoms through this enzyme is generally determined by the availability of $NADP^+$.

$$\text{Glucose-6-P} + 2NADP^+ + H_2O \rightarrow \text{Ribulose-5-P} + 2NADPH + 2H^+ + CO_2 \qquad 16.1$$

When ribose is required to supply other pathways, the process described in Equation 16.1 dominates the pentose phosphate pathway, with the ribulose-5-P being converted to ribose-5-P. However, it is often the case—for example in tissues in which large quantities of acetyl-CoA are converted to fatty acids—that the main product required from the pentose phosphate pathway is NADPH for reductive biosynthesis. The latter requires that excess ribose-5-phosphate be prevented from accumulating, since that would cause product inhibition and limit the production of NADPH. The second phase of the pathway consumes 3 equivalents of ribulose-5-P, producing 2 equivalents of fructose-6-P and 1 equivalent of glyceraldehyde-3-P. All of the latter are further metabolized via glycolysis, thus preventing the accumulation of ribulose-5-P. The reactions of the second phase of the pathway are outlined in Figure 16–1.

Under conditions of high cell energy charge, when vigorous NADPH-dependent biosynthesis is proceeding, glucose is used to generate large quantities of NADPH with minimal glucose waste. For example, 6 glucose molecules that are passed through the pentose phosphate pathway give rise to the stoichiometry shown in Equation 16.2.

$$6\ \text{Glucose-6-P} + 12NADP^+ + 6H_2O \rightarrow 6CO_2 + 12NADPH + 4\ \text{Fructose-6-P}$$
$$+\ 2\ \text{Glyceraldehyde-3-P} + 12H^+ \qquad 16.2$$

When the energy charge is high, the glyceraldehyde-3-P and the fructose-6-P can be converted to a total of 5 glucose-6-P equivalents via gluconeogenesis. The result is that large quantities of NADPH are produced and glucose is conserved by reconverting products of the pentose phosphate pathway back to glucose.

Metabolism of Fructose and Galactose

Fructose and galactose are two important sugars in human nutrition. The disaccharide, sucrose, is comprised of 1 molecule each of glucose and fructose. Fructose, abundant in fruits and often used as a sweetener in place of sucrose, is metabolized mainly by the fructose-1-phosphate pathway illustrated in Figure 16–2. However, fructose can also be phosphorylated by hexokinase with the production of fructose-6-P (Figure 16–2) but the affinity of hexokinase for fructose is about 20 times lower than it is for glucose, and relatively little fructose is metabolized by this path.

A. Fructose Intolerance: Studies on humans have established that many people with dietary fructose intolerance lack the liver enzyme known as **fructose-1-phosphate aldolase (also aldolase B)**, which is required for the breakdown of fructose-1-P (see Figure 16–2). In afflicted individuals, a diet high in fructose leads to the accumulation of fructose-1-P and a severe loss of free inorganic phosphate (which becomes sequestered as the hexose phosphate). The phosphate deficiency, in turn, leads to an inability to carry out liver oxidative phosphorylation and ultimately to the death of hepatocytes due to low cell-energy charge.

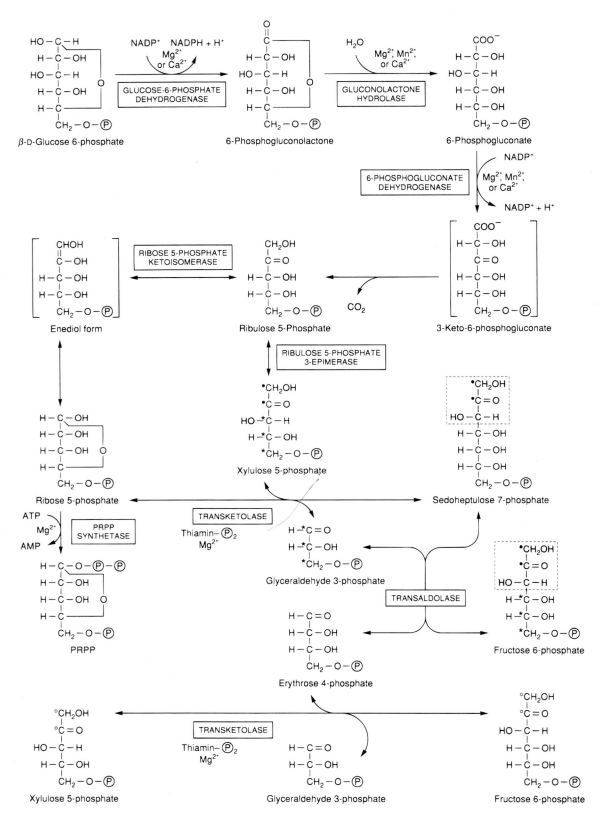

Figure 16–1. The pentose phsophate pathway. $\text{P} - \text{Po}_3^{2-}$, PRPP, 5-phosphoribosyl-1-pyrophosphate. (Reproduced, with permission, from Murray, RK: *Harper's Biochemistry*, 23rd ed. Appleton & Lange, 1993.)

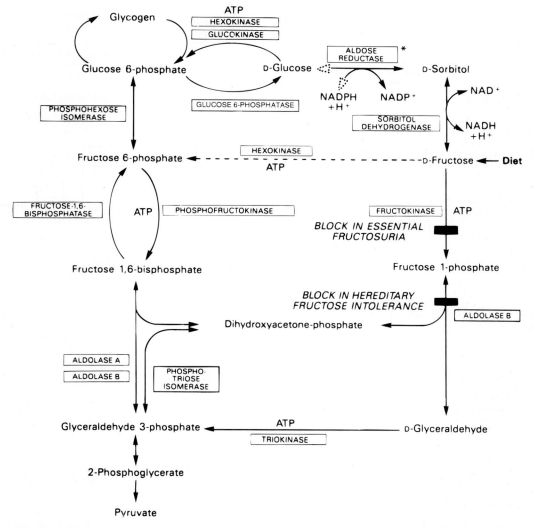

Figure 16–2. Metabolism of fructose. Aldolase A is found in all tissues except the liver, where only aldolase B is present. (*not found in liver). (Reproduced, with permission, from Murray, RK: *Harper's Biochemistry*, 23rd ed. Appleton & Lange, 1993.)

B. Galactose Metabolism and Hypergalactosemia: Lactose, the disaccharide of milk, is made of glucose plus galactose and is the main source of dietary galactose. The intestinal conversion of lactose to its monomers—glucose and galactose—takes place via the mucosal enzyme **lactase**, which often disappears from the intestine after the age of 5 years. Individuals lacking lactase are said to be **lactose-intolerant**; they experience gastric disturbances when consuming a diet rich in lactose. Lactase deficiency eliminates the main source of dietary galactose, but galactose synthesis via the UDP-hexose epimerase (see Figure 16–3A) provides adequate galactose for metabolic purposes.

The catabolism of galactose in normal tissue involves phosphorylation by galactose kinase (Figure 16.3), followed by UMP transfer from UDP-glucose to galactose-1-P, forming UDP-galactose. The UDP-galactose is epimerized by a UDP-hexose epimerase to regenerate UDP-glucose. During lactation, UDP-galactose is used to form mammary tissue lactose as illustrated in Figure 16–3B.

The enzyme lactose synthase is a heterodimer, with 1 polypeptide being known as protein A or galactosyltransferase and the other as protein B or α-lactalbumin, the major milk protein.

The inability to metabolize galactose by normal means is usually due to defective galactose kinase or uridylyl transferase. Individuals who lack the function of one of these enzymes exhibit hypergalac-

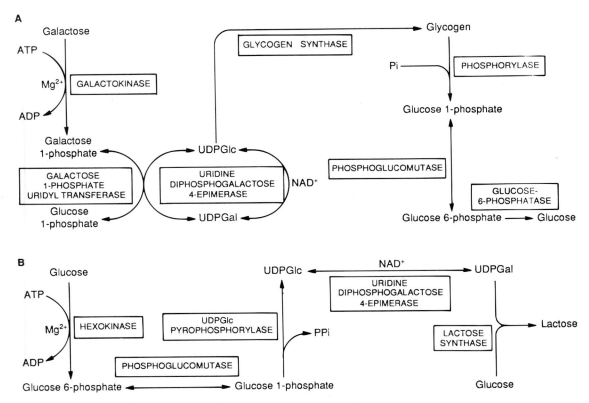

Figure 16–3. Pathway of conversion of **A:** Galactose to glucose in the liver. **B:** Glucose to lactose in the lactating mamary gland. (Reproduced, with permission, from Murray, RK: *Harper's Biochemistry*, 23rd ed. Appleton & Lange, 1993.)

tosemia. Absence of the enzyme galactose-1-P uridylyl transferase is the more severe of the two malfunctions. When affected newborns continue to consume milk, they develop galactosemia, urinary galacitol, and ultimately neurological deficiencies. In addition they experience frequent vomiting, diarrhea, jaundice and hepatomegaly. Unless it is controlled by exclusion of galactose from the diet, the disease can go on to produce blindness and fatal liver damage. Even on a galactose-restricted diet, transferase-deficient individuals exhibit urinary galacitol excretion and persistently elevated erythrocyte galactose-1-P levels. Blindness is due to the conversion of circulating galactose to the sugar alcohol galacitol, by an NADPH-dependent **galactose reductase** that is present in neural tissue and in the lens of the eye. At normal circulating levels of galactose this enzyme activity causes no pathological effects. However, a high concentration of galacitol in the lens causes osmotic swelling, with the resultant formation of cataracts and other symptoms.

Questions

DIRECTIONS (item 1–12): Each numbered item or incomplete statement is followed by answers or by completions of the statement. Select the ONE lettered answer or completion that is BEST in each case.

1. The transketolase enzyme requires the following for maximal activity:
    a. TPP.
    b. Biotin.
    c. CoA.

    d. Dihydroxyacetone phosphate.
    e. Schiff base formation.

2. The rate-limiting step in the formation of ribose-5-phosphate occurs at:
    a. The transketolase catalyzed reaction.
    b. The glucose-6-phosphate dehydrogenase catalyzed reaction.
    c. The transaldolase catalyzed reaction.
    d. The 6-phosphogluconate dehydrogenase catalyzed reaction.
    e. The ribulose-5-phosphate catalyzed reaction.

3. Enzymes of the pentose phosphate pathway that are directly responsible for interconversions of 3-, 4-, 5-, 6-, or 7-carbon sugars include:
    a. Isomerase and transketolase.
    b. Epimerase and transaldolase.
    c. Glucose-6-phosphate dehydrogenase.
    d. Transketolase and transaldolase.
    e. Isomerase and epimerase.

4. Principle roles of the pentose phosphate pathway include:
    a. The production of high-energy phosphate energy when the energy charge is low.
    b. The production of NADPH for use in reductive biosynthetic reactions.
    c. The production of NADH for use in making ATP when the energy charge is low.
    d. The production of ribose phosphate for use in nucleotide synthesis.
    e. Two of the above are correct.

5. Enzymatic reactions involved in the metabolism of excess sucrose include:
    a. Hexokinase, with the production of fructose-6-phosphate.
    b. Hexokinase, with the production of glucose-6-phosphate.
    c. Fructokinase, with the production of fructose-1-phosphate.
    d. All of the above are correct.

Questions 6 and 7 below are based on the following case history. A child born and raised in Chicago planned to spend the summer on a relative's fruit farm and to help with the harvest. The summer passed uneventfully, but several days after the harvest began the child became jaundiced and very sick. Upon admission to the hospital the following clinical findings were made: In addition to the expected hyperbilirubinemia the patient was hypoglycemic, had a markedly elevated rise in blood fructose concentration, and was hyperlacticacidemic. Further history-taking revealed that during the harvest it was customary for the family to indulge in fruit-filled meals and to snack freely on fruit while carrying out the harvest. The following conclusions were reached:

6. The elevated blood fructose was due to:
    a. Defective liver hexokinase.
    b. Defective liver fructokinase.
    c. An allergic reaction to constituents in the fruit diet.
    d. Defective liver fructose-1,6-bis aldolase.
    e. Defective liver fructose-1-phosphate aldolase.

7. The hyperlacticacidemia was attributed to:
    a. An inability to carry out liver oxidative phosphorylation and gluconeogenesis, secondary to a severe deficiency of available inorganic phosphate in hepatocytes.
    b. Excess production of pyruvate from dietary fructose sources.
    c. Inability to metabolize pyruvate because of a dietary vitamin insufficiency leading to inhibition of the pyruvate dehydrogenase complex.
    d. An allergic reaction to a constituent in the fruit diet.
    e. A diet-induced insufficiency of niacin, leading to low cell levels of available NAD/NADH.

DIRECTIONS (items 8–12): Match the phrases with the correct lettered responses from the list below. Each lettered response may be used more than once or not at all.

    a. lactose synthase
    b. Galactosyl transferase
    c. α-lactalbumin
    d. Lactase
    e. Lactose
    f. Galactokinase
    g. Galactose-1-phosphate uridylyl transferase
    h. Galactose-1-phosphate
    i. UDP-hexose epimerase
    j. Serum albumin
    k. Galacitol
    l. An aldose reductase

8. Missing or defective in lactose intolerance.

9. Involved in producing needed galactose in cases of diet restricted from containing galactose or lactose.

10. Defective or missing in the most severe form of galactosemia.

11. The major milk protein in human milk.

12. Responsible for galacitol formation.

Answers

1. a	**5.** d	**9.** i
2. b	**6.** e	**10.** g
3. d	**7.** a	**11.** c
4. e	**8.** d	**12.** l

17

The Pyruvate Dehydrogenase Complex and the Krebs Cycle

Objectives

- To review the conversion of pyruvate to acetyl-CoA by the mitochondrial PDH complex and the further oxidation of the acetyl carbons to CO_2 and H_2O via Krebs cycle activity.
- To define the two main metabolic products of intramitochondrial pyruvate metabolism and the metabolic pathways that can be traversed by these products.
- To name the proteins, tightly bound coenzymes, and product carrier coenzymes of the PDH complex.
- To identify the products of pyruvate dehydrogenase activity.
- To define the role of transacyclase in the PDH complex; to identify the cofactor of transacyclase.
- To identify the oxidized and reduced forms of the cofactor of dihydrolipoyl transacyclase.

- To write the net reaction representing the activity of the PDH complex on pyruvate.
- To define the allosteric effectors of PDH and PDH kinase; to define the role of PDH kinase in regulation of PDH and PDH complex activity.
- To identify, in a cell with high energy charge, the regulatory features that result in carbon atoms from pyruvate being stored in glycogen and fat.
- To identify the cofactor of each enzyme of the Krebs cycle and the structure of each intermediate; to compare the standard free energy charge for the citrate synthase reaction with that for the hydrolysis of ATP to ADP and Pi.
- To write the reaction catalyzed by citrate synthase, identifying the substrate atoms that condense to yield citrate.
- To identify the cofactor of aconitase and the origin of the chiral carbon of isocitrate.
- To identify the components of the non-heme iron complexes.
- To define the effectors on the IDH associated with the Krebs cycle and the nature of the regulation induced by each of the effectors.
- To compare the composition, mechanism of action, and regulation of the αKG dehydrogenase complex with that of the PDH complex.
- To identify the pathways, other than the Krebs cycle, in which succinyl-CoA and αKG are important metabolites.
- To identify the covalent intermediates formed during the substrate level phosphorylation catalyzed by succinate synthetase.
- To identify the flavin associated with SDH and the role of the SDH iron sulfur center.
- To identify the structure of oxidized and reduced CoQ_{10}.
- To identify the substrate, product, and cofactor of the MDH-catalyzed reaction.
- To identify the role of the citrate synthase reaction on the conversion of malate to OAA.
- To write an equation describing the net chemical changes associated with the oxidation of one equivalent of pyruvate by the PDH complex and Krebs cycle activity.

Concepts

The bulk of ATP used by many cells to maintain homeostasis is produced by the oxidation of pyruvate to CO_2 and H_2O, with generation of reduced nicotinamide adenine dinucleotide (NADH). The NADH is principally used to drive the processes of oxidative phosphorylation, which are responsible for converting the reducing potential of NADH to the high energy phosphate in ATP.

The PDH Complex

The fate of pyruvate depends on the cell energy charge. In cells or tissues with a high energy charge pyruvate is directed toward gluconeogenesis, but when the energy charge is low pyruvate is preferentially oxidized within the mitochondrion to CO_2 and H_2O, with generation of 15 equivalents of ATP per pyruvate. When transported into the mitochondrion, pyruvate encounters two principal metabolizing enzymes: pyruvate carboxylase and **pyruvate dehydrogenase (PDH)**, the first enzyme of the PDH complex. With a high cell-energy charge, coenzyme A (CoA) is highly acylated, principally as acetyl-CoA, and able allosterically to activate pyruvate carboxylase, directing pyruvate toward gluconeogenesis as described earlier. When the energy charge is low CoA is not acylated, pyruvate carboxylase is inactive, and pyruvate is preferentially metabolized via the PDH complex and the Krebs cycle to CO_2 and H_2O. Reduced NADH generated during the oxidative reactions is used to drive ATP synthesis via oxidative phosphorylation.

As summarized in Table 17–1, the PDH complex is comprised of 3 separate enzymes and requires 5 different coenzymes. Each PDH complex contains multiple copies of each enzyme; as a result, the molecular weight of PDH complexes is in the range of 6 to 8 million Daltons and the structure can be visualized by electron microscopy. Three of the coenzymes of the complex are tightly bound to enzymes of the complex and two are employed as carriers of the products of PDH complex activity. The processes carried out by the PDH complex are illustrated in Figure 17–1.

The first enzyme of the complex is PDH itself which oxidatively decarboxylates pyruvate. During the course of the reaction the acetyl group derived from decarboxylation of pyruvate is bound to thiamine pyrophosphate, TPP (Figure 17–2).

		Table 17–1. Components of the PDH complex.	
Enzymes of the PDH Complex	**Number of Copies Per Complex**	**Tightly Bound Coenzymes, One Per Enzyme**	**Soluble Carrier Coenzymes**
Pyruvate dehydrogenase	20–30	Thiamin pyrophosphate	
Dihydrolipoyl transacetylase	60	Lipoic acid	Coenzyme A
Dihydrolipoyl dehydrogenase	6	Flavin adenine dinucleotide	NAD⁺

The next reaction of the complex (Figure 17–1), is the transfer of the 2-carbon acetyl group from acetyl-TPP to lipoic acid (LA), the covalently bound coenzyme of lipoyl transacylase. The structure of LA, bound to an ε amino of the protein, and the structure of the acylated-LA product of the reaction are illustrated in Figure 17–3.

The final disposition of the original pyruvate carbon atoms by the PDH complex is their transfer to CoA as the acetyl group, as shown in Figure 17–1.

The transfer of the acetyl group from acyl-lipoamide to CoA results in the formation of 2 sulfhydryl (SH) groups in lipoate requiring reoxidation to the disulfide (S-S) form to regenerate lipoate as a competent acyl acceptor. The enzyme dihydrolipoyl dehydrogenase, with flavin adenine dinucleotide (FAD) as a cofactor, catalyzes the oxidation as shown in Figure 17–4.

The final activity of the PDH complex is the transfer of reducing equivalents from the $FADH_2$ of dihydrolipoyl dehydrogenase to NAD^+ (Figure 17–1). The fate of the NADH is oxidation via mitochondrial electron transport, to produce 3 equivalents of ATP.

Equation 17.1 represents the net chemical reaction carried out by the PDH complex.

$$\text{Pyruvate} + \text{CoA} + \text{NAD}^+ \rightarrow \text{CO}_2 + \text{acetyl} - \text{CoA} + \text{NADH} + \text{H}^+ \qquad 17.1$$

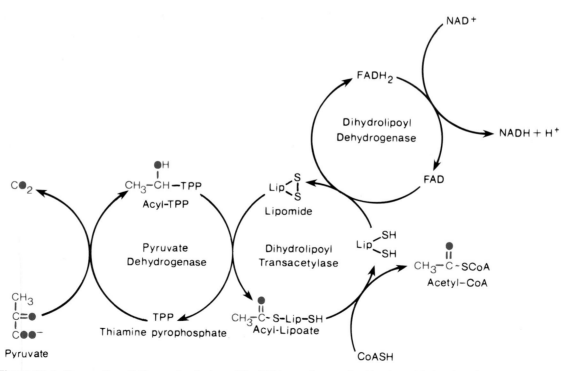

Figure 17–1. Enzymatic activities and cofactors of the PDH complex required for the oxidative decarboxylation of pyruvate to acetyl-CoA. Flow diagram depicting the overall activity of the PDH complex.

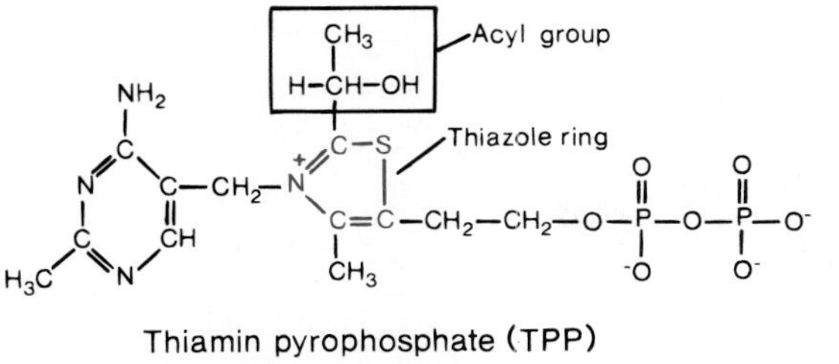

Figure 17–2. Structure of acetyl-TPP, the covalent coenzyme intermediate of pyruvate dehydrogenase in the PDH complex.

Regulation of the PDH Complex

The reactions of the PDH complex connect several of the main metabolic pathways of the cell: glycolysis/gluconeogenesis, fatty acid synthesis, and the Krebs or citric acid cycle. As a consequence, **the activity of the PDH complex is highly regulated by a variety of allosteric effectors and by covalent modification**. The importance of the PDH complex to the maintenance of homeostasis is evident from the fact that although diseases associated with deficiencies of the PDH complex have been observed, affected individuals often do not survive to maturity. Since the energy metabolism of highly aerobic tissues such as the brain is dependent on normal conversion of pyruvate to acetyl-CoA, aerobic tissues are most sensitive to deficiencies in components of the PDH complex. Most genetic diseases associated with PDH complex deficiency are due to mutations in PDH. The main pathologic result of such mutations is moderate to severe cerebral lactic acidosis and encephalopathies.

This discussion will focus on regulation of the normal PDH complex as it is understood in liver. The main regulatory features of the PDH complex are summarized in Figure 17–5. Two products of the complex, NADH and acetyl-CoA, are negative allosteric effectors on *PDH-a,* the non-phosphorylated, active form of PDH. These effectors reduce the affinity of the enzyme for pyruvate, thus limiting the flow of carbon through the PDH complex. In addition, NADH and acetyl-CoA are powerful positive effectors on PDH kinase, the enzyme that inactivates PDH by converting it to the phosphorylated *PDH-b* form. Since NADH and acetyl-CoA accumulate when the cell energy charge is high, it is not surprising that high ATP levels also up-regulate PDH kinase activity, reinforcing down-regulation of PDH activity in energy-rich cells. Note, however, that **pyruvate is a potent negative effector on PDH kinase**, with the result that when pyruvate levels rise, *PDH-a* will be favored even with high levels of NADH and acetyl-CoA.

Concentrations of pyruvate which maintain PDH in the active *a*-form are sufficiently high so that, in energy-rich cells, the allosterically down-regulated, high K_M form of PDH is nonetheless capable of

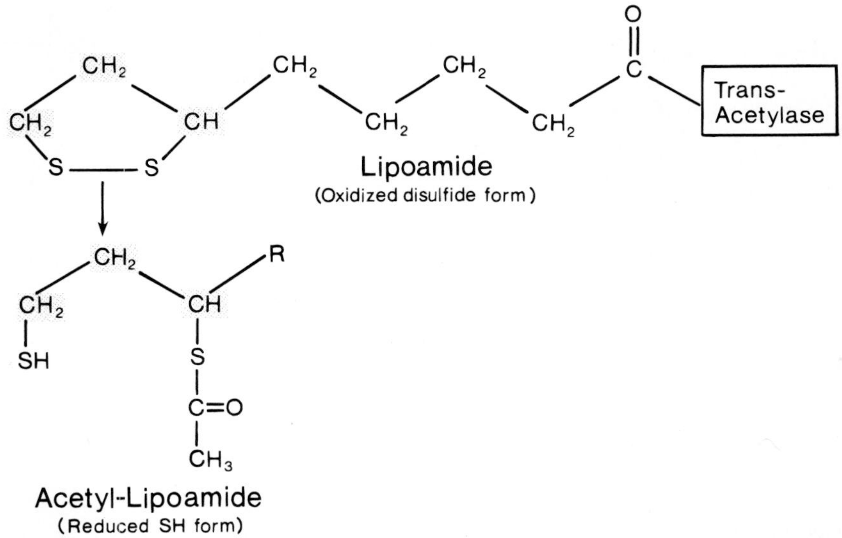

Figure 17–3. Formation of acetyl lipoamide (LA). The S-S bond of LA of LA is shown opened as a result of reduction of LA and addition of the acetyl group from acetyl-TPP.

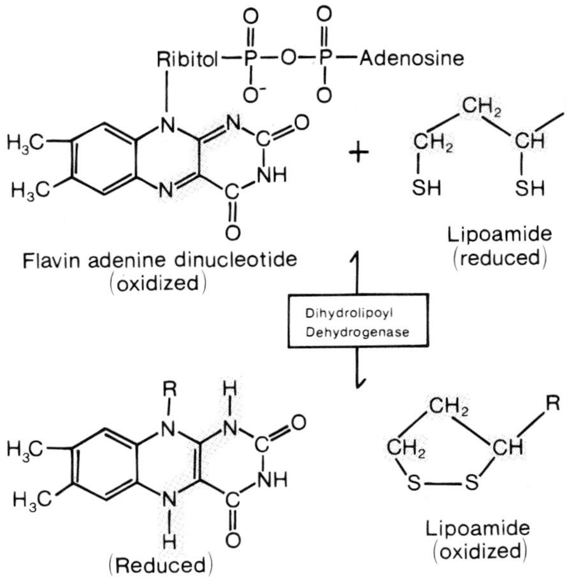

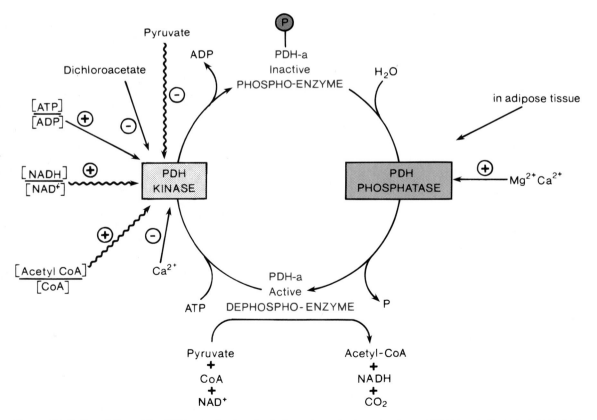

Figure 17–4. Regeneration of the oxidized (S-S) form of lipoic acid by FAD, the coenzyme of dihydrolipoyl dehydrogenase. The product is reduced FAD (FADH$_2$).

Figure 17–5. Regulation of the PDH complex by allosteric and covalent mechanisms. PDH kinase phosphorylates and inactivates PDH. Substrates and products of the PDH complex are potent allosteric regulators of PDH kinase and of PDH itself.

converting pyruvate to acetyl-CoA. With large amounts of pyruvate in cells having high energy charge and high NADH, pyruvate carbon will be directed to the 2 main storage forms of carbon—glycogen via gluconeogenesis and fat production via fatty acid synthesis—where acetyl-CoA is the principal carbon donor.

Although the regulation of *PDH-b* phosphatase is not well understood, it is quite likely regulated to maximize pyruvate oxidation under energy-poor conditions and to minimize PDH activity under energy-rich conditions.

ENZYMES OF THE KREBS CYCLE

Citrate Synthase (Condensing Enzyme)

The overall pathway of the Krebs cycle is illustrated in Figure 17–6. The first reaction of the cycle is condensation of the methyl carbon of acetyl-CoA with the keto carbon (C-2) of oxaloacetate (OAA), as illustrated in Figure 17–7. The standard free energy of the reaction, -8.0 kcal/mol, drives it strongly in the forward direction. Since the formation of OAA from its precursor is thermodynamically unfavorable, the highly exergonic nature of the citrate synthase reaction is of central importance in keeping the entire cycle going in the forward direction, since it "pulls" oxaloacetate formation by mass action principals.

Aconitase, a Non-Heme-Iron Protein

Citrate is a prochiral molecule: it can become chiral with a single rearrangement, as shown in Figure 17–8. The isomerization of citrate to isocitrate by aconitase is stereospecific, with the migration of the $-OH$ from the central carbon of citrate (formerly the keto carbon of OAA) being always to the adjacent carbon which is derived from the methylene (CH_2) of OAA. **The stereospecific nature of the isomerization determines that the CO_2 lost, as isocitrate is oxidized to succinyl-CoA, is derived from the oxaloacetate used in citrate synthesis.**

Aconitase is one of several mitochondrial enzymes known as **non-heme-iron** proteins. These proteins contain inorganic iron and sulfur, known as iron sulfur centers, in a coordination complex with cysteine sulfurs of the protein. There are two prominent classes of non-heme-iron complexes, those containing two equivalents each of inorganic iron and sulfur Fe_2S_2, and those containing 4 equivalents of each Fe_4S_4. Aconitase is a member of the Fe_4S_4 class. Its iron sulfur centers are often designated as $Fe_4S_4Cys_4$, indicating that 4 cystine sulfur atoms are involved in the complete structure of the complex. In iron sulfur compounds the iron is generally involved in oxidation reduction events.

Isocitrate Dehydrogenase With NAD$^+$ as Coenzyme

There are two different isocitrate dehydrogenase (IDH) enzymes. The IDH of the Krebs cycle uses NAD^+ as a cofactor, whereas the other IDH uses $NADP^+$ as a cofactor. Unlike the NAD^+-requiring enzyme, which is located only in the mitochondrial matrix, the $NADP^+$-requiring enzyme is found in both the mitochondrial matrix and the cytosol. IDH catalyzes the rate-limiting step, as well as the first NADH-yielding reaction of the Krebs cycle (Figure 17–9). It is generally considered that control of carbon flow through the cycle is regulated at IDH by the powerful negative allosteric effectors NADH and ATP and by isocitrate ADP and AMP, which are potent positive effectors of the enzyme. From the latter it is clear that cell energy charge is a key factor in regulating carbon flow through the Krebs cycle.

The CO_2 produced by the IDH reaction is the original C-1 of the oxaloacetate used in the citrate synthase reaction.

α-Ketoglutarate Dehydrogenase Complex

As illustrated in Figure 17–10, α-KG is oxidatively decarboxylated by the α-ketoglutarate dehydrogenase (αKG dehydrogenase) complex, with the production of the second Krebs cycle equivalents of CO_2 and NADH. This multienzyme complex is very similar to the PDH complex in the intricacy of its protein makeup, cofactors, and its mechanism of action. Also, as with the PDH complex, the reactions of the αKG dehydrogenase complex proceed with a large negative standard free energy change (approximately -8 Kcal/mol). Although the αKG dehydrogenase enzyme is not subject to covalent modification, allosteric regulation is quite complex, with activity being regulated by energy charge, the NAD$^+$/NADH ratio, and effector activity of substrates and products.

Succinyl-CoA and α-ketoglutarate are also important metabolites outside the Krebs cycle. In particular, α-ketoglutarate represents a key anapleurotic metabolite linking the entry and exit of carbon

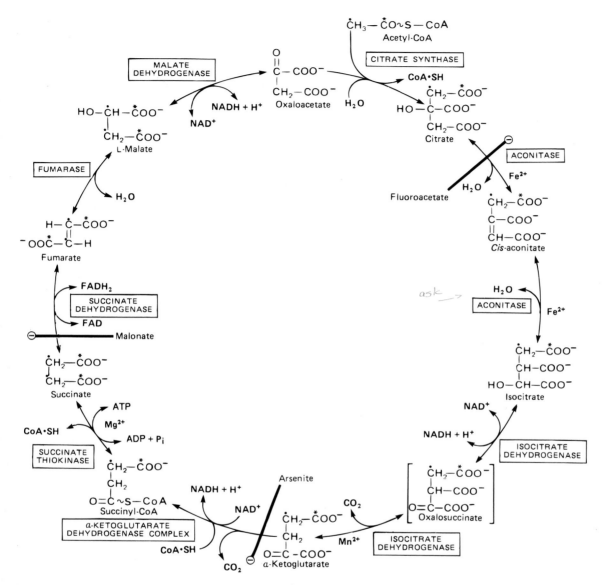

Figure 17–6. The citric acid (Krebs) cycle. Oxidation of NADH and FADH$_2$ in the respiratory chain leads to the generation of ATP via oxidative phosphorylation. In order to follow the passage of acetyl-CoA through the cycle, the 2 carbon atoms of the acetyl radical are shown labeled on the carboxyl carbon (using the designation [*]) and on the methyl carbon (using the designation [•]). Although 2 carbon atoms are lost as CO$_2$ in one revolution of the cycle, these atoms are not derived from the acetyl-CoA that has immediately entered the cycle but from that portion of the citrate molecule which derived from oxaloacetate. However, on completion of a single turn of the cycle, the oxaloacetate that is regenerated is now labeled, which leads to labeled CO$_2$ being evolved during the second turn of the cycle. Because succinate is a symmetric compound and because succinate dehydrogenase does not differentiate between its 2 carboxyl groups, "randomization" of label occurs at this step such that all 4 carbon atoms of oxaloacetate appear to be labeled aafter one turn of the cycle. During gluconeogenesis, some of the label in oxaloacetate is incorporated into glucose and glycogen. The sites of inhibition by (⊖) by fluoroacetate, malonate, and arsenite are indicated.

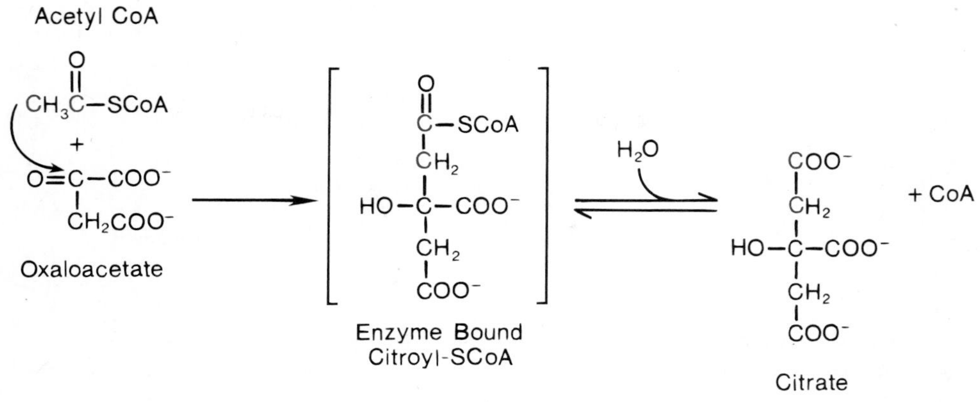

CITRATE SYNTHASE

Figure 17–7. The condensation, by citrate cynthase, of the methyl acetate of acetyl-CoA with the keto carbon of OAA to produce citric acid.

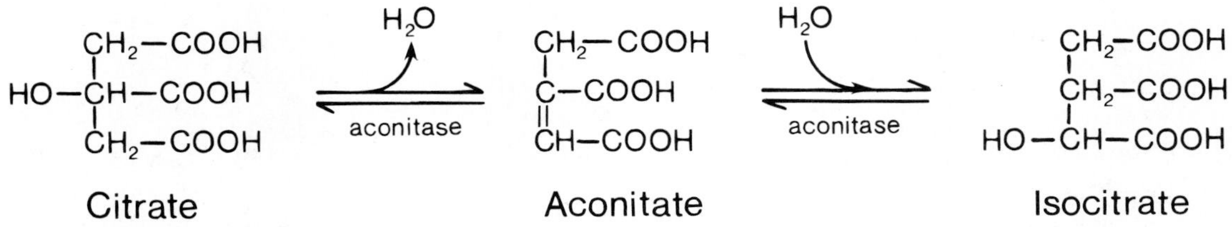

Figure 17–8. Conversion of prochiral citrate to the enzyme-bound intermediate aconitate and finally production of the chiral product of aconitase, isocitrate.

from the Krebs cycle to pathways involved in amino acid metabolism (Table 17–2). Succinyl-CoA, along with glycine, contributes all the carbon and nitrogen atoms required for the synthesis of protoporphyrin in heme biosynthesis.

Succinyl-CoA Synthetase (Succinyl Thiokinase)

The conversion of succinyl-CoA to succinate by succinyl-CoA synthetase involves use of the high-energy thioester of succinyl-CoA to drive synthesis of a high-energy nucleotide phosphate, by a process known as **substrate-level phosphorylation**. In this process a high energy enzyme—phosphate intermediate is formed, with the phosphate subsequently being transferred to GDP as illustrated in Figure 17–11. Mitochondrial GTP is used in a trans-phosphorylation reaction catalyzed by the mitochondrial enzyme **nucleoside diphosphokinase** to phosphorylate ADP, producing ATP and regenerating GDP for the continued operation of succinyl-CoA synthetase.

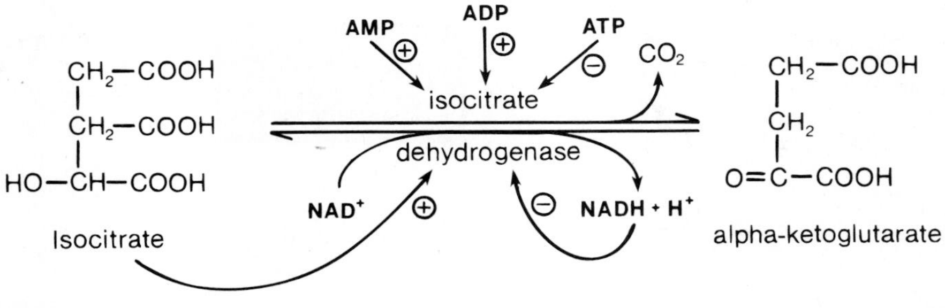

Figure 17–9. The reaction catalyzed by the NAD-requiring isocitrate dehydrogenase. The products are NADH, CO_2, and α-ketoglutarate. Also shown are allosteric regulators and their effects on enzyme activity.

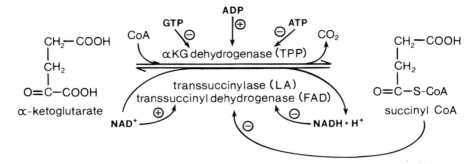

Figure 17–10. The α-ketoglutarate dehydrogenase complex is a multienzyme complex composed of α-ketoglutarate dehydrogenase, lipoyl transsuccinylase and dihydrolipoyl dehydrogenase. The enzyme complex has similar cofactors as their respective counterparts in the PDH complex. The effects of various regulatory molecules is also shown.

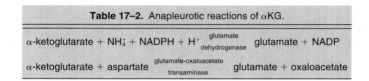

Table 17–2. Anapleurotic reactions of αKG.
α-ketoglutarate + NH₄⁺ + NADPH + H⁺ $\xrightarrow[\text{dehydrogenase}]{\text{glutamate}}$ glutamate + NADP
α-ketoglutarate + aspartate $\xrightarrow[\text{transaminase}]{\text{glutamate-oxaloacetate}}$ glutamate + oxaloacetate

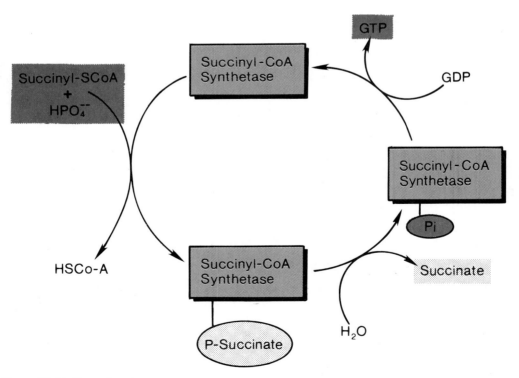

Figure 17–11. Formation of succinate from succinyl-CoA by succinyl-CoA synthetase. A high energy phospho-enzyme intermediate is formed and the phosphate is used to phosphorylate GDP to GTP.

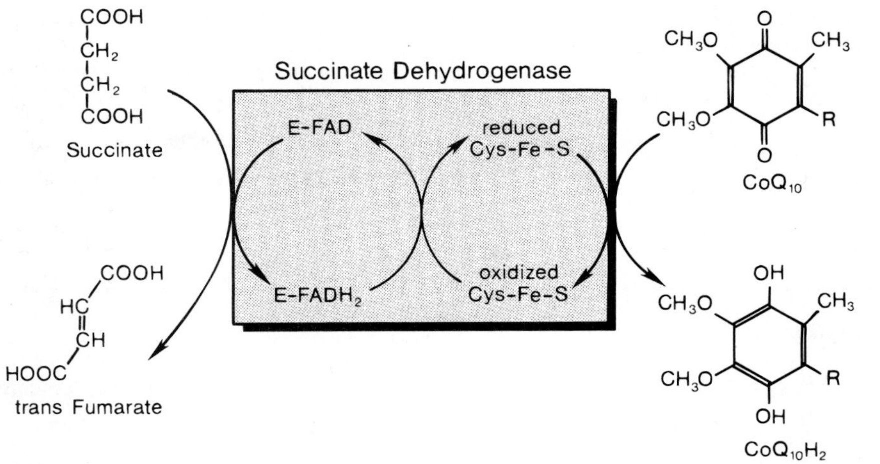

Figure 17–12. Oxidation of succinate to *trans*-fumarate by mitochondrial succinate dehydrogenase (SDH). Coenzyme Q is the final acceptor of reducing equivalents from SDH. In the CoQ structure R is a 10 unit poly-isoprenoid structure.

Succinate Dehydrogenase (SDH)

Succinate dehydrogenase catalyzes the oxidation of succinate to fumarate with the sequential reduction of enzyme-bound FAD and non-heme-iron, as shown in Figure 17–12. In mammalian cells the final electron acceptor is coenzyme Q_{10} (CoQ_{10}), a mobile carrier of reducing equivalents that is restricted by its lipophilic nature to the lipid phase of the mitochondrial membrane. The enzyme has a number of important features, which are summarized in Table 17–3.

Fumarase (Fumarate Hydratase)

The fumarase-catalyzed reaction shown in Equation 17.2 is specific for the trans form of fumarate.

$$\text{trans-Fumarate} + \text{H}_2\text{O} \rightarrow \text{L-Malate} \qquad \textbf{17.2}$$

The result is that the hydration of fumarate proceeds stereospecifically with the production of L-malate.

Malate Dehydrogenase (MDH)

L-malate is the specific substrate for MDH, the final enzyme of the Krebs cycle. The forward reaction of the cycle, the oxidation of malate, is illustrated in Equation 17.3.

$$\text{L-malate} + \text{NAD}^+ \rightarrow \text{oxaloacetate} + \text{NADH} + \text{H}^+ \qquad \textbf{17.3}$$

In the forward direction the reaction has an equilibrium constant of about 10^{-5}M, equivalent to a standard free energy of about +7 kcal/mol, indicating the very unfavorable nature of the forward direction. As noted earlier, the citrate synthase reaction that condenses oxaloacetate with acetyl-CoA has a standard free energy of about −8 kcal/mol and is responsible for pulling the MDH reaction in the

Table 17–3. Key characteristics of succinic dehydrogenase.
1. An integral protein of the inner mitochondrial membrane
2. A marker enzyme for inner mitochondrial membranes
3. The only oxidative enzyme of the Krebs cycle *not* requiring NAD
4. Its coenzyme is covalently bound FAD
5. It is a non-heme-iron containing protein
6. The reduced flavin (FADH2) is oxidized by membrane localized coenzyme Q_{10}
7. Malonate (HOOC—CH₂—COOH) is a well known competitive inhibitor

forward direction. The overall change in standard free energy change is about -1 kcal/mol for the conversion of malate to oxaloacetate and on to succinate.

Equation 17.4 represents the stoichiometric changes associated with passage of 1 equivalent of acetate from acetyl-CoA through the Krebs cycle.

$$\text{acetyl-SCoA} + 3\text{NAD}^+ + \text{FAD} + \text{GDP} + \text{Pi} + 2\text{H}_2\text{O} \rightarrow$$
$$2\text{CO}_2 + 3\text{NADH} + \text{FADH}_2 + \text{GTP} + 2\text{H}^+ + \text{HSCoA} \qquad 17.4$$

Questions

DIRECTIONS (item 1): The incomplete statement is followed by answers or completions of the statement. Select the ONE lettered completion that is BEST in this case.

1. In glycolytic reactions leading to formation of Acetyl-CoA, the C-2 and C-5 of glucose are.
   a. Phosphorylated and oxidized to the level of a carboxylic acid.
   b. Phosphorylated and oxidized to CO_2.
   c. Phosphorylated, oxidized to the level of a carboxylic acid, and covalently linked to a sulfur atom.
   d. Phosphorylated, oxidized to CO_2, and covalently linked to a sulfur atom.

DIRECTIONS (items 2–5): Match the following lettered mitochondrial enzymes with the properties identified In the numbered list. Each enzyme may be used more than once or not at all.

   a. Succinate dehydrogenase
   b. Isocitrate dehydrogenase
   c. NADH dehydrogenase
   d. Cytochrome c oxidase
   e. Pyruvate dehydrogenase

2. This copper-containing constituent of the electron transport chain is found in the inner mitochondrial membrane.

3. This TCA cycle enzyme is found in the mitochondrial matrix.

4. This FAD-linked dehydrogenase of the Krebs cycle is found in the inner mitochondrial membrane.

5. This enzyme is part of a very large complex (about 8×10^6 MW) which is found in the mitochondrial matrix.

DIRECTIONS (items 6–16): Each numbered item or incomplete statement is followed by answers or by completions of the statement. Select the ONE lettered answer or completion that is BEST in each case.

6. In conversion to succinate, glucose carbons:
   a. C1 and C6 become oxidized to carboxylic acids.
   b. C1 and C4 are lost as CO_2.
   c. C2 and C5 become carboxylate groups.
   d. C3 and C4 become involved in thioester bonding to CoA.
   e. C2 and C5 become CH_2 groups of succinate.

7. All Nicotinamide nucleotides:
   a. Are carriers of reducing equivalents that are generally used to drive ATP production via a chemiosmotic potential.
   b. Go from the oxidized state to the reduced state by the addition of two protons and two electrons to one of their heterocyclic rings.

    c. Are key carriers of reducing equivalents that are generally used for driving reductive biosynthetic reactions.

    d. Are unlike adenosine triphospho nucleotides in that they lack phosphoanhydride bonds.

    e. Are reduced by mitochondrial isocitrate dehydrogenases.

8. Constituents involved in pyruvate dehydrogenase complex activity include:
   a. TTP as a cofactor of the decarboxylating enzyme.
   b. FAD and NADP as cofactors for a "disulfhydryl" dehydrogenase.
   c. FMN and NAD as cofactors for dihydrolipoyl dehydrogenase.
   d. Lipoic acid, NADP, and FAD.
   e. Lipoic acid, NADP, FAD and TPP.

9. Carbon atoms from acetyl-CoA entering citrate are not racemized in Krebs cycle intermediates arising prior to the α-keto glutarate dehydrogenase step because:
   a. The product of citrate synthase is chiral.
   b. The 1 and 6 groups of citrate are chemically identical.
   c. Isocitrate is prochiral.
   d. Citrate is prochiral.
   e. Isocitrate has a 3-point attachment to isocitrate dehydrogenase.

10. Which of the following represent the reaction product of the citrate-condensing enzyme?
    a. OAA.
    b. Citrate.
    c. Acetyl-CoA.
    d. Isocitrate.
    e. Aconitate.

11. Which of the following enzymes requires TPP, FAD, and lipoic acid as cofactors?
    a. Succinyl-CoA synthetase.
    b. Malate dehydrogenase.
    c. Aconitase.
    d. Fumarase.
    e. α-Ketoglutarate dehydrogenase.

12. Which enzyme causes production of NADH in a reaction with a standard free energy of more than +5 kcal/mol.
    a. Malate dehydrogenase.
    b. α-Ketoglutarate dehydrogenase.
    c. Citrate synthase.
    d. Aconitase.
    e. Succinate dehydrogenase.

13. Which enzyme requires coenzyme A as a cofactor and catalyzes oxidative decarboxylation of a transaminase product?
    a. Succinate dehydrogenase.
    b. Succinyl-CoA synthetase.
    c. Malate dehydrogenase.
    d. Aconitase.
    e. α-Ketoglutarate dehydrogenase.

14. Which of the following enzymes has an iron sulfur center and catalyzes a reaction that does not lead to a net transfer of electrons?
    a. α-Ketoglutarate dehydrogenase.
    b. Malate dehydrogenase.
    c. Aconitase.
    d. Succinate dehydrogenase.
    e. Malate dehydrogenase.

DIRECTIONS (item 15): With reference to the following case history, select the ONE response that BEST completes the statement.

15. A candidate for colostomy was malnourished due to the intestinal tract disorder. The patient was maintained on total parenteral nutrition prior to the surgery, and post-surgical complications necessitated a continuation of total parenteral nutrition. After a period of seemingly normal recovery, but while still maintained on total parenteral nutrition, the patient experienced an episode of severe lactic acidosis with a minimal increase in blood pyruvate. A number of support measures failed to control the condition, with the result that blood pH values as low as 6.70 were observed. When it was discovered that the parenteral nutrition did not include vitamins the attending physician administered a pharmacological dose of all water soluble vitamins and supplied vitamins along with the other parenteral nutrients. The patient soon responded favorably and recovered without further incident. From the following list of possibilities the most likely cause of the patient's difficulty was:

   a. Lack of niacin for production of nicotinamide nucleotides.
   b. Inability to synthesize a number of citric acid cycle enzymes, including pyruvate dehydrogenase, because of the absence of vitamin-derived cofactors.
   c. Inability to carry out the reactions of the PDH complex because of the absence of thiamine.
   d. Inability to carry out the succinic dehydrogenase reaction because of the lack of riboflavin.

Answers

1. c	**6.** c	**11.** e
2. d	**7.** e	**12.** a
3. b	**8.** a	**13.** e
4. a	**9.** d	**14.** c
5. e	**10.** b	**15.** c

18

Biological Oxidations

Objectives

- To review the principles of physiologically important oxidation/reduction reactions and the role these reactions play in production of ATP via oxidative phosphorylation.
- To identify the component of a redox half cell that is common to all half cells.
- To identify the symbols used to designate standard reduction potential and standard biological reduction potential.
- To define the relationship between difference in standard electrode potentials of a reaction comprised of 2 electrochemical half cells and the standard free energy of the same reaction.
- To identify the terms in the Nernst equation that equate standard free energy with the standard electrode potential difference.
- To estimate, by calculation, the efficiency of the mitochondrial electron transport system in converting substrate energy into ATP energy.
- To identify the nature of the primary transmembrane gradient generated by mitochondrial electron transport.
- To identify the cofactors/coenzymes and metal ions of the electron transport assembly and the path of electron flow through the system.

- To identify the coenzymes of Complex I and II and the difference in free energy for the reactions catalyzed by the 2 complexes.
- To identify the constituents of Complexes III and IV and the carrier that transfers electrons between these complexes.
- To identify the processes by which the obligatory 2-electron reactions of NADH dehydrogenase and succinic dehydrogenase are connected to the single-electron reactions of the cytochromes.
- To identify characteristics that differentiate Heme, Heme A, and Heme C.
- To identify the 2 forces used in calculating the chemiosmotic potential.
- To compare the ATP-generating potential of mitochondrial NADH oxidation via NADH dehydrogenase with that of succinate oxidation.
- To identify the process by which cell energy charge modulates the PMF and regulates mitochondrial electron transport.
- To identify 2 pathways by which cytosolic NADH can be used to support the process of oxidative phosphorylation and compare the efficiency of the 2 processes.
- To identify the components of the malate aspartate shuttle system.
- To identify the difference in reactions catalyzed by oxidases and oxygenases.
- To identify the components of the P450 system of hydroxylating enzymes and their intracellular location.
- To name the enzymes responsible for protection against oxygen free radicals that form in tissues; to identify the reactants and products of each enzyme.

Concepts

While the large quantity of NADH resulting from Krebs cycle activity can be used for reductive biosynthesis, the reducing potential of mitochondrial NADH is most often used to raise the cell energy charge by supplying the energy for ATP synthesis via oxidative phosphorylation. Oxidation of NADH with phosphorylation of ADP to form ATP are processes supported by the mitochondrial electron transport assembly and ATP synthase, which are integral protein complexes of the inner mitochondrial membrane. The electron transport assembly is comprised of a series of protein complexes that catalyze sequential oxidation reduction reactions; some of these reactions are thermodynamically competent to support ATP production via ATP synthase provided a coupling mechanism, such as a common intermediate, is available. Proton translocation and the development of a transmembrane proton gradient provides the required coupling mechanism. This chapter reviews the principles of oxidation reduction reactions and biological reactions involving oxygen and the use of biological reducing potential to form ATP via oxidative phosphorylation.

PRINCIPLES OF OXIDATION REDUCTION REACTIONS

Redox reactions involve the transfer of electrons from one chemical species to another. The oxidized plus the reduced form of each chemical species is referred to as an electrochemical **half cell**. Two half cells having at least one common intermediate comprise a complete, coupled, oxidation reduction reaction. Coupled electrochemical half cells have the thermodynamic properties of other coupled chemical reactions. If one half cell is far from electrochemical equilibrium, its tendency to achieve equilibrium (ie, to gain or lose electrons) can be used to alter the equilibrium position of a coupled half cell. Typical electrochemical half cells are illustrated by Equations 18.1 and 18.2 and the sum of the half cell reactions in Equation 18.3. Note that electrons are common to both half cells and typically provide the coupling between electrochemical half cells.

$$\frac{1}{2}O_2 + 2e^- + 2H^+ \Leftrightarrow H_2O \qquad\qquad 18.1$$

$$NADH + H^+ \Leftrightarrow NAD^+ + 2H^+ + 2e^- \qquad\qquad 18.2$$

$$NADH + \frac{1}{2}O_2 + 3H^+ + 2e^- \Leftrightarrow NAD^+ + H_2O + 2H^+ + 2e^- \qquad\qquad 18.3$$

Equation 18.3 (the sum of 18.1 and 18.2) simplifies to give the more familiar expression, shown in Equations 18.4 and 18.7, which describe the overall reaction involved in oxidation of one equivalent of NADH by the mitochondrial electron transport system:

$$\text{NADH} + \tfrac{1}{2}O_2 + H^+ \longrightarrow \text{NAD}^+ + H_2O \qquad\qquad \textbf{18.4}$$

The thermodynamic potential of a *chemical reaction* is calculated from equilibrium constants and concentrations of reactants and products, as described in Chapter 2. Because it is not practical to measure electron concentrations directly, the electron energy potential of a redox system is determined from the electrical potential or voltage of the individual half cells, relative to a standard half cell. When the reactants and products of a half cell are in their standard state and the voltage is determined relative to a standard hydrogen half cell (whose voltage, by convention, is zero), the potential observed is defined as the **standard electrode potential**, E_o. If the pH of a standard cell is in the biological range, pH 7, its potential is defined as the **standard biological electrode potential** and designated E_o'. By convention, standard electrode potentials are written as potentials for reduction reactions of half cells. The reduction potentials for the half cells of Equations 18.1 and 18.2 and the difference in standard electrode potentials for the coupled reactions are given in Equations 18.5, 18.6, 18.7, and 18.8. By convention, potential is in volts and $\delta E_o'$ is calculated from the following algebraic difference: [(electron acceptor half cell potential) − (electron donor half cell potential)]. Note that algebraic subtraction of both the electrode potentials and the chemical expressions leads to Equation 18.8.

$$\tfrac{1}{2}O_2 + 2H^+ + 2e^- \Leftrightarrow 2H_2O \qquad E_o' = +0.82V \qquad\qquad \textbf{18.5}$$

$$\text{NAD}^+ + H^+ + 2e^- \Leftrightarrow \text{NADH} \qquad E_o' = -0.32V \qquad\qquad \textbf{18.6}$$

$$\tfrac{1}{2}O_2 - \text{NAD}^+ + H^+ \Leftrightarrow 2H_2O - \text{NADH} \qquad \Delta E_o' = +1.14V \qquad\qquad \textbf{18.7}$$

or

$$\text{NADH} + \tfrac{1}{2}O_2 + H^+ \Leftrightarrow \text{NAD}^+ + H_2O \qquad \Delta E_o' = +1.14V \qquad\qquad \textbf{18.8}$$

The free energy of a reaction such as that in Equation 18.8 is calculated directly from its $\Delta E_o'$ by the Nernst expression as shown in Equation 18.9, where n is the number of electrons involved in the reaction and F is the Faraday constant, 23.06 kcal/volt/mol or 94.4 kJ/volt/mol.

$$\Delta G^{\circ\prime} = -n F \Delta E_o' \qquad\qquad \textbf{18.9}$$

For the oxidation of NADH, η is 2 and the standard biological reduction potential, E_o' is −52.6 kcal/mol.

If the concentration of half cell components is other than that of the standard state, the half cell will have an electrode potential proportionately different from that of the standard state. The electrode potential for a redox reaction at pH 7, but away from the standard state, is defined as $\delta e'$. The free energy calculated from the Nernst expression is equal to $\delta G'$, as shown in Equation 18.10.

$$\Delta G' = -n F \Delta E' \qquad\qquad \textbf{18.10}$$

With a free energy change of −52.6 kcal/mol, it is clear that NADH oxidation has the potential for driving the synthesis of a number of ATPs: the standard free energy for the reaction $\text{ADP} + P_i \Leftrightarrow \text{ATP}$ is +7.3 kcal/mole. Direct chemical analysis has shown that for every 2 electrons transferred from NADH to oxygen, 3 equivalents of ATP are synthesized. From these considerations we can calculate that 21.9 kcal of energy are conserved in ATP ($3 \times 7.3 = 21.9$), while 52 kcal of free energy are available. The efficiency of the process is $21.9/52 \times 100 = 42\%$, a much greater conservation of energy than is found in most mechanical devices.

NADH OXIDATION

NADH is oxidized by a series of catalytic redox carriers that are integral proteins of the inner mitochondrial membrane. The free energy change in several of these steps is very exergonic (large and negative). Coupled to these oxidation reduction steps is a transport process in which protons (H^+) from the mitochondrial matrix are translocated to the cytosol. The redistribution of protons leads to formation of a proton gradient across the mitochondrial membrane (high proton concentration outside). The size of the gradient is proportional to the free energy change of the electron transfer reactions. The result of these reactions is that the redox energy of NADH is converted to the energy of the proton gradient. In the presence of ADP, protons flow down their thermodynamic gradient from outside the mitochondrion back into the mitochondrial matrix. This process is facilitated by a proton carrier in the inner mitochondrial membrane known as **ATP synthase**. As its name implies, this carrier is coupled to ATP synthesis. The following material reviews the electron transport reactions that generate the proton gradient and the use of the proton gradient to synthesize ATP via ATP synthase.

THE ELECTRON TRANSPORT ASSEMBLY

Electron flow through the mitochondrial electron transport assembly is illustrated in Figure 18–1. The process is diagrammed for passage of 2 electrons from NADH (top) and succinate (bottom) to molecular oxygen via the electron transferring cofactors of the pathway. With the exception of NADH, succinate, and CoQ, all of the components of the pathway are integral proteins of the inner mitochondrial membrane whose cofactors undergo redox reactions. NADH and succinate are soluble in the mitochondrial matrix, while CoQ is a small, mobile carrier that transfers electrons between the primary dehydrogenases and cytochrome *b*. CoQ is also restricted to the membrane phase because of its hydrophobic character.

COMPLEXES I AND II

The mitochondrial electron transport proteins are clustered in groups in the inner membrane; the proteins in each cluster, as shown in Figure 18–1, tend to copurify. The clusters are known as Complexes I, II, III, and IV. Complex I, also known as NADH:CoQ oxidoreductase, is composed of NADH dehydrogenase with FMN as cofactor, plus non-heme-iron proteins having at least 1 iron sulfur center. Complex I is responsible for transferring electrons from NADH to CoQ. The $\Delta E_0'$ for the latter transfer is 0.42 V, corresponding to a $\Delta G'$ of -19 kcal/mol of electrons transferred (Figure 18–2). With its highly exergonic free energy change, the flow of electrons through Complex I is more than adequate

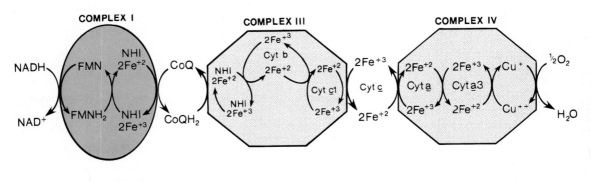

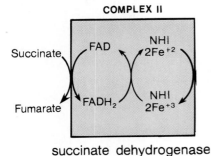

succinate dehydrogenase

Figure 18–1. Top. Electron flow from NADH to oxygen through complexes I, III, and IV. Bottom. Flow of electrons from succinate into the oxidative phosphorylation pathway through complex II.

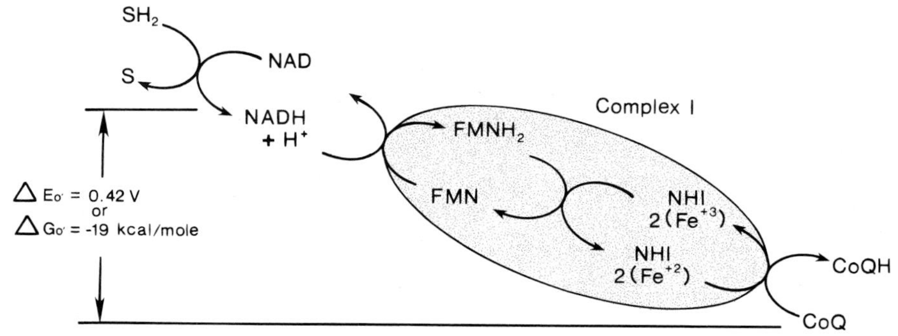

Figure 18–2. Energy change as electrons flow from NADH to CoQ.

to drive ATP synthesis. Complex II is also known as succinate dehydrogenase or succinate: CoQ oxidoreductase. The $\Delta E'_o$ for electron flow through Complex II is about 0.05 V, corresponding to a $\Delta G'$ of -2.3 kcal/mol of electrons transferred, which is insufficient to drive ATP synthesis. The difference in free energy of electron flow through Complexes I and II accounts for the fact that a pair of electrons originating from NADH and passing to oxygen supports production of 3 equivalents of ATP, while 2 electrons from succinate support the production of only 2 equivalents of ATP.

COMPLEXES III AND IV

Reduced CoQ (CoQH$_2$) diffuses in the lipid phase of the membrane and donates its electrons to Complex II, whose principal components are the heme proteins known as cytochromes b and c_1, and a non-heme-iron protein, known as the Reskie iron sulfur protein. In contrast to the heme of hemoglobin and myoglobin, the heme iron of all cytochromes participates in the cyclic redox reactions of electron transport, alternating between the oxidized (Fe^{+3}) and reduced (Fe^{+2}) forms. The electron carrier from Complex III to Complex IV is the smallest of the cytochromes, cytochrome c (molecular weight 12,000). Complex IV, also known as **cytochrome oxidase**, contains the hemeproteins known as cytochrome a and cytochrome a_3, as well as copper-containing proteins in which the copper undergoes a transition from Cu$^+$ to Cu^{++} during the transfer of electrons through the complex to molecular oxygen. Oxygen is the final electron acceptor, with water being the final product of oxygen reduction.

COENZYMES OF THE ELECTRON TRANSPORT PATHWAY

FAD and FMN

Normal oxidation of NADH or succinate is always a 2-electron reaction, with the transfer of 2 hydride ions to a flavin. A hydride ion is composed of 1 proton and 1 electron and is written as **$H\dot{V}$**. Unlike NADH and succinate, flavins can participate in either 1-electron or 2-electron reactions; thus, flavin that is fully reduced by the dehydrogenase reactions can subsequently be oxidized by 2 sequential 1-hydride reactions. Figure 18–3 illustrates the structure and the reactions of FMN's 3-member, heterocyclic **isoalloxazine** ring. The oxidation reduction reactions of FAD's isoalloxazine ring are identical. The fully reduced form of a flavin is known as the quinol form and the fully oxidized form is known

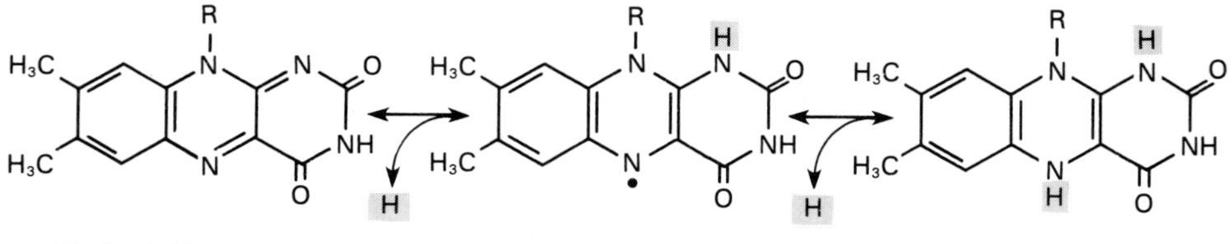

Figure 18–3. Oxidoreduction of isoalloxazine ring in flavin nucleotides. (Reproduced, with permission, from Murray, RK: *Harper's Biochemistry*, 23rd ed. Appleton & Lange, 1993.)

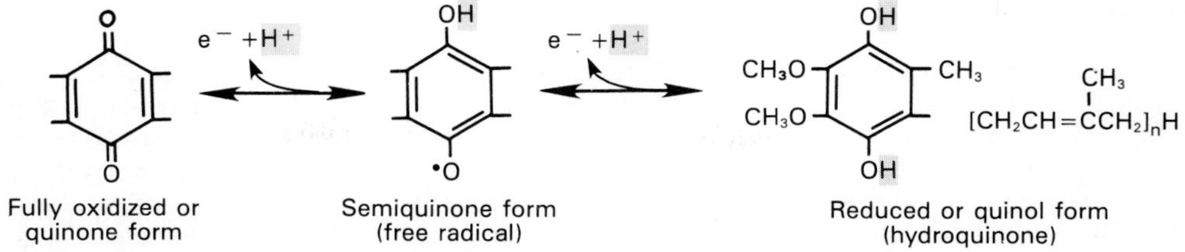

Fully oxidized or quinone form → (e⁻ +H⁺) → **Semiquinone form (free radical)** → (e⁻ +H⁺) → **Reduced or quinol form (hydroquinone)**

Figure 18–4. Structure of ubiquinone (Q). n = Number of isoprenoid units, which is 10 in higher animals, ie, Q_{10-}. (Reproduced, with permission, from Murray, RK: *Harper's Biochemistry*, 23rd ed. Appleton & Lange, 1993.)

as the quinone form; the intermediate containing a single electron is known as the semiquinone or semiquinol form.

Coenzyme Q

Like flavins, CoQ (also known as ubiquinone) can undergo either 1- or 2-electron reactions leading to formation of the reduced quinol, the oxidized quinone, and the semiquinone intermediate. Figure 18–4 illustrates the fully oxidized and reduced structures and the long hydrophobic polyisoprenoid substitution on the ring. In higher animals, 10 isoprene units, covalently aligned head-to-tail, constitute the polyisoprenoid substituent. The ability of flavins and CoQ to form semiquinone intermediates is a key feature of the mitochondrial electron transport systems, since these cofactors link the obligatory 2-electron reactions of NADH and succinate with the obligatory 1-electron reactions of the cytochromes.

The cytochromes are heme proteins. Like hemoglobin and myoglobin, the cytochromes generally contain 1 heme group per polypeptide—except for cytochrome *b*, which has 2 heme residues in 1 polypeptide chain. Cytochromes vary in the structure of the heme and in its binding to apoprotein. Illustrated in Figure 18–5 is iron protoporphyrin IX, often simply known as heme. This is the heme of hemoglobin, myoglobin, and all *b* type cytochromes. It is always bound to its apoprotein noncovalently.

Cytochromes of the *c* type contain a modified iron protoporphyrin IX known as heme C. In heme C the 2 vinyl (C=C) side chains are covalently bonded to cysteine sulfhydryl residues of the apoprotein. Only cytochromes of the *c* type contain covalently bound heme. Heme *a* is again a modified iron protoporphyrin IX. The modifications include addition of a long poly isoprenoid side chain added to 1 of the vinyl groups, and oxidation of 1 methyl group to an aldehyde. Heme *a* is found in cytochromes of the *a* type and in the chlorophyll of green plants.

An important property of heme proteins involved in redox reactions is the marked spectral change which they all exhibit at specific wavelengths when their iron atoms undergo oxidation and reduction. Typically, reduced heme proteins have a higher absorption coefficient than their oxidized counterpart. The sequence of hemeproteins in the electron transport assembly has been deduced from the sequential spectral changes observed when the components of the electron transport system undergo their characteristic sequential oxidation or reduction.

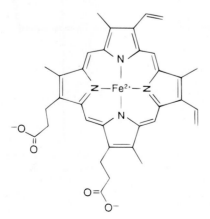

Figure 18–5. Structure of heme. (Reproduced, with permission, from Murray, RK: *Harper's Biochemistry*, 23rd ed. Appleton & Lange, 1993.)

OXIDATIVE PHOSPHORYLATION

The free energy available as a consequence of transferring 2 electrons from NADH or succinate to molecular oxygen is −57 and −36 kcal/mol, respectively. Oxidative phosphorylation traps this energy as the high-energy phosphate of ATP. In order for oxidative phosphorylation to proceed, two principal conditions must be met. First, the inner mitochondrial membrane must be physically intact so that protons can only reenter the mitochondrion by a process coupled to ATP synthesis. Second, a high concentration of protons must be developed on the outside of the inner membrane (Figure 18–6).

The energy of the proton gradient is known as the **chemiosmotic potential, or proton motive force (PMF)**. This potential is the sum of the concentration difference of protons across the membrane and

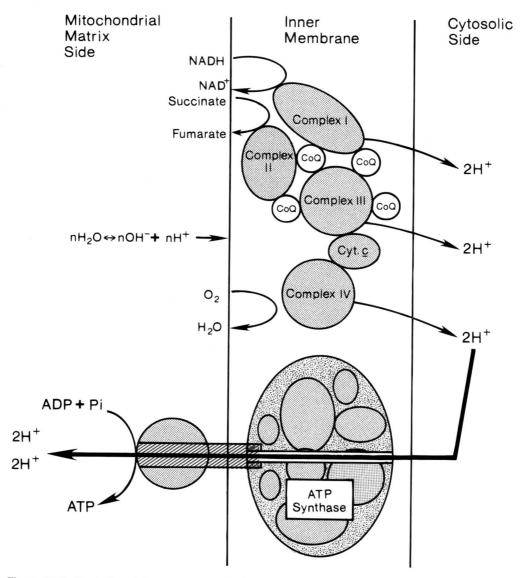

Figure 18–6. Illustration of the events associated with oxidative phosphorylation. A pair of electrons from NADH (or succinate) are transferred to oxygen forming H_2. The energy of the redox reactions is used to pump pairs of protons out of the mitochondrion at 3 steps of the electron transport system. The protons arise from dissociation reactions within the mitochondrion with each of the proton translocation reactions being coupled to a redox reaction having a large negative free energy change.

the difference in electrical charge across the membrane. The free energy of these 2 potentials can be calculated from the basic chemiosmotic equation shown in Equation 18.11.

$$\Delta G' = n F \Delta \Psi - 2.303 RT \Delta p \qquad \textbf{18.11}$$

The Faraday constant is F; the voltage difference, $\Delta \Psi$, is typically -150 mV (inside negative), and ΔpH is in the range of 1.5 (more acidic outside). With the latter values, the free energy of moving one mole of protons, with charge $n = +1$, from the exterior to the interior of the mitochondrion is about -5.2 kcal.

Figure 18–6 depicts 2 electrons from NADH as generating a 6-proton gradient. Thus, oxidation of 1 mole of NADH leads to the availability of a PMF with a free energy of about -31.2 kcal (6×-5.2 kcal). The energy of the gradient is used to drive ATP synthesis: protons are transported back down their thermodynamic gradient into the mitochondrion through the integral membrane protein known as ATP synthase (or Complex V). ATP synthase is a multiple subunit complex that binds ADP and inorganic phosphate at its catalytic site inside the mitochondrion, and requires a proton gradient for activity in the forward direction. ATP synthase is composed of 3 fragments: F_o, which is localized in the membrane; F_1, which protrudes from the inside of the inner membrane into the matrix; and oligomycin sensitivity-conferring protein (OSCP), which connects F_o to F_1. The reaction catalyzed by ATP synthase is shown in Equation 18.12.

In damaged mitochondria, permeable to protons, the ATP synthase reaction is active in the reverse direction acting as a very efficient ATP hydrolase or ATPase.

$$nH^+_{outside} + ADP_{inside} + Pi_{inside} \underset{\text{Synthase}}{\overset{\text{ATP}}{\rightleftharpoons}} ATP_{inside} + nH^+_{inside} \qquad \textbf{18.12}$$

Stoichiometry of Oxidative Phosphorylation

For each pair of electrons originating from NADH, 3 equivalents of ATP are synthesized, requiring 22.4 kcal of energy (ATP:2e$-$ = 3). Thus, with 31.2 kcal of available energy, it is clear that the proton gradient generated by electron transport contains sufficient energy to drive normal ATP synthesis. Electrons from succinate have about ⅔ the energy of NADH electrons: they generate PMFs that are about ⅔ as great as NADH electrons and lead to the synthesis of only 2 moles of ATP per mole of succinate oxidized. (ATP:2e$-$ = 2)

Regulation of Oxidative Phosphorylation

Since electron transport is directly coupled to proton translocation, the flow of electrons through the electron transport system is regulated by the magnitude of the PMF. The higher the PMF the lower the rate of electron transport, and vice versa. Under resting conditions, with a high cell energy charge, the demand for new synthesis of ATP is limited and, although the PMF is high, flow of protons back into the mitochondria through ATP synthase is minimal. When energy demands are increased, such as during vigorous muscle activity, cytosolic ADP rises and is exchanged with intramitochondrial ATP via the transmembrane adenine nucleotide carrier **ADP/ATP translocase**. Increased intramitochondrial concentrations of ADP cause the PMF to become discharged as protons pour through ATP synthase, regenerating the ATP pool. Thus, while the rate of electron transport is dependent on the PMF, the magnitude of the PMF at any moment simply reflects the energy charge of the cell. In turn the energy charge, or more precisely ADP concentration, **normally determines the rate of electron transport by mass action principles**. The rate of electron transport is usually measured by assaying the rate of oxygen consumption and is referred to as the **cellular or mitochondrial respiratory rate**. The respiratory rate is known as the **state 4** rate when the energy charge is high, the concentration of ADP is low, and electron transport is limited by ADP. When ADP levels rise and inorganic phosphate is available, the flow of protons through ATP synthase is elevated and higher rates of electron transport are observed; the resultant respiratory rate is known as the **state 3** rate. Thus, under physiological conditions mitochondrial respiratory activity cycles between state 3 and state 4 rates.

The pathway of electron flow through the electron transport assembly, and the unique properties of the PMF, have been determined through the uses of a number of important antimetabolites. Some of these agents are inhibitors of electron transport at specific sites in the electron transport assembly, while others stimulate electron transport by discharging the proton gradient. For example, antimycin A is a specific inhibitor of cytochrome b. In the presence of antimycin A cytochrome b can be reduced but not oxidized. As expected, in the presence of cytochrome c remains oxidized in the presence of antimycin A, as do the downstream cytochromes a and a_3. An important class of antimetabolites are the uncou-

Table 18–1. Antimetabolites and their effects of oxidative phosphorylation.

Name	Function	Site of Action
Rotenone	e-transport inhibitor	Complex I
Amytal	e-transport inhibitor	Complex I
Antimycin A	e-transport inhibitor	Complex III
Cyanide	e-transport inhibitor	Complex IV
Carbon Monoxide	e-transport inhibitor	Complex IV
Azide	e-transport inhibitor	Complex IV
2,4,-dinitrophenol	Uncoupling agent	Trans membrane H^+ carrier
Pentachlorophenol	Uncoupling agent	Trans membrane H^+ carrier
Oligomycin	Inhibits ATP synthase	OSCP fraction of ATP synthase

pling agents exemplified by 2,4-dinitrophenol (DNP). Uncoupling agents act as lipophilic weak acids, associating with protons on the exterior of mitochondria, passing through the membrane with the bound proton, and dissociating the proton on the interior of the mitochondrion. These agents cause maximum respiratory rates but the electron transport generates no ATP, since the translocated protons do not return to the interior through ATP synthase. Table 18–1 contains a list of antimetabolites that have been useful in elucidating the process of oxidative phosphorylation.

ENERGY CONSERVATION FROM CYTOSOLIC NADH

In contrast to oxidation of mitochondrial NADH, **cytosolic NADH** gives rise to 2 equivalents of ATP if it is oxidized by the **glycerol phosphate shuttle** and 3 ATPs if it proceeds via the **malate aspartate shuttle**. As illustrated in Figure 18–7, the glycerol phosphate shuttle is coupled to an inner mitochondrial membrane, FAD-linked dehydrogenase, of low energy potential like that found in Complex II.

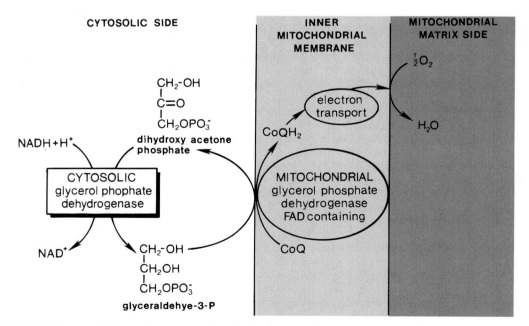

Figure 18–7. Glycerophosphate shuttle for transfer of reducing equivalents from the cytosol into the mitochondrion. (Reproduced, with permission, from Murray, RK: *Harper's Biochemistry*, 23rd ed. Appleton & Lange, 1993.)

Thus, cytosolic NADH oxidized by this pathway can generate only 2 equivalents of ATP. The shuttle involves two different glycerol-3-phosphate dehydrogenases: one is cytosolic, acting to produce glycerol-3-phosphate, and one is an integral protein of the inner mitochondrial membrane that acts to oxidize the glycerol-3-phosphate produced by the cytosolic enzyme. The net result of the process is that reducing equivalents from cytosolic NADH are transferred to the mitochondrial electron transport system. The catalytic site of the mitochondrial glycerol phosphate dehydrogenase is on the outer surface of the inner membrane, allowing ready access to the product of the second, or cytosolic, glycerol-3-phosphate dehydrogenase.

In some tissues, such as that of heart and muscle, mitochondrial glycerol-3-phosphate dehydrogenase is present in very low amounts, and the malate aspartate shuttle is the dominant pathway for aerobic oxidation of cytosolic NADH. In contrast to the glycerol phosphate shuttle, the malate aspartate shuttle generates 3 equivalents of ATP for every cytosolic NADH oxidized. The key components of the malate aspartate shuttle are illustrated in Figure 18–8. It is important to recognize that a fundamental reason for the pathway to operate as shown is that oxaloacetate is impermeable to the mitochondrial membrane.

In action, NADH efficiently reduces OAA to malate via cytosolic malate dehydrogenase (MDH). Malate is transported to the interior of the mitochondrion via the αKG/malate antiporter. Inside the mitochondrion, malate is oxidized by the MDH of the Krebs cycle, producing OAA and NADH. In this step the **cytosolic, NADH-derived reducing equivalents** become available to the NADH dehydrogenase of the inner mitochondrial membrane and are oxidized, giving rise to 3 ATPs as described earlier. The mitochondrial transaminase uses glutamate to convert membrane-impermeable OAA to aspartate and αKG. This provides a pool of αKG for the αKG/malate antiporter, and the Asp is translocated out of the mitochondrion via the Asp/Glu antiporter.

OXYGEN, OXIDASES, AND OXYGENASES

Oxidase complexes, like cytochrome oxidase, transfer electrons directly from NADH and other substrates to oxygen, producing water. *Oxygenases,* widely localized in membranes of the endoplasmic reticulum, catalyze the addition of molecular oxygen to organic molecules. There are 2 kinds of oxygenase complexes, monooxygenases and dioxygenases. Dioxygenases add the 2 atoms of molecular oxygen (O_2) to carbon and nitrogen of organic compounds. Monooxygenase complexes play a key role in detoxifying drugs and other compounds (eg, PCBs and dioxin) and in the normal metabolism of steroids, fatty acids and fat soluble vitamins. As illustrated in Figure 18–9, monooxygenases act by se-

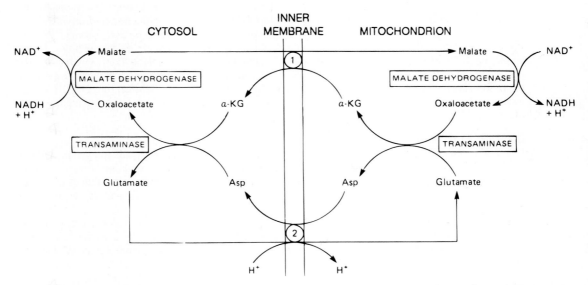

Figure 18–8. Malate shuttle for transfer of reducing equivalents from the cytosol into the mitochondrion. ① Ketoglutarate transporter; ② glutamate-aspartate transporter (note the proton symport with glutamate). (Reproduced, with permission, from Murray, RK: *Harper's Biochemistry,* 23rd ed. Appleton & Lange, 1993.)

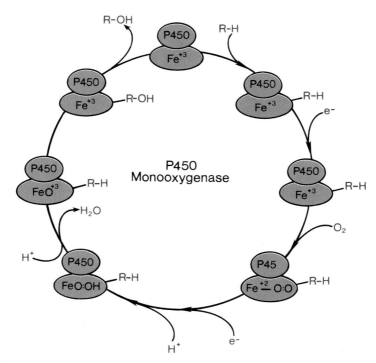

Figure 18–9. P450 hydroxylase (Monooxygenase) activity. Substrates to be hydroxylated bind to oxidized P450 with subsequent reduction of the heme iron and addition of O_2. Addition of a second reducing equivalent (e^-) and H^+ leads to production of water as the first product and leaving the reactive heme complex FeO^{3+}. The latter oxidizes the bound RH generating the hydroxylated second product, ROH. Fe signifies the redox state of the heme iron of P450.

quentially transferring 2 electrons from NADH or NADPH to 1 of the 2 atoms of oxygen in O_2, generating H_2O from 1 oxygen atom and incorporating the other oxygen atom into an organic compound as a hydroxyl group (ROH). The hydroxylated products are markedly more water-soluble than their precursors and are much more readily excreted from the body. Widely used synonyms for the monooxygenases are: mixed function oxidases, hydroxylases, and mixed function hydroxylases.

The chief components of monooxygenase complexes include cytochrome b_5, cytochrome P450, and cytochrome P450 reductase, which contains FAD plus FMN. There are many P450 isozymes; for example, up to 50 different P450 gene products can be found in liver, where the bulk of drug metabolism occurs. Some of these same gene products are also found in other tissues, where they are responsible for tissue-specific oxygenase activities. P450 reducing equivalents arise either from NADH via cytochrome b_5 or from NADPH via cytochrome P450 reductase, both of which are associated with cytochrome P450 in the membrane-localized complexes.

Oxygen Free Radicals

Enzymatic reactions involving molecular oxygen usually produce water or organic oxygen in well regulated reactions having specific products. However, under some metabolic conditions (eg, reperfusion of anaerobic tissues) unpaired electrons gain access to molecular oxygen in unregulated, non-enzymatic reactions. The products, called free radicals, are quite toxic. Their names and the reactions that produce them are summarized in Table 18–2.

These free radicals, especially hydroxy radical, randomly attack all cell components, including proteins, lipids and nucleic acids, potentially causing extensive cellular damage. Tissues are replete with enzymes to protect against the random chemical reactions that these free radicals initiate. The system of free radical scavengers is presented below.

Superoxide Dismutase (SOD)

Animals have Zn^{++} and Cu^{++} containing SOD. This converts superoxide to peroxide (Equation 18.13) and thereby minimizes production of hydroxy radical, the most potent of the oxygen free radicals

Table 18–2. Generation of various oxygen free radicals.

Oxygen Free Radicals		
	Nonenzymatic Reaction	**Product**
1.	$O_2 + e^- \rightarrow O_2^{\bullet}$	Superoxide
2.	$O_2^{\bullet} + H^+ \rightarrow HO_2^{\bullet}$	Hydroxy peroxy radical
3.	$H^+ + O_2^{\bullet} + HO_2^{\bullet} \rightarrow H_2O_2 + O_2$	Hydrogen peroxide
4.	$H_2O_2 + e^- \rightarrow OH^- + OH^{\bullet}$	Hydroxide + hydroxy radical

GLUTATHIONE

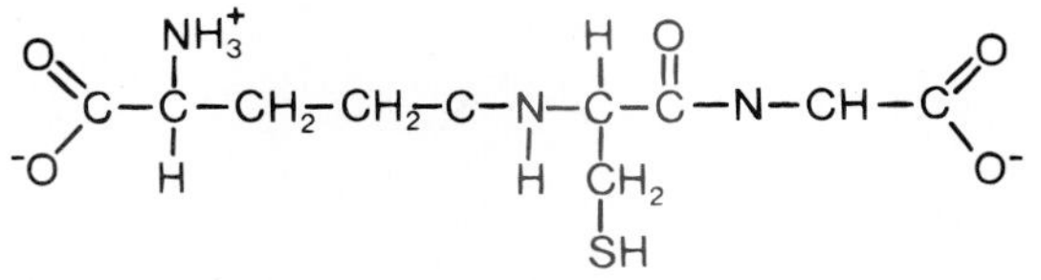

Figure 18–10. Glutathione (γ-glytamyl-cysteinyl-glycine). (Reproduced, with permission, from Murray, RK: *Harper's Biochemistry*, 23rd ed. Appleton & Lange, 1993.)

$$O_2^{\bullet} + O_2^{\bullet} \quad ^{\text{SOD}} \quad H_2O_2 + O_2 \qquad \qquad 18.13$$

Peroxidases

Peroxides produced by SOD are also toxic. They are detoxified by conversion to water via the enzyme peroxidase, which requires a source of reducing equivalents, as shown in Equation 18.14.

$$H_2O_2 + SH_2 \quad ^{\text{Peroxidase}} \quad 2H_2O + S \qquad \qquad 18.14$$

The best known mammalian peroxidase is glutathione peroxidase, which contains selenium as a prosthetic group.

Glutathione (Figure 18–10) is important in maintaining the normal reduction potential of cells and provides the reducing equivalents for glutathione peroxidase to convert hydrogen peroxide to water. In red blood cells the lack of glutathione leads to extensive peroxide attack on the plasma membrane, producing fragile red blood cells that readily undergo hemolysis.

Catalase

Localized in peroxisomes, catalase provides a redundant route for the degradation of hydrogen peroxide. Mammalian catalase has one of the highest turnover numbers of any documented enzyme.

$$2H_2O_2 \quad ^{\text{Catalase}} \quad 2H_2O + O_2 \qquad \qquad 18.15$$

Antioxidant Properties of Vitamin E (Alpha-tocopherol)

Membrane phospholipids containing polyunsaturated fatty acids (PUFA) are continuously under attack by free radicals. Vitamin E solubilized in membrane lipid bilayers acts as a free radical scavenger, protecting sensitive membrane lipids from free radical damage.

Questions

DIRECTIONS: Select the ONE lettered phrase that offers the BEST completion of this statement.

1. Components of the mitochondrial electron transport system are not found in:
   a. Cytochrome C.
   b. Complex I.
   c. Complex II.
   d. Complex III.
   e. $F_1/F_0/OSCP$ complex.

DIRECTIONS (items 2–5): Each numbered sentence describes a compound that alter mitochondrial activities. Select the ONE lettered answer corresponding to the compound that BEST fits the description. Each compound may be used more than once or not at all.

    a. Dinitrophenol
    b. Carbon monoxide
    c. Antimycin A
    d. Oligomycin

2. This compound prevents electron transport through the cytochrome b-cytochrome c_1 complex.

3. This compound makes the mitochondrial membrane permeable to protons, stimulates O_2 consumption, and inhibits ATP synthesis.

4. This compound binds to heme iron of cytochrome oxidase.

5. This compound binds and inhibits ATP synthase.

DIRECTIONS (items 6–13): Each numbered item or incomplete statement is followed by answers or by completions of the statement. Select the ONE lettered answer or completion that is BEST in each case.

6. In cases of acute cyanide poisoning, cyanide binds to the Fe^{3+} of a cytochrome. Which statement best describes this cytochrome?
    a. It directly oxidizes cytochrome b.
    b. It reduces cytochrome c_1.
    c. It is directly reduced by cytochrome c.
    d. It is localized in the mitochondrial matrix.
    e. It binds carbon monoxide.

7. Which of the following statements best characterizes ATP Synthase?
    a. It is found dissolved in the mitochondrial matrix.
    b. Its catalytic function is to synthesize ATP in a reaction driven by a chemiosmotic potential.
    c. It couples ATP export from the mitochondrial matrix to ATP synthesis.
    d. For activity it requires a low H^+ ion concentration outside the mitochondria.
    e. Oligomycin binds to ATP synthase, directly preventing ATP export from the mitochondria.

8. Nicotinamide nucleotides:
    a. Are carriers of reducing equivalents that are generally used to drive ATP production via a chemiosmotic potential.
    b. Go from the oxidized state to the reduced state by the addition of 2 protons and 2 electrons to one of their heterocyclic rings.
    c. Are key carriers of reducing equivalents used for driving reductive biosynthetic reactions.
    d. Are unlike adenosine triphospho nucleotides in that they lack phosphoanhydride bonds.
    e. Are best described by 2 of the above.

9. Regarding normal physiological oxidative phosphorylation, it can be stated that:
    a. The ATP:2e$^-$ is always very close to 3.0.
    b. Electron transport generates a "pH gradient" that provides all the energy used to form phosphoanhydride bonds.
    c. An intact inner mitochondrial membrane is required for oxidative phosphorylation to occur.
    d. ATP synthase function depends solely on the electrical potential generated by electron transport.
    e. The ATP:2e$^-$ depends on the H^+:2e$^-$.

10. How many ATPs are formed for each extramitochondrial NADH that is oxidized to NAD+ by O_2 via the glycerol-phosphate shuttle?

    a. 1.
    b. 2.
    c. 3.
    d. 4.
    e. None of the above.

11. Coenzyme Q and flavin nucleotides:
    a. Can exist as free radical species.
    b. Are derived from vitamins.
    c. Are found freely soluble in the lipid phase of the inner mitochondrial membrane.
    d. Contain heterocyclic rings.
    e. Are generally found in the mitochondrial matrix.

12. The following hemeproteins have the 6th coordinate position of the iron available for chemical reaction:
    a. Cytochrome a_3.
    b. Cytochrome a_3, and Cytochrome c_1.
    c. Cytochrome a_3, Cytochrome c_1, and hemoglobin.
    d. Cytochrome a_3, myoglobin, and hemoglobin.
    e. Cytochrome c_1, hemoglobin, and myoglobin.

13. Cytosolic reducing equivalents cannot gain access to the mitochondrial electron transport chain via:
    a. The malate/aspartate shuttle.
    b. A FAD-linked dehydrogenase.
    c. NADH transport across the inner membrane.
    d. The glycerol phosphate shuttle.

DIRECTIONS (items 14–16): Match the following numbered reactions with the correct enzyme from the lettered list.

    a. P450
    b. Peroxidase
    c. Catalase
    d. Cytochrome oxidase $(a + a_3)$
    e. Superoxide dismutase

14. $Fe^{+2} + O_2 + 4H^+ \xrightarrow{E} H_2O + Fe^{+3}$.

15. $2H_2O_2 \xrightarrow{E} 2H_2O + O_2$.

16. $\overset{\cdot\cdot}{{}_2} + O_2^{\cdot} + H^+ \xrightarrow{E} H_2O_2 + O_2$.

DIRECTIONS (items 17–18): Each numbered item or incomplete statement is followed by answers or by completions of the statement. Select the ONE lettered answer or completion that is BEST in each case.

17. Mitochondrial heme proteins have the following characteristics:
    a. With two exceptions, the 6th coordinate position of heme prosthetic groups are normally blocked from interaction with molecular oxygen.
    b. The most frequently found ligand in the 5th and 6th coordinate position of mitochondrial heme proteins is a cysteine sulfur.
    c. The redox transitions in mitochondrial heme proteins are accounted for by iron undergoing transitions from Fe^{+3} to Fe^{+4}.
    d. With one exception, in mitochondrial heme proteins, the ratio of heme prosthetic group to apoprotein is 1.
    e. The heme of all mitochondrial hemeproteins has the same structure as the heme of myoglobin or hemoglobin.

18. The drug detoxifying P450 electron transport system of the endoplasmic reticulum is characterized by all of the following except:
    a. Its oxygen binding protein reduces each equivalent of O_2 to 2 equivalents of H_2O in a single process involving a 4-electron transition.
    b. The system includes P450 and 2 different electron donors.
    c. A product of its activity is H_2O.
    d. Its action most often produces an oxygen-containing molecule that is not water.
    e. P450 is a heme protein.

Answers

1. e	7. b	13. c
2. c	8. e	14. d
3. a	9. c	15. c
4. b	10. b	16. e
5. d	11. a	17. d
6. e	12. d	18. a

19   Amino Sugars and Glycoconjugates

Objectives

- To describe the physiologically relevant amino sugars.
- To describe the clinical significance of glucuronic acid.
- To describe the composition of glycoproteins.
- To describe O- and N-linkages of sugars in glycoproteins.
- To define the three major classes of N-linked sugar structures in glycoproteins.
- To explain the clinical significance of carbohydrate modification of proteins and the role of glycoproteins in lysosomal storage diseases.

Concepts

The amino sugars—**N-acetylgalactosamine (GalNAc), N-acetylglucosamine (GlcNAc) and N-acetylneuraminic acid (NANA)**—are important constituents of the glycoproteins, gangliosides and glycosaminoglycans (Chapters 10 and 20). The complex glycoconjugates are needed for proper cell-cell adhesion and extracellular matrix construction to ensue tissue integrity. Equally important is the role of glycoconjugates in cell-cell and immunological self-recognition. Another important carbohydrate is glucuronic acid, which is incorporated into proteoglycans and is necessary for the clearance of bilirubin, steroid hormones, and certain drugs.

    The carbohydrate found in membranes exists in one form only: oligosaccharides covalently attached to proteins to form glycoproteins, and to a lesser extent covalently attached to lipid and forming the glycolipids (see Chapter 10). Glycoproteins consist of proteins covalently linked to carbohydrate. The distinction between proteoglycans (Chapter 20) and glycoproteins lies in the level and types of carbohydrate modification. Glycoproteins contain less complex carbohydrate modification.

Physiologically Important Amino Sugars

Glucose can serve as the carbon skeleton for all other sugars. During the synthesis of the amino sugars, **fructose-6-phosphate** serves as the immediate acceptor of the amide nitrogen (donated from glutamine). This irreversible reaction, which represents the committed step in the synthesis of amino sugars, generates **glucosamine-6-phosphate**; it is catalyzed by **glutamine:fructose-6-phosphate aminotransferase**. The glucosamine 6-phosphate is then converted to **UDP-GlcNAc** by acetyl-CoA mediated acetylation and a **mutase** catalyzedreaction that converts the 6-phosphate to 1-phosphate. Finally, UDP-GlcNAc serves as the precursor for the synthesis of **GalNAc** and the sialic acids, particularly **NANA**. The interrelationships of the different amino sugars are depicted in Figure 19–1.

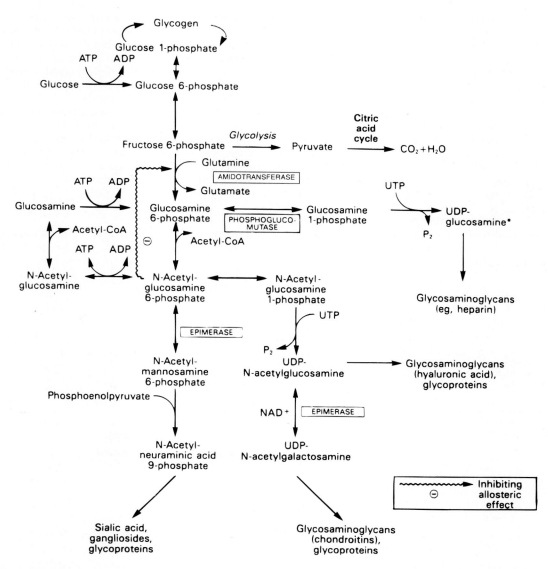

Figure 19–1. A summary of the interrelationships in metabolism of amino sugar. *Analogous to UDPGlc. Other purine or pyrimidine necleotides may be similarly linked to sugars or amino sugars. Examples are thymidine diphosphate (TDP)-glucosamine and TDP-N-acetylglucosamine. (Reproduced, with permission, from Murray, RK: *Harper's Biochemistry*, 23rd ed. Appleton & Lange, 1993.)

Glucuronic Acid

Glucuronate is a highly polar molecule which is incorporated into proteoglycans as well as combining with bilirubin and steroid hormones; it can also be combined with certain drugs to increase their solubility. Glucuronate is synthesized in the **uronic acid pathway**, an alternative pathway for the oxidation of glucose that does not provide a means of producing ATP, but is utilized for the generation of the activated form of glucuronate, **UDP-glucuronate**. The conversion of glucose to glucuronate begins through the action of several hepatic enzymes that have been discussed earlier (Chapter 14). Glucose-6-phosphate is converted to glucose-1-phosphate by phosphoglucomutase, and then activated to UDP-glucose by **UPD-glucose pyrophosphorylase**. UDP-glucose is oxidized to **UDP-glucuronate** by the NAD$^+$ requiring enzyme, **UDP-glucose dehydrogenase** (Figure 19–2).

CLINICAL SIGNIFICANCE

In the human adult, a significant number of erythrocytes die each day. This turnover releases significant amounts of the iron-free portion of heme, **porphyrin**, which is subsequently degraded (Chapter 26). The primary sites of porphyrin degradation are found in the reticuloendothelial cells of the liver, spleen and bone marrow. The breakdown of porphyrin yields **bilirubin**, a product that is non-polar and therefore insoluble. In the liver, to which it is transported in the plasma bound to albumin, bilirubin is solubilized by conjugation to glucuronate. The soluble **conjugated bilirubin diglucuronide** is then secreted into the bile. An inability to conjugate bilirubin, for instance in hepatic disease or when the level of bilirubin production exceeds the capacity of the liver, is the root cause of jaundice.

The conjugation of glucuronate to certain non-polar drugs is important for their solubilization in the liver; glucuronate conjugated drugs are more easily cleared from the blood by the kidneys for excretion in the urine. The glucuronate-drug conjugation system can, however, lead to drug resistance: chronic exposure to certain drugs, such as barbiturates and AZT, leads to an increase in the synthesis of the **UDP-glucuronosyl transferases** in the liver that are involved in glucuronate-drug conjugation. The increased levels of these hepatic enzymes results in a higher rate of drug clearance leading to a reduction in the effective dose of glucuronate cleared drugs.

Glycoproteins

Membranes associated carbohydrate is exclusively in the form of oliogsaccharides covalently attached to proteins forming **glycoproteins**, and to a lesser extent covalently attached to lipid forming the glycolipids (see Chapter 10). Glycoproteins consist of proteins covalently linked to carbohydrate. The pre-

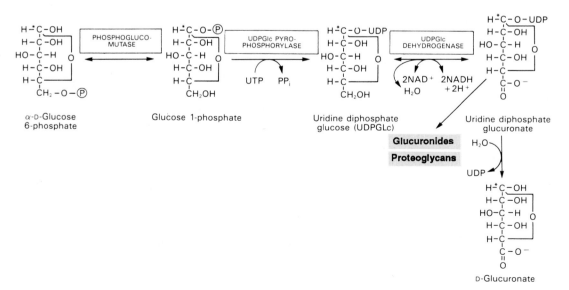

Figure 19–2. Synthesis of D-Glucuronate from α-D-Glucose-6-phosphate. (Reproduced, with permission, from Murray, RK: *Harper's Biochemistry*, 23rd ed. Appleton & Lange, 1993)

dominant sugars found in glycoproteins are **glucose, galactose, mannose, fucose, GalNAc, GlcNAc and NANA**. The distinction between proteoglycans (Chapter 9) and glycoproteins is in the level and types of carbohydrate modification. The carbohydrate modifications found in glycoproteins are rarely comples: carbohydrates are linked to the protein component through either **O-glycosidic or N-glycosidic** bonds (Figure 19–3). The O-glycosidic linkage is to the hydroxyl of S, T, or hydroxylysine. The linkage of carbohydrate to hydroxylysine is generally only found in the collagens. The linkage of carbohydrate to 5-hydroxylysine is either the single sugar galactose or the disaccharide glucosylgalactose. The N-glycosidic linkage is through the amide group of N (Figure 19–3).

In S- and T-type O-linked glycoproteins, the carbohydrate directly attached to the protein is GalNAc. In N-linked glycoproteins, it is GlcNAc. The predominant carbohydrate attachment in glycoproteins of mammalian cells is via N-glycosidic linkage. The site of carbohydrate attachment to N-linked glycoproteins is found within a consensus sequence of amino acids, N-X-S(T), where X is any amino acid except P. N-linked glycoproteins all contain a common core of carbohydrate attached to the polypeptide. This core consists of three mannose residues and 2 GlcNAc. A variety of other sugars are attached to this core and comprise three major N-linked families (illustrated in Figure 19–4):

1. High-mannose type contains all mannose outside the core in varying amounts.
2. Hybrid type contains various sugars and amino sugars.
3. Complex type is similar to the hybrid type but, in addition, contains sialic acids to varying degrees.

Most proteins that are secreted or bound to the plasma membrane are modified by carbohydrate attachment. The part that is modified, in plasma membrane-bound proteins, is the extracellular portion. Intracellular proteins are less frequently modified by carbohydrate attachment.

Mechanism of Carbohydrate Linkage to Proteins

The protein component of all glycoproteins is synthesized from polyribosomes that are bound to the endoplasmic reticulum (ER). The processing of the sugar groups occurs **cotranslationally** in the lumen

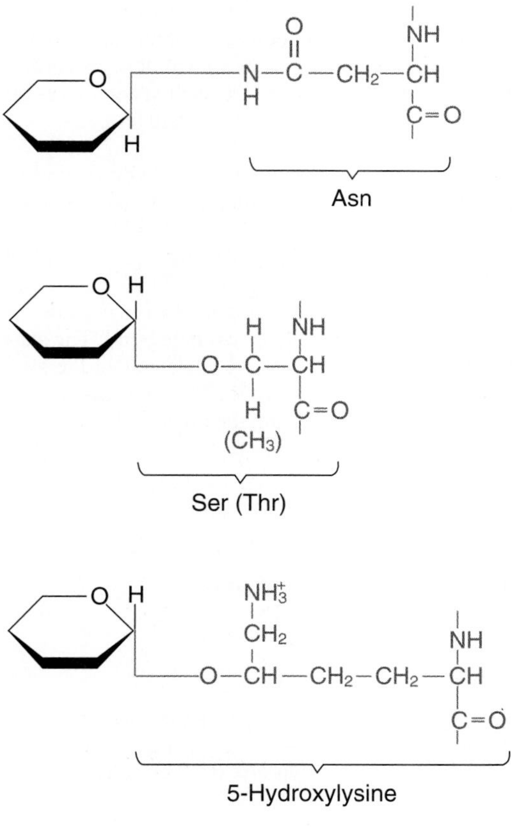

Figure 19–3. Types of glycosidic bonds in glycoproteins. (a) N-linked glycosidic bond to an Asn residue in the sequence Asn-X-Ser/Thr. (b) O-linked glycosidic bond to a Ser (or Thr) residue. (c) O-linked glycosidic bond to a 5-hydroxylysine residue in collagen. (Reproduced, with permission from Voet, D. and Voet, JG: *Biochemistry*, 1st ed. John Wiley and Sons, Inc. 1990.)

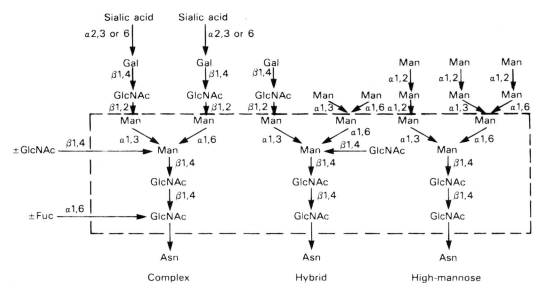

Figure 19–4. Structures of the major types of asparagine-linked oligosaccharides. The boxed area encloses the pentasaccharide core common to all N-linked glycoproteins. (Reproduced, with permission, from Kornfeld R, Kornfeld S: Assembly of asparagine-linked oligosaccharides. *Annu Rev Biochem* 1985;54:631.)

of the ER and continues in the Golgi apparatus for N-linked glycoproteins. Attachment of sugars in O-linked glycoproteins occurs **post-translationally** in the Golgi apparatus.

A. O-linked Sugars: The synthesis of O-linked glycoproteins occurs via the stepwise addition of nucleotide-activated sugars directly onto the polypeptide. The nucleotide-activated sugars are coupled to either UDP, GDP (as with mannose) or CMP (for instance, NANA). The attachment of sugars is catalyzed by specific **glycoprotein glycosyltransferases**. Evidence indicates that each specific type of carbohydrate linkage in O-linked glycoproteins is the result of a different glycosyltransferase.

B. N-linked Sugars: As indicated earlier, the three major classes of N-linked carbohydrate modifications are high-mannose, hybrid and complex (Figure 19–4). The major distinguishing feature of the complex class is the presence of sialic acid, whereas the hybrid class contains no sialic acid.

In contrast to the step-wise addition of sugar groups to the O-linked class of glycoproteins, N-linked glycoprotein synthesis requires a lipid intermediate: **dolichol phosphate** (Figure 19–5). Dolichols are polyprenols (C80–C100 containing 16 to 20 isoprene units, in which the terminal unit is saturated.

The oligosaccharide unit is attached to dolichol phosphate through a pyrophosphate bond. The sugars used for N-linked glycoprotein synthesis are activated by coupling to nucleotides, as in the synthesis of O-linked glycoproteins. GlcNAc is coupled to UDP, and Man to GDP. The first reaction involves the formation of **GlcNAc-P-P-dolichol**, with the release of UMP from the nucleotide-activated

$$HO-CH_2-CH_2-\overset{\overset{\displaystyle H}{|}}{\underset{\underset{\displaystyle CH_3}{|}}{C}}-CH_2\left[CH_2-CH=\overset{\overset{\displaystyle CH_3}{|}}{C}-CH_2\right]_n CH_2-CH=\overset{\overset{\displaystyle CH_3}{|}}{C}-CH_3$$

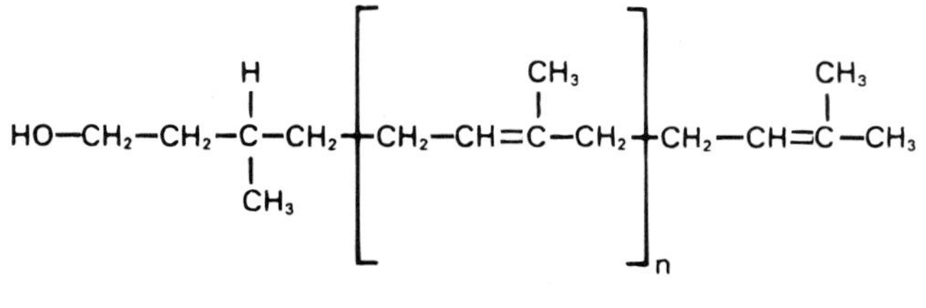

Figure 19–5. The structure of the dolichol. The phospate in dolichol phosphate is attached to the primary alcohol group at the left-hand end of the molecule. The group within the brackets in an isoprene unit (n = 17 − 20 isoprenoid units. (Reproduced, with permission, from Murray, RK: *Harper's Biochemistry*, 23rd ed. Appleton & Lange, 1993.)

sugar, UDP-GlcNAc (Figure 19–6). The second GlcNAc and Man transferase reactions proceed via sugar transfer from the nucleotide-activated sugar directly to GlcNAc-P-P-dolichol unit (Figure 19–6). From this point, additional mannose are added to the Man-GlcNAc-GlcNAc-P-P-dolichol by transfer from Man-P-dolichol; formed from dolichol phosphate and GDP-Man (Figure 19–6). Once the oligosaccharide core unit is complete, it is transferred to an N residue in the protein. As indicated above, this N residue is found within N-X-S(T) consensus sequence.

After the oligosaccharide core is transferred to the protein, additional modifications take place through the action of glycosyltransferases as well as through the removal of certain glycosyl residues. These modifications occur as the protein migrates through the Golgi apparatus to the cell surface (Figure 19–7).

Enzymes that are destined for the **lysosomes** (lysosomal enzymes) are directed there by a specific carbohydrate modification. During transit through the Golgi apparatus, a residue of GlcNAc-1-P is added to carbon 6 of one or more specific Man residues that have been incorporated into these enzymes (Figure 19–7). The GlcNAc is activated by coupling to UDP and is transferred by a **GlcNAc phosphotransferase**, yielding **GlcNAc-1-P-6-Man-protein**. A second reaction removes the GlcNAc, leaving Man residues phosphorylated in the 6 position: **P-6-Man-protein**. A specific Man-6-phosphate receptor is present in the membranes of the Golgi apparatus. It is the binding of Man-6-phosphate to this receptor that targets proteins to the lysosomes.

Two distinct Man-6-P receptors have been identified. The receptor present in the Golgi apparatus that is responsible for lysosomal targeting of enzymes is a 46k Da protein. The cellular receptor for the growth factor, **insulin-like growth factor II (IGF-II)**, is also a Man-6-P receptor. The IGF-II-M6P receptor is a 300 kDa protein.

CLINICAL SIGNIFICANCE

Glycoproteins on cell surfaces are important for communication between cells, for maintaining cell structure and for self-recognition by the immune system. The alteration of cell-surface glycoproteins can, therefore, produce profound physiological effects, of which several are listed below.

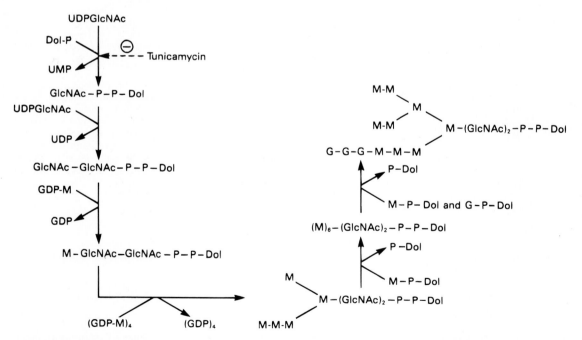

Figure 19–6. Pathways of biosynthesis of dolichol-P-P-oligosaccharide. Note that the internal mannose residues are donated by GDP-mannose, whereas the more external mannose residues and the glucose residues are donated by dolichol-P-mannose and dolichol-P-glucose. UDP, uridine diphosphate; Dol, dolichol; P, phosphate; UMP, uridine monophosphate; GDP, guanosine diphosphate, M, mannose; G, glucose. (Reproduced, with permission, from Murray, RK: *Harper's Biochemistry*, 23rd ed. Appleton & Lange, 1993.)

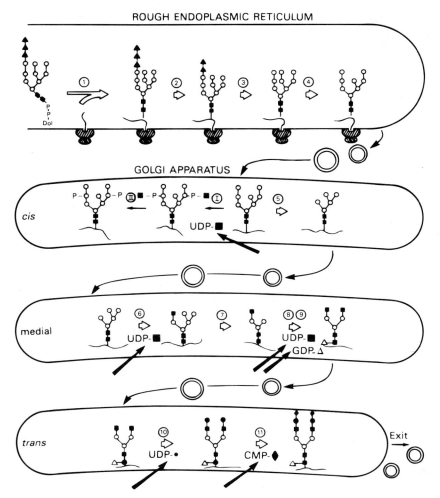

ROUGH ENDOPLASMIC RETICULUM

GOLGI APPARATUS

Figure 19–7. Schmatic pathways of oligosaccharide processing. The reactions are catalyzed by the follow-ing enzymes: ①, oligosacchayltransferase; ②, α-glucosidase I; ③, α-glucosidase II; ④, endoplasmic reticu-lum α1, 2 mannosidase; I, N acetylglucosaminylphosphotransferase; II, N-acetylglucosamine-1-phosphodi-ester α-N-Acetylglucosaminidase; ⑤, Golgi apparatus αmannosidase I; ⑥, N-acetylglucosaminyltransferase I; ⑦, Golgi apparatus αmannosidase II; ⑧, N-acetyl-glucosaminyltransferase II; ⑨, fucosyltransferase; ⑩, galactosyltransferase; ⑪, sialyltransferase. ■ N-acetylglucosamine; O, mannose; ▲, glucose; △, fucose; ●, galactose; ◆, sialic acid. (Reproduced, with permission, from Kornfeld R, Kornfeld S: Assembly of as-paragine-linked oligosaccharides. *Annu Rev Biochem* 1985:631.)

A. Truncation: The truncation of erythrocyte surface glycoproteins leads to cell clumping, as in HEMPAS (congenital dyserythropoietic anemia type II).

B. Infiltration: Several viruses, bacteria and parasites have exploited the presence of cell-surface glycoproteins, using attachment to them as portals of entry into the cell.

1. Rhinoviruses–Rhinoviruses utilize attachment to **ICAM-1** (intracellular adhesion molecule-1) to gain entry into cells.

2. B19–The pathogenic human parvovirus B19, attaches to the erythrocyte-specific cell surface glo-boside identified as **erythrocyte P antigen** to infect erythrocytes.

3. Plasmodium vivax–The malarial parasite *Plasmodium vivax* binds to the erythrocyte **chemokine receptor** known as the **Duffy blood group antigen** (also known as the erythrocyte receptor for interleukin-8) to infect erythrocytes.

4. MN blood group system–This system is a well-characterized set of erythrocyte surface antigens that represent the variable carbohydrate modifications of the trans-membrane glycoprotein, **glycophorin**. Glycophorin is the cellular receptor for influenza virus as well as the receptor for erythrocyte invasion by the malarial parasite *Plasmodium falciparum*.

5. Helicobacter pylori–This is the bacterium responsible for chronic active gastritis and gastric and duodenal ulcers; it is also the causative agent for one of the most common forms of cancer in humans, adenocarcinomas. This bacterium attaches to the **Lewis blood group antigen** on the surface of gastric mucous cells.

6. Fibroblast growth factor–The receptor for fibroblast growth factor (**FGF**) has been reported to be the portal of entry for human **herpes virus Type I**.

C. Lipid Linkage: Some glycoproteins are tethered to the membrane by a lipid linkage; the protein is attached to the carbohydrate through phosphatidylethanolamine (PE) linkage, and the carbohydrate is in turn attached to the membrane via linkage to phosphatidylinositol (PI), which anchors the structure within the membrane. The linkage is called a glycosylphosphotidylinositol (GPI) anchor, and proteins that are anchored in this way are termed **glypiated proteins**. The disease of **paroxysmal nocturnal hemoglobinuria** results from the presence of glypiated proteins free in serum. One such protein, **decay-accelerating factor**, is present on the surface of erythrocytes and prevents erythrocyte lysis by complement. When this factor is lost from the erythrocyte surface, abnormal hemolysis occurs, with the end result of hemoglobin accumulation in the urine.

D. Glycoprotein Degradation: Degradation of the glycoproteins is as medically relevant as is their synthesis (see Table 19–1). Degradation occurs within lysosomes and requires specific lysosomal hydrolases termed **glycosidases**. **Exoglycosidases** remove sugars sequentially from the non-reducing end and exhibit restricted substrate specificities. In contrast, **endoglycosidases** cleave carbohydrate linkages from within and exhibit broader substrate specificities. Several inherited disorders involving the abnormal storage of glycoprotein degradation products have been identified in humans. These disorders result from defects in the genes encoding specific glycosidases, leading to incomplete degradation and subsequent over-accumulation of partially degraded glycoproteins. As a general class, such

Table 19–1. Enzyme defects in degradation of Asn-GlcNAc-type glycoproteins.

Disease	Enzyme Deficiency	Symptoms
Aspartylglycosaminuria	Aspartylglycosaminidase	Progressive mental retardation, delayed speech and motor development, coarse facial features
β-Mannosidosis	β-Mannosidase	Primarily neurological defects, speech impairment
α-Mannosidosis	α-Mannosidase	Mental retardation, dystosis multiplex, hepatosplenomegaly, hearing loss, delayed speech
G$_{M2}$ Gangliosidosis (Sandhoff-Jatzkewitz disease)	β-N-acetylhexosaminidases A and B	See Chapter 10
G$_{M1}$ Gangliosidosis	β-Galactosidase	See Chapter 10
Sialidosis (also identified as Mucolipidosis 1)	Neuraminidase (sialidase)	Myoclonus, congenital ascites, hepatosplenomegaly, coarse facial features, delayed mental and motor development
Fucosidosis	α-Fucosidase	Progressive motor and mental deterioration, growth retardation, coarse facial features, recurrent sinus and pulmonary infections

disorders are known as **lysosomal storage diseases**. Since glycoproteins may share carbohydrate structures similar to those found in glycolipids, the defective degradation of these carbohydrates can lead to accumulation of both glycolipids and glycoproteins.

E. Glycoprotein Targeting Defects: Defects in the proper targeting of glycoproteins to the lysosomes can also lead to clinical complications. Deficiencies in the enzyme responsible for the transfer of GlcNAc-1-P to Man residues (**GlcNAc phosphotransferase**) in lysosomal enzymes leads to the formation of dense **inclusion bodies** in the fibroblasts. Two disorders related to deficiencies in the targeting of lysosomal enzymes are termed **I-cell disease (mucolipidosis II)** and **pseudo-Hurler polydystrophy (mucolipidosis III, also called mucolipidosis-HI)**. I-cell disease is characterized by severe psychomotor retardation, skeletal abnormalities, coarse facial features, painful restricted joint movement, and early mortality. Pseudo-Hurler polydystrophy is less severe: It progresses more slowly, and afflicted individuals live to adulthood.

Questions

DIRECTIONS (items 1–11): Each numbered item or incomplete statement is followed by answers or by completions of the statement. Select the ONE lettered answer or completion of statement that is BEST in each case.

1. Glycoproteins are characterized by:
    a. Carbohydrate as the dominant component.
    b. Carbohydrate as the dominant component with attached peptides.
    c. Carbohydrate as the dominant component with attached peptides in N-or O-glycosidic linkages to asparagine and serine.
    d. Protein as the dominant component.
    e. Protein as the dominant component with oligosaccharides attached in N-and O-glycosidic linkages to asparagine and serine.

2. Predominant carbohydrates found in mammalian glycoproteins include:
    a. NANA, glucose and mannose.
    b. Fructose, galactose and xylulose.
    c. NANA, galactose and fructose.
    d. GlcNAc, fructose and xylulose.
    e. GalNAc, ribose and galactose.

3. Hyperbilirubinemia may be associated with:
    a. Defective conjugation of bilirubin to glucuronate in the liver.
    b. Defective conjugation of bilirubin to glucuronate in the intestines.
    c. Excess resorption of bilirubin as a result of an increase in glucuronate conjugation.
    d. Excess renal excretion of bilirubin.
    e. Excess renal excretion of glucuronate.

4. A 13-year-old male is presented with the following symptoms: mental retardation, hepatosplenomegaly, coarse facial features and dystosis multiplex. These are most diagnostic of which disease?
    a. Aspartylglycosaminuria.
    b. Sandhoff-Jatzkewitz disease.
    c. Fucosidosis.
    d. α-Mannosidosis.
    e. Sialidosis.

5. I-cell disease is characterized by:
    a. Mental retardation, hepatosplenomegaly, and skeletal abnormalities.
    b. Psychomotor retardation, coarse facial features, and hepatomegaly.
    c. Coarse facial features, kidney failure, and skin rashes.

    d. Skeletal abnormalities, severe mental retardation, and blindness.
    e. Coarse facial features, skeletal abnormalities, and myelin deficiencies.

6. Enzymes that are destined for packaging within the lysosomes (lysosomal enzymes) are modified in such a way that they contain:
    a. GlcNAc-6-phosphate.
    b. GalNAc.
    c. Mannose-6-phosphate.
    d. Fucose.
    e. UDP-mannose.

7. The carbohydrate found attached directly to protein via an N-glycosidic linkage is:
    a. Fucose.
    b. Mannose.
    c. GlcNAc.
    d. GalNAc.
    e. NANA.

8. The precursor nitrogen source for the synthesis of the amino sugars is:
    a. Arginine.
    b. Aspartate.
    c. Glutamate.
    d. Glutamine.
    e. Asparagine.

9. Which of the following most closely describes the composition of the carbohydrate portion of glycoproteins of the complex type, excluding the common core sugars?
    a. Galactose, GalNAc and fructose.
    b. Galactose, GlcNAc and NANA.
    c. Mannose and GlcNAc.
    d. Mannose and GalNAc.
    e. Mannose, GlcNAc, GalNAc and fucose.

10. The role of dolichol phosphate is:
    a. To anchor proteins to membranes of the endoplasmic reticulum in order for carbohydrate modifications to be carried out.
    b. To act as a "scaffold" upon which the initial core of sugars of N-linked glycoproteins is first assembled before transfer to the protein.
    c. To anchor proteins to Golgi apparatus in order for carbohydrate modifications to be carried out.
    d. To "activate" proteins for initiation of carbohydrate modification.
    e. To act as a "scaffold" upon which the initial core of sugars of O-linked glycoproteins is first assembled before transfer to the protein.

11. I-cell disease is the result of a deficiency in:
    a. Neuraminidase, leading to an accumulation of G_{M3} gangliosides.
    b. β-galactosidase, leading to an accumulation of G_{M1} gangliosides.
    c. GlcNAc phosphotransferase, leading to abnormal targeting of lysosomal enzymes.
    d. Neuraminidase, leading to an accumulation of G_{M2} gangliosides.
    e. GlcNAc phosphotransferase, leading to defective targeting of lysosomal enzymes.

Answers

1. e	**5.** b	**9.** b
2. a	**6.** c	**10.** b
3. a	**7.** c	**11.** e
4. d	**8.** d	

20

Lipid Metabolism

Objectives

- To describe the origin of cytoplasmic acetyl-CoA used for the synthesis of fatty acids.
- To describe the relevant features of the biosynthesis of fatty acids.
- To describe how the body elongates and desaturates fatty acids.
- To describe the pathway of fatty acid oxidation.
- To describe the alternative pathways used for the oxidation of unsaturated and odd-chain length fatty acids.
- To describe the mobilization of fatty acids from stores within adipose tissue.
- To describe the mechanisms used to regulate fatty acid metabolism.
- To describe the clinical significance of altered fatty acid metabolism.
- To describe the pathway and regulation of ketogenesis.
- To describe the clinical significance of ketogenesis.
- To describe the metabolism of the complex lipids including triacylglycerols, phospholipids, plasmalogens and sphingolipids.
- To describe the synthesis of the eicosanoids.
- To define the clinical properties of the major eicosanoids.

Concepts

Fatty acids constitute biomolecules of major importance. They are oxidized by all cells, leading to the production of significant quantities of energy per mole. They also serve as precursors for the synthesis of the acylglycerols and glycosphingolipids, molecules that are the main lipid constituents of biological membranes.

The reactions of fatty acid synthesis and their oxidation are remarkably similar, forming two pathways that are reversals of each other. There are only two differences in the chemistry of these pathways. First of all, fatty acid synthesis occurs in the cytoplasm, whereas oxidation takes place in the mitochondria. Second, the two processes differ in their use of nucleotide cofactors: the synthesis of fats involves the oxidation of NADPH, whereas oxidation involves the reduction of $FADH^+$ and NAD^+. Both oxidation and synthesis of fats require an activated two-carbon intermediate, acetyl-CoA. In synthesis, though, the acetyl-CoA is temporarily bound to the enzyme complex as malonyl-CoA.

Origins of Acetyl-CoA for Synthesis

Acetyl-CoA is generated in the mitochondria primarily from two sources, the PDH reaction and fatty acid oxidation. In order for these acetyl units to be used for fatty acid synthesis, they must be present and available in the cytoplasm. The shift from fatty acid oxidation and glycolytic oxidation occurs when the cell's need for energy diminishes. This results in reduced oxidation of acetyl-CoA in the TCA cycle and the oxidative phosphorylation pathway. Under these conditions the mitochondrial acetyl units can be stored as fat for future energy demands.

Acetyl-CoA enters the cytoplasm in the form of citrate, via the **tricarboxylate transport** system (Figure 20–1). In the cytoplasm, citrate is converted to oxaloacetate (OAA) and acetyl-CoA by the **ATP-citrate lyase** reaction, which is driven by ATP. The acetyl-CoA is then available as a substrate for fat synthesis. The cycle continues with the conversion of OAA to malate by the action of **malate dehydrogenase**. The malate then undergoes oxidative decarboxylation to pyruvate by **malic enzyme**. The co-enzyme for this reaction is $NADP^+$, which generates NADPH. An advantage of this series of reactions for converting mitochondrial acetyl-CoA into cytoplasmic acetyl-CoA is that the NADPH produced by the malic enzyme reaction can be a major source of reducing co-factor for the fatty acid synthesis. The other major source of NADPH is the pentose phosphate pathway (Chapter 16). By this

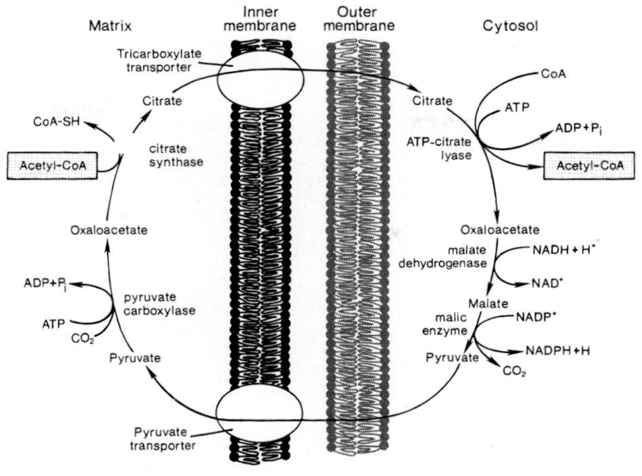

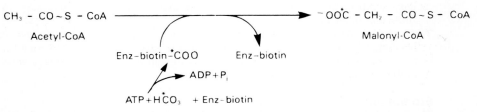

Figure 20–1. Pathway for the movement of acetyl-CoA from inside the mitochondria to the cytosol of the cell for use in fatty acid (and cholesterol) biosynthesis.

pathway, the pyruvate can re-enter the mitochondria, where it is converted to OAA through the action of pyruvate carboxylase. The cycle can then continue.

Synthetic Pathway for Saturated Fatty Acids

The synthesis of malonyl-CoA through the carboxylation of acetyl-CoA is the first and the **committed step** of fatty acid synthesis (Figure 20–2). This reaction is catalyzed by **acetyl-CoA carboxylase (ACC)**, the major site of regulation of fatty acid synthesis.

The rate of fatty acid synthesis is controlled by the equilibrium between monomeric ACC and polymeric ACC. The activity of ACC requires polymerization. This conformational change is enhanced by citrate and inhibited by long-chain fatty acids. ACC also is controlled through hormone mediated phosphorylation (see below).

The synthesis of fatty acids from malonyl-CoA is carried out by **fatty acid synthase, FAS** (Figure 20–3). The active forms of FAS is a dimer of identical subunits. All the reactions of fatty acid synthesis are carried out by the multiple enzymatic activities of FAS. The full process involves four domains of enzymatic activity: **3-ketoacyl synthase (also called condensing enzyme), 3-ketoacyl reductase, 3-OH ketoacyl dehydratase and enoyl-CoA reductase.** The two reduction reactions require NADPH as a co-factor (Figure 20–4).

The carrier of the elongating acyl groups during fatty acid synthesis is a **phosphopantetheine** prosthetic group attached to a serine hydroxyl within a specific domain of FAS. The carrier portion of the enzyme is called **acyl carrier protein, ACP**; however, ACP is actually not a separate polypeptide but a distinct region of activity in the single FAS polypeptide.

The synthesis of fatty acids by FAS proceeds through the following steps (Figure 20–4):

1. The initiation of fatty acid synthesis begins when FAS is "charged" by the attachment of an acetyl group from acetyl-CoA to a cysteine thiol group within the 3-ketoacyl synthetase domain, with the subsequent release of CoA.

Figure 20–2. Biosynthesis of malonyl-CoA. Enz, acetyl-CoA carboxytase. (Reproduced, with permission, from Murray, RK: *Harper's Biochemistry*, 23rd ed. Appleton & Lange, 1993.)

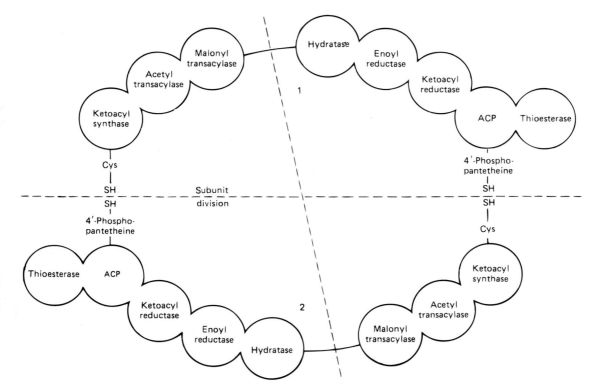

Figure 20–3. Fatty acid synthase multienzyme complex. The complex is a dimer to 2 identical polypeptide monomers, 1 and 2, each consisting of 7 enzyme activities and the acyl carrier protein (ACP). Cys-SH, cysteine thiol. The -SH of the 4'-phosphopantetheine of one monomer is in close proximity to the − SH of the cysteine residue of the ketoacyl synthasee of the other monomer, suggesting a "head-to-tail" arrangement of the 2 monomers. The detailed sequence of the enzymes in each monomer is tentative (based on Wakil). Though each monomer contains all the partial activities of the reaction sequence, the actual functional unit consists of one-half of a monomer interacting with the complementary half of the other. Thus, 2 acyl chains are produced simultaneously. (Reproduced, with permission, from Murray, RK: *Harper's Biochemistry*, 23rd ed. Appleton & Lange, 1993.)

2. Once FAS is activated, a malonyl group from malonyl-CoA is attached to the pantetheine sulfhydryl group of ACP (with the release of CoA). The complex is now primed to carry out a first round of synthesis.

3. The initial acetyl group is transferred from the cysteine sulfhydryl of the 3-ketoacyl synthetase domain and condensed with the malonyl group on ACP, with the concomitant release of CO_2. The release of CO_2 at this point propels the reaction in the forward direction and renders the reaction incapable of incorporating net CO_2 into fats.

4. The condensed product (an acetoacetyl group at this stage) is reduced, dehydrated and reduced again by the remaining enzymatic activities of FAS.

5. This cyclic process is completed when the resultant acyl group is transferred from ACP to the cysteine sulfhydryl of the condensing enzyme domain.

6. All subsequent steps of fatty acid synthesis involve the addition of acetyl units to the elongating fatty acid from malonyl-CoA. As in the initial step, the malonyl group is added to the pantetheine sulfhydryl of ACP.

7. The cycles continue by the addition of acetyl units until the 16-carbon, fully saturated fatty acid, palmitate, is synthesized. Next, palmitoyl-CoA is released from the FAS complex through the action of the **thioesterase** domain of FAS. The palmitate can then undergo separate elongation and/or unsaturation to yield other fatty acid molecules.

Elongation of Fatty Acids

The fatty acid product released from FAS is palmitoyl-CoA (via the action of **palmitoyl thioesterase**), a 16:0 fatty acid. Elongation and unsaturation of fatty acids occurs primarily in the endoplasmic retic-

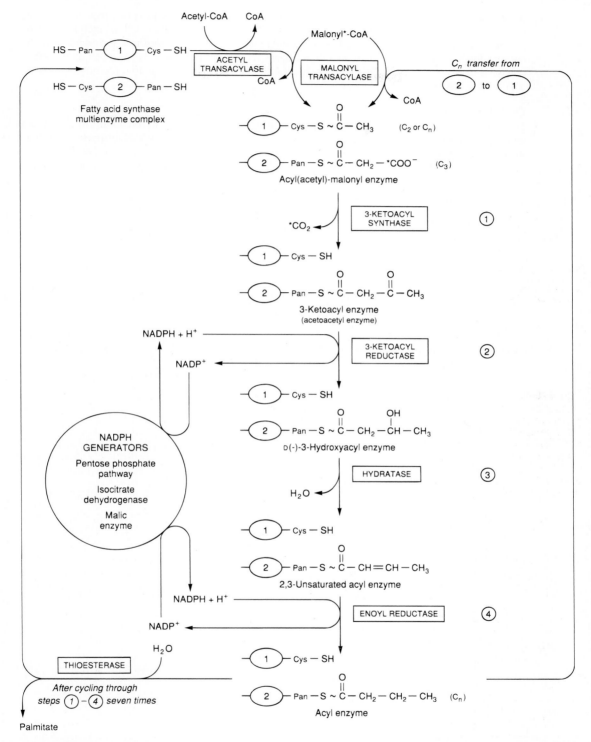

Figure 20–4. Biosynthesis of long-chain fatty acids. Details of how addition of a malonyl residue causes the acyl chain to grow by 2 carbon atoms. Cys-, cysteine residue; pan 4′-phosphopantetheine. Details of the fatty acid synthase dimer are shown in Figure 20–3. ① and ② represent the individual monomers of fatty acid synthase. (Reproduced, with permission, from Murray, RK: *Harper's Biochemistry*, 23rd ed. Appleton & Lange, 1993.)

ulum, and to some degree also in the mitochondria (Figure 20–5). Elongation involves the condensation of acyl-CoA groups with malonyl-CoA, with release of CO_2 as in the reaction catalyzed by FAS. The product undergoes reduction, dehydration and reduction, yielding a saturated fatty acid that is two carbons longer than before. The reduction reactions require NADPH as a co-factor, as do the similar reactions catalyzed by FAS. When elongation takes place in the mitochondria, the process involves acetyl-CoA instead of malonyl-CoA.

Desaturation of Fatty Acids

Desaturation of fatty acids also occurs in membranes of the endoplasmic reticulum. In mammalian cells, these reactions are catalyzed by four broad-specificity **fatty acyl-CoA desaturases** (non-heme iron containing enzymes). These enzymes introduce unsaturation at C_4, C_5, C_6, or C_9. The electrons re-

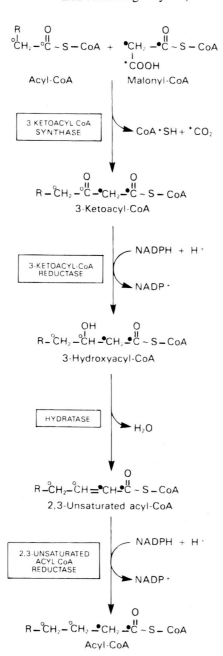

Figure 20–5. Microsomal system for fatty acid chain elongation (elongase). (Reproduced, with permission, from Murray, RK: *Harper's Biochemistry*, 23rd ed. Appleton & Lange, 1993.)

moved from the oxidized fatty acids during desaturation are transferred from the desaturases to **cytochrome b₅** and then **NADH-cytochrome b₅ reductase**. These electrons are uncoupled from mitochondrial oxidative-phosphorylation and, therefore, do not yield ATP.

Since these mammalian enzymes cannot introduce sites of unsaturation beyond C_9, they are incapable of generating *de novo* either **linoleate** (18:2, cis-9, 12) or **linolenate** (18:3, cis-9, 12, 15). These fatty acids must be acquired from the diet and are therefore referred to as **essential fatty acids** (see Chapter 10). Linoleic acid is especially important in that it is required for the synthesis of **arachidonic acid**, a precursor for the **eicosanoids** (the prostaglandins and thromboxanes, see below), and is a constituent of epidermal cell sphingolipids that form a barrier to water permeability in the skin.

Fatty Acid Oxidation

Fatty acids must be activated in the cytoplasm before being oxidized in the mitochondria. Activation is catalyzed by **fatty acyl-CoA ligase** (also called **acyl-CoA synthetase** or **thiokinase**). The net result of this activation process is the consumption of 2 molar equivalents of ATP.

$$\text{Fatty acid} + \text{ATP} + \text{CoA} \rightarrow \text{Acyl-CoA} + \text{PP}_i + \text{AMP} \qquad \textbf{20.1}$$

Oxidation of fatty acids occurs in the mitochondria. The transport of fatty acyl-CoA into the mitochondria is accomplished via an **acyl-carnitine** intermediate, which itself is generated by the action of **carnitine acyltransferase I**, an enzyme that resides in the outer mitochondrial membrane. The acyl-carnitine molecule then is transported into the mitochondria, where **carnitine acyltransferase II** catalyzes the regeneration of the fatty acyl-CoA molecule (Figure 20–6).

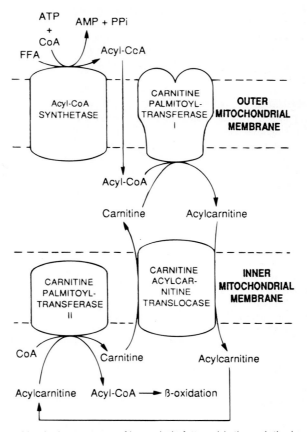

Figure 20–6. Role of carnitine in the transport of long-chain fatty acids through the inner miochondrial membrane. Long chainacyl-CoA cannot pass through the inner mitochondrial membrane, but its metabolic product, acylcarnitine, can. (Reproduced, with permission, from Murray, RK: *Harper's Biochemistry*, 23rd ed. Appleton & Lange, 1993.)

The process of fatty acid oxidation is termed β-**oxidation**, since it occurs through the sequential re-moval of 2-carbon units by oxidation at the β-carbon position of the fatty acyl-CoA molecule (Figure 20–7).

Each round of β-oxidation produces one mole of NADH, one mole of FADH$_2$ and one mole of acetyl-CoA (Figure 20–8). The acetyl-CoA—the end product of each round of β-oxidation—then en-ters the TCA cycle, where it is further oxidized to CO$_2$ with the concomitant generation of three moles of NADH, one mole of FADH$_2$ and one mole of ATP (see Chapter 17). The NADH and FADH$_2$ gen-erated during the fat oxidation and acetyl-CoA oxidation in the TCA cycle then can enter the respira-tory pathway for the production of ATP (see Chapter 18).

The oxidation of fatty acids yields significantly more energy per carbon atom than does the oxida-tion of carbohydrates. The net result of the oxidation of one mole of oleic acid (an 18-carbon fatty acid) is 146 moles of ATP (2 mole equivalents are used during the activation of the fatty acid), as compared with 114 moles from an equivalent number of glucose carbon atoms.

Alternative Oxidation Pathways

The majority of natural lipids contain an even number of carbon atoms. A small proportion contain odd numbers; upon complete β-oxidation, these yield acetyl-CoA units plus a single mole of propi-onyl-CoA. The propionyl-CoA is converted, in an ATP-dependent pathway, to succinyl-CoA. The suc-cinyl-CoA can then enter the TCA cycle for further oxidation.

The oxidation of unsaturated fatty acids is essentially the same process as for saturated fats, except when a double bond is encountered. In such a case, the bond is isomerized by a specific **enoyl-CoA isomerase** and oxidation continues. In the case of linoleate, the presence of the Δ^{12} unsaturation results in the formation of a dienoyl-CoA during oxidation (Figure 20–9). This molecule is the substrate for an additional oxidizing enzyme, the NADPH requiring **2,4-dienoyl-CoA reductase**.

Phytanic acid is a fatty acid present in the tissues of ruminants and in dairy products and is, there-fore, an important dietary component of fatty acid intake. Because phytanic acid is methylated, it can-not act as a substrate for the first enzyme of the β-oxidation pathway (acyl-CoA dehydrogenase). An additional mitochondrial enzyme, α-**hydroxylase**, adds a hydroxyl group to the α-carbon of phytanic acid, which then serves as a substrate for the remainder of the normal oxidative enzymes. This process is termed α-**oxidation**.

Lipid Mobilization

The primary sources of fatty acids for oxidation are diet and mobilization from cellular stores. Fatty acids from the diet can are delivered from the gut to cells via transport in the blood (see Chapter 21). Fatty acids are stored in the form of **triacylglycerols**, primarily within adipocytes of adipose tissue. In response to energy demands, the fatty acids of stored triacylglycerols can be mobilized for use by pe-ripheral tissues. The release of metabolic energy, in the form of fatty acids, is controlled by a complex series of interrelated cascades that result in the activation of **hormone-sensitive lipase** (Figure 20–10).

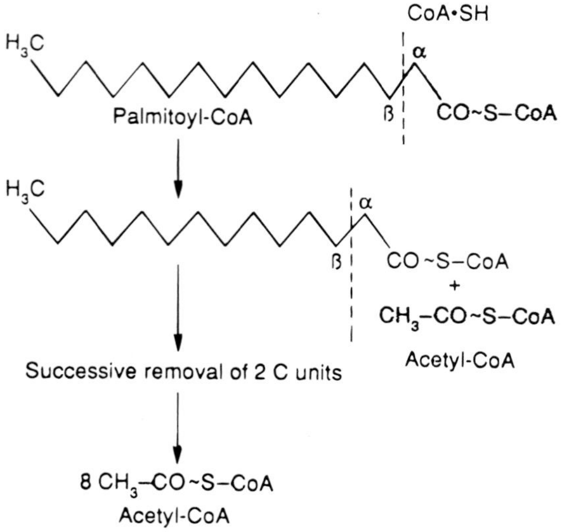

Figure 20–7. Overview of β-oxidation of fatty acid. (Re-produced, with permission, from Murray, RK: *Harper's Biochemistry*, 23rd ed. Appleton & Lange, 1993.)

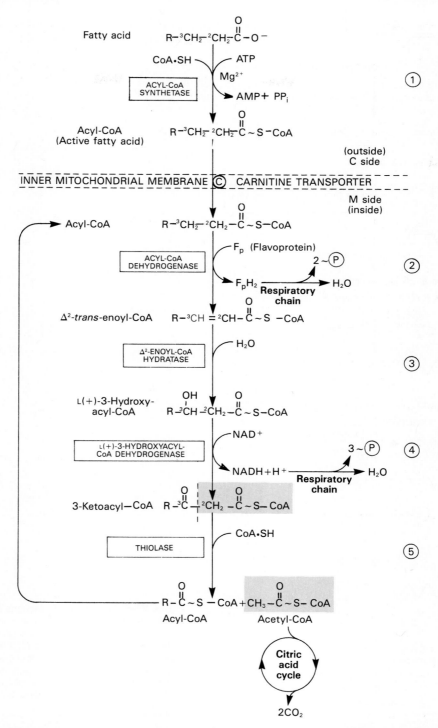

Figure 20–8. β-Oxidation fo fatty acids. Long-chain acyl-CoA is cycled through reactions 2–5, acetyl-CoA being split off each cycle by thiolase (reaction 5). When the acyl radicals is only 4 carbon atoms in length, 2 acetyl-CoA molecules are formed in reaction 5. (Reproduced, with permission, from Murray, RK: *Harper's Biochemistry*, 23rd ed. Appleton & Lange, 1993.)

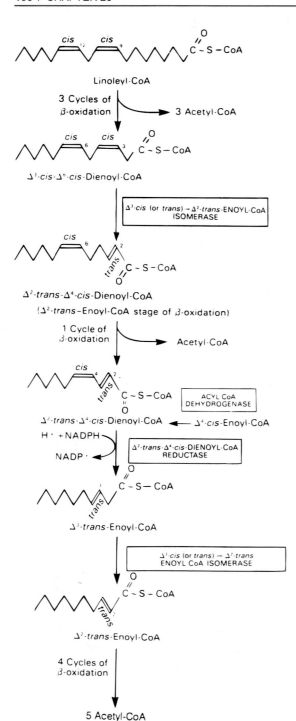

Figure 20–9. Sequence of reactions in the oxidation of unsaturated fatty acids, eg, linoleic acid. δ⁴-*cis*-fatty acids or fatty acids forming δ⁴-*cis*enoyl-CoA enter the pathway at the position shown. NADPH for the dienoyl-CoA reductase step is supplied by intramitochondrial sources such as glutamate dehydrogenase, isocitrate dehydrogenase, and NAD(P)H transhydrogenase. (Reproduced, with permission, from Murray, RK: *Harper's Biochemistry*, 23rd ed. Appleton & Lange, 1993.)

The stimulus to activate this cascade, in adipocytes, can be glucagon, epinephrine or β-corticotropin. These hormones bind cell-surface receptors that are coupled to the activation of adenylate cyclase upon ligand binding. The resultant increase in cAMP leads to activation of PKA, which in turn phosphorylates and activates hormone-sensitive lipase. This enzyme hydrolyzes fatty acids from carbon atoms 1 or 3 of triacylglycerols. The resultant diacylglycerol is then a substrate for diacylglycerol lipases and finally for monoacylglycerol lipases. The net result of the action of these three enzymes is three moles

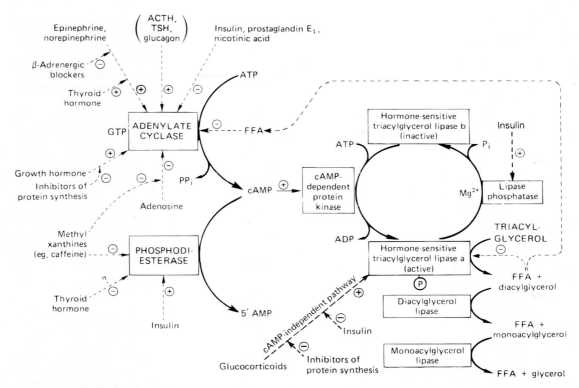

Figure 20–10. Control of adipose tissue lipolysis. TSH, thyroid-stimulating hormone; FFA, free fatty acids. Note the cascade sequence of ractions affording amplification of each step. The llipolytic stimulus is "switched off" by (1) removal of the stimulating hormone; (2) the action of lipase phosphatase; (3) the inhibition of the lipase and adenylate cyclase by high concentration of FFA; (4) the inhibition of adenylate cyclase by adenosine; and (5) the removal of cAMP by the action of phosphodiesterase. ACTH, TSH, and glucagon may not activate adenylate cyclase in vivo, since the concentration of each hormone required in vivo is much higher than is found in the circulation. Positive ($\oplus$) and negative ($\ominus$) regulatory effects are represented by broken lines and substrate flow by solid lines. (Reproduced, with permission, from Murray, RK: *Harper's Biochemistry*, 23rd ed. Appleton & Lange, 1993.)

of free fatty acid and one mole of glycerol. The free fatty acids diffuse from adipose cells, combine with albumin in the blood, and are thereby transported to other tissues, where they passively diffuse into cells.

In contrast to the hormonal activation of adenylate cyclase and (subsequently) hormone-sensitive lipase in adipocytes, the mobilization of fat from adipose tissue is inhibited by numerous stimuli. The most significant inhibition is that exerted upon adenylate cyclase by insulin. When an individual is well fed, insulin released from the pancreas prevents the inappropriate mobilization of stored fat. Instead, any excess fat and carbohydrate are incorporated into the triacylglycerol pool within adipose tissue (see Chapter 28).

Regulation of Fatty Acid Metabolism

In order to understand how the synthesis and degradation of fats needs to be exquisitely regulated, one must consider the energy requirements of the organism as a whole. The blood is the carrier of triacylglycerols in the form of VLDLs and chylomicrons (see Chapter 21), fatty acids bound to albumin, amino acids, lactate, ketone bodies and glucose. The pancreas is the primary organ involved in sensing the organism's dietary and energetic states by monitoring glucose concentrations in the blood. Low blood glucose stimulates the secretion of glucagon, whereas, elevated blood glucose calls for the secretion of insulin.

The metabolism of fat is regulated by two distinct mechanisms. One is **short-term regulation**, which can come about through events such as substrate availability, allosteric effectors and/or enzyme modification. The other mechanism, **long-term regulation**, is achieved by alteration of the rate of enzyme synthesis and turn-over.

ACC is the rate-limiting (committed) step in fatty acid synthesis. This enzyme is activated by citrate and inhibited by palmitoyl-CoA and other long-chain fatty acyl-CoAs. ACC activity can also be affected by phosphorylation. For instance, glucagon-stimulated increases in PKA activity result in the phosphorylation of certain serine residues in ACC, leading to decreased activity of the enzyme. By contrast, insulin leads to PKA-independent phosphorylation of ACC at sites distinct from glucagon, which bring about increased ACC activity. Both these reaction chains are examples of **short-term regulation**.

Insulin, a product of the well-fed state, stimulates ACC and FAS synthesis, whereas starvation leads to a decrease in the synthesis of these enzymes. The levels of lipoprotein lipase in the capillaries of adipose tissue also are increased by insulin and decreased by starvation. However, the effects of insulin and starvation on lipoprotein lipase in the heart are just the inverse of those in adipose tissue. This sensitivity allows the heart to absorb any available fatty acids in the blood in order to oxidize them for energy production. Starvation also leads to increases in the levels of cardiac enzymes of fatty acid oxidation, and to decreases in FAS and related enzymes of synthesis.

Adipose tissue contains **hormone-sensitive lipase**, which is activated by PKA-dependent phosphorylation; this activation increases the release of fatty acids into the blood. This in turn leads to the increased oxidation of fatty acids in other tissues such as muscle and liver. In the liver, the net result (due to increased acetyl-CoA levels) is the production of ketone bodies (see below). This would occur under conditions in which the carbohydrate stores and gluconeogenic precursors available in the liver are not sufficient to allow increased glucose production. The increased levels of fatty acid that become available in response to glucagon or epinephrine are assured of being completely oxidized, because PKA also phosphorylates ACC; the synthesis of fatty acid is thereby inhibited.

Insulin has the opposite effect to glucagon and epinephrine: it increases the synthesis of triacylglycerols (and glycogen). One of the many effects of insulin is to lower cAMP levels, which leads to increased dephosphorylation through the enhanced activity of protein phosphatases such as PP-1. With respect to fatty acid metabolism, this yields dephosphorylated and inactive hormone-sensitive lipase. Insulin also stimulates certain phosphorylation events. This occurs through activation of several cAMP-independent kinases, one of which phosphorylates and thereby stimulates the activity of ACC.

Fat metabolism can also be regulated by malonyl-CoA-mediated inhibition of **carnitine acyltransferase I**. Such regulation serves to prevent *de novo* synthesized fatty acids from entering the mitochondria and being oxidized.

CLINICAL SIGNIFICANCE

The majority of clinical problems related to fatty acid metabolism are associated with processes of oxidation. These disorders fall into four main groups.

A. Deficiencies in Carnitine: Deficiencies in carnitine lead to an inability to transport fatty acids into the mitochondria for oxidation. This can occur in newborns and particularly in preterm infants. Carnitine deficiencies also are found in patients undergoing hemodialysis or exhibiting organic aciduria. Carnitine deficiencies may manifest systemic symptomatology or may be limited to only muscles. Symptoms can range from mild occasional muscle cramping to severe weakness or even death. Treatment is by oral carnitine administration.

B. Deficiencies in Carnitine Acyltransferase: Deficiencies in carnitine acyltransferase affect primarily the liver and lead to reduced fatty acid oxidation and ketogenesis. **Carnitine acyltransferase II** deficiencies affect skeletal muscle, resulting in recurrent muscle pain and fatigue and myoglobinuria following strenuous exercise. Carnitine acyltransferases may also be inhibited by sulfonylurea drugs such as **tolbutamide** and **glyburide**.

C. Deficiencies in Acyl-CoA Dehydrogenases: A group of inherited diseases that impair β-oxidation result from deficiencies in acyl-CoA dehydrogenases. The enzymes affected may belong to one of three categories: **long-chain acyl-CoA dehydrogenase (LCAD), medium-chain acyl-CoA dehydrogenase (MCAD) or short-chain acyl-CoA dehydrogenase (SCAD)**. MCAD deficiency is the most common form of this disease. In the first years of life, this deficiency will become apparent following a prolonged fasting period. Symptoms include vomiting, lethargy and frequently coma. Excessive urinary excretion of medium-chain dicarboxylic acids as well as their glycine and carnitine esters is diagnostic of this condition. In the case of this enzyme deficiency, taking care to avoid prolonged fasting is sufficient to prevent clinical problems.

D. Refsum's Disease: Refsum's disease is a rare inherited disorder in which patients lack the mitochondrial α-oxidizing enzyme. As a consequence, they accumulate large quantities of phytanic acid in their tissues and serum. This leads to severe symptoms, including cerebellar ataxia, retinitis pigmentosa, nerve deafness and peripheral neuropathy. As expected, the restriction of dairy products and ruminant meat from the diet can ameliorate the symptoms of this disease.

Ketogenesis

During high rates of fatty acid oxidation, primarily in the liver, large amounts of acetyl-CoA are generated. These exceed the capacity of the TCA cycle, and one result is the synthesis of ketone bodies, or **ketogenesis** (Figure 20–11). The ketone bodies are **acetoacetate**, β-**hydroxybutyrate**, and **acetone**.

The formation of acetoacetyl-CoA occurs by condensation of two moles of acetyl-CoA through a reversal of the **thiolase-** catalyzed reaction of fat oxidation. Acetoacetyl-CoA and an additional acetyl-CoA are converted to β-**hydroxy-β-methylglutaryl-CoA** (HMG-CoA) by **HMG-CoA synthase**, an enzyme found in large amounts only in the liver (Figure 20–12). Some of the HMG-CoA leaves the mitochondria, where it is converted to mevalonate (the precursor for cholesterol synthesis; see Chapter 21) by **HMG-CoA reductase**. HMG-CoA in the mitochondria is converted to acetoacetate by the action of **HMG-CoA lyase** (Figure 20–12). Acetoacetate can undergo spontaneous decarboxylation to acetone, or it can be enzymatically converted to β-hydroxybutyrate through the action of β-**hydroxybutyrate dehydrogenase** (Figure 20–11). When the level of glycogen in the liver is high, the production of β-hydroxybutyrate increases.

When carbohydrate utilization is low or deficient, the level of oxaloacetate will also be low, resulting in a reduced flux through the TCA cycle. This in turn leads to increased release of ketone bodies from the liver for use as fuel by other tissues. In early stages of starvation, when the last remnants of fat are oxidized, the brain will receive a majority of its fuel in the form of ketone bodies. Acetoacetate

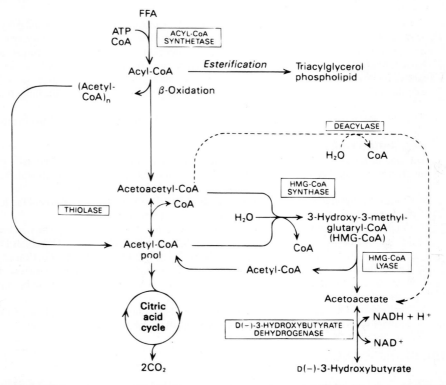

Figure 20–11. Pathways of ketogenesis in the liver. FFA, free fatty acid; HMG, 3-hydroxy-3-methylglutaryl. (Reproduced, with permission, from Murray, RK: *Harper's Biochemistry*, 23rd ed. Appleton & Lange, 1993.)

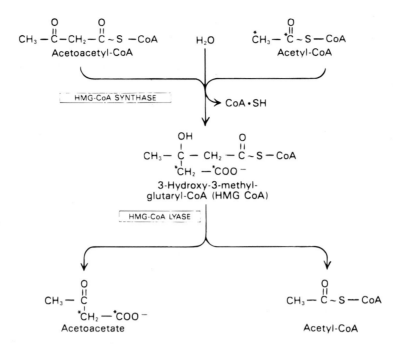

Figure 20–12. Formation of acetoacetate through intermediate production of HMG-CoA. (Reproduced, with permission, from Murray, RK: *Harper's Biochemistry*, 23rd ed. Appleton & Lange, 1993.)

and β-hydroxybutyrate, in particular, also serve as major substrates for the biosynthesis of neonatal cerebral lipids.

Ketone bodies are utilized by extrahepatic tissues through the conversion of β-hydroxybutyrate to acetoacetate and of acetoacetate to acetoacetyl-CoA. The first step involves the reversal of the β-hydroxybutyrate dehydrogenase reaction, and the second involves the action of **acetoacetate:succinyl-CoA transferase** (reaction 20.2). The latter enzyme is present in all tissues except the liver. Importantly, its absence allows the liver to produce ketone bodies but not to utilize them. This ensures that extrahepatic tissues have access to ketone bodies as a fuel source during prolonged starvation.

$$\text{Acetoacetate} + \text{Succinyl-CoA} \Leftrightarrow \text{Acetoacetyl-CoA} + \text{Succinate} \qquad 20.2$$

Regulation of Ketogenesis

The fate of the products of fatty acid metabolism is determined by an individual's physiological status. Ketogenesis takes place primarily in the liver and may be affected by several factors:

1. Control in the release of free fatty acids from adipose tissue directly affects the level of ketogenesis in the liver. This is, of course, substrate-level regulation.
2. Once fats enter the liver, they have two distinct fates. They may be activated to acyl-CoAs and oxidized, or esterified to glycerol in the production of triacylglycerols. If the liver has sufficient supplies of glycerol-3-phosphate, most of the fats will be turned to the production of triacylglycerols.
3. The generation of acetyl-CoA by oxidation of fats can be completely oxidized in the TCA cycle. Therefore, if the demand for ATP is high, the fate of acetyl-CoA is likely to be further oxidation to CO_2.
4. The level of fat oxidation is regulated hormonally through the phosphorylation of ACC, which may activate it (in response to insulin) or inhibit it (in the case of glucagon).

CLINICAL SIGNIFICANCE

The production of ketone bodies occurs at a relatively low rate during normal feeding and under conditions of normal physiological status. Normal physiological responses to carbohydrate shortages cause the liver to increase the production of ketone bodies from the acetyl-CoA generated from fatty acid oxidation. This allows the heart and skeletal muscles primarily to use ketone bodies for energy, thereby preserving the limited glucose for use by the brain.

The most significant disruption in the level of ketosis, leading to profound clinical manifestations, occurs in untreated insulin-dependent diabetes mellitus. This physiological state, **diabetic ketoacidosis (DKA)**, results from a reduced supply of glucose (due to a significant decline in circulating insulin) and a concomitant increase in fatty acid oxidation (due to a concomitant increase in circulating glucagon). The increased production of acetyl-CoA leads to ketone body production that exceeds the ability of peripheral tissues to oxidize them. Ketone bodies are relatively strong acids ($pK_a \sim 3.5$), and their increase lowers the pH of the blood. This acidification of the blood is dangerous chiefly because it impairs the ability of hemoglobin to bind oxygen.

METABOLISM OF COMPLEX LIPIDS

Triglycerides

Fatty acids are stored for future use as triacylglycerols in all cells, but primarily in adipocytes of adipose tissue. Triacylglycerols constitute molecules of glycerol to which three fatty acids have been esterified. The fatty acids present in triacylglycerols are predominantly saturated. The major building block for the synthesis of triacylglycerol is glycerol. Dihydroxyacetone, produced during glycolysis, can also serve as a backbone precursor but does so to a much lesser extent than glycerol.

The glycerol backbone of triacylglycerols is activated by phosphorylation at the C-3 position by **glycerol kinase** (Figure 20–13). The utilization of dihydroxyacetone phosphate for the backbone is carried out through the action of **glycerol-3-phosphate dehydrogenase**, a reaction that requires NADH (Figure 20–13). The fatty acids incorporated into triacylglycerols are activated to acyl-CoAs through the action of **acyl-CoA synthetase**. Two molecules of acyl-CoA are esterified to glycerol-3-phosphate to yield **1,2-diacylglycerol phosphate (phosphatidic acid)**. The phosphate is then removed to yield **1,2-diacylglycerol**, the substrate for addition of the third fatty acid (Figure 20–13). Intestinal monoacylglycerols, derived from the hydrolysis of dietary fats, can also serve as substrates for the synthesis of 1,2-diacylglycerols (Figure 20–13).

Phospholipids

Phospholipids are synthesized by the addition of a basic group (predominantly a nitrogenous base) to phosphatidic acid or 1,2-diacylglycerol. Most phospholipids have a saturated fatty acid on C-1 and an unsaturated fatty acid on C-2 of the glycerol backbone. The major classifications of phospholipids (see Chapter 10) are **phosphatidylcholine (PC), phosphatidylethanolamine (PE), phosphatidylserine (PS), phosphatidylinositol (PI), phosphatidylglycerol (PG), and diphosphatidylglycerol (DPG**, an important form being **cardiolipin)**. Phospholipids can be synthesized by two mechanisms. One utilizes a CDP-activated polar head group for attachment to the phosphate of phosphatidic acid. The other utilizes CDP-activated 1,2-diacylglycerol and an inactivated polar head group.

A. PC: Choline is activated first by phosphorylation and then by coupling to CDP prior to attachment to phosphatidic acid (Figure 20–13). PC is also synthesized by the addition of choline to CDP-activated 1,2-diacylglycerol. A third pathway to PC synthesis, as shown in Figure 20–13, involves the conversion of either PS or PE to PC. The conversion of PS first requires decarboxylation to yield PE; this then undergoes a series of three methylation reactions, utilizing S-adenosylmethionine (SAM) as methyl group donor.

B. PE: Synthesis of PE can occur by two pathways. The first requires that ethanolamine is activated by phosphorylation and then by coupling to CDP. The ethanolamine is then transferred from CDP-ethanolamine to phosphatidic acid to yield PE. The second involves the decarboxylation of PS (Figure 20–13).

C. PS: The pathway for PS synthesis involves an exchange reaction of serine for ethanolamine in PE. This exchange occurs when PE is in the lipid bilayer of the membrane (Figure 20–13). As indicated above, PS can serve as a source of PE through a decarboxylation reaction (Figure 20–13).

D. PI: The synthesis of PI involves CDP-activated 1,2-diacylglycerol condensation with *myo*inositol (Figure 20–13). PI subsequently undergoes a series of phosphorylations of the hydroxyls of inos-

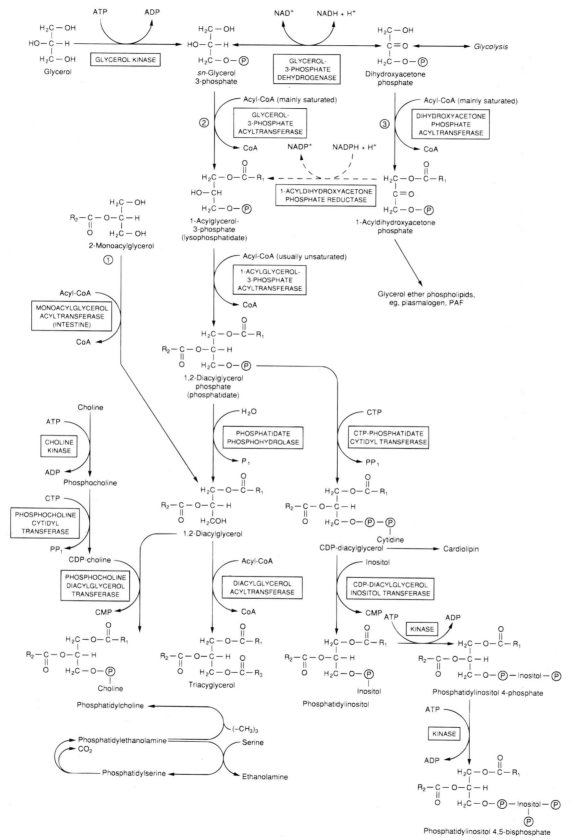

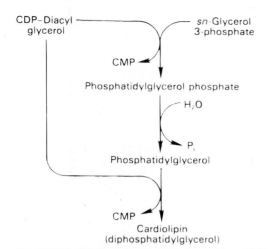

Figure 20–14. Biosynthesis of cardiolipin. (Reproduced, with permission, from Murray, RK: *Harper's Biochemistry*, 23rd ed. Appleton & Lange, 1993.)

itol, leading to the production of polyphosphoinositides. One polyphosphoinositide **(phosphatidylinositol 4,5-bisphosphate, PIP$_2$)** is a critically important membrane phospholipid involved in the transmission of signals for cell growth and differentiation from outside the cell to inside (see Chapter 34).

E. PG: PG is synthesized from CDP-diacylglycerol and glycerol-3-phosphate (Figure 20–13). The vital role of PG is to serve as the precursor for the synthesis of diphosphatidylglycerols (DPGs).

F. DPG: One important class of diphosphatidylglycerols is the **cardiolipin**. These molecules are synthesized by the condensation of CDP-diacylglycerol with PG (Figure 20–14).

The fatty acid distribution at the C-1 and C-2 positions of glycerol within phospholipids is continually in flux, owing to phospholipid degradation and the continuous phospholipid remodeling that occurs while these molecules are in membranes. Phospholipid degradation results from the action of **phospholipases**. The various phospholipases each exhibit substrate specificities for different positions in phospholipids (Figure 20–15).

In many cases the acyl group which was initially transferred to glycerol, by the action of the acyl transferases, is not the same acyl group present in the phospholipid when it resides within a membrane. The remodeling of acyl groups in phospholipids is the result of the action of **phospholipase A$_1$** and **phospholipase A$_2$**. The products of these enzymes are called **lysophospholipids** and can be substrates for acyl transferases utilizing different acyl-CoA groups. Lysophospholipids can also accept acyl groups from other phospholipids in an exchange reaction catalyzed by **lysolecithin:lecithin acyltransferase (LLAT)**.

Phospholipase A$_2$ is also an important enzyme, whose activity is responsible for the release of arachidonic acid from the C-2 position of membrane phospholipids. The released arachidonate is then a substrate for the synthesis of the prostaglandins and leukotrienes (see below).

Plasmalogens

Plasmalogens are glycerol ether phospholipids. They are of two types, alkyl ether and alkenyl ether. Dihydroxyacetone phosphate serves as the glycerol precursor for the synthesis of glycerol ether phospholipids. Three major classes of plasmalogens have been identified: **choline, ethanolamine, and serine plasmalogens** (see Figure 10–9). One particular choline plasmalogen (1-alkyl, 2-acetyl phosphatidylcholine) has been identified as an extremely powerful biological mediator, capable of inducing cellular responses at concentrations as low as 10^{-11} M. This molecule is called **platelet activating factor, PAF** (see Figure 10–10).

Figure 20–13. Biosynthesis of triacylglycerol and phospholipids. ①, Monoacylglycerol pathway; ②, glycerol phosphate pathway; ③, dihydroxyacetone pathway. Phosphatidylethanolamine may be formed from ethanolamine by a similar pathway to that shown for the formation of phosphatidylcholine form choline. PAF, platelet-activating factor. (Reproduced, with permission, from Murray, RK: *Harper's Biochemistry*, 23rd ed. Appleton & Lange, 1993.)

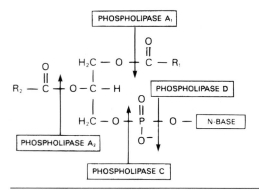

Figure 20–15. Sites of the hydrolytic activity of phospho-lipases on a phospholipid substrate. (Reproduced, with permission, from Murray, RK: *Harper's Biochemistry*, 23rd ed. Appleton & Lange, 1993.)

Sphingolipids

The sphingolipids contain a backbone of **ceramide**, which is derived from **sphingosine** (Figure 20–16) by fatty acylation of the amino group. The sphingolipids include the **sphingomyelins** and **glyco-sphingolipids** (the cerebrosides, sulfatides, globosides and gangliosides). Sphingomyelin is the only sphingolipid that is a phospholipid. Sphingolipids are a component of all membranes but are particularly abundant in the myelin sheath.

A. Sphingomyelins: The sphingomyelins are synthesized by the transfer of phosphorylcholine from phosphatidylcholine to a ceramide, in a reaction catalyzed by **sphingomyelin synthase**.

B. Cerebrosides: Cerebrosides are glycosphingolipids containing a monohexoside, predominantly galactose. Galactocerebrosides are synthesized from ceramide and UDP-galactose. The glucose-containing cerebrosides are not normally found in membranes and represent an intermediate in the synthesis or degradation of more complex glycosphingolipids. Excess accumulation of glucocerebro-sides is observed in **Gaucher's disease** (see Chapter 10).

C. Sulfatides: The sulfuric acid esters of galactocerebrosides are the **sulfatides**. Sulfatides are synthesized from galactocerebrosides and activated sulfate, **3′-phosphoadenosine 5′-phosphosulfate (PAPS)**. Excess accumulation of sulfatides is observed in **sulfatide lipidosis** (metachromatic leukodystrophy; see Chapter 10).

D. Globosides: Globosides represent cerebrosides that contain additional carbohydrates, predominantly galactose, glucose or GalNAc. Lactosyl ceramide is a globoside found in erythrocyte plasma membranes. Globotriaosylceramide (also called ceramide trihexoside) contains glucose and two moles of galactose and accumulates, primarily in the kidneys, of patients suffering from **Fabry's disease** (see Chapter 10).

E. Gangliosides: Gangliosides are very similar to globosides except that they also contain NANA in varying amounts. Deficiencies in lysosomal enzymes, which normally are responsible for the degradation of the carbohydrate portions of various gangliosides, underlie the symptoms observed in rare autosomally inherited diseases termed **lipid storage diseases** (see also Chapter 10).

Synthesis of Eicosanoids

The eicosanoids consist of the **prostaglandins (PGs), thromboxanes (TXs)** and **leukotrienes (LTs)**. The PGs and TXs are collectively identified as **prostanoids**. Prostaglandins were originally shown to

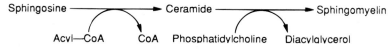

Figure 20–16. Biosynthesis of ceramide and sphingomyelin. (Reproduced, with permission, from Murray, RK: *Harper's Biochemistry*, 23rd ed. Appleton & Lange, 1993.)

be synthesized in the prostate gland, thromboxanes from platelets (thrombocytes) and leukotrienes from leukocytes, hence the derivation of their names.

The eicosanoids produce a wide range of biological effects on inflammatory responses (predominantly those of the joints, skin, and eyes), on the intensity and duration of pain and fever, and on reproductive function (including the induction of labor). They also play important roles in inhibiting gastric acid secretion, regulating blood pressure through vasodilation or constriction, and inhibiting or activating platelet aggregation and thrombosis (see Table 20–1).

The principal eicosanoids of biological significance to humans are a group of molecules derived from the C_{20} fatty acid, arachidonic acid. Minor eicosanoids are derived from dihomo-γ-linoleic acid and eicosopentaenoic acid. The major source of arachidonic acid is through its release from cellular stores. Within the cell, it resides predominantly at the C-2 position of membrane phospholipids and is released from there upon the activation of phospholipase A_2. The immediate dietary precursor of arachidonate is linoleate. Linoleate is also the precursor for dihomo-γ-linoleic acid and eicosopentaenoic acid synthesis. Therefore, the absence of linoleic acid from the diet would seriously threaten the body's ability to synthesize eicosanoids.

All mammalian cells except erythrocytes synthesize eicosanoids. These molecules are extremely potent, able to cause profound physiological effects at very dilute concentrations. All eicosanoids function locally at the site of synthesis, through receptor-mediated G-protein linked signaling pathways leading to an increase in cAMP levels.

Two main pathways are involved in the biosynthesis of eicosanoids. The prostaglandins and thromboxanes are synthesized by the **cyclic pathway** (Figure 20–17) and the leukotrienes by the **linear pathway** (Figure 20–18).

The cyclic pathway is initiated through the action of **prostaglandin endoperoxide synthetase**. This enzyme possesses two activities, **cyclooxygenase** and **peroxidase** (Figure 20–17). The linear pathway is initiated through the action of **lipoxygenases**. It is the enzyme **5-lipoxygenase** that gives rise to the leukotrienes (Figure 20–18).

A widely used class of drugs, the nonsteroidal anti-inflammatory drugs (NSAIDs) such as ibuprofen, indomethacin, naproxen, phenylbutazone and aspirin, all act upon the cyclooxygenase activity. Another class, the corticosteroidal drugs, act to inhibit phospholipase A_2, thereby inhibiting the release of arachidonate from membrane phospholipids and the subsequent synthesis of eicosinoids.

Table 20–1. Properties of significant eicosanoids.

Eicosanoid	Major Site(s) of Synthesis	Major Biological Activities
PGD_2	Mast cells	Vasodilation
PGE_2	Kidney, spleen, heart	Vasodilation, enhancement of the effects of bradykinin and histamine, induction of uterine contractions, and of platelet aggregation, maintaining the open passageway of the fetal ductus arteriosus
PGF_2	Kidney, spleen, heart	Vasoconstriction, smooth muscle contraction
PGH_2		Precursor to thromboxanes A_2 and B_2, induction of platelet aggregation and vasoconstriction
PGI_2	Heart, vascular endothelial cells	Inhibits platelet aggregation, induces vasodilation
TXA_2	Platelets	Induces platelet aggregation and vasoconstriction
TXB_2	Platelets	Induces vasoconstriction
LTB_4	Monocytes, basophils, neutrophils, eosinophils, mast cells, epithelial cells	Induces leukocyte chemotaxis and aggregation
LTC_4	Monocytes and alveolar macrophages, basophils, eosinophils, mast cells, epithelial cells	Component of SRS-A[a], induces vasodilation and bronchoconstriction
LTD_4	Monocytes and alveolar macrophages, eosinophils, mast cells, epithelial cells	Predominant component of SRS-A[a], induces vasodilation and bronchoconstriction
LTE_4	Mast cells and basophils	Component of SRS-A[a], induces vasodilation and bronchoconstriction

[a]SRS-A = Slow-reactive substance of anaphylaxis

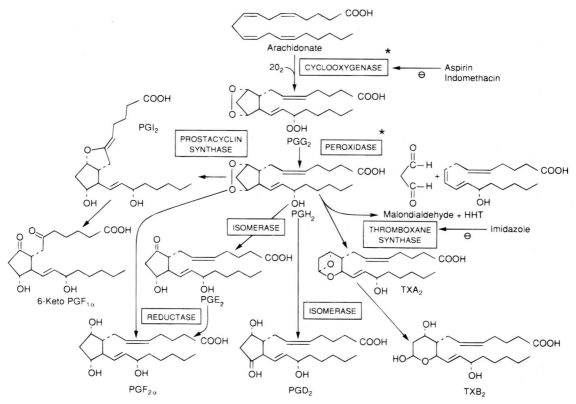

Figure 20–17. Conversion of arachidonic acid to prostaglandins and thromboxanes of series 2. PG, prostaglandin; TX, thromboxane; PGI prostacyclin; HHT, hydroxyheptadecatrienoate. ˙Both of these activities are attributed to one enzyme—prostaglandin endoperoxide synthase. Similar conversions occur in prostaglandins and thromboxanes of series 1 and 3. (Reproduced, with permission, from Murray, RK: *Harper's Biochemistry*, 23rd ed. Appleton & Lange, 1993.)

Questions

DIRECTIONS (items 1–20): Each numbered item or incomplete statement is followed by answers or by completions of the statement. Select the ONE lettered answer or completion that is BEST in each case.

1. A 2-year-old male is presented with progressive joint deformity, hoarse voice, skin granulomatous lesions and mental retardation. Biopsy of the liver indicates an accumulation of acidic ceramides. These symptoms and the results of the liver biopsy are indicative of which disease?
   a. Metachromatic leukodystrophy.
   b. Farber's lipogranulomatosis.
   c. Sandhoff-Jatzkewitz disease.
   d. Fucosidosis.
   e. Gaucher's disease.

2. Triacylglycerol degradation generates free glycerol, which can be utilized for the:
   a. Synthesis of ATP.
   b. Synthesis of glycogen.
   c. Synthesis of platelet activating factor (PAF).
   d. Synthesis of triacylglycerols.
   e. All of the above.

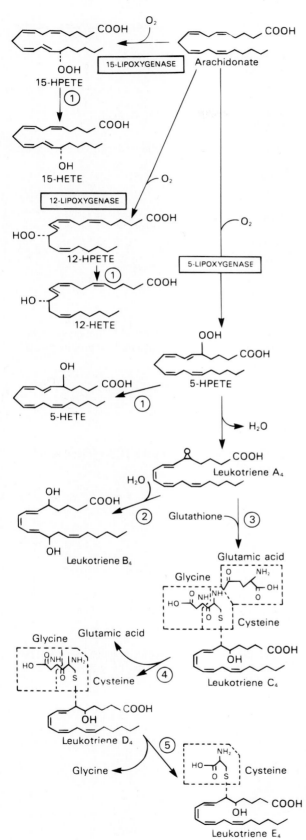

Figure 20–18. Conversion of arachidonic acid to leukotrienes of series 4 via the lipoxygenase pathway. HPETE, hydroperoxyeicosatetraenoate; HETE, hydroxyeicosatetraenoate. Some similar conversions occur in series 3 and 5 leukotrienes. ①, Peroxidase; ②, leukotriene A₄ epoxide hydrolase; ③, glutathione S-transferase; ④, γ-glutamyltransferase; ⑤, cysteinyl-glycine dipeptidase. (Reproduced, with permission, from Murray, RK: *Harper's Biochemistry*, 23rd ed. Appleton & Lange, 1993.)

3. Fatty acids enter the mitochondrial matrix for oxidation:
   a. In the form of acyl-CoAs directly by active transport.
   b. As free fatty acid.
   c. Following conversion to carnitine.
   d. Complexed with carnitine as acyl-carnitines.
   e. In the form of acyl-CoAs by passive diffusion.

4. The reducing equivalents necessary for fatty acid biosynthesis are produced during conversion of:
   a. Glyceraldehyde-1,3-bisphosphate to 3-phosphoglycerate.
   b. 6-phosphogluconate to ribulose-5-phosphate.
   c. Glucose-6-phosphate to 6-phosphogluconate.
   d. Pyruvate to malate.
   e. Glucose-6-phosphate to fructose-6-phosphate.

5. During fasting or starvation, the brain receives energy in the form of:
   a. Acetyl-CoA.
   b. Acetoacetyl-CoA.
   c. Hydroxymethylglutaryl-CoA.
   d. γ-hydroxybutyrate.
   e. Glucose.

6. The propionyl-CoA generated from β-oxidation of fatty acids containing odd numbers of carbon atoms enters the TCA cycle as succinyl-CoA. The net yield of ATP from a mole of propionyl-CoA is:
   a. 4.
   b. 6.
   c. 5.
   d. 7.
   e. 3.

7. When excess acetyl-CoA, produced by the liver, cannot be utilized by the TCA cycle, it accumulates in the body as:
   a. Glucose.
   b. Triglycerides.
   c. Acetyl-CoA.
   d. β-Hydroxybutyrate.
   e. Glycogen.

8. Synthesis of fatty acids requires:
   a. NADH.
   b. $FADH_2$.
   c. NADPH.
   d. NAD^+.
   e. $NADP^+$.

9. A 5-year-old female is presented with hepatosplenomegaly, abnormal bleeding disorders, and defects in long bone development and neurological dysfunction. Liver biopsy reveals an accumulation of glucocerebrosides. These symptoms indicate the child is suffering from:
   a. Fabry's disease.
   b. Niemann-Pick disease.
   c. Gaucher's disease.
   d. Krabbe's disease.
   e. Sandhoff-Jatzkewitz disease.

10. Human fatty acid oxidation is controlled by:
    a. The activity of lipoprotein lipase in all tissues.
    b. Malonyl-CoA inhibition of carnitine acyltransferase I in the liver.
    c. Simple substrate availability.
    d. Allosteric regulation on two of the enzymes of the beta oxidation pathway.
    e. Phosphorylation of ACC.

11. The major biological action of the prostaglandins PGI_2 is:
    a. Vasoconstriction.
    b. Promoting smooth-muscle contraction.
    c. Bronchodilation.
    d. Inhibition of platelet aggregation.
    e. Bronchoconstriction.

12. The leukotriene LTB_4:
    a. Mediates neutrophil chemotaxis.
    b. Is a component of SRS-A.
    c. Induces aggregation of platelets.
    d. Increases vasodilation.
    e. Increases vascular permeability.

13. Regulation of fatty acid synthesis occurs:
    a. By phosphorylation of inactive fatty acid synthase.
    b. By phosphorylation of inactive ACC.
    c. By dephosphorylation of inactive fatty acid synthase.
    d. By dephosphorylation of inactive ACC.
    e. Fatty acid synthesis is essentially unregulated.

14. The site of action of the nonsteroidal anti-inflammatory drugs is:
    a. Thromboxane synthase.
    b. Prostaglandin cyclooxygenase.
    c. Lipoxygenase.
    d. Prostaglandin hydroperoxidase.
    e. Prostacyclin synthase.

15. Which of the following is a true organismal response to starvation?
    a. Liver ACC will be phosphorylated and activated.
    b. Adipose hormone-sensitive lipase will be inhibited.
    c. Cardiac lipoprotein lipase will be inhibited.
    d. Adipose glycerol kinase will be activated.
    e. Liver pyruvate kinase will be phosphorylated and activated.

16. Liver cells contribute to the overall bodily content of ketone bodies primarily because:
    a. They lack the form of the β-ketothiolase necessary to hydrolyze acetoacetyl-CoA.
    b. They lack the enzyme hydroxymethylglutaryl-CoA-lyase.
    c. They lack the enzyme hydroxymethylglutaryl-CoA synthetase.
    d. They lack the enzyme acetoacetate succinyl-CoA transferase.
    e. They lack the enzyme β-hydroxybutyrate dehydrogenase.

17. During the synthesis of one mole of palmitate, _____ moles of NADPH can be derived from the pathway utilized in the transport of the acetyl-CoA molecules out of the mitochondria.
    a. 8.
    b. 10.
    c. 14.
    d. 6.
    e. 12.

18. A patient on tolbutamide therapy for adult-onset diabetes suffers severe muscle pain and fatigue after his daily jog. The most likely explanation for these symptoms is:
    a. Tolbutamide poisoning within muscle cells, leading to reduced creatine kinase levels.
    b. Tolbutamide inhibition of carnitine acyltransferase activity, resulting in rreduced fat oxidation.
    c. Tolbutamide inhibition of FAS, leading to decreased availability of free fatty acids for oxidation in muscle cells.
    d. Tolbutamide inhibition of glucose uptake by skeletal muscle.
    e. Tolbutamide poisoning within the liver, leading to decreased utilization of the lactate produced by skeletal muscle.

19. Inhibiting the intake of the essential fatty acids or increasing the intake of fatty acids from fish oil has an anti-inflammatory effect. This is most likely due to:
   a. Lowering the level of histamine release from mast cells, due to alterations in the membrane phospholipid composition in cells.
   b. Decreasing the level of cholesterol biosynthesis and subsequently the level of potential vascular injury in response to atherosclerotic plaque formation.
   c. Altering the fatty acid composition within membrane phospholipids, thereby altering the availability of arachidonate for prostaglandin synthesis.
   d. Inhibiting the level of circulating leukocytes by decreasing production of chemoattractants by neutrophils.
   e. Preventing monocyte adherence to endothelial cell membranes, thereby decreasing invasion into tissues.

20. The net moles of ATP equivalents available from the complete β-oxidation of palmitic acid is:
   a. 131.
   b. 139.
   c. 121.
   d. 119.
   e. 129.

Answers

1. b	8. c	15. a
2. e	9. c	16. d
3. d	10. b	17. a
4. c	11. d	18. b
5. d	12. a	19. c
6. c	13. b	20. e
7. d	14. b	

21

Lipid Digestion and Lipoproteins

Objectives

- To define the terms: Chylomicron, VLDL, IDL, LDL and HDL.
- To describe the mechanisms of intestinal absorption of lipids.
- To describe the different apolipoproteins and their roles in lipid mobilization and metabolism.
- To describe the functions of chylomicrons, VLDLs, IDLs, LDLs, and HDLs in lipid metabolism.
- To describe the role of LDL receptors in lipid mobilization.
- To describe the clinical significances of altered lipid mobilization.
- To describe the major pharmacologic interventions used in lowering plasma lipid levels.

Concepts

INTESTINAL UPTAKE OF LIPIDS

For the body to make use of dietary lipids, the lipids must first be absorbed from the small intestine. Since these molecules are oils, they are essentially insoluble in the aqueous environment of the intestine. The solubilization (or emulsification) of dietary lipids is therefore accomplished by means of bile salts, which are synthesized from cholesterol in the liver and then stored in the gallbladder; they are secreted following the ingestion of fat (Figure 21–1).

The emulsification of dietary fats renders them accessible to **pancreatic lipases**, primarily lipase and phospholipase A₂. These enzymes, secreted into the intestine from the pancreas, generate free fatty acids and a mixture of mono- and diacylglycerols from dietary triacylglycerols. Pancreatic lipase degrades triacylglycerols at the 1 and 3 positions sequentially to generate 1, 2-diacylglycerols and 2-acylglycerols. Phospholipids are degraded at the 2 position by pancreatic phospholipase A₂ releasing a free fatty acid and the lysophospholipid. The products of pancreatic lipases then diffuse into the intestinal epithelial cells, where resynthesis of triacyglycerols occurs.

Dietary triacylglycerols and cholesterol as well as triacylglycerols and cholesterol synthesized by the liver are solubilized in lipid-protein complexes identified as lipoproteins (Figure 21–2). These complexes contain triacylglycerol lipid droplets and cholesteryl esters surrounded by the polar phospholipids and proteins. The lipid and protein composition of lipoprotein complexes is variable depending upon site of synthesis and plasma life-span (see Table 21–1). The protein components of lipoprotein particles are identified as apolipoproteins and serve specific functions within the lipoprotein complex (Table 21–2).

Circulating Lipid Carriers

A. Chylomicrons: Chylomicrons are assembled in the intestinal mucosa as a means to transport dietary cholesterol and triacylglycerols to the rest of the body (Figure 21–1). Chylomicrons are, therefore, the molecules formed to mobilize dietary (exogenous) lipids. The predominant lipids of chylomicrons are triacylglycerols (Table 21–1). The apolipoproteins that predominate before the chylomicrons enter the circulation include apo-B-48 and apo-A-I, -II and -IV. Apo-B-48 combines only with chylomicrons.

Chylomicrons leave the intestine via the lymphatic system and enter the circulation at the left subclavian vein. In the bloodstream, chylomicrons acquire apo-C-II and apo-E from plasma HDLs (Figure 21–3). In the capillaries of adipose tissue and muscle, the fatty acids of chylomicrons are removed from the triacylglycerols by the action of **lipoprotein lipase (LPL)**, which is found on the

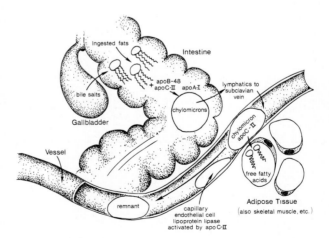

Figure 21–1. Intestinal absorption of fats. Dietary triacylglycerols are packaged into chylomicrons along with the apolipoproteins A-I, B-48 and C-II where they are delivered to the subclavian vein via the lymphatics. Apolipoprotein C-II activates capillary endothelial cell lipoprotein lipase leading to release of free fatty acids from the chylomicrons.

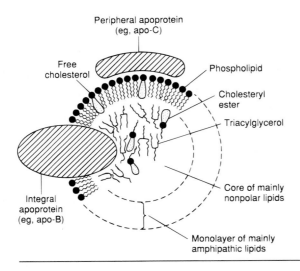

Figure 21–2. Generalized structure of a plasma lipoprotein. The similarities with the structure of the plasma membrane are to be noted. A small amount of cholesteryl ester and triacylglycerol are to be found in the surface layer and a little free cholesterol in the core. (Reproduced, with permission, from Murray, RK: *Harper's Biochemistry*, 23rd ed. Appleton & Lange, 1993.)

surface of the endothelial cells of the capillaries (Figure 21–1). **The apo-C-II in the chylomicrons activates LPL** in the presence of phospholipid (see Figure 21–1). The free fatty acids are then absorbed by the tissues and the glycerol backbone of the triacylglycerols is returned, via the blood, to the liver and kidneys. Glycerol is converted to the glycolytic intermediate DHAP and oxidized or used for glucose synthesis. During the removal of fatty acids, a substantial portion of the phospholipid and apo-As and apo-Cs is transferred to HDLs. The loss of apo-C-II prevents LPL from further degrading the **chylomicron remnants** (Figure 21–3).

Chylomicron remnants—containing primarily cholesterol, apo-E and apo-B-48—are then delivered to, and taken up by, the liver through interaction with the **chylomicron remnant receptor** (Figure 21–3). The recognition of chylomicron remnants by the **hepatic remnant receptor requires apo-E.** Chylomicrons function to deliver dietary triacylglycerols to adipose tissue and muscle and dietary cholesterol to the liver (Figure 21–3).

B. Very Low-Density Lipoproteins (VLDLs): The dietary intake of both fat and carbohydrate, in excess of the needs of the body, leads to their conversion into triacylglycerols in the liver. These triacylglycerols are packaged into VLDLs and released into the circulation for delivery to various tissues (primarily muscle and adipose tissue) for storage or production of energy through oxidation (Figure 21–4). VLDLs are, therefore, the molecules formed to transport endogenously derived triacylglycerols to extra-hepatic tissues. In addition to triacylglycerols, VLDLs contain some cholesterol and choles-

Table 21–1. Composition of the major lipoprotein complexes.

Complex	Source	Density (g/ml)	% Protein	% TG[a]	% PL[b]	% CE[c]	% C[d]	% FFA[e]
Chylomicron	Intestine	< 0.95	1–2	85–88	8	3	1	0
VLDL	Liver	0.95–1.006	7–10	50–55	18–20	12–15	8–10	1
IDL	VLDL	1.006–1.019	10–12	25–30	25–27	32–35	8–10	1
LDL	VLDL	1.019–1.063	20–22	10–15	20–28	37–48	8–10	1
*HDL$_2$	Intestine, liver (chylomicrons and VLDLs)	1.063–1.125	33–35	5–15	32–43	20–30	5–10	0
*HDL$_3$	Intestine, liver (chylomicrons and VLDLs)	1.125–1.21	55–57	3–13	26–46	15–30	2–6	6
Albumin-FFA	Adipose tissue	> 1.281	99	0	0	0	0	100

[a]Triacylglycerols, [b]Phospholipids, [c]Cholesteryl esters, [d]Free cholesterol, [e]Free fatty acids
*HDL$_2$ and HDL$_3$ derived from nascent HDL as a result of the acquisition of cholesteryl esters

Table 21–2. Apoprotein classifications.

Apoprotein-MW(Da)	Lipoprotein Association	Function and Comments
apo-A-I–29,016	Chylomicrons, HDL	Major protein of HDL, activates LCAT
apo-A-II–17,400	Chylomicrons, HDL	Primarily in HDL, enhances hepatic lipase actiivty
apo-A-IV–46,000	Chylomicrons and HDL	Present in triacylglycerol-rich lipoproteins
apo-B-48–241,000	Chylomicrons	Exclusively found in chylomicrons; derived from apo-B-100 gene by RNA editing in intestinal epithelium; lacks the LDL receptor-binding domain of apo-B-100
apo-B-100–513,000	VLDL, IDL, and LDL	Major protein of LDL; binds to LDL receptor; one of the longest known proteins in humans
apo-C-I–7,600	Chylomicrons, VLDL, IDL, and HDL	May also activate LCAT
apo-C-II–8,916	Chylomicrons, VLDL, IDL, and HDL	Activates lipoprotein lipase
apo-C-III–8,750	Chylomicrons, VLDL, IDL, and HDL	Inhibits lipoprotein lipase
apo-D–20,000; also called cholesterol ester transfer protein, CETP	HDL	Exclusively associated with HDL, cholesteryl ester transfer
apo-E–-34,000 (at least 3 alleles [E2, E3, E4] each of which may have multiple isoforms)	Chylomicron remnants, VLDL, IDL, and HDL	Binds to LDL receptor, apo-E-ϵ4 allele amplification associated with late-onset Alzheimer's disease
apo-H–50,000 (also known as β-2-glyco-protein I)	Chylomicrons	Triacylglycerol metabolism
apo(a)–at least 19 different alleles; protein ranges in size from 300,000–800,000	LDL	Disulfide bonded to apo-B-100, forms a complex with LDL identified as lipoprotein(a), Lp(a); strongly resembles plasminogen; may deliver cholesterol to sites of vascular injury; high-risk association with premature coronary artery disease and stroke

teryl esters and the apoproteins apo-B-100, apo-C-I, -II, -III and apo-E (Table 21–1). Like nascent chylomicrons, newly released VLDLs acquire apo-Cs and apo-E from circulating HDLs.

The fatty acid portion of VLDLs is released to adipose tissue and muscle in the same way as for chylomicrons, through the action of lipoprotein lipase. The action of lipoprotein lipase coupled to a loss of certain apoproteins (the apo-Cs) converts VLDLs to IDLs, also termed VLDL remnants. The apo-Cs are transferred to HDLs. The predominant remaining proteins are apo-B-100 and apo-E. Further loss of triacylglycerols converts IDLs to LDLs (Figure 21–4).

C. Intermediate-Density Lipoproteins (IDLs): IDLs are formed as triacylglycerols are removed from VLDLs. The fate of IDLs is either conversion to LDLs or direct uptake by the liver. Conversion of IDLs to LDLs occurs as more triacylglycerols are removed. The liver takes up IDLs after they have interacted with the LDL receptor to form a complex, which is endocytosed by the cell (see below, Figure 21–6). For LDL receptors in the liver to recognize IDLs requires the presence of both apo-B-100 and apo-E (the LDL receptor is also called the apo-B-100/apo-E receptor). The importance of apo-E in cholesterol uptake by LDL receptors has been demonstrated in transgenic mice lacking functional apo-E genes. These mice develop severe atherosclerotic lesions at 10 weeks of age.

D. Low-Density Lipoproteins (LDLs): The cellular requirements for cholesterol as a membrane component is satisfied in one of two ways: Either it is synthesized *de novo* within the cell, or it is supplied from extra-cellular sources, namely, chylomicrons and LDLs. As indicated above, the dietary cholesterol that goes into chylomicrons is supplied to the liver by the interaction of chylomicron remnants with the remnant receptor. In addition, cholesterol synthesized by the liver can be transported to extra-hepatic tissues if packaged in VLDLs. In the circulation VLDLs, are converted to LDLs through the action of lipoprotein lipase (Figure 21–4). LDLs are the primary plasma carriers of cholesterol for delivery to all tissues.

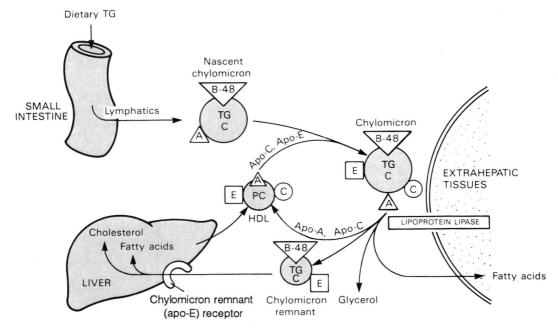

Figure 21–3. Metabolic fate of chylomicrons. A, apolipoprotein A; B-48, apolipoprotein B-48; ©, apolipoprotein C; E, apolipoprotein E; HDL, high-density lipoprotein; TG, triacylglycerol; C, cholesterol and cholesteryl ester; P, phospholipid. Only the predominant lipids are shown. (Reproduced, with permission, from Murray, RK: *Harper's Biochemistry*, 23rd ed. Appleton & Lange, 1993.)

The exclusive apolipoprotein of LDLs is apo-B-100. LDLs are taken up by cells via LDL receptor-mediated endocytosis, as described above for IDL uptake. The uptake of LDLs occurs predominantly in liver (75%), adrenals and adipose tissue. As with IDLs, the interaction of LDLs with LDL receptors requires the presence of apo-B-100 (Figure 21–4). The endocytosed membrane vesicles (endosomes) fuse with lysosomes, in which the apoproteins are degraded and the cholesterol esters are hydrolyzed to yield free cholesterol (Figure 21–6). The cholesterol is then incorporated into the plasma membranes as necessary. Excess intracellular cholesterol is re-esterified by **acyl-CoA-cholesterol acyltransferase (ACAT)**, for intracellular storage. The activity of ACAT is enhanced by the presence of intracellular cholesterol.

Insulin and tri-iodothyronine (T3) increase the binding of LDLs to liver cells, whereas glucocorticoids (eg, dexamethasone) have the opposite effect. The precise mechanism for these effects is unclear but may be mediated through the regulation of apo-B degradation. The effects of insulin and T3 on hepatic LDL binding may explain the hypercholesterolemia and increased risk of atherosclerosis that have been shown to be associated with uncontrolled diabetes or hypothyroidism.

An abnormal form of LDL, identified as lipoprotein-X (Lp-X), predominates in the circulation of patients suffering from LCAT deficiency (see Table 21–3) or cholestatic liver disease. In both cases there is an elevation in the level of circulating free cholesterol and phospholipids.

E. High-Density Lipoproteins (HDLs): HDLs are synthesized *de novo* in the liver and small intestine, as primarily protein-rich disc-shaped particles (Figure 21–5). These newly formed HDLs are nearly devoid of any cholesterol and cholesteryl esters. The primary apoproteins of HDLs are apo-A-I, apo-C-I, apo-C-II and apo-E. In fact, a major function of HDLs is to act as circulating stores of apo-C-I, apo-C-II and apo-E.

HDLs are converted into spherical lipoprotein particles through the accumulation of cholesteryl esters (Figure 21–5). This accumulation converts nascent HDLs to HDL_2 and HDL_3 (see Table 21–1). Any free cholesterol present in chylomicron and VLDL remnants (IDLs) can be esterified through the action of the HDL-associated enzyme, **lecithin-cholesterol acyl transferase, LCAT** (Figure 21–5). LCAT is synthesized in the liver. This enzyme is so named because it transfers a fatty acid from the C-2 position of lecithin (phosphatidylcholine; see Chapter 10) to the C-3-OH of cholesterol, generat-

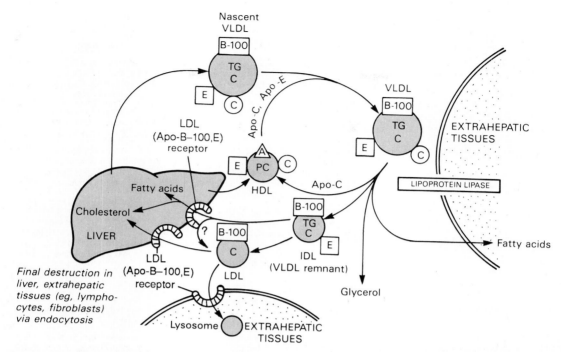

Figure 21–4. Metabolic fate of very low density lipoproteins (VLDL) and production of low-density lipoproteins (LDL). A, apolipoprotein A; B-100, apolipoprotein B-100; ©, apolipoprotein C; E, apolipoprotein E; HDL, high-density lipoprotein; TG, triacylglycerol; IDL, intermediate density lipoprotein; C, cholesterol and cholesteryl ester; P, phospholipid. Only the predominant lipids shown. It is possible that some IDL is also metabolized via the chylomicron remnant (apo-E) receptor. (Reproduced, with permission, from Murray, RK: *Harper's Biochemistry*, 23rd ed. Appleton & Lange, 1993.)

ing a cholesteryl ester and lysolecithin. The activity of LCAT requires interaction with apo-A-I, which is found on the surface of HDLs.

Cholesterol-rich HDLs return to the liver, where they are endocytosed. Hepatic uptake of HDLs, or **reverse cholesterol transport**, may be mediated through an HDL-specific apo-A-I receptor or through lipid-lipid interactions. Macrophages also take up HDLs through apo-A-I receptor interaction. HDLs can acquire cholesterol and apo-E from the macrophages; the cholesterol-enriched HDLs are then secreted from the macrophages. The increased apo-E in these HDLs leads to an increase in their uptake and catabolism by the liver.

HDLs also acquire cholesterol by extracting it from cell surface membranes. This process has the effect of lowering the level of intracellular cholesterol, since the cholesterol stored within cells as cholesteryl esters will be mobilized to replace the cholesterol removed from the plasma membrane (Figure 21–5).

The cholesterol esters of HDLs can also be transferred to VLDLs and LDLs through the action of the HDL-associated enzyme, **cholesterol ester transfer protein (CETP**, also identified as **apo-D)**. This has the added effect of allowing the excess cellular cholesterol to be returned to the liver through the LDL-receptor pathway as well as the HDL-receptor pathway.

LDL Receptors

The LDLs are the principal plasma carriers of cholesterol, delivering cholesterol from the liver (via hepatic synthesis of VLDLs) to peripheral tissues, primarily the adrenals and adipose tissue. LDLs also return cholesterol to the liver. The cellular uptake of cholesterol from LDLs occurs following the interaction of LDLs with the LDL receptor (also called the apo-B-100/apo-E receptor). The sole apoprotein present in LDLs is apo-B-100, which is required for interaction with the LDL receptor.

The LDL receptor is a polypeptide of 839 amino acids that spans the plasma membrane (Figure 21–6). An extracellular domain is responsible for apo-B-100/apo-E binding. The intracellular domain is responsible for the clustering of LDL receptors into regions of the plasma membrane termed "coated pits." Once LDL binds the receptor, the complexes are rapidly internalized (endocytosed). ATP-

Table 21–3. Hyperlipoproteinemias.

Disorder	Defect	Comments
Type I (familial LPL deficiency, familial hyperchylomicronemia)	(a) deficiency of LPL; (b) production of abnormal LPL; (c) apo-C-II deficiency	Slow chylomicron clearance, reduced LDL and HDL levels; treated by low fat/complex carbohydrate diet; no increased risk of coronary artery disease
Type II (FH)	Four classes of LDL receptor defect (see text above)	Reduced LDL clearance leads to hypercholesterolemia, resulting in athersclerosis and coronary artery disease
Type III (familial dysbetalipoproteinemia, remnant removal disease, broad beta disease)	Hepatic remnant clearance impaired due to apo-E abnormality; patients only express the apo-E2 isoform that interacts poorly with the apo-E receptor	Causes xanthomas, hypercholesterolemia and athersclerosis in peripheral and coronary arteries due to elevated levels of chylomicrons and VLDLs
Type IV (familial hypertriacylglcerolemia)	Elevated production of VLDL associated with glucose intolerance and hyperinsulinemia	Frequently associated with type-II (non-insulin dependent) diabetes mellitus, obesity, alcoholism or administration of progestational hormones; elevated cholesterol as a result of increased VLDLs
Type V familial	Elevated chylomicrons and VLDLs due to unknown cause	Hypertriacylglycerolemia and hypercholesterolemia with decreased LDLs and HDLs
Familial hyperalphalipoproteinemia	Increased level of HDLs	A rare condition that is beneficial for health and longevity
Familial hyperapobetalipoproteinemia	Increased LDL production and delayed clearance of triacylglycerols and fatty acids	Strongly associated with increased risk of coronary artery disease
Familial defective apo-B-100	Gln for Arg mutation in exon 26 of apo-B gene (amino acid 3500); leads to reduced affinity of LDL for LDL receptor	Dramatic increase in LDL levels; no affect on HDL, VLDL or plasma triglyceride levels; significant cause of hypercholesterolemia and premature coronary artery disease
Familial LCAT deficiency	Absence of LCAT leads to inability of HDLs to take up cholesterol (reverse cholesterol transport)	Decreased levels of plasma cholesteryl esters and lysolecithin; abnormal LDLs (Lp-X) and VLDLs; symptoms also found associated with cholestasis
Wolman's disease (cholesteryl ester storage disease)	Defect in lysosomal cholesteryl ester hydrolase, affects metabolism of LDLs	Reduced LDL clearance leads to hypercholesterolemia resulting in athersclerosis and coronary artery disease
Hormone-releasable hepatic lipase deficiency	Deficiency of the lipase leads to accumulation of triacylglycerol-rich HDLs and VLDL remnants (IDLs)	Causes xanthomas and coronary artery disease

dependent proton pumps lower the pH in the endosomes, which results in dissociation of the LDL from the receptor. The portion of the endosomal membranes harboring the receptor are then recycled to the plasma membrane, and the LDL-containing endosomes fuse with lysosomes. Acid hydrolases of the lysosomes degrade the apoproteins and release free fatty acids and cholesterol (Figure 21–6). As indicated above, the free cholesterol is either incorporated into plasma membranes or esterified (by ACAT) and stored within the cell.

The level of intracellular cholesterol is regulated through cholesterol-induced suppression of LDL receptor synthesis and cholesterol-induced inhibition of cholesterol synthesis. The increased level of intracellular cholesterol that results from LDL uptake has the additional effect of activating ACAT, thereby allowing the storage of excess cholesterol within cells. However, the effect of cholesterol-induced suppression of LDL receptor synthesis is a *decrease* in the rate at which LDLs and IDLs are removed from the serum. This can lead to excess circulating levels of cholesterol and cholesteryl esters when the dietary intake of fat and cholesterol exceeds the needs of the body. The excess cholesterol tends to be deposited in the skin, tendons and (more gravely) within the arteries, leading to atherosclerosis.

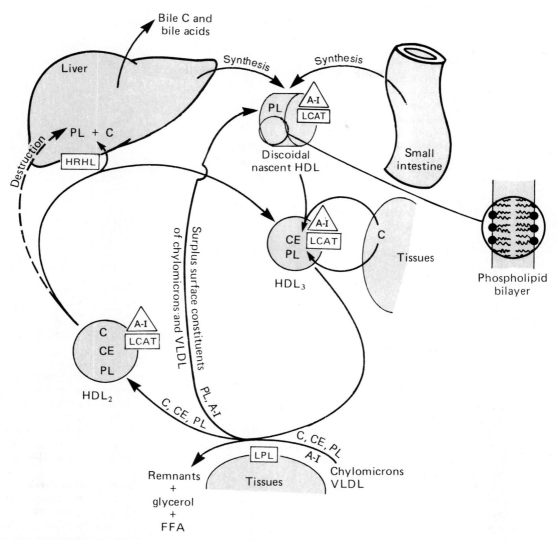

Figure 21–5. Metabolism of high-density lipoprotein (HDL). HRHL, heparin-releasable hepatic lipase; LCAT, lecithin: cholesterol acyltransferase; LPL, lipoprotein lipase; C, cholesterol; CE, cholesteryl ester, PL, phospholipid; FFA, free fatty acids; A-I, apoprotein A-I. The figure illustrates the role of the 3 enzymes HRHL, LCAT, and LPL in the postulated HDL cycle for the transport of cholesterol from the tissues to the liver. HDL_2, HDL_3—see Table 27–2. In addition to triacylglycerol, HRHL hydrolyzes phospholipid on the surface of HDL_2, releasing cholesterol for uptake into the liver, allowing formation of smaller and more dense HDL_3. HRHL activity is increased by androgens and decreased by estrogens, which may account for higher concentrations of plasma HDL_2 in women. (Reproduced, with permission, from Murray, RK: *Harper's Biochemistry*, 23rd ed. Appleton & Lange, 1993.)

CLINICAL SIGNIFICANCE

Fortunately, few individuals carry the inherited defects in lipoprotein metabolism that lead to **hyper-** or **hypolipoproteinemias** (Tables 21–3 and 21–4). Persons suffering from diabetes mellitus, hypothyroidism and kidney disease may often exhibit abnormal lipoprotein metabolism as a result of secondary effects of their disorders. For example, because lipoprotein lipase (LPL) synthesis is regulated by insulin, LPL deficiencies leading to type I hyperlipoproteinemia may occur as a secondary outcome of diabetes mellitus. Additionally, insulin and thyroid hormones positively affect hepatic LDL-receptor interactions; therefore, the hypercholesterolemia and increased risk of athersclerosis as-

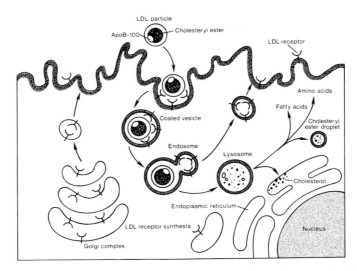

Figure 21–6. Cycle of LDL receptor-LDL particle internalization. In the endosomes the LDLs are degraded and the free LDL-receptor is recycled to the plasma membrane.

sociated with uncontrolled diabetes or hypothyroidism is likely due to decreased hepatic LDL uptake and metabolism.

Of the disorders listed in Tables 21–3 and 21–4, familial hypercholesterolemia (FH) may be the most prevalent in the general population. Heterozygosity at the FH locus occurs in 1:500 individuals, whereas homozygosity is observed in 1:1,000,000 individuals. FH is an inherited disorder comprising four different classes of mutation in the LDL receptor gene. The class 1 defect (the most common) results in a complete loss of receptor synthesis. The class 2 defect results in the synthesis of a receptor protein that is not properly processed in the Golgi apparatus, and therefore is not transported to the plasma membrane. The class 3 defect results in an LDL receptor that is incapable of binding LDLs. The class 4 defect results in receptors that bind LDLs but do not cluster in coated pits and are therefore not internalized.

FH sufferers may be either heterozygous or homozygous for a particular mutation in the receptor gene. Homozygotes exhibit grossly elevated serum cholesterol (primarily in LDLs). The elevated levels of LDLs result in their phagocytosis by macrophages. These lipid-laden phagocytic cells tend to deposit within the skin and tendons, leading to **xanthomas**. A greater complication results from cholesterol deposition within the arteries, leading to **atherosclerosis**—the major contributing factor of nearly all cardiovascular diseases.

Table 21–4. Hypolipoproteinemias.		
Disorder	**Defect**	**Comments**
Abetalipoproteinemia (acanthocytosis, Bassen-Kornzweig syndrome)	No chylomicrons, VLDLs or LDLs, due to defect in apo-B expression	Rare defect; intestine and liver accumulate, malabsorption of fat, retinitis pigmentosa, ataxic neuropathic disease; erythrocytes have "thorny" appearance
Familial hypobetalipoproteinemia	At least 20 different apoB gene mutations identified; LDL concentrations 10–20% of normal; VLDL slightly lower, HDL normal	Mild or no pathological changes
Familial alpha-lipoprotein deficiency (Tangier disease, Fish-eye disease, apo-A-1 and -C-III deficiencies)	All of these related syndromes have reduced HDL concentrations, no effect on chylomicron or VLDL production	Tendency to hypertriacylglycerolemia; some elevation in VLDLs; Fish-eye disease characterized by severe corneal opacity

Pharmacologic Intervention

Drug treatment to lower plasma lipoproteins and/or cholesterol is primarily aimed at reducing the risk of athersclerosis and subsequent coronary artery disease that exists in patients with elevated circulating lipids. Drug therapy usually is considered as an option only if nonpharmacologic interventions (altered diet and exercise) have failed to lower plasma lipids.

A. Mevinolin, Mevastatin, Lovastatin: These drugs are fungal HMG-CoA reductase (see Chapter 22) inhibitors. The net result of treatment is an increased cellular uptake of LDLs, since the intracellular synthesis of cholesterol is inhibited and cells are therefore dependent on extracellular sources of cholesterol. However, since mevalonate (the product of the HMG-CoA reductase reaction) is required for the synthesis of other important isoprenoid compounds besides cholesterol, long-term treatments carry some risk of toxicity.

B. Nicotinic Acid: Nicotinic acid reduces the plasma levels of both VLDLs and LDLs by inhibiting hepatic VLDL secretion, as well as suppressing the flux of fatty acid release from adipose tissue by inhibiting lipolysis. Because of its ability to cause large reductions in circulating levels of cholesterol and triacylglycerol, nicotinic acid is used to treat Type II, III, IV, and V hyperlipoproteinemias.

C. Clofibrate, Gemfibrozil, Fenofibrate: These compounds are derivatives of fibric acid and promote rapid VLDL turnover by activating lipoprotein lipase. They also induce the diversion of hepatic free fatty acids from esterification reactions to those of oxidation, thereby decreasing the liver's secretion of triacylglycerol and cholesterol.

D. Probucol: Probucol increases the rate of LDL metabolism and may block the intestinal transport of cholesterol. The net result is a significant reduction in plasma cholesterol levels.

E. Cholestyramine or Colestipol: These compounds are nonabsorbable resins that bind bile acids, which are then not reabsorbed by the liver but excreted. The drop in hepatic reabsorption of bile acid releases a feedback mechanism that had been inhibiting bile acid synthesis (see Chapter 22). As a result, a greater amount of cholesterol is converted to bile acids to maintain a steady level in circulation. Additionally, the synthesis of LDL receptors increases to allow increased cholesterol uptake for bile acid synthesis, and the overall effect is a reduction in plasma cholesterol. (This treatment is ineffective in homozygous FH patients, since they are completely deficient in LDL receptors.)

Questions

DIRECTIONS (items 1–11): Each numbered item or incomplete statement is followed by answers or by completion of the statement. Select ONE lettered answer or completion that is BEST in each case.

1. The principal plasma carrier of dietary triacylglycerols is:
    a. Serum albumins.
    b. LDLs.
    c. HDLs.
    d. Chylomicrons.
    e. VLDLs.

2. One major function of the HDLs is to:
    a. Deliver cholesterol from the liver to non-hepatic tissues.
    b. Transfer cholesterol from peripheral tissues to the liver.
    c. Transfer cholesterol from chylomicrons to VLDLs.
    d. Catalyze the synthesis of ACAT.
    e. Catalyze the synthesis of LCAT.

3. In order for chylomicron remnants to be taken up by the liver, which apoprotein in the remnant must interact with the remnant receptor?
   a. Apo-E.
   b. Apo-B-100.
   c. Apo-B-48.
   d. Apo-A-I.
   e. Apo-C-II.

4. The level of the abnormal LDL identified as Lp-X is elevated in the plasma of patients suffering from:
   a. Familial hyperalphalipoproteinemia.
   b. Wolman disease.
   c. Familial defective apo-B-100.
   d. Familial LCAT deficiency.
   e. Familial hypertriacylglycerolemia.

5. Which of the following drugs exerts its effects on serum cholesterol levels by causing an increased synthesis of LDL receptors?
   a. Cholestyramine.
   b. Probucol.
   c. Lovastatin.
   d. Nicotinic acid.
   e. Gemfibrozil.

6. Activation of lipoprotein lipase requires which apoprotein?
   a. Apo-C-I.
   b. Apo-A-I.
   c. Apo-E.
   d. Apo-B-100.
   e. Apo-C-II.

7. The major function of apo-D is:
   a. To interact with the LDL receptor to allow hepatic uptake of LDLs.
   b. To interact with the hepatic remnant receptor to allow uptake of chylomicron remnants.
   c. To transfer cholesterol from HDLs to VLDLs and LDLs.
   d. To inhibit the uptake of LDLs by macrophages.
   e. To activate esterification of free cholesterol in LDLs and VLDLs.

8. Patients suffering from familial defective apolipoprotein B-100 have significantly elevated levels of plasma LDLs. The major reason for this is:
   a. The inability of LDLs to interact with the LDL receptor.
   b. The inability of LDLs to activate lipoprotein lipase.
   c. The inability of LDLs to activate cholesterol transfer to HDLs.
   d. The genetic defect additionally results in an elevated synthesis of LDLs.
   e. The LDL interaction with LDL receptors failing to stimulate endocytosis.

9. Familial hypercholesterolemia is characterized by:
   a. An increase in cholesterol biosynthesis beyond the needs of the body.
   b. Defects in LDL receptor structure and/or function.
   c. Defects in the synthesis of the apoprotein responsible for activation of lipoprotein lipase.
   d. An increase in the hepatic synthesis of VLDLs.
   e. A decrease in the ability of liver to take up chylomicron remnants.

10. Which apoprotein is found exclusively associated with chylomicrons?
    a. Apo-B-100.
    b. Apo-B-48.
    c. Apo-E.
    d. Apo-A-I.
    e. Apo-A-IV.

11. What is meant by the term "reverse cholesterol transport"?
    a. Transport of cholesterol from LDLs to VLDLs.
    b. Transport of cholesterol from macrophages to chylomicron remnants.
    c. Transport of cholesterol from HDLs to chylomicron remnants.
    d. Transport of cholesterol from peripheral tissue to the liver.
    e. Transport of cholesterol from the liver to the intestine.

DIRECTIONS (items 12–18): Match the following apoproteins with either the most correct function or the most correct lipoprotein particle association. Each answer is to be used only once.

12. Apo-C-II		a. Resembles plasminogen.
13. Apo-B-48		b. Activates lipoprotein lipase.
14. Apo-B-100 D		c. Exclusively with chylomicrons.
15. Apo-A-I		d. Binds the LDL receptor.
16. Apo-D		e. Exclusively with HDLs.
17. Apo(a)		f. Activates hepatic lipase.
18. Apo-A-II		g. Activates LCAT.

Answers

1. d	**7.** c	**13.** c
2. b	**8.** a	**14.** d
3. a	**9.** b	**15.** g
4. d	**10.** b	**16.** e
5. a	**11.** d	**17.** a
6. e	**12.** b	**18.** f

Cholesterol Metabolism

22

Objectives

- To describe the major steps in the biosynthesis of cholesterol.
- To describe the mechanisms used to control cholesterol biosynthesis.
- To describe the pathways through which cholesterol is utilized.
- To describe the synthesis of the bile acids.
- To describe the clinical significances of bile acid synthesis.

Concepts

Cholesterol is an extremely important biological molecule that has roles in membrane structure as well as being a precursor for the synthesis of the steroid hormones and bile acids. Both dietary cholesterol and that synthesized *de novo* are transported through the circulation in lipoprotein particles. The same is true of cholesteryl esters, the form in which cholesterol is stored within cells.

The synthesis and utilization of cholesterol must be tightly regulated in order to prevent over-accumulation and abnormal deposition within the body (see Chapter 21). Of particular importance clinically is the abnormal deposition of cholesterol and cholesterol-rich lipoproteins in the coronary arteries. Such deposition, eventually leading to atherosclerosis, is the leading contributory factor in diseases of the coronary arteries.

Biosynthesis of Cholesterol

Slightly less than half of the cholesterol in the body derives from biosynthesis *de novo*. Biosynthesis in the liver accounts for approximately 10%, and in the intestines approximately 15%, of the amount produced each day. Cholesterol synthesis occurs in the cytoplasm and microsomes from the two-carbon acetate group of acetyl-CoA. The process has five major steps: (1) acetyl-CoAs are converted to 3-hydroxy-3-methylglutaryl-CoA (HMG-CoA); (2) HMG-CoA is converted to mevalonate; (3) mevalonate is converted to the isoprene-based molecule, isopentenyl pyrophosphate (IPP), with the concomitant loss of CO_2; (4) IPP is converted to squalene; (5) squalene is converted to cholesterol.

The acetyl-CoA utilized for cholesterol biosynthesis is derived from an oxidation reaction (eg, fatty acids or pyruvate) in the mitochondria and is transported to the cytoplasm by the same process as that described for fatty acid synthesis (see Chapter 20). Acetyl-CoA can also be derived from cytoplasmic oxidation of ethanol by acetyl-CoA synthetase. All the reduction reactions of cholesterol biosynthesis use NADPH. The isoprenoid intermediates of cholesterol biosynthesis can be diverted to other synthesis reactions, such as those for dolichol, coenzyme Q or the side chain of heme A.

Acetyl-CoA units are converted to mevalonate by a series of reactions that begins with the formation of **HMG-CoA** (Figure 22–1). Unlike the HMG-CoA formed during ketone body synthesis in the mitochondria (see Chapter 20), this form is synthesized in the cytoplasm. However, the pathway and

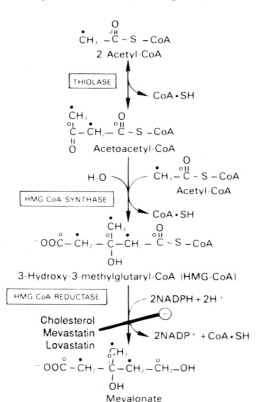

Figure 22–1. Biosynthesis of mevalonate, HMG, 3-hydroxy-3-methylglutaryl, HMG-CoA reductase is inhibited by cholesterol and the fungal metabolites mevastatin (Compactin) and lovastatin (mevinolin), which are competitive with HMG-CoA. (Reproduced, with permission, from Murray, RK: *Harper's Biochemistry*, 23rd ed. Appleton & Lange, 1993.)

the necessary enzymes are the same as those in the mitochondria. Two moles of acetyl-CoA are condensed in a reversal of the **thiolase** reaction, forming acetoacetyl-CoA. Acetoacetyl-CoA and a third mole of acetyl-CoA are converted to HMG-CoA by the action of **HMG-CoA synthase**. HMG-CoA is converted to mevalonate by **HMG-CoA reductase** (an enzyme bound to the endoplasmic reticulum). HMG-CoA reductase absolutely requires NADPH as a cofactor, and 2 moles of NADPH are consumed during the conversion of HMG-CoA to mevalonate (Figure 22–1). **The reaction catalyzed by HMG-CoA reductase is the rate-limiting step of cholesterol biosynthesis, and this enzyme is subject to complex regulatory controls.**

Mevalonate is then activated by three successive phosphorylations, yielding **5-pyrophospho-mevalonate** (Figure 22–2). In addition to activating mevalonate, the phosphorylations maintain its solubility, since otherwise it is insoluble in water. After phosphorylation, an ATP-dependent decarboxylation yields **isopentenyl pyrophosphate**, IPP, an activated isoprenoid molecule (Figure 22–2). Isopentenyl pyrophosphate is in equilibrium with its isomer, **dimethylallyl pyrophosphate, DMPP**. One molecule of IPP condenses with one molecule of DMPP to generate **geranyl pyrophosphate, GPP**. GPP further condenses with another IPP molecule to yield **farnesyl pyrophosphate, FPP**. Finally, the NADPH-requiring enzyme, **squalene synthase**, catalyzes the head-to-tail condensation of two molecules of FPP, yielding squalene (squalene synthase also is tightly associated with the endoplasmic reticulum). Squalene undergoes a two-step cyclization to yield **lanosterol.** The first reaction is catalyzed by **squalene monooxygenase**. This enzyme uses NADPH as a cofactor to introduce molecular oxygen as an epoxide at the 2,3 position of squalene. Through a series of 19 additional reactions, lanosterol is converted to cholesterol (Figure 22–2).

Regulating Cholesterol Metabolism

Normal healthy adults synthesize cholesterol at a rate of approximately 1g/day and consume approximately 0.3g/day. A relatively constant level of cholesterol in the body (150–200 mg/dL) is maintained primarily by controlling the level of *de novo* synthesis. The level of cholesterol synthesis is regulated in part by the dietary intake of cholesterol. Cholesterol from both diet and synthesis is utilized in the formation of membranes (Chapter 12) and in the synthesis of the steroid hormones and bile acids (see below). The greatest proportion of cholesterol is used in bile acid synthesis.

The cellular supply of cholesterol is maintained at a steady level by three distinct mechanisms regulation of HMG-CoA reductase activity and levels; regulation of excess intracellular free cholesterol through the activity of acyl-CoA:cholesterol acyltransferase, ACAT (see Chapter 21); and regulation of plasma cholesterol levels via LDL receptor-mediated uptake and HDL-mediated reverse transport (see Chapter 21). These mechanisms are illustrated in Figure 22–3.

Regulation of HMG-CoA reductase activity is the primary means for controlling the level of cholesterol biosynthesis. This regulation takes place both through covalent modification and through control of the absolute level of the enzyme within cells. HMG-CoA reductase is a single polypeptide embedded in microsomal membranes; it is most active in its unmodified form. Phosphorylation of the enzyme decreases its activity (Figure 22–4).

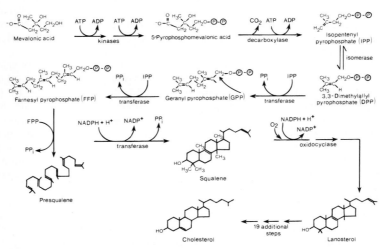

Figure 22–2. Major steps in the synthesis of cholesterol from mevalonic acid.

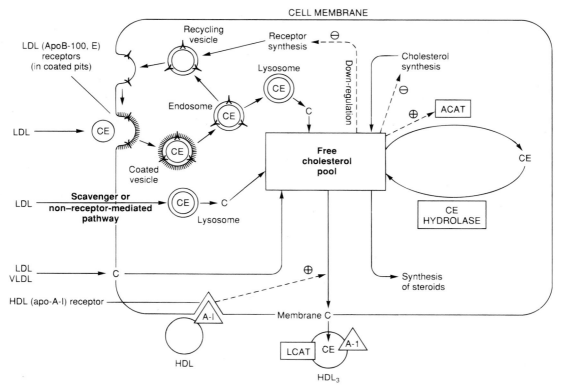

Figure 22–3. Factors affecting cholesterol balance at the cellular level. Reverse cholesterol transport may be initiated by HDL binding to an HDL (ap0-A-I-) receptor, which via protein kinase C, stimulates translocation of cholesterol to the plasma membrane. C, cholesterol; CE, choesteryl ester; ACAT, acyl-CoA: cholesterol acultransferase; LCAT, lecithin: cholesterol acyltransferase; A-1, apoprotein A-1; LDL, low-density lipoprotein; VLDL, very low density lipoprotein. LDL and HDL are not shown to scale. (Reproduced, with permission, from Murray, RK: *Harper's Biochemistry*, 23rd ed. Appleton & Lange, 1993.)

HMG-CoA reductase is phosphorylated by **reductase kinase, RK**. RK itself is activated via phosphorylation. The phosphorylation of RK is catalyzed by **reductase kinase kinase, RKK**. There are two isoforms of RKK, one independent of cAMP and one dependent upon cAMP. The cAMP-dependent RKK is activated in the presence of cAMP. Since the intracellular level of cAMP is regulated by hormonal stimuli, regulation of cholesterol biosynthesis is hormonally controlled. Insulin leads to a decrease in cAMP, which in turn activates cholesterol synthesis. Alternatively, glucagon and epinephrine—which increase the level of cAMP—inhibit cholesterol synthesis.

The activity of HMG-CoA reductase is further controlled by the cAMP signaling pathway. Increases in cAMP lead to phosphorylation and activation of **phosphoprotein phosphatase inhibitor-1 (PPI-1)**. HMG-CoA reductase is dephosphorylated and, thereby, activated by **phosphoprotein phosphatase-1 (PP-1)**. At the same time, PP-1 dephosphorylates and inactivates RK. The cAMP-induced activation of PPI-1 results in a decrease in the level of active PP-1, resulting in a reduced ability of PP-1 to dephosphorylate HMG-CoA reductase and RK. This maintains RK in the phosphorylated and active state, and HMG-CoA reductase in the phosphorylated and inactive state. As the stimulus leading to increased cAMP production is removed, the level of phosphorylations decreases and that of dephosphorylations increases. The net result is a return to a higher level of HMG-CoA reductase activity.

The ability of insulin to stimulate, and glucagon to inhibit, HMG-CoA reductase activity is consistent with the effects of these hormones on other metabolic pathways. The basic function of these two hormones is to control the availability and delivery of energy to all cells of the body (see Chapter 28).

Long-term control of HMG-CoA reductase activity is exerted primarily through control over the synthesis and degradation of the enzyme. When levels of cholesterol are high, the level of expression

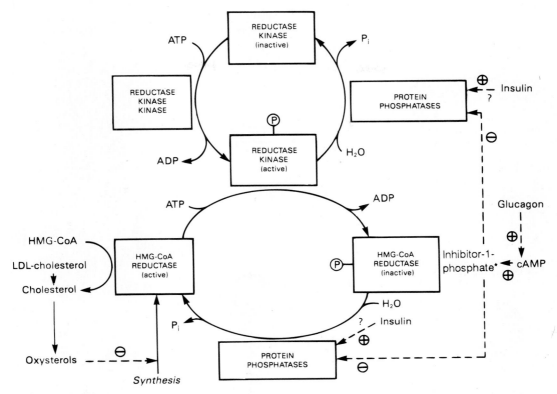

Figure 22–4. Possible mechanisms in the regulation of cholesterol synthesis by HMG-CoA reductase. Insulin has a dominant role compared with glucagon. (Reproduced, with permission, from Murray, RK: *Harper's Biochemistry*, 23rd ed. Appleton & Lange, 1993.)

of the HMG-CoA gene is reduced. Conversely, reduced levels of cholesterol activate expression of the gene. Insulin also brings about long-term regulation of cholesterol metabolism by increasing the level of HMG-CoA reductase synthesis. The rate of HMG-CoA turnover is also regulated by the supply of cholesterol. When cholesterol is abundant, the rate of HMG-CoA reductase degradation increases.

The Utilization of Cholesterol

Cholesterol is transported in the plasma predominantly as cholesteryl esters associated with lipoproteins (Figure 22–5; see also Chapter 21). Dietary cholesterol is transported from the small intestine to the liver within chylomicrons. Cholesterol synthesized by the liver, as well as any dietary cholesterol in the liver that exceeds hepatic needs, is transported in the serum within LDLs. As described in Chapter 21, the liver synthesizes VLDLs and these are converted to LDLs through the action of endothelial cell-associated lipoprotein lipase. Cholesterol found in plasma membranes can be extracted by HDLs and esterified by the HDL-associated enzyme LCAT. The cholesterol acquired from peripheral tissues by HDLs can then be transferred to VLDLs and LDLs via the action of cholesteryl ester transfer protein (apo-D), which is associated with HDLs. As explained in Chapter 21, "reverse cholesterol transport" allows peripheral cholesterol to be returned to the liver in LDLs. Ultimately, cholesterol is excreted in the bile as free cholesterol or as bile acids, after conversion in the liver.

Bile Acids

The end products of cholesterol utilization are the bile acids (Figure 22–6), synthesized in the liver. Synthesis of bile acids is the predominant mechanisms for the excretion of excess cholesterol. However, the excretion of cholesterol in this form is insufficient to compensate for an excess dietary intake of cholesterol.

The most abundant bile acids in human bile are **chenodeoxycholic acid** (45%) and **cholic acid** (31%). The carboxyl group of bile acids is conjugated via an amide bond to either glycine or taurine

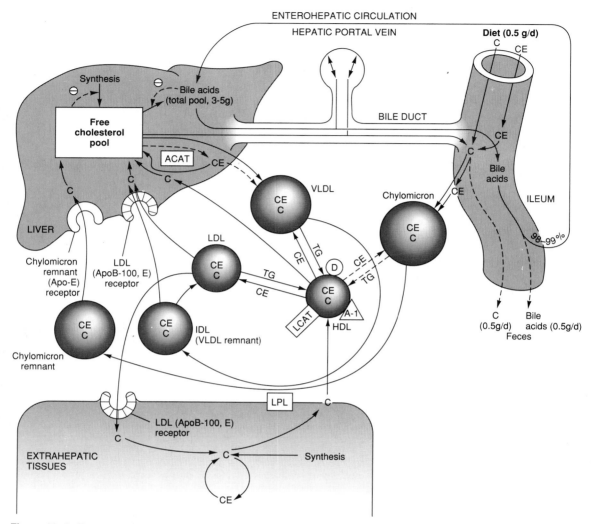

Figure 22–5. Transport of cholesterol between the tissues in humans. C, free cholesterol; CE, cholesteryl ester; VLDL, very low density lipoprotein; IDL, intermediate-density lipoprotein; LDL, low-density lipoprotein; HDL, high-density lipoprotein; ACAT, acyl-CoA:cholesterol acyltransferase; LCAT, lecithin:cholesterol acyltransferase; A-1, apoprotein A-1: D, cholesteryul ester transfer protein; LPL lipoprotein lipase. (Reproduced, with permission, from Murray, RK: *Harper's Biochemistry*, 23rd ed. Appleton & Lange, 1993.)

before their secretion into the bile canaliculi (Figure 22–5). These conjugation reactions yield **glycocholic acid** and **taurocholic acid**, respectively. The bile canaliculi join with the bile ductules, which then form the bile ducts. Bile acids are carried from the liver through these ducts to the gallbladder, where they are stored for future use. The ultimate fate of bile acids is secretion into the intestine, where they aid in the emulsification of dietary lipids. In the gut the glycine and taurine residues are removed and the bile acids are either excreted (only a small percentage) or reabsorbed by the gut and returned to the liver. This process is termed the **enterohepatic circulation** (see Figure 22–5).

CLINICAL SIGNIFICANCE

Bile acids perform four physiologically significant functions:

1. Their synthesis and subsequent excretion in the feces represent the only significant mechanism for the elimination of excess cholesterol.

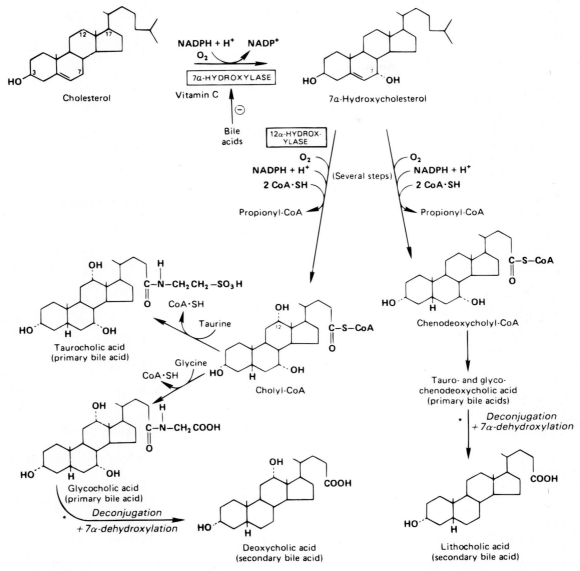

Figure 22–6. Biosynthesis and degradation of bile acids. * Catalyzed by microbial enzymes. (Reproduced, with permission, from Murray, RK: *Harper's Biochemistry*, 23rd ed. Appleton & Lange, 1993.)

2. Bile acids and phospholipids solubilize cholesterol in the bile, thereby preventing the precipitation of cholesterol in the gallbladder.

3. They facilitate the digestion of dietary triacylglycerols by acting as emulsifying agents that render fats accessible to pancreatic lipases.

4. They facilitate the intestinal absorption of fat-soluble vitamins.

Questions

DIRECTIONS (items 1–12): Each numbered item or incomplete statement is followed by answers or by completions of the statement. Select ONE lettered answer or completion that is BEST in each case.

1. The commitment to cholesterol biosynthesis occurs during the action of which enzyme?
   a. Squalene synthase.
   b. HMG-CoA synthetase.
   c. HMG-CoA reductase.
   d. HMG-CoA dehydrogenase.
   e. Mevalonate kinase.

2. The rate-limiting step in cholesterol biosynthesis:
   a. Occurs during production of mevalonic acid.
   b. Is regulated by the hormones glucagon and insulin.
   c. Is allosterically affected by isopentenyl pyrophosphate.
   d. Occurs during the production of squalene.
   e. Depends upon the level of active acetyl-CoA synthase.

3. VLDLs are synthesized in the liver for transport of which of the following to peripheral tissues:
   a. Triacylglycerols.
   b. LDL receptors.
   c. Apoproteins for HDL synthesis.
   d. Apolipoprotein-B-100.
   e. Phospholipids.

4. High dietary intake of cholesterol leads to:
   a. Increased HMG-CoA reductase synthesis.
   b. Decreased apo-B-100 synthesis.
   c. Increased apo-E receptor synthesis.
   d. Decreased ACAT activity.
   e. Decreased LDL receptor synthesis.

5. Inhibition of HMG-CoA reductase as a means to lower plasma cholesterol may have detrimental long-term side effects. This is most likely due to the fact that:
   a. Long-term reduction in cholesterol affects cell membrane integrity.
   b. The product of the reaction is required for the synthesis of physiologically important molecules other than cholesterol.
   c. Long-term reduction in cholesterol synthesis affects the ability of the body to emulsify dietary lipids.
   d. The product of the reaction is required in the production of HDLs by the liver.
   e. The product of the reaction also binds to dolichol phosphate and activates it during the synthesis of glycoproteins.

6. The hypercholesterolemia that is associated with familial hypercholesterolemia is due to:
   a. Abnormally reduced production of HDLs, resulting in a diminished capacity of reverse cholesterol transport.
   b. Abnormally elevated production of LDLs.
   c. Defective LDL receptors leading to reduced uptake of plasma LDLs and IDLs.
   d. Abnormally elevated production of VLDLs by the liver.
   e. Defective remnant receptors on hepatic cells, leading to reduced uptake of chylomicron remnants.

7. The pharmacological benefit of Lovastatin is in the fact that this drug:
   a. Inhibits HMG-CoA reductase, which leads to a reduction in cholesterol synthesis.
   b. Inhibits the intestinal absorption of dietary cholesterol.
   c. Inhibits hepatic production of VLDLs.
   d. Increases the rate of LDL metabolism.
   e. Absorbs bile acids, increasing their rate of elimination in the feces.

8. The major function of bile acids is to:
   a. Form the core of HDLs, allowing them to remove cholesterol from membranes.
   b. Form the surface of VLDLs in order to prevent loss of cholesterol.
   c. Provide precursor carbons for cholesteryl ester synthesis.

    d. Aid in the emulsification of dietary lipids.
    e. Form the core of VLDLs to which cholesteryl esters attach.

9. Regulation of the intracellular levels of cholesterol is predominantly effected by:
    a. The rate of LDL receptor synthesis.
    b. The total intake of cholesterol in the diet.
    c. The rate of lipoprotein lipase synthesis.
    d. The activity of LCAT in the liver.
    e. The rate of bile acid transport in the enterohepatic circulation.

10. The hypercholesterolemia that is associated with type III hyperlipoproteinemia is due to:
    a. Increased hepatic synthesis of VLDLs.
    b. Decreased hepatic synthesis of HDLs.
    c. Loss of LDL receptor synthesis.
    d. Increased hepatic cholesterol synthesis, due to elevated activity of HMG-CoA reductase.
    e. Impaired hepatic clearance of chylomicron remnants.

11. The ability of insulin to increase cholesterol biosynthesis is due primarily to:
    a. Insulin-stimulated phosphorylation of HMG-CoA reductase.
    b. Insulin-stimulated reduction in the intracellular level of cAMP.
    c. Insulin-stimulated production of intracellular cAMP.
    d. Insulin-stimulated phosphorylation of HMG-CoA reductase kinase.
    e. Insulin-stimulated synthesis of HMG-CoA reductase kinase.

12. Several of the enzymes of cholesterol biosynthesis require which of the following co-factors?
    a. $FADH_2$.
    b. Biotin.
    c. Thiamine pyrophosphate.
    d. NADPH.
    e. NADH.

Answers

1. a	**5.** b	**9.** b
2. b	**6.** c	**10.** e
3. a	**7.** a	**11.** b
4. e	**8.** d	**12.** d

Nitrogen Metabolism
23

Objectives

- To identify the reactions and enzymes that allow glutamate and glutamine to act as collectors and carriers of nitrogen and as a reservoir of carbon for anapleurotic Krebs cycle reactions.
- To identify the amino acids that are usually considered to be essential in the diet and to explain how some nonessential amino acids may become dietarily essential.

- To define positive nitrogen balance and negative nitrogen balance; to identify physiological conditions that may produce these states.
- To identify the role of ubiquitin in protein degradation.
- To identify the role of the class of enzymes known as aminotransferases.
- To to distinguish a transamidinase from a transaminase.
- To identify the role of S-adenosyl methionine in the synthesis of creatine.
- To identify the role of glutamate and glutamine in preventing blood alkalization and neurotoxicity.
- To compare the reactions involved in introducing amido (amidine) groups into aspartate and glutamate.
- To identify the substrates and products of amino acid oxidase activity that involves oxygen as electron acceptor.
- To identify the amino carrier of the urea cycle; the compounds that donate urea nitrogen to the urea cycle; and the common compounds of the urea cycle and the Krebs cycle. To explain how these two cycles interact to maximize elimination of nitrogen through the urea cycle.
- To identify the urea cycle reactions that take place in the mitochondria.

Concepts

Humans are totally dependent on other organisms for converting atmospheric nitrogen into forms available to the body. This chapter begins, therefore, with nitrogen fixation and the entry of reduced nitrogen into organic compounds. The chapter also will discuss the entry and elimination of dietary nitrogen by the body. Subsequent chapters review the pathways by which nonessential and essential amino acids are metabolized and converted into other essential nitrogenous compounds.

Production of Ammonia, the Precursor to Amino Acids

The most abundant forms of nitrogen in the biosphere are gaseous nitrogen (N_2) and the oxides of nitrogen, nitrate (NO_3^-) plus nitrite (NO_2^-). None of these provide nitrogen in a form that is usable by the human body, although the closely related compound nitric oxide (NO) is a potent and physiologically important vasodilator. The oxides in the environment are converted to ammonia by bacterial nitrate reductase and nitrite reductase, while N_2 is converted to ammonia primarily by the bacterial nitrogenase complex. The reductases and the nitrogenase complex are multiple subunit protein assemblies, which use ATP and the reducing power of metabolites to convert nitrogen to ammonia. The reaction of the nitrogenase complex, in which N_2 (valence state = 0) is converted to the valance state of -3 in NH_3 and NH_4^+, is shown in Equation 23.1. Note the great requirement for metabolic energy in the form of ATP and reducing equivalents.

$$N_2 + 12ATP + 3NADH + 3H^+ \xrightarrow[\text{NITROGENASE}]{\text{BACTERIAL}} 2NH_4 + 12ADP + 12Pi + 3NAD^+ \qquad \textbf{23.1}$$

Reduced nitrogen, in the form of ammonia, is used by all organisms to form amino acids. Reduced nitrogen enters the human body as dietary free amino acids, protein, and the ammonia produced by intestinal tract bacteria. A pair of principal enzymes, glutamate dehydrogenase and glutamine synthetase, are found in all organisms and effect the conversion of ammonia into the amino acids glutamate and glutamine, respectively. Amino and amide groups from these 2 substances are freely transferred to other carbon skeletons by transamination and transamidation reactions (which will be reviewed later).

$$NH_4^+ + \alpha KG + NAD(P)H + H^+ \underset{\text{dehydrogenase}}{\overset{\text{glutamate}}{\rightleftharpoons}} L - glutamate + NAD(P)^+ + H_2O \qquad \textbf{23.2}$$

$$glutamate + NH_4^+ + ATP \xrightarrow{\text{glutamine synthetase}} glutamine + ADP + Pi + H^+ \qquad \textbf{23.3}$$

GLUTAMATE DEHYDROGENASE

In the forward reaction, shown in Equation 23.2, **glutamate dehydrogenase** (which can utilize either NADH or NADPH) is important in converting free ammonia and α-ketoglutarate to glutamate, forming one of the 20 amino acids required for protein synthesis. However, it should be recognized that the reverse reaction is a key anapleurotic process linking amino acid metabolism with Krebs cycle activity. In a reversal of the reaction in Equation 23.2, glutamate dehydrogenase provides an oxidizable carbon source used for the production of energy. As expected for a branch point enzyme with an important link to energy metabolism, glutamate dehydrogenase is regulated by the cell energy charge. ATP and GTP are negative effectors, whereas ADP and GDP are positive allosteric effectors. Thus, when the level of ATP is high, conversion of glutamate to α-KG and other Krebs cycle intermediates is limited; when the cellular energy charge is low, glutamate is converted to ammonia and oxidizable Krebs cycle intermediates. **Glutamate is also a principal amino donor to other amino acids in transamination reactions.**

The glutamine synthetase reaction is also important in several respects. First it produces glutamine, one of the 20 major amino acids. Second, in animals, **glutamine is the major amino acid found in the circulatory system**. Its role there is to carry ammonia to and from various tissues but principally from peripheral tissues to the kidney, where the amide nitrogen is hydrolyzed by the enzyme glutaminase (Equation 23.4); this process regenerates glutamate and free ammonium ion, which is excreted in the urine.

$$\text{glutamine} + \text{H}_2\text{O} \xrightarrow[\text{glutaminase}]{\text{kidney}} \text{glutamate} + \text{NH}_3 \qquad\qquad \textbf{23.4}$$

Note that, in this function, ammonia arising in peripheral tissue is carried in a nonionizable form which has none of the neurotoxic or alkalosis-generating properties of free ammonia.

Nitrogenous Compounds of the Digestive Tract

While glutamine, glutamate, and the remaining nonessential amino acids can be made by animals, the majority of the amino acids found in human tissues necessarily come from dietary sources (about 400 g of protein per day). Protein digestion begins in the stomach, where a proenzyme called pepsinogen is secreted, autocatalytically converted to Pepsin A, and used for the first step of proteolysis. However, most proteolysis takes place in the duodenum as a consequence of enzyme activities secreted by the pancreas. All of the serine proteases and the Zn peptidases listed in Table 23–1 are pancreatic secretions and, like pepsin, they are produced in the form of their respective proenzymes. These proteases are both endopeptidase and exopeptidase, and their combined action in the intestine leads to the production of amino acids, dipeptides, and tripeptides, all of which are taken up by enterocytes of the mucosal wall.

A circuitous regulatory pathway leading to the secretion of proenzymes into the intestine is triggered by the appearance of food in the intestinal lumen. Special mucosal endocrine cells secret the peptide hormones **cholecystokinin (CCK)** and **secretin** into the circulatory system. Together, CCK and secretin cause contraction of the gall bladder and the exocrine secretion of a bicarbonate-rich, alkaline fluid, containing protease proenzymes from the pancreas into the intestine. A second, paracrine role of CCK is to stimulate adjacent intestinal cells to secrete enteropeptidase, a protease that cleaves trypsinogen to produce trypsin. Trypsin also activates trypsinogen as well as all the other proenzymes in the pancreatic secretion, producing the active proteases and peptidases that hydrolyze dietary polypeptides.

Table 23–1. Proteases of the digestive tract, their proenzyme precursors, and the nature of the peptide bond specificity of each protease.

Enzyme	Proenzyme	Activator	Cleavage Sites
Carboxyl Proteases			
Pepsin A	Pepsinogen A	Autoactivation, pepsin	Tyr, Phe, Leu
Serine Proteases			
Trypsin	Trypsinogen	Enteropeptidase, trypsin	Arg, Lys
Chymotrypsin	Chymotrypsinogen	Trypsin	Tyr, Trp, Phe, Met, Leu
Elastase	Proelastase	Trypsin	Ala, Gly, Ser
Zn-Proteases			
Carboxypeptidase A	Procarboxypeptidase A	Trypsin	Val, Leu, Ile, Ala
Carboxypeptidase B	Procarboxypeptidase B	Trypsin	Arg, Lys

Subsequent to luminal hydrolysis, small peptides and amino acids are transferred through entero-cytes to the portal circulation by diffusion, facilitated diffusion, or active transport. A number of Na^+-dependent amino acid transport systems with overlapping amino acid specificity have been described. In systems such as that illustrated in Figure 23–1, Na^+ and amino acids at high luminal concentrations are co-transported down their concentration gradient to the interior of the cell. The ATP-dependent Na^+/K^+ pump exchanges the accumulated Na^+ for extracellular K^+, reducing intracellular Na^+ levels and maintaining the high extracellular Na^+ concentration (high in the intestinal lumen, low in entero-cytes) required to drive this transport process.

Transport mechanisms of this nature are ubiquitous in the body. Small peptides are accumulated by a proton (H^+) driven transport process and hydrolyzed by intracellular peptidases, as illustrated in Figure 23–1. Amino acids in the circulatory system and in extracellular fluids are transported into cells of the body by at least 7 different ATP-requiring active transport systems with overlapping amino acid specificities.

Many other nitrogenous compounds are found in the intestine. Most, such as those listed in Table 23–2, are bacterial products of protein degradation. Some have powerful pharmacological (vasopres-sor) effects.

Prokaryotes such as *E. coli* can make the carbon skeletons of all 20 amino acids and transaminate those carbon skeletons with nitrogen from glutamine or glutamate to complete the amino acid struc-tures. Humans cannot synthesize the branched carbon chains found in branched-chain amino acids or the ring systems found in phenylalanine and the aromatic amino acids; nor can we incorporate sulfur into covalently bonded structures. Therefore, the 9 so-called "essential" amino acids (Table 23–3) must be supplied from the diet. Nevertheless, it should be recognized that—depending on the composition of the diet and physiological state of an individual—one or another of the non-essential amino acids may also become a required dietary component. For example, arginine is not usually considered to be essential, because enough for adult needs is made by the urea cycle. However, the urea cycle gener-ally does not provide sufficient arginine for the needs of a growing child.

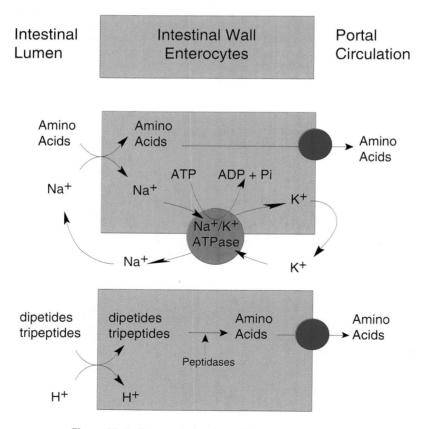

Figure 23–1. Transport of amino acids across the intestinal wall.

Table 23–2. Products of bacterial activity. Intestinal bacteria degrade dietary proteins, producing the deaminated vasopressor amines that are collectively known as **ptomaines**. Along with indole and skatole, the intestinal ptomaines are excreted in the feces.

Substrates	Products	
	Vasopressor amines	Other
Lysine	Cadaverene	
Arginine	Agmatine	
Tyrosine	Tyramine	
Ornithine	Putrescine	
Histidine	Histamine	
Tryptophan		Indole and skatole
All amino acids		NH4+

To take a different type of example, cysteine and tyrosine are considered non-essential but are formed from the essential amino acids methionine and phenylalanine, respectively. If sufficient cysteine and tyrosine are present in the diet, the requirements for methionine and phenylalanine are markedly reduced; conversely, if methionine and phenylalanine are present in only limited quantities, cysteine and tyrosine can become essential dietary components. Finally, it should be recognized that if the α-keto acids corresponding to the carbon skeleton of the essential amino acids are supplied in the diet, aminotransferases in the body will convert the keto acids to their respective amino acids, largely supplying the basic needs.

Unlike fats and carbohydrates, nitrogen has no designated storage depots in the body. Since the half-life of many proteins is short (on the order of hours), insufficient dietary quantities of even one amino acid can quickly limit the synthesis and lower the body levels of many essential proteins. The result of limited synthesis and normal rates of protein degradation is that the balance of nitrogen intake and nitrogen excretion is rapidly and significantly altered. Normal, healthy adults are generally in **nitrogen balance**, with intake and excretion being very well matched. Young growing children, adults recovering from major illness, and pregnant women are often in **positive nitrogen balance**. Their intake of nitrogen exceeds their loss as net protein synthesis proceeds. When more nitrogen is excreted than is incorporated into the body, an individual is in **negative nitrogen balance**. Insufficient quantities of even one essential amino acid is adequate to turn an otherwise normal individual into one with a negative nitrogen balance.

The biological value of dietary proteins is related to the extent to which they provide all the necessary amino acids. Proteins of animal origin generally have a high biological value; plant proteins have

Table 23–3. A list of the 22 amino acids that are conventionally considered to be required in the diet and those nonessential amino acids which are generally considered to be adequately made by humans.

Nonessential	Essential
Alanine	Histidine
Arginine	Isoleucine
Asparagine	Leucine
Aspartate	Lysine
Cysteine	Methionine
Glutamate	Phenylalanine
Glutamine	Threonine
Glycine	Tryptophan
Proline	Valine
Serine	
Tyrosine	

a wide range of values from almost none to quite high. In general, plant proteins are deficient in lysine, methionine, and tryptophan and are much less concentrated and less digestible than animal proteins. The absence of lysine in low-grade cereal proteins, used as a dietary mainstay in many underdeveloped countries, leads to an inability to synthesize protein (because of missing essential amino acids) and ultimately to a syndrome known as **kwashiorkor**, common among children in these countries.

REMOVAL OF NITROGEN FROM AMINO ACIDS

Intracellular Proteases

Nitrogen elimination begins intracellularly with protein degradation. There are two main routes for converting intracellular proteins to free amino acids: a lysosomal pathway, by which extracellular and some intracellular proteins are degraded, and cytosolic pathways that are important in degrading proteins of intracellular origin. In 1 cytosolic pathway a protein known as ubiquitin is activated by conversion to an AMP derivative, and cytosolic proteins that are damaged or otherwise destined for degradation are enzymically tagged with the activated ubiquitin. Ubiquitin-tagged proteins are then attacked by cytosolic ATP-dependent proteases that hydrolyze the targeted protein, releasing the ubiquitin for further rounds of protein targeting.

Aminotransferases

The dominant reactions involved in removing amino acid nitrogen from the body are known as transaminations. This class of reactions funnels nitrogen from all free amino acids into a small number of compounds; then, either they are oxidatively deaminated, producing ammonia, or their amine groups are converted to urea by the urea cycle. Transaminations involve moving an α amino group from a donor α amino acid to the keto carbon of an acceptor α keto acid, as illustrated in Figure 23–2.

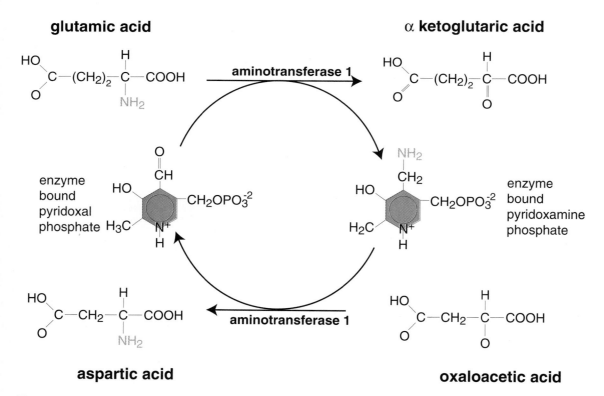

Figure 23–2. The mechanism of a typical aminotransferase enzyme illustrated for the ubiquitously occurring oxalate glutamate aminotransferase.

The latter reversible reactions are catalyzed by a group of intracellular enzymes known as **aminotransferases**, which generally employ covalently bound pyridoxal phosphate as a cofactor. However, some amino transferases employ **pyruvate** as a cofactor.

Aminotransferases exist for all amino acids except threonine and lysine. The most common compounds involved as a donor/acceptor pair in transamination reactions are glutamic acid and α-ketoglutaric acid, which participate in reactions with many different aminotransferases. Serum aminotransferases such as **serum glutamate-oxaloacetate-aminotransferase** (Figure 23–2) have been used as clinical markers of tissue damage, with increasing serum levels indicating an increased extent of damage.

Transamidinases

A small but clinically important amount of **creatinine** is excreted in the urine daily, and the **creatinine clearance rate** is often used as an indicator of kidney function. The first reaction in creatinine formation is the transfer of the amido (or amidine) group of arginine to glycine, forming guanidinoacetate as illustrated in Figure 23–3.

Subsequently, a methyl group is transferred from the ubiquitous 1-carbon-donor S-adenosylmethionine to guanidinoacetate to produce creatine and phosphocreatine, some of which spontaneously cyclizes to creatinine (Figure 23–3), and is eliminated in the urine. The quantity of urine creatinine is generally constant for an individual and approximately proportional to muscle mass. In individuals with damaged muscle cells, creatine leaks out of the damaged tissue and is rapidly cyclized, greatly increasing the quantity of circulating and urinary creatinine.

Glutamate Dehydrogenase and Glutaminase

Because of the participation of α-ketoglutarate in numerous transaminations, glutamate is a prominent intermediate in nitrogen elimination as well as in anabolic pathways (see Figure 23–2). Glutamate formed in the course of nitrogen elimination is either oxidatively deaminated (Equation 23.2) by liver glutamate dehydrogenase, forming ammonia, or converted to glutamine by glutamine synthase and transported to kidney tubule cells. There the glutamine is sequentially **deamidated** by glutaminase (Equation 23.3) and **deaminated** by kidney glutamate dehydrogenase.

The ammonia produced in the latter two reactions is excreted as NH_4^+ in the urine, where it helps maintain urine pH in the normal range of pH 4 to pH 8. The extensive production of ammonia by peripheral or liver glutamate dehydrogenase is not feasible because of the highly toxic effects of

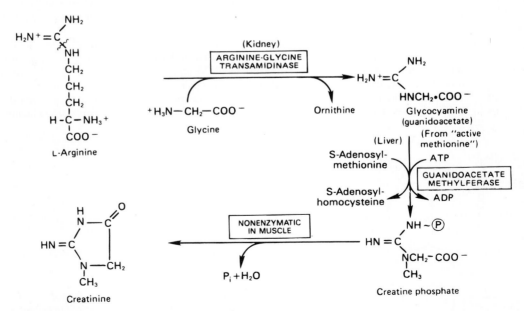

Figure 23–3. Biosynthesis of creatine and creatinine. (Reproduced, with permission, from Murray, RK: *Harper's Biochemistry*, 23rd ed. Appleton & Lange, 1993.)

circulating ammonia. Normal serum ammonium concentrations are in the range of 20–40 μmol, and an increase in circulating ammonia to about 400 μmol causes alkalosis and neurotoxicity.

A final, therapeutically useful amino acid–related reaction is the amidation of aspartic acid to produce asparagine. The enzyme asparagine synthase catalyzes the ATP-requiring the transamidation reaction shown in Equation 23.5.

$$\text{Aspartate} + \text{Glutamine} + \text{ATP} \xrightarrow[\text{synthase}]{\text{asparagine}} \text{Glutamate} + \text{Asparagine} + \text{AMP} + \text{PPi} \quad \textbf{23.5}$$

Most cells perform this reaction well enough to produce all the asparagine they need. However, some leukemia cells require exogenous asparagine, which they obtain from the plasma. Chemotherapy using the enzyme asparaginase takes advantage of this property of leukemic cells by hydrolyzing serum asparagine to ammonia and aspartic acid, thus depriving the neoplastic cells of the asparagine that is essential for their characteristic rapid growth.

Amino Acid Oxidases

In the peroxisomes of mammalian tissues, especially liver, there are 2 stereospecific amino acid oxidases involved in elimination of amino acid nitrogen. D-amino acid oxidase is an FAD-linked enzyme, and while there are few D-amino acids that enter the human body the activity of this enzyme in liver is quite high. L-amino acid oxidase is FMN-linked and has broad specificity for the L-amino acids. The reactions catalyzed by the amino acid oxidases are chemically similar and can be characterized as illustrated in Figure 23–4.

A number of substances, including oxygen, can act as electron acceptors from the flavoproteins. If oxygen is the acceptor the product is hydrogen peroxide, which is then rapidly degraded by the catalases found in liver and other tissues.

Missing or defective biogenesis of peroxisomes or L-amino acid oxidase causes **generalized hyper-aminoacidemia** and **hyper-aminoaciduria**, generally leading to neurotoxicity and early death.

THE UREA CYCLE

Earlier it was noted that kidney glutaminase was responsible for converting excess glutamine from the liver to urine ammonium. However, **about 80% of the excreted nitrogen is in the form of urea** which is also largely made in the liver, in a series of reactions that are distributed between the mitochondrial matrix and the cytosol. The series of reactions that form urea is known as the urea cycle or the Krebs-Henseleit cycle, and is illustrated in Figure 23–5.

The essential features of the urea cycle reactions and their metabolic regulation are as follows: Arginine from the diet or from protein breakdown is cleaved by the cytosolic enzyme **arginase**, generating urea and ornithine. In subsequent reactions of the urea cycle a new urea residue is built on the ornithine, regenerating arginine and perpetuating the cycle.

Ornithine arising in the cytosol is transported to the mitochondrial matrix, where **ornithine transcabamoylase** catalyzes the condensation of ornithine with carbamoyl phosphate, producing citrulline.

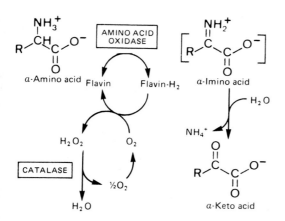

Figure 23–4. Oxidative deamination catalyzed by L-amino acid oxidase (L-α-amino acid: O_2 oxidoreductase). The α-imino acid, shown in brackets, is not a stable intermediate. (Reproduced, with permission, from Murray, RK: *Harper's Biochemistry*, 23rd ed. Appleton & Lange, 1993.)

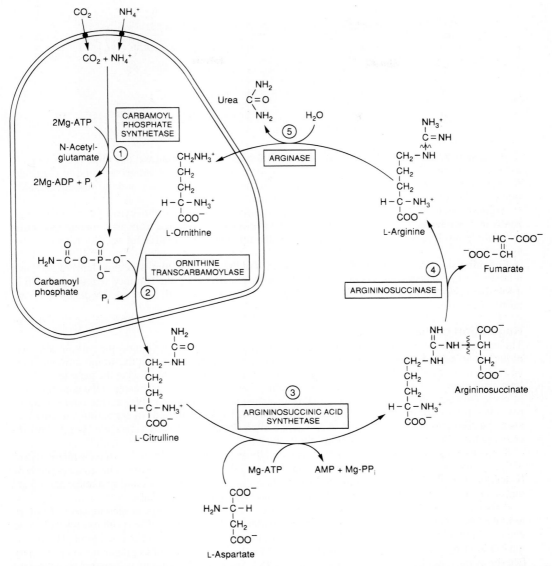

Figure 23–5. Reactions and intermediates of urea biosynthesis. The amines contributing to the formation of urea are shaded. Reactions ① and ② occur in the matrix of liver mitochondria and reactions ③, ④, and ⑤ in liver cytosol. CO_2 (as bicarbonate), ammonium ion, and ornithine and citrulline traverse the mitochondrial matrix via specific carriers (·) present in the inner membrane of liver mitochondria. (Reproduced, with permission, from Murray, RK: *Harper's Biochemistry*, 23rd ed. Appleton & Lange, 1993.)

The energy for the reaction is provided by the high-energy anhydride of carbamoyl phosphate. The product, citrulline, is then transported to the cytosol, where the remaining reactions of the cycle take place.

The synthesis of citrulline requires *a prior* activation of carbon and nitrogen as carbamoyl phosphate (CP). As illustrated in Figure 23–5, the activation requires 2 equivalents of ATP and the mitochondrial matrix enzyme carbamoyl phosphate synthetase-I (CPS-I). There are two forms of CPS: a mitochondrial enzyme, CPS-I, which forms CP destined for inclusion in the urea cycle, and a cytosolic CPS synthetase (CPS-II), which is involved in pyrimidine synthesis (see Chapter 27). CPS-I is positively regulated by the allosteric effector N-acetylglutamate, while the cytosolic enzyme is acetylglutamate independent.

Table 23–4. The summed reactions of the urea cycle.

$CO_2 + 2NH_4^+ + 2ATP \rightarrow$ carbamoyl phosphate $+ 2ADP + Pi$
ornithine + carbamoyl phosphate $\rightarrow$ citrulline + Pi
citrulline + aspartate + ATP $\rightarrow$ argininosuccinate + AMP + PPi
argininosuccinate $\rightarrow$ fumarate + arginine
arginine $\rightarrow$ urea + ornithine

$CO_2 + 2NH_4^+ +$ aspartate $+ 3ATP \rightarrow$ urea + fumarate $+ 2ADP + AMP + PPi + 2Pi$

In a 2-step reaction, catalyzed by cytosolic **argininosuccinate synthetase**, citrulline is converted to aspartate. The reaction involves the addition of AMP (from ATP) to the amido carbonyl of citrulline, forming an activated intermediate on the enzyme surface (AMP-citrulline), and the subsequent addition of aspartate to form argininosuccinate (see Figure 23–5).

Arginine and fumarate are produced from argininosuccinate by the cytosolic enzyme **argininosuccinate lyase**. In the final step of the cycle **arginase** cleaves urea from aspartate, regenerating cytosolic ornithine, which can be transported to the mitochondrial matrix for another round of urea synthesis.

Beginning and ending with ornithine, the reactions of the cycle consumes 3 equivalents of ATP and a total of 4 high-energy nucleotide phosphates, as demonstrated by the summed reactions shown in Table 23–4. Urea is the only new compound generated by the cycle; all other intermediates and reactants are recycled.

Regulation of the Urea Cycle

The urea cycle operates only to eliminate excess nitrogen. On high-protein diets the carbon skeletons of the amino acids are oxidized for energy or stored as fat and glycogen, but the amino nitrogen must be excreted. To facilitate this process, enzymes of the urea cycle are controlled at the gene level. With long-term changes in the quantity of dietary protein, changes of 20-fold or greater in the concentration of cycle enzymes are observed. When dietary proteins increase significantly, enzyme concentrations rise. On return to a balanced diet, enzyme levels decline. Under conditions of starvation, enzyme levels rise as proteins are degraded and amino acid carbon skeletons are used to provide energy, thus increasing the quantity of nitrogen that must be excreted.

Short-term regulation of the cycle occurs principally at carbamoyl synthetase, which is relatively inactive in the absence of its allosteric activator N-acetylglutamate. The steady-state concentration of N-acetylglutamate is set by the concentration of its components acetyl-CoA and glutamate and by arginine, which is a positive allosteric effector of N-acetylglutamine synthetase.

Earlier it was noted that ammonia was neurotoxic. Marked brain damage is seen in cases of failure to make urea via the urea cycle or to eliminate urea through the kidneys. The result of either of these events is a buildup of circulating levels of ammonium ion. Aside from its effect on blood pH, ammonia readily traverses the brain blood barrier and in the brain is converted to glutamate via glutamate dehydrogenase, depleting the brain of α-ketoglutarate. As the α-ketoglutarate is depleted oxaloacetate falls correspondingly, and ultimately Krebs cycle activity comes to a halt. In the absence of aerobic oxidative phosphorylation and Krebs cycle activity, irreparable cell damage and neural cell death ensue.

Questions

DIRECTIONS (items 1–4): Each numbered item or incomplete statement is followed by answers or by completions of the statement. Referring to the diagram below, select the ONE lettered answer or completion that is BEST in each case.

    a. Reactions catalyzed by B and A.
    b. Reactions catalyzed by B and C.
    c. Reactions catalyzed by C and D.
    d. Reactions catalyzed by D and E.

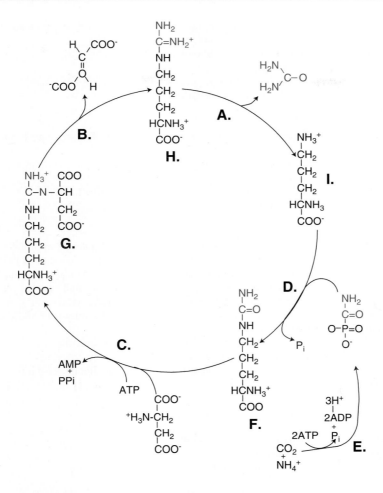

1. Mitochondrial reactions include:

2. Citrulline is represented by structure:

3. Argininosuccinate is represented by structure:

4. The reaction catalyzed by Arginase is:

DIRECTIONS (items 5–14): Each numbered item or incomplete statement is followed by answers or by completions of the statement. Select the ONE lettered answer or completion that is BEST in each case.

5. Some of the nitrogen atoms required for animal metabolism can be obtained directly from:
   a. Atmospheric nitrogen (N_2).
   b. NO_2.
   c. Ammonia or ammonium.
   d. NO.

6. Aside from urea, the main circulatory system carrier of nitrogen atoms destined for excretion by the kidney is:
   a. Glutamate.
   b. Alanine.
   c. Asparagine.
   d. Glutamine.

7. In situations of extrahepatic biliary obstruction, including obstruction of the pancreatic duct, one might expect the following postprandial findings:
   a. Abnormally low levels of intestinal tract chymotrypsinogen.
   b. Abnormally low levels of gastric pepsin A.
   c. Abnormally low levels of duodenal chymotrypsin.
   d. Abnormally low plasma levels of cholecystokinin.
   e. 2 of the above.

8. A deficiency in the ability to secrete enteropeptidase would lead to the following postprandial condition(s):
   a. Abnormally low levels of intestinal zymogens postprandially.
   b. Abnormally low levels of active intestinal proteases.
   c. Abnormally high levels of undigested protein in the lower digestive tract.
   d. Increased plasma levels of serum amylase.
   e. 2 of the above.

9. A pregnant woman presented at the emergency room with the following symptoms. She was sweating profusely, had been vomiting, and was near-comatose. Her family related that she had had no previous illness that could related to this problem and that she had been careful to maintain a well-balanced diet during her pregnancy. Among other laboratory findings, her plasma ammonia levels were 3 times greater than normal. She was administered arginine and glucose in an intravenously saline infusion. In several hours her condition markedly improved and she was released. The cause of her difficulty was:
   a. A reduced blood volume, which was corrected by the bulk saline infusion.
   b. A deficiency of blood glucose for aerobic brain glycolysis.
   c. A deficiency of dietary arginine, which is usually included in the diet because it is considered an essential amino acid.
   d. None of the above.

10. Transamidation reactions include:
   a. Formation of liver glutamate from α-ketoglutarate.
   b. Formation of liver glutamine from glutamate.
   c. Formation of liver asparagine from aspartate.

11. The cofactor(s) for aminotransferase enzymes include:
   a. Pyridoxine phosphate.
   b. Carbamoyl phosphate.
   c. Pyruvate.
   d. Flavin mononucleotide (FMN).
   e. 2 of the above.

12. Abnormal neural development, generalized aminoaciduria, and an absence of peroxisomes in a liver biopsy were among the findings used to diagnose a case of Zellweger cerebrohepatorenal syndrome (CHRS). The aminoaciduria in this instance was most likely due to:
   a. A deficiency in an enzyme of the urea cycle.
   b. A deficiency in the serum enzyme responsible for oxalic acid, glutamic acid aminotransferase (OGAT) activity.
   c. A deficiency of acetyl coenzyme A required for the synthesis of N-acetylglutamate, the positive effector of carbamoylphosphate synthetase and the urea cycle.
   d. Deficient amino acid oxidase activity.

13. Upregulation of urea cycle activity is attained by:
   a. Altered dietary habits that include increased dietary protein.
   b. Elevated cytoplasmic aspartate.
   c. Increased dietary and cytoplasmic arginine.
   d. Increased dietary N-acetylglutamate.
   e. 2 of the above.

Answers

1. d.
2. f.
3. g.
4. a.
5. c.
6. d.
7. e.
8. e. Enteropeptidase is required to cleave trypsinogen, and trypsin is required to activate the other protease proenzymes. Thus zymogen levels will be normal, active enzymes low (B), and undigested protein high (C).
9. d. The arginine is required to make ornithine, but normally humans make enough arginine to supply ornithine for the urea cycle. In rapidly metabolizing people the need for ornithine can outstrip the ability of the body to make arginine. Thus *arginine, which is not normally considered an essential amino acid*, can become a dietary requirement.)
10. c. The glutamine amido group is transferred to aspartate from asparagine, but the glutamine amido group is formed from ammonia via glutamine synthatase.
11. e. (a and c are correct).
12. d. Amino acid oxidase is a peroxisomal enzyme and is responsible for oxidative amino acid deamination.
13. e. N-acetylglutamate will be degraded before it reaches the mitochondrial matrix. Aspartate-dependent increases require up-regulation of citrulline formation; this is effected by arginine-mediated up-regulation of N-acetylglutamate synthesis, which is the first step in a regulatory cascade culminating in the increased synthesis of citrulline. Persistently high levels of dietary protein increases biosynthesis of all urea cycle enzymes.

Nonessential Amino Acid Synthesis 24

Objectives

- To identify the pathways that lead to the synthesis of the nonessential amino acids by humans.
- To identify the role of the alanine cycle in the carbon and nitrogen economy of the body.
- To identify the source of nitrogen used in alanine synthesis.
- To identify the intermediates between methionine and cysteine.
- To explain how SAdoMet acts as a methyl donor and as a nitrogen donor.
- To identify the clinical consequences of genetic defects in cystathionine synthase and cystathioninase, and the clinical effects of deficient phenylalanine hydroxylase or deficient tetrahydrobiopterin reductase.
- To define the metabolic relationship between glutamate, glutamate semialdehyde, ornithine, and proline.
- To define the relationship between the pathway to serine synthesis and the pathway to glycolysis.
- To define the relationship between one-carbon metabolism in the body and glycine biosynthesis.
- To identify the one-carbon conjugates of THF and the enzymes that interconvert the various THF derivatives.
- To compare asparagine biosynthesis with glutamine biosynthesis.

Concepts

All tissues have some capability for synthesis of the nonessential amino acids, amino acid remodeling, and conversion of nonamino-acid carbon skeletons into amino acids and other derivatives that contain nitrogen. However, the liver is the major site of nitrogen metabolism in the body. The multiple roles of the liver in nitrogen metabolism are listed in Table 24–1.

This chapter focuses on item 2 in Table 24–1, reviewing the formation of nonessential amino acids and the flow of amino acids between tissues.

Glutamate, Aspartate, and Alanine Biosynthesis and the Alanine Cycle

Three amino acids—glutamate, aspartate, and alanine—are synthesized from their widely distributed α-keto acid precursors by simple 1-step transamination reactions. Chapter 23 reviewed the importance of glutamate as a common intracellular amino donor for transamination reactions and of aspartate as a precursor of ornithine for the urea cycle. Aside from its role in protein synthesis, alanine is second only to glutamine in prominence as a circulating amino acid. In this capacity it serves a unique role in the transfer of nitrogen from peripheral tissue to the liver. Alanine is transferred to the circulation by many tissues, but mainly by muscle, in which alanine is formed from pyruvate at a rate proportional to intracellular pyruvate levels. Liver accumulates plasma alanine, reverses the transamination that occurs in muscle, and proportionately increases urea production. The pyruvate is either oxidized or converted to glucose via gluconeogenesis (Figure 24–1). When alanine transfer from muscle to liver is coupled with glucose transport from liver back to muscle, the process is known as the **alanine cycle**. The key feature of the cycle is that in 1 molecule, alanine, peripheral tissue exports pyruvate and ammonia (which are potentially rate-limiting for metabolism) to the liver, where the carbon skeleton is recycled and most nitrogen eliminated.

There are 2 main pathways to production of muscle alanine: directly from protein degradation, and via the transamination of pyruvate by pyruvate:glutamate aminotransferase. The various reactions leading to alanine production are summarized in Equations 24.1, 24.2, and 24.3. The metabolic production of glutamate is presented in Equations 24.1 and 24.2; Equation 24.3 represents the glutamate:pyruvate aminotransferase reaction.

$$\alpha\text{-ketoglutarate} + NH_4^+ + NAD(P)H + H^+ \xrightarrow[\text{dehydrogenase}]{\text{glutamate}} \text{glutamate} + NAD(P)^+ + H_2O \quad 24.1$$

$$\alpha\text{-ketoglutarate} + \alpha\text{-amino acid (Asp)} \xrightarrow[\text{transferase}]{\text{amino}} \text{glutamate} + \alpha\text{-keto acid (OAA)} \quad 24.2$$

$$\text{glutamate} + \text{pyruvate} \xrightarrow[\text{transferase}]{\text{amino}} \alpha\text{-ketoglutarate} + \text{alanine} \quad 24.3$$

Cysteine Synthesis

As illustrated in Figure 24–2, the key features of cysteine synthesis are as follows. The sulfur for cysteine synthesis comes from the essential amino acid methionine. Homocysteine, the product of a transmethylation reaction employing S-adenosylmethionine (SAdoMet), condenses with serine to produce cystathionine, which is subsequently cleaved by cystathionase to produce cysteine and α-ketobutyrate. The sum of the latter 2 reactions is known as **transsulfuration**. Cysteine is used for protein synthesis and other body needs, while the α-ketobutyrate is decarboxylated and converted to propionyl-CoA. While cysteine readily oxidizes in air to form the disulfide cystine, the cells contain little if any free cystine, because the ubiquitous reducing agent glutathione effectively reverses the formation of cystine by a non-enzymatic reduction reaction.

Transmethylation reactions employing the general methylating agent SAdoMet are extremely important, but in this case the role of SAdoMet in transmethylation is secondary to the production of

Table 24–1. Role of the liver in nitrogen metabolism.

1.	Disposal of the surplus amino acid nitrogen and some carbon atoms as urea and kidney glutamine.
2.	Delivery of a balanced amino acid supply to the circulation for use by other tissues including synthesis of nonessential amino acids.
3.	Synthesis of nonprotein, nitrogen containing compounds such as heme, pyrimidines, creatinine.
4.	Protein synthesis, for liver cell proteins and for serum proteins such as serum albumin.

ALANINE CYCLE

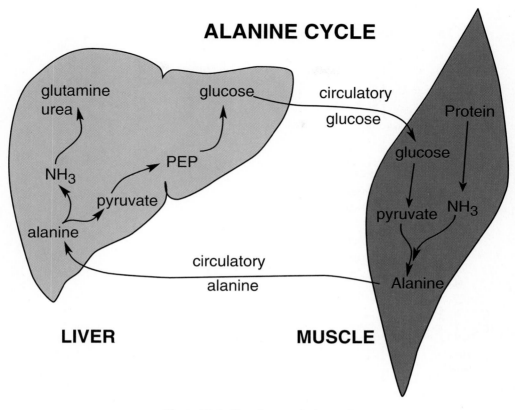

Figure 24–1. The glucose-alanine cycle.

homocysteine (essentially a by-product of transmethylase activity). Notice that in the production of SAdoMet all phosphates of an ATP are lost: 1 as P_i and 2 as PP_i, and that it is adenosine which is transferred to methionine and not AMP (Figure 24–3).

The 2 key enzymes of this pathway, cystathionine synthase and cystathionase (cystathionine lyase), both use pyridoxal phosphate as a cofactor, and both are under regulatory control. Cystathionase is under negative allosteric control by cysteine, as well cysteine inhibits the expression of the cystathionine synthase gene. Genetic defects are known for both the synthase and the lyase. Missing or impaired cystathionine synthase leads to homocystinuria and is often associated with mental retardation, although the complete syndrome is multifaceted and many individuals with this disease are mentally normal. Some instances of genetic homocystinuria respond favorably to pyridoxine therapy, suggesting that in these cases the defect in cystathionine synthase is a decreased affinity for the cofactor. Missing or impaired cystathionase leads to excretion of cystathionine in the urine but does not have any other untoward effects. Rare cases are known in which cystathionase is defective and operates at a low level. This genetic disease leads to methioninuria with no other consequences.

Tyrosine Biosynthesis

Tyrosine is produced in cells by hydroxylating the essential amino acid phenylalanine. This relationship is much like that between cysteine and methionine. Half of the phenylalanine required goes into the production of tyrosine; if the diet is rich in tyrosine itself, the requirements for phenylalanine are reduced by about 50%.

Phenylalanine hydroxylase is a classic P450-linked, mixed-function hydroxylase: 1 atom of oxygen is incorporated into water and the other into a hydroxyl of the product tyrosine, as illustrated in Figure 24–4. The reductant is the tetrahydrofolate-related cofactor **tetrahydrobiopterin**, which is maintained in the reduced state by the NADPH-dependent enzyme **dihydrobiopterin reductase**. Missing or deficient phenylalanine hydroxylase leads to the genetic disease known as **phenlyketonuria** (PKU), which if untreated leads to severe mental retardation. As illustrated in Figure 24–5,

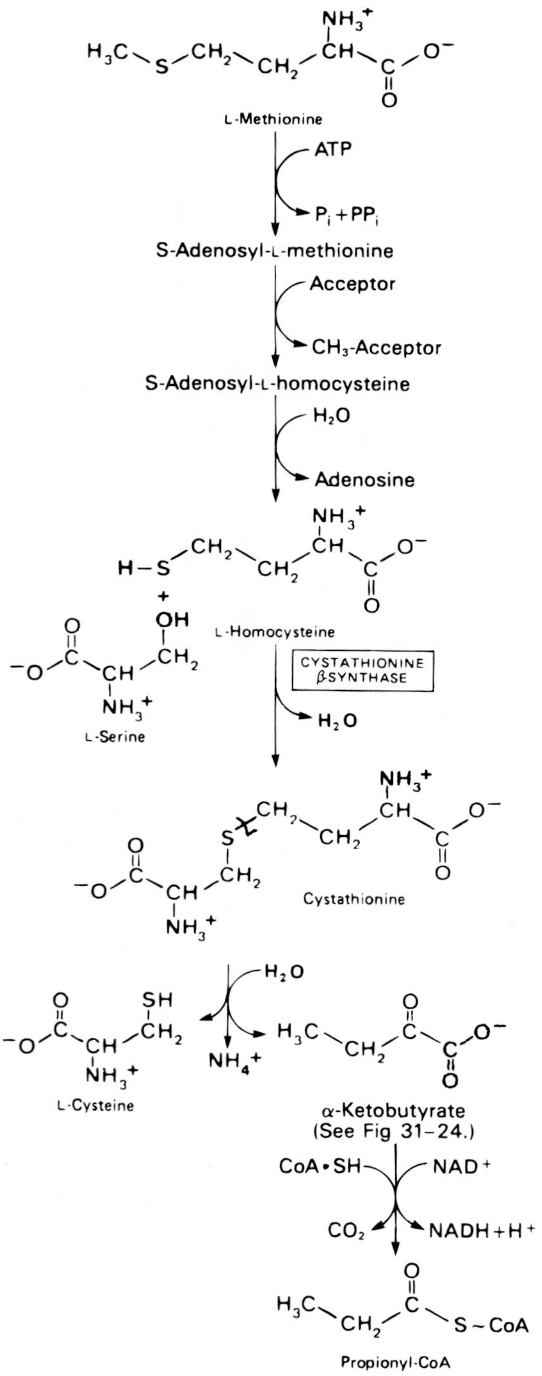

Figure 24–2. Role of methione in the synthesis of cysteine.

the mental retardation is caused by the accumulation of phenylalanine, which becomes a major donor of amino groups in aminotransferase activity and depletes neural tissue of α-ketoglutarate. This absence of α-ketoglutarate in the brain shuts down the Krebs cycle and the associated production of aerobic energy, which is essential to normal brain development.

The product of phenylalanine transamination, phenylpyruvic acid, is reduced to phenylacetate and phenyl lactate, and all 3 compounds appear in the urine. If the problem is diagnosed early, the addi-

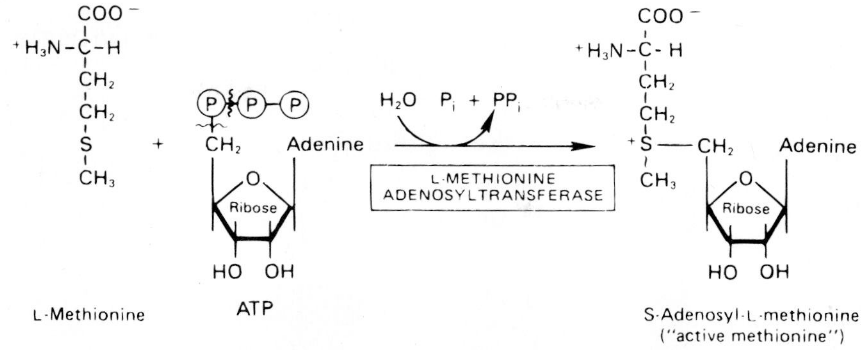

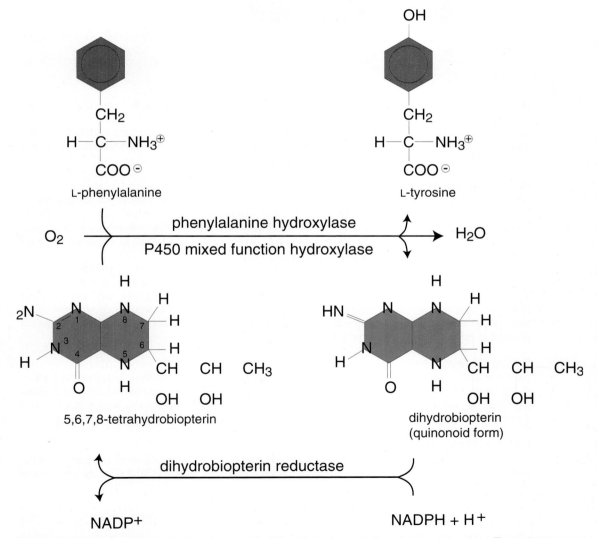

Figure 24–3. Formation of S-adenosylmethionine. ~CH₃ represents the high energy transfer potential of the active methyl group in methionine.

Figure 24–4. The phenylalanine hydroxylase reaction. Two distinct enzymatic activities are involved. The reductase catalyzes reduction of dihydrobiopterin by NADPH, and the hydroxylase the reduction of O_2 to H_2O and of phenylalanine to tyrosine. This reaction is associated with several defects of phenylalanine metabolism.

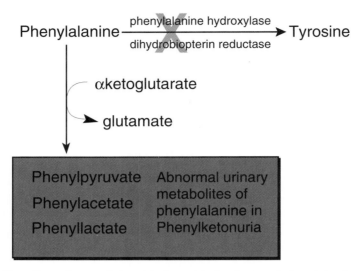

Figure 24–5. Alternative pathways of phenylalanine catabolism in phenylketonuria leading to the production of abnormal urinary metabolites.

tion of tyrosine and restriction of phenylalanine from the diet can minimize the extent of mental retardation.

In other pathways, tetrahydrobiopterin is a cofactor. Thus the effects of missing or defective dihydrobiopterin reductase cause even more severe neurological difficulties than those usually associated with PKU resulting from deficient hydroxylase activity.

Proline Biosynthesis

Glutamate is the precursor of both proline and ornithine, with glutamate semialdehyde being a branch point intermediate leading to one or the other of these 2 products. While ornithine is not one of the 20 amino acids used in protein synthesis, it plays a significant role as the acceptor of carbamoyl phosphate in the urea cycle (see Chapter 23). The production of ornithine from glutamate is important when dietary arginine, the other principal source of ornithine, is limited.

The fate of glutamate semialdehyde depends on prevailing cellular conditions. Ornithine production occurs from the semialdehyde via a simple glutamate-dependent transamination, producing ornithine. When arginine concentrations become elevated, the ornithine contributed from the urea cycle plus that from glutamate semialdehyde inhibit the aminotransferase reaction, with accumulation of the semialdehyde as a result. The semialdehyde cyclizes spontaneously to pyrroline-5-carboxylate and is then reduced to proline by an NAD(P)H-dependent reductase (Figure 24–6).

Serine Biosynthesis

As illustrated in Figure 24–7, the main pathway to serine starts with the glycolytic intermediate 3-phosphoglycerate. An NADH-linked dehydrogenase converts 3-phosphoglycerate into a keto acid, 3-phosphopyruvate, suitable for subsequent transamination. Aminotransferase activity with glutamate as a donor produces 3-phosphoserine, which is converted to serine by phosphoserine phosphatase.

Glycine Biosynthesis

As shown in Equation 24.4, the main pathway to glycine is a 1-step reaction involving transfer of the hydroxymethyl group from serine to the cofactor tetrahydrofolic acid (THF), producing glycine and $N^{5,10}$-methylene THF.

$$\text{serine} + \text{THF} \xrightarrow[\text{transferase}]{\text{serine hydroxymethyl}} \text{glycine} + N^{5,10}\text{methylene THF} + H_2O \qquad 24.4$$

Glycine is involved in many anabolic reactions other than protein synthesis. A brief list reflecting glycine's central role in anabolic process is presented in Table 24–2.

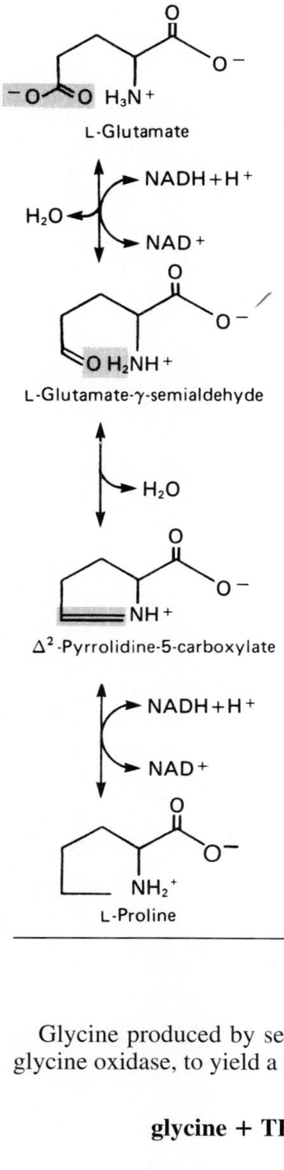

Figure 24–6. Biosynthesis of proline from glutamate by reversal of the reactions of proline catabolism.

Glycine produced by serine hydroxymethyl transferase or from the diet can also be oxidized by glycine oxidase, to yield a second equivalent of $N^{5,10}$-methylene THF, as illustrated in Equation 24.5.

$$\text{glycine} + \text{THF} + \text{NAD}^+ \xrightarrow[\text{oxidase}]{\text{glycine}} CO_2 + NH_4^+ + N^{5,10}\text{methylene THF} \qquad 24.5$$

The Role of Tetrahydrofolic Acid as a One-Carbon Atom Carrier

A key feature of THF is that it can carry carbon at 3 different levels of oxidation, as indicated in Table 24–3. The carbon carried in the various oxidation states is always associated with atoms N^5 or N^{10}, or else it bridges the N^5 and N^{10} atoms (see Figure 36-10). While numerous compounds can contribute to the THF pool, the main reaction(s) providing carbon elements to THF is serine conversion to glycine and the subsequent degradation of glycine to CO_2 and H_2O.

Once THF contains a 1-carbon element in the 5, 10, or 5-10 position, all the other oxidation levels of the 1-carbon unit can be readily formed by the conversions summarized in Figure 24–8. These conversions are catalyzed by the 6 enzyme activities listed in Table 24–4.

Aspartate/Asparagine and Glutamate/Glutamine Biosynthesis

As described in Chapter 23, glutamate is synthesized in all organisms, including humans, by a reductive amination of the α-keto acid, α-ketoglutarate (Figure 24–8); it is thus a nitrogen-fixing reaction. In addition, glutamate arises by aminotransferase reactions, with the amino nitrogen being donated by a number of different amino acids. Thus glutamate is a general collector of amino nitrogen, as was also

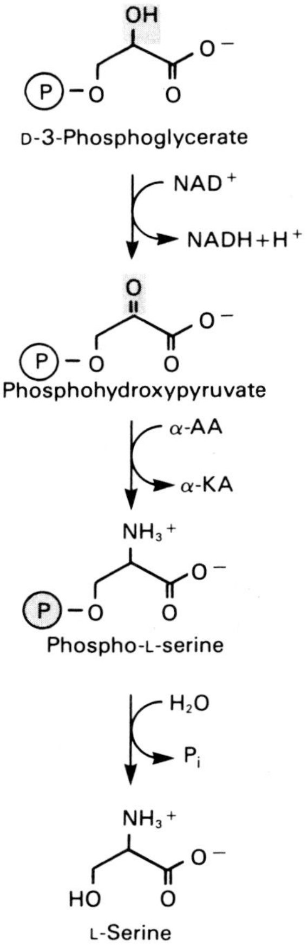

Figure 24–7. Serine biosynthesis. α-AA, α-amino acids, α-KA, α-keto acids.

Table 24–2. Metabolites requiring glycine as a precursor.

Proteins
Nucleic acids (purines)
Heme
Glutathione
Creatinine
Serine

Table 24–3. Oxidation states of carbon atoms carried by tetrahydrofolate.

One Carbon Oxidation State	Carbon Source	THF Derivative	Anabolic Product
1. Most reduced (methanol)	$N^{5,10}$-Methylene THF	N^5-Methyl-THF	Methionine
2. Intermediate (formaldehyde)	Glycine, serine, formaldehyde	$N^{5,10}$-methylene-THF	Thymidine
3. Most oxidized (formic acid)	Formic acid	N^{10}-Formyl-THF	Purines
	Histidine	Formimino-THF	
	Formimino-THF	$N^{5,10}$-Methenyl	Purines

Table 24–4. Enzymes that interconvert tetrahydrofolate derivatives.

	Enzyme	Action
1.	N^{10} Formyltetrahydrofolate Synthetase	Adds activated formate to THF
2.	N^5,N^{10}-Methylene THF Dehydrogenase	Interconverts N^5,N^{10} methylene and methenyl derivatives
3.	N^5,N^{10}-Methenyl THF Cyclohydrolase	Converts the N^{10}-formyl THF to the cyclic methenyl derivative
4.	N^5,N^{10}-methylene THF Reductase	Converts the N^5,N^{10}-methylene derivative to the N^5-methyl derivative
5.	Formimino transferase	Produces N^5-formimino THF
6.	Formiminotetrahydrofolate cyclodeaminase	Converts N^5-formimino THF to the central methenyl derivative in a deamination reaction followed by cyclization

discussed in Chapter 23. In contrast, aspartate is formed only from its α-keto acid analog, oxaloacetate, by aminotransferase reactions, generally with glutamate as the amino donor.

Aspargine synthetase and glutamine synthetase, although similarly named, catalyze the production of asparagine and glutamine from their respective α amino acids by very different reactions. Glutamine is produced from glutamate by the direct incorporation of ammonia (see Equation 23.3); and this can be considered another nitrogen fixing reaction. Asparagine, however, is formed by an amidotransferase reaction, as illustrated in Figure 24–9.

Aminotransferase reactions are readily reversible. The direction of any individual transamination depends principally on the concentration ratio of reactants and products. By contrast, transamidation reactions—which are dependent on ATP—are considered irreversible. As a consequence, the degradation of asparagine and glutamine take place by a hydrolytic pathway rather than by a reversal of the pathway by which they were formed.

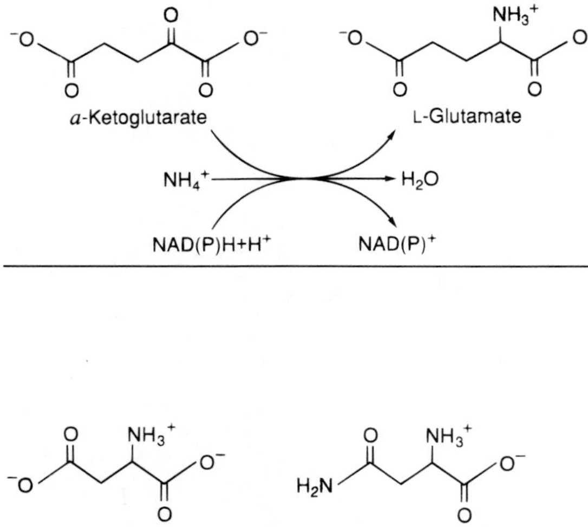

Figure 24–8. The glutamate dehydrogenase reaction. Reductive amination of α-ketoglutarate by NH_4^+ proceeds at the expense of NAD(P)H.

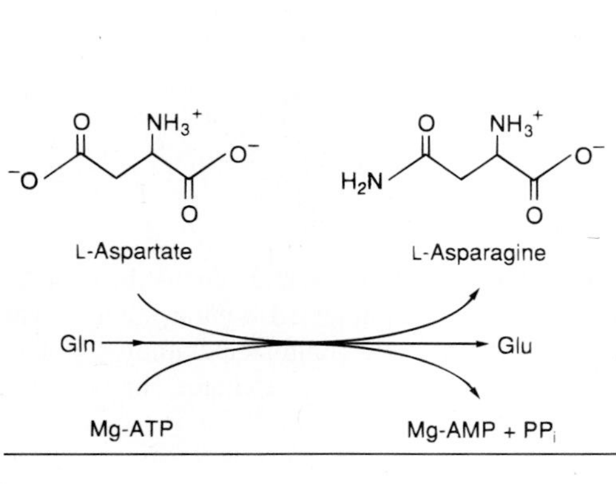

Figure 24–9. The asparagine synthetase reaction.

Questions

DIRECTIONS (items 1–12): Each numbered item or incomplete statement is followed by answers or by completions of the statement. Select the ONE lettered answer or completion that is BEST in each case.

1. The production of alanine by peripheral tissue, and especially by muscle, is primarily dependent on:
   a. Allosteric effectors of pyruvate:glutamate aminotransferase.
   b. Intracellular levels of pyruvate.
   c. Intracellular levels of glutamate.
   d. Intracellular levels of NADH.

2. The role of the alanine cycle is:
   a. To deliver muscle pyruvate to liver in exchange for glucose.
   b. To deliver liver nitrogen to muscle.
   c. To deliver muscle alanine to the kidneys.
   d. To provide all tissues with an equivalent homeostatic level of alanine.

3. Nitrogen of muscle alanine can be derived from tissue ammonia through the action of the enzyme:
   a. Glutamate dehydrogenase.
   b. Alanine dehydrogenase.
   c. Glutamine synthetase.
   d. Alanine synthetase.

4. Sulfur for cysteine synthesis is derived from:
   a. Sulfites.
   b. H_2S.
   c. Coenzyme A.
   d. Methionine.
   e. Tetrahydrobiopterin.

5. The nucleotide required in the main pathway to cysteine synthesis is:
   a. NADH.
   b. NADPH.
   c. FADH.
   d. ATP.
   e. GTP.

6. The carbon atoms of cysteine are derived from:
   a. Methionine.
   b. Serine.
   c. Homocysteine.
   d. Pyruvate.
   e. α-ketobutyrate.

7. The number of phosphate atoms in S-adenosylmethionine:
   a. None.
   b. One.
   c. Two.
   d. Three.

8. A mildly retarded 18-year-old male was treated several times for arterial and venous thromboses and was suspected of having Marfan syndrome. However, a pronounced homocystinuria was among the diagnostic findings. The patient was treated with multiple vitamins including high levels of pyridoxine. After pyridoxine therapy, no further thromboses were reported. The probable cause of the patient's difficulty was:

    a. A genetic defect leading to elevated levels of blood clotting factors.
    b. A genetic defect leading to excess cysteine production.
    c. A genetic defect in the enzyme cystathionine synthase.
    d. A genetic defect in the enzyme cystathioninase.
    e. A genetic defect in the enzyme SAdoMet synthetase.

9. Severe phenylketonuria and shortened life span is a result of:
    a. Excess dietary phenylalanine.
    b. Excess dietary tyrosine.
    c. Deficient or missing dihydrobiopterin reductase.
    d. Deficiency in tetrahydrofolate.

10. The biochemical basis for development-related mental retardation that is caused by an enzymatic defect in the pathway to tyrosine synthesis is most likely to be:
    a. Excess production of tyrosine, causing ammonia neurotoxicity.
    b. Insufficient production of tyrosine, causing deficient production of brain proteins.
    c. Ammonemia leading to depletion of α-ketoglutarate in brain tissue.
    d. Pathologically severe, chronic alkalosis.

11. Glutamate semialdehyde is:
    a. An intermediate on the pathway to proline synthesis.
    b. An intermediate on a pathway that interconverts proline and ornithine.
    c. Formed by glutamate oxidation.
    d. All of the above.
    e. None of the above.

12. Serine is synthesized by:
    a. Transamination of an amino acid with α-ketoglutarate as an amino acceptor.
    b. By transamination of an α-keto acid derived from a Krebs cycle intermediate.
    c. By transamination of an α-keto acid derived from an intermediate of glycolysis.
    d. Principally by modification of an essential amino acid.

DIRECTIONS (items 13–18): For each numbered description, select the BEST response from among the lettered list of compounds.

    a. Glycine
    b. Proline
    c. Phenyl lactate
    d. Tyrosine
    e. Cysteine
    f. Glutamine
    g. Tryptophan
    h. Valine
    i. Threonine

13. Abnormal sulfur containing compound formed from an essential amino acid.

14. Essential amino acid that can be dietarily spared by its α-keto analog.

15. Amino acid that spares the dietary requirement for phenylalanine.

16. Amino acid that spares the dietary requirement for methionine.

17. Carrier of peripheral tissue ammonia to liver and kidney.

18. Amino acid produced from glutamate via glutamate γ semialdehyde.

Answers

1. b	**7.** a	**13.** c
2. a	**8.** c	**14.** h
3. a	**9.** c	**15.** d
4. d	**10.** c	**16.** e
5. d	**11.** d	**17.** f
6. b	**12.** c	**18.** b

Metabolic Fate of Amino Acids

25

Objectives

- To identify the main pathways of amino acid catabolism, including the disposition of nitrogen atoms and the carbon skeletons.
- To define the terms "ketogenic" and "glucogenic" as they apply to amino acids and list the ketogenic and the glucogenic amino acids.
- To identify the common intermediate in the catabolism of proline, arginine, and ornithine.
- To identify 2 prominent paths for serine catabolism, and the pathway through which most carbon flows.
- To identify the reaction paths leading from threonine to glucogenic and ketogenic metabolites and the pathway through which most carbon flows. To list the key intermediates that arise in the path between α-ketobutyrate and succinate.
- To illustrate 2 reaction paths for the catabolism of glycine.
- To identify the major cysteine catabolic pathway; the structure of cysteine sulfinate, and the product of cysteine metabolism that is a precursor to bile salts.
- To identify the enzymes that are common to the catabolic pathway of the branched-chain amino acids; to identify the glucogenic and/or ketogenic product of each catabolic pathway.
- To identify the significance of homogentisic acid in the urine in amounts sufficient to cause formation of dark-colored urine.
- To discuss the consequence of defective tyrosine transaminase and list the abnormal metabolic products that appear in body fluids.
- To identify how aminotransferase activity leading to the production of saccharopine is different from most other aminotransferases; to identify the reaction that forms the basis for lysine's characterization as an essential amino acid.
- To identify the result of failure of the dibasic amino acid transport system.
- To identify the basis for classifying arginine as a glucogenic amino acid.
- To identify the unusual features of removing the α-amino group of histidine.
- To identify the product of histidine decarboxylation and the physiological role of the product.
- To identify the glucogenic and ketogenic products of tryptophan catabolism.
- To identify the branch-point intermediate in tryptophan catabolism that leads to nicotinic acid.
- To identify the enzymes and intermediates that are common to leucine and tryptophan catabolism.

Concepts

This chapter reviews the catabolic fate of dietary amino acids as well as those produced during normal breakdown of body proteins. In times of dietary surplus, the potentially toxic nitrogen of amino acids is eliminated via transaminations, deamination, and urea formation; the carbon skeletons are generally conserved as carbohydrate, via gluconeogenesis, or as fatty acid via fatty acid synthesis pathways. In this respect amino acids fall into three categories: **glucogenic, ketogenic, or glucogenic and ketogenic**. Glucogenic amino acids are those that give rise to a net production of pyruvate or Krebs Cycle intermediates, such as α-ketoglutarate or oxaloacetate, all of which are precursors to glucose via gluconeogenesis. **All amino acids except lysine and leucine are at least partly glucogenic.** Lysine and leucine are the only amino acids that are solely ketogenic, giving rise only to acetylCoA or acetoacetylCoA, neither of which can bring about net glucose production.

A small group of amino acids comprised of isoleucine, phenylalanine, threonine, tryptophan, and tyrosine give rise to both glucose and fatty acid precursors and are thus characterized as being glucogenic and ketogenic. The amphibolic nature of the carbon skeltons of the amino acids is illustrated in Figure 25–1. Finally, it should be recognized that amino acids have a third possible fate. During times of starvation the reduced carbon skeleton is used for energy production, with the result that it is degraded to CO_2 and H_2O.

Alanine, Glutamine/Glutamate, and Asparagine/Aspartate

Glutaminase is an important kidney tubule enzyme involved in converting glutamine (from liver and from other tissue) to glutamate and NH_3^+ (Equation 25.1), with the NH_3^+ being excreted in the urine. Glutaminase activity is present in many other tissues as well, although its activity is not nearly as prominent as in the kidney. The glutamate produced from glutamine is converted to α-ketoglutarate, making glutamine a glucogenic amino acid.

$$\text{glutamine} \xrightarrow{\text{glutaminase}} NH_3^+ + \text{glutamate} \xrightarrow[\text{transferase}]{\text{amino}} \alpha - \text{ketoglutarate} \qquad 25.1$$

$$\text{asparagine} \xrightarrow{\text{asparaginase}} NH_3 + \text{aspartate} \xrightarrow[\text{transferase}]{\text{amino}} \text{oxaloacetate} \qquad 25.2$$

Asparaginase is also widely distributed within the cell, where it converts asparagine into ammonia and aspartate (Equation 25.2). Aspartate transaminates to oxaloacetate, which follows the gluconeogenic pathway to glucose.

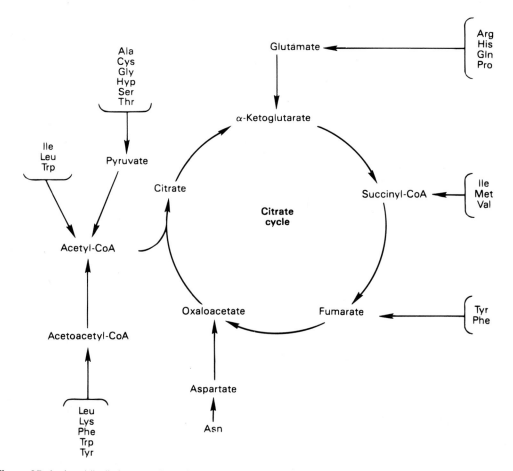

Figure 25–1. Amphibolic intermediates formed from the carbon skeletons of amino acids. (Reproduced, with permission, from Murray, RK: *Harper's Biochemistry*, 23rd ed. Appleton & Lange, 1993.)

Important metabolic reactions of this group of amino acids have been reviewed earlier. To restate briefly, glutamate and aspartate are important in collecting and eliminating amino nitrogen via glutamine synthetase and the urea cycle, respectively. The catabolic path of the carbon skeletons, illustrated in Equations 25.1 and 25.2, involves simple 1-step aminotransferase reactions that directly produce net quantities of a Krebs cycle intermediate. The glutamate dehydrogenase reaction operating in the direction of α-ketoglutarate production provides a second avenue leading from glutamate to gluconeogenesis.

Alanine is also important in intertissue nitrogen transport as part of the alanine cycle (Chapter 24). As shown in Equation 25.3, alanine's catabolic pathway involves a simple aminotransferase reaction that directly produces pyruvate. Generally pyruvate produced by this pathway will result in the formation of oxaloacetate, although when the energy charge of a cell is low the pyruvate will be oxidized to CO_2 and H_2O via the PDH complex and the Krebs cycle.

$$\text{alanine} \xrightarrow{\text{aminotransferase}} \text{pyruvate} \qquad \textbf{25.3}$$

Arginine, Ornithine, Proline, and Hydroxyproline

As shown in Figure 25–2, the catabolism of arginine begins with hydrolytic removal of urea by arginase with formation of ornithine. Excess ornithine transaminates to form glutamate semialdehyde; proline is converted to pyrroline-5-carboxylate by a P450-coupled mitochondrial enzyme, **proline oxidase**, and then also rearranges, to glutamate semialdehyde. Glutamate semialdehyde is oxidized to glutamate by an ATP-independent glutamate semialdehyde dehydrogenase, which is converted to α-ketoglutarate in a transamination reaction. Thus arginine, ornithine and proline, are glucogenic.

The hydroxyproline catabolic pathway is illustrated by Equation 25.4 and initially undergoes the same oxidation and rearrangement reactions as proline producing hydroxyglutamate semialdehyde. The latter is oxidized to γ-hydroxyglutamate and the amino nitrogen is removed by transamination, resulting in the formation of α-keto-γ-hydroxyglutaric acid. At this point the pathways diverge as α-keto-γ-hydroxyglutaric acid is cleaved to yield pyruvate and glyoxylic acid, both of which are glucogenic.

$$\textbf{Hydroxyproline} \xrightarrow[\text{mitochondial P450}]{\text{proline oxidase}} \gamma\text{-hydroxyglutamate} \xrightarrow{\text{aminotransferase}}$$

$$\alpha\text{-keto-}\gamma\text{-hydroxyglutarate} \xrightarrow{H_2O} \text{pyruvate} + \text{glyoxylate} \qquad \textbf{25.4}$$

Serine

The conversion of serine to glycine and then to CO_2 and NH_3, with the production of 2 equivalents of methyleneTHF, was described in Chapter 24. However, the bulk of serine degradation is by the pyridoxal phosphate-requiring enzyme known as **glycine synthase** (Figure 25–3).

Threonine Catabolism

There are at least 3 pathways for threonine catabolism. One involves a pathway initiated by threonine dehydrogenase yielding α-amino-β-ketobutyrate. The α-amino-β-ketobutyrate is either converted to acetyl-CoA and glycine or spontaneoulsy degrades to aminoacetone which is converted to pyruvate. The second pathway involves serine/threonine dehydratase yielding α-ketobutyrate which is further catabolized to propionyl-CoA and finally the Krebs cycle intermediate, succinyl-CoA. The third pathway utilizes threonine aldolase (Figure 25–4). The products of this reaction are both ketogenic (acetyl-CoA) and glucogenic (pyruvate).

Glycine

Glycine is classified as a glucogenic amino acid, since it can be converted to serine by a reversal of the reaction that forms methylene-THF (Figure 25–3), and serine can be converted to pyruvate by serine/threonine dehydratase (Figure 25–4). Nevertheless, the main glycine catabolic pathway leads to the production of CO_2, ammonia, and 1 equivalent of methyleneTHF by the mitochondrial enzyme glycine synthase, as illustrated in Figure 25–3.

Cysteine

There are 3 pathways for cysteine catabolism. The simplest, but least important pathway runs via a liver desulfurase and produces H_2S and pyruvate. The more important catabolic pathway is via a P450-coupled dioxygenase that oxidizes the cysteine sulfhydryl to sulfinate, producing the intermedi-

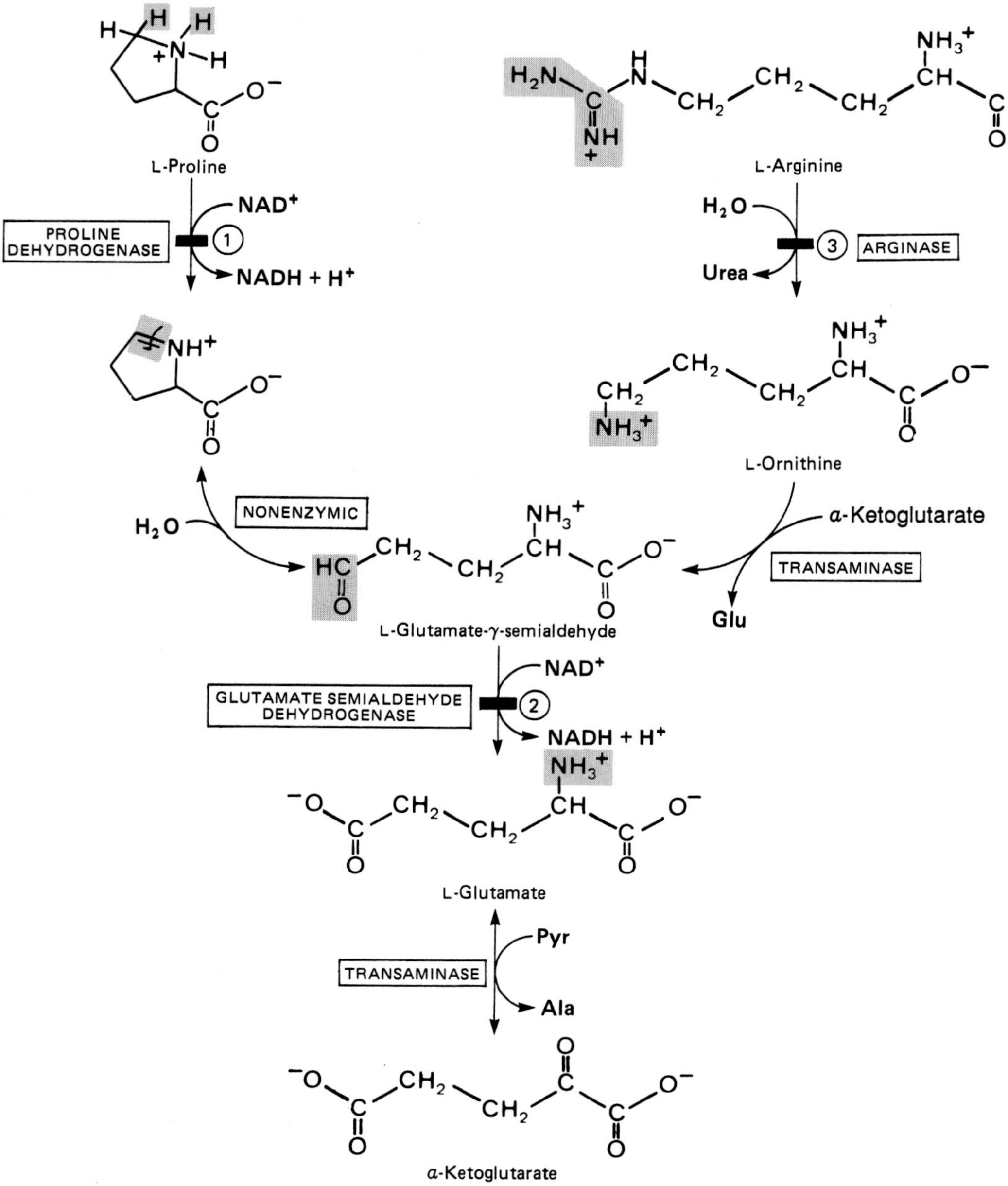

Figure 25–2. Catabolism of L-proline (*left*) and of L-arginine (*right*) to α-ketoglutarate. Circled numerals mark the sites of the metabolic defects in 1, type I hyperprolinemia; 2, type II hyperprolinemia; and 3, hyperargininemia. (Reproduced, with permission, from Murray, RK: *Harper's Biochemistry*, 23rd ed. Appleton & Lange, 1993.)

ate cysteine sulfinate (Figure 25–5). The glucogenic product pyruvate is formed from cysteine sulfinate by a transamination, followed by the removal of SO_3^-. Other than protein, the most important product of cysteine metabolism is the bile salt precursor **taurine**, which is used to form the bile acid conjugates **taurocholate** and **taurochenodeoxycholate** (see Chapter 22).

Methionine

The principal fates of the essential amino acid methionine are incorporation into polypeptide chains, and use in the production of α-ketobutyrate and cysteine via S-adenosyl methionine as described in

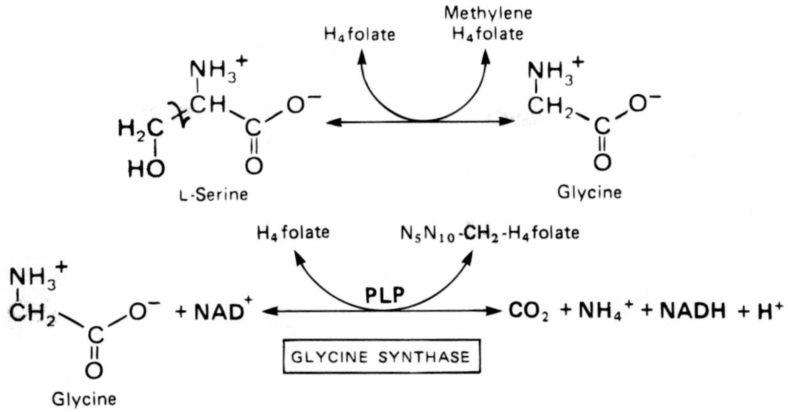

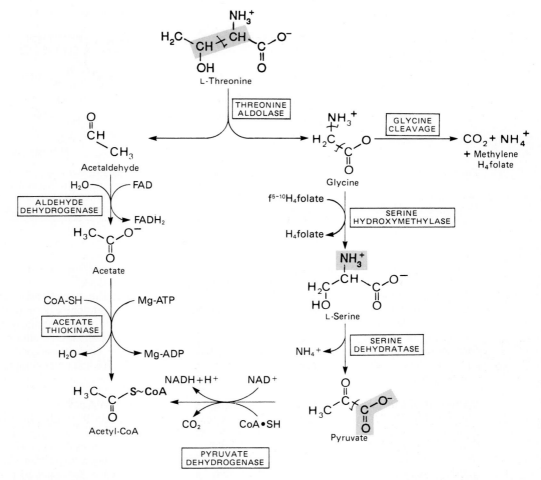

Figure 25–3. The freely reversible serine hydroxymethyltransferase reaction (*left*). H$_4$, tetrahydrofolate. Reversible cleavage of glycine by the mitochondrial glycine synthase comple (*right*). PLP, pyridoxal phosphate. (Reproduced, with permission, from Murray, RK: *Harper's Biochemistry*, 23rd ed. Appleton & Lange, 1993.)

Figure 25–4. Conversion of threonine and glycine to serine, pyruvate, and acetyl-CoA. f$^{5-10}$H$_4$folate, formyl [5–10] tetrahydrofolic acid. (Reproduced, with permission, from Murray, RK: *Harper's Biochemistry*, 23rd ed. Appleton & Lange, 1993.)

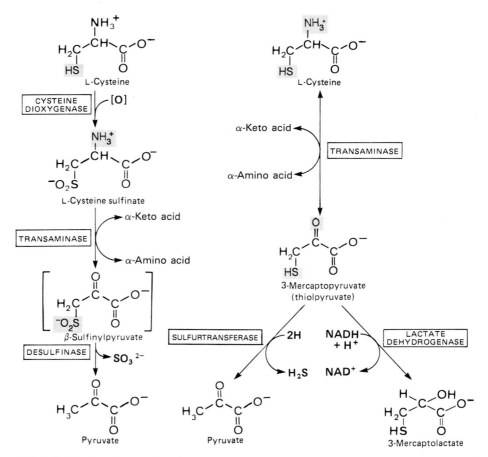

Figure 25–5. Catabolism of L-cysteine via the direct oxidative (cysteine sulfinate) pathway (*left*) and by the transamination (3-mercaptopyruvate) pathway (*right*). β-Sulfinylpyruvate is a putative intermediate. Oxidation of the sulfite produced in the last reaction of the direct oxidative pathway is catalyzed by sulfite oxidase. (Reproduced, with permission, from Murray, RK: *Harper's Biochemistry*, 23rd ed. Appleton & Lange, 1993.)

Chapter 24 (Figure 24–2). The transulfuration reactions that produce cysteine from homocysteine and serine also produce α-ketobutyrate, the latter being converted to succinyl-CoA.

Regulation of the methionine metabolic pathway is based on the availability of methionine and cysteine. If both amino acids are present in adequate quantities, S-adenosyl methionine accumulates and is a positive effector on cystathionine synthase, encouraging the production of cysteine and α-ketobutyrate (both of which are glucogenic). However, if methionine is scarce, S-adenosyl methionine will form only in small quantities, thus limiting cystathionine synthase activity. Under these conditions accumulated homocysteine is remethylated to methionine, using methyl-THF and other compounds as methyl donors.

Valine, Isoleucine, and Leucine: The Branched Chain Amino Acids

This group of essential amino acids all contain saturated, branched-chain R groups. Because this arrangement of carbon atoms cannot be made by humans, these amino acids are an essential element in the diet. As illustrated in Figure 25–6, the catabolism of all 3 compounds uses the same enzymes in the first 2 catabolic steps. The first step in each case is a transamination using a single branched-chain transaminase, with α-ketoglutarate as amine acceptor. As a result, 3 different α-keto acids are produced and are oxidized using a common **branched-chain keto acid dehydrogenase**, yielding the 3 different CoA derivatives shown in Figure 25–6. Subsequently the metabolic pathways diverge, producing many intermediates (which are not considered in this review). The principal product from valine is pro-

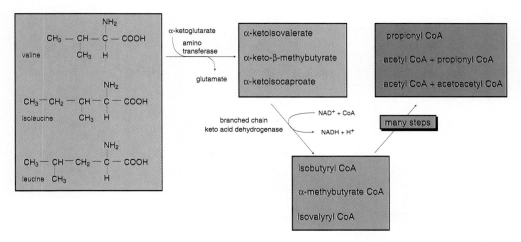

Figure 25–6. Catabolism of the branched chain amino acids.

pionyl-CoA, the glucogenic precursor of succinate. Isoleucine catabolism terminates with production of acetyl-CoA and propionyl-CoA; thus isoleucine is both glucogenic and ketogenic. Leucine gives rise to acetyl-CoA and acetoacetyl-CoA, and is thus classified as ketogenic only.

GENETIC DISEASES

There are a number of genetic diseases associated with faulty catabolism of the branched-chain keto acids. The most common defect is in the branched-chain keto acid dehydrogenase. Since there is only one dehydrogenase enzyme for all 3 amino acids, all 3 α-keto acids accumulate and are excreted in the urine. The disease is known as **maple sugar urine disease** because of the characteristic odor of the urine in afflicted individuals. Mental retardation in these cases is extensive. Unfortunately, since these are essential amino acids, they cannot be heavily restricted in the diet; ultimately, the life of afflicted individuals is short and development is abnormal. The main neurological problems are due to poor formation of myelin in the CNS.

Phenylalanine and Tyrosine Catabolism

Phenylalanine normally has only 2 fates: incorporation into polypeptide chains, and production of tyrosine via the tetrahydrobiopterin-requiring phenylalanine hydroxylase (see Figure 24–4). Thus, phenylalanine catabolism always follows the pathway of tyrosine catabolism illustrated in Figure 25–7. The main pathway for tyrosine degradation involves conversion to fumarate and acetoacetate, allowing phenylalanine and tyrosine to be classified as both glucogenic and ketogenic. A number of physiologically important metabolites obtained from tyrosine are discussed in Chapter 26

As in phenylketonuria (deficiency of phenylalanine hydroxylase), deficiency of tyrosine transaminase leads to urinary excretion of tyrosine and the intermediates between phenylalanine and tyrosine. The adverse neurological symptoms are the same for the two diseases.

Genetic diseases are also associated with other defective enzymes of the tyrosine catabolic pathway. Some of these diseases are indicated in Figure 25–7. The first genetic disease ever recognized, **alcaptonuria**, is caused by defective **homogentisic acid oxidase**. Homogentisic acid accumulation is relatively innocuous, causing urine to darken on standing, but no life-threatening effects accompany the disease. The other genetic deficiencies lead to more severe symptoms, most of which are associated with abnormal neural development, mental retardation, and shortened life span.

Lysine Catabolism

Lysine catabolism is unusual in the way that the epsilon amino group is transferred to α-ketoglutarate and into the general nitrogen pool. The reaction is a transamination in which the epsilon amino group is transferred to the α-keto carbon of α-ketoglutarate forming the metabolite, **saccharopine** (Figure 25–8). Unlike the majority of transamination reactions, this one does not employ pyridoxal phosphate as a cofactor. Saccharopine is immediately hydrolyzed by the enzyme α-**aminoadipic semialdehyde**

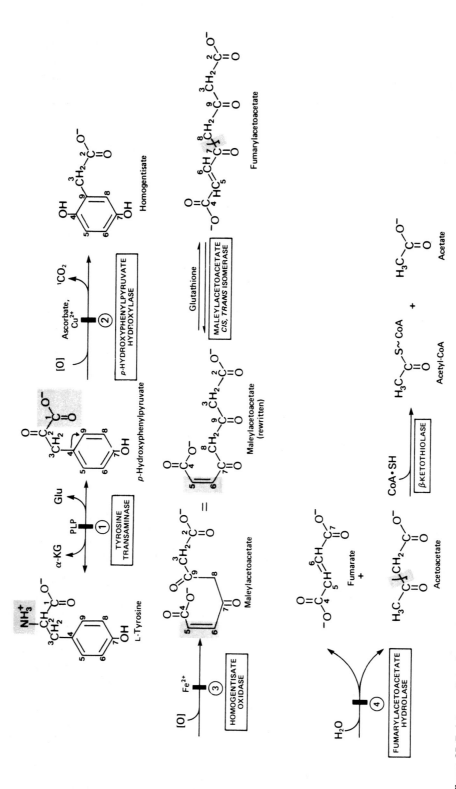

Figure 25–7. Intermediates in tyrosine catabolism. With the exception of β-ketothiolase, reactions are discussed in the text. Carbon atoms of intermediates are numbered to assist readers in determining the ultimate fate of each carbon. α-KG, α-ketoglutarate; Glu, glutamate; PLP, pyridoxal phosphate. The circled numerals represent the probable sites of the metabolic defects in ①, type II tyrosinemia; ②, neonatal tyrosinemia; ③, alkaptonuria; and ④, type I tyrosinemia, or tyrosinosis.

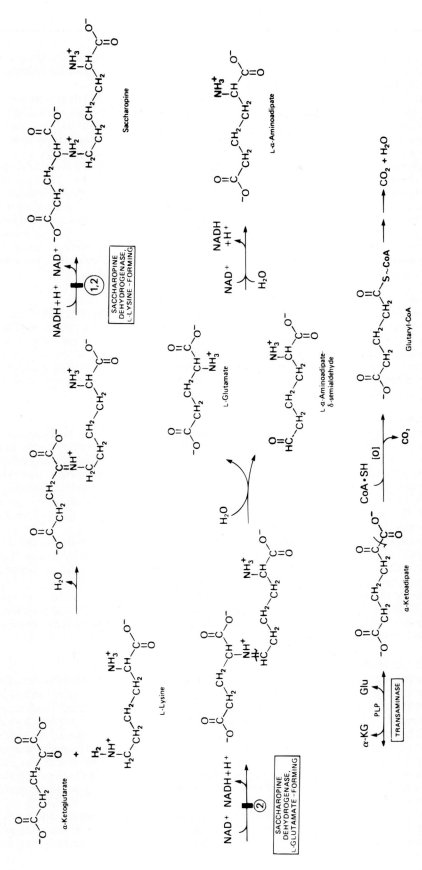

Figure 25–8. Catabolism of L-lysine. (α-KG, α-ketoglutarate; Glu, glutamate; PLP, pyridoxal phosphate.) The circled numerals indicate the probable sites of the metabolic defects in ①, periodic hyperlysinemia with associated hyperammonemia; and ②, persistent hyperlysinemia without associated hyperammonemia. (Reproduced, with permission, from Murray, RK: *Harper's Biochemistry*, 23rd ed. Appleton & Lange, 1993.)

synthase in such a way that the amino nitrogen remains with the α-carbon of α-ketoglutarate, producing glutamate and α-aminoadipic semialdehyde. **Because this transamination reaction is not reversible, lysine is an essential amino acid.**

There are many intermediates on the path from α-aminoadipic semialdehyde to acetoacetylCoA, but only the key features of the reaction pathway are noted here. The α-amino of aminoadipic semialdehyde is removed by a standard transamination involving α-ketoglutarate as the amino acceptor, and both adipic acid carboxylates are oxidized to CO_2. Thus, the 6-carbon aminoadipic semialdehyde is reduced to a 4-carbon molecule that finally appears as the ketogenic acetoacetyl-CoA.

Genetic deficiencies in the enzyme α-aminoadipic semialdehyde synthase have been observed in individuals who excrete large quantities of urinary lysine and some saccharopine. The lysinemia and associated lysinuria are benign. Other serious disorders associated with lysine metabolism are due to failure of the transport system for lysine and the other dibasic amino acids across the intestinal wall. Lysine is essential for protein synthesis; a deficiency of its transport into the body can cause seriously diminished levels of protein synthesis. Probably more significant however, is the fact that arginine is transported on the same dibasic amino acid carrier, and resulting arginine deficiencies limit the quantity of ornithine available for the urea cycle. The result is severe hyperammonemia after a meal rich in protein. The addition of citrulline to the diet prevents the hyperammonemia.

Arginine

The main aspects of arginine catabolism were considered in Chapter 23 during review of the urea cycle, and earlier in this chapter during the consideration of proline catabolism. Consequently, the two key issues associated with arginine catabolism are reiterated only briefly here. First arginase cleaves urea from arginine, producing ornithine, which is the essential nitrogen carrier of the urea cycle. Second, the ornithine produced from arginine can be transaminated to produce glutamate semialdehyde, which is readily converted to α-ketoglutarate. Thus, the main pathway of arginine catabolism leads to α-ketoglutarate, making arginine glucogenic in nature.

Histidine Catabolism

Histidine catabolism begins with release of the α-amino group catalyzed by **histidase**, introducing a double bond into the molecule. As a result the deaminated product, urocanate, is not the usual α-keto acid associated with loss of α-amino nitrogens. As illustrated in Figure 25–9, the end product of histidine catabolism is glutamate, making histidine one of the glucogenic amino acids. However, another key feature of histidine catabolism is that the ring opening of the 4-imidazolone-5-propionate intermediate makes a ring nitrogen available to combine with tetrahydrofolic acid (THF), producing the 1-carbon THF intermediate known as N^5-formimino THF. The latter reaction is 1 of 2 routes to N^5-formimino THF (see Figure 24–8).

The principal genetic deficiency associated with histidine metabolism is absence or deficiency of the first enzyme of the pathway, histidase. The resultant histidinemia is relatively benign. The disease, which is of relatively high incidence (1 in 10,000), is most easily detected by the absence of urocanate from skin and sweat, where it is normally found in relative abundance.

Decarboxylation of histidine in the intestine by bacteria gives rise to histamine, as described in Chapter 23. Similarly, histamine arises in many tissues by the decarboxylation of histidine, which in excess causes constriction or dilation of various blood vessels. The general symptoms are those of asthma and various allergic reactions.

Tryptophan Metabolism

The principal pathway of tryptophan catabolism is illustrated in Figure 25–10. However, a number of important side reactions occur from intermediates on the pathway to acetoacetate. The first enzyme of the catabolic pathway is an iron porphyrin oxygenase that opens the indole ring. The latter enzyme is highly inducible, its concentration rising almost 10-fold on a diet high in tryptophan. **Kynurenine** is the first key branch point intermediate in the pathway. The main path continues to 2-amino-3-carboxymuconic acid, but kynurenic acid produced along one side branch is an important antagonist of excitatory amino acids in the brain. A second side branch produces anthranilic acid plus alanine. Another equivalent of alanine is produced further along the main catabolic pathway, and it is the production of these alanine residues that allows tryptophan to be classified among the glucogenic and ketogenic amino acids.

The second important branch point metabolite is **2-amino-3-carboxymuconic acid**. The main flow of carbon elements from this intermediate is to α-ketoadipic acid. An important side reaction in liver is a transamination and several rearrangements to produce limited amounts of nicotinic acid, which leads

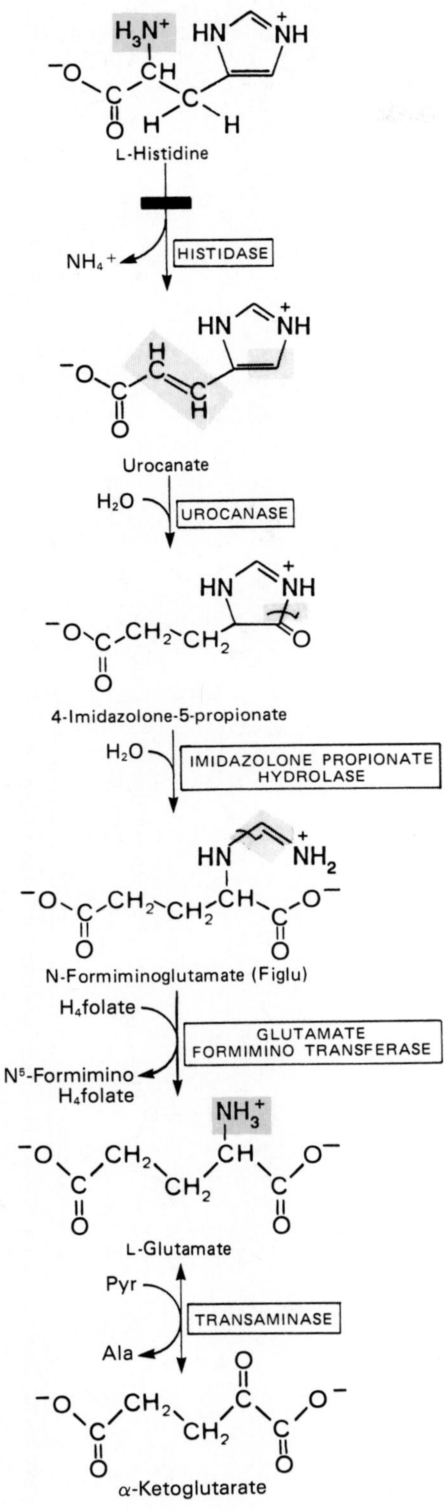

Figure 25–9. Catabolism of L-histidine to α-ketoglutarate. H₄ folate, tetrahydrofolate. The reaction catalyzed by histidase represents the site of the probable metabolic defect in histidinemia. (Reproduced, with permission, from Murray, RK: *Harper's Biochemistry*, 23rd ed. Appleton & Lange, 1993.)

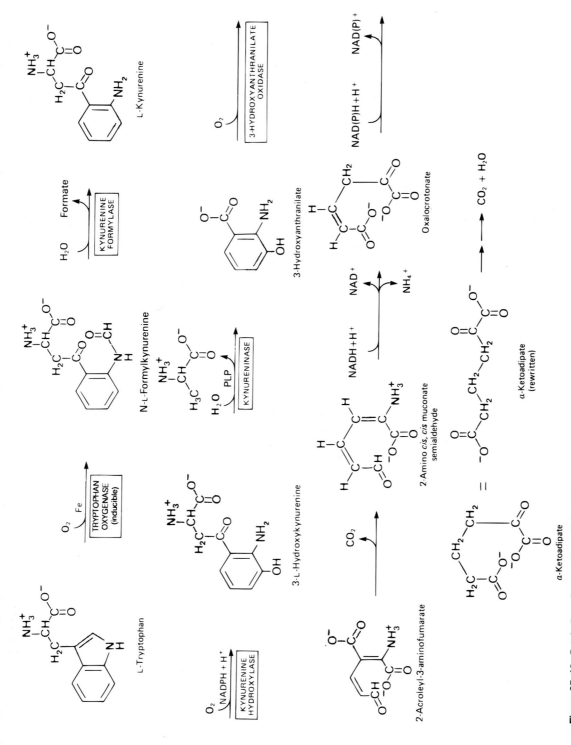

Figure 25–10. Catabolism of L-tryptophan. PLP, pyridoxal phosphate. (Reproduced, with permission, from Murray, RK: *Harper's Biochemistry*, 23rd ed. Appleton & Lange, 1993.)

to production of a small amount of NAD^+ and $NADP^+$. Earlier, α-ketoadipic acid was encountered as a principal intermediate in leucine metabolism, and thus tryptophan and leucine metabolism converge at this compound. The α-ketoadipate is converted to the ketogenic compound acetoacetyl-CoA, providing the final identification of tryptophan as both glucogenic and ketogenic in nature.

Questions

DIRECTIONS (items 1–4): The following statements refer to the compounds A, B, C, and D shown in the diagram below and are either true or false. Select "A" if the statement is TRUE; select "B" if it is FALSE.

A. $^-OOC-CH-CH_2-COO^-$
  NH_3^+

C. $^-OOC-CH-CH_2-CH_2-COO^-$
  NH_3^+

B. $^-OOC-CH-CH_2-CH_2-S-CH_3$
  NH_3^+

D. $^-OOC-CH-CH_2-CH-CH_3$
  NH_3^+ CH_3

1. Compounds A and B are glucogenic.

2. Compounds B, C, and D are essential in the human diet.

3. Compound D is ketogenic.

4. Compound D is degraded to pyruvate after deamination.

DIRECTIONS (items 5–11): Match the following lettered compounds with the properties described. Some answers may be used once or not at all.

  a. Glycine
  b. Proline
  c. Phenyl lactate
  d. Tyrosine
  e. Cysteine
  f. Glutamine
  g. Tryptophan
  h. Valine
  i. Threonine

5. Non-sulfur-containing compound formed from an essential amino acid.

6. Compound whose amino group is removed by a dehydration reaction.

7. Essential amino acid that can be dietarily spared by its α-keto analog.

8. Amino acid that spares the dietary requirement for phenylalanine.

9. Amino acid that spares the dietary requirement for methionine.

10. Carrier of peripheral tissue NH_3 to liver and kidney.

11. Amino acid produced from glutamate via glutamate γ-semialdehyde.

DIRECTIONS (items 12–15): Each numbered item or incomplete statement is followed by answers or completions of the statement. Select the ONE lettered response that is BEST in each case.

12. Lysine is an essential amino acid because:
    a. Animals cannot make the lysine carbon skeleton.
    b. The α-amino group of lysine cannot be formed by an aminotransferase.
    c. The ε-amino of lysine cannot be added to urocanate.
    d. More lysine is required for protein synthesis than can be made in the body.

13. The amino acids that are strictly ketogenic are:
    a. Lysine and leucine.
    b. Valine and isoleucine.
    c. Leucine and isoleucine.
    d. Lysine, leucine and isoleucine.
    e. Tyrosine and tryptophan.

14. Glutamate semialdehyde is a branch point metabolite in the catabolism of:
    a. Proline.
    b. Proline and ornithine.
    c. Proline, ornithine, and arginine.
    d. Proline, ornithine arginine, and hydroxyproline.

15. Catabolites of cysteine include:
    a. α-Ketobutyrate.
    b. Taurine.
    c. Cholate.
    d. S-adenosylmethionine.

Answers

1. a		**6.** i		**11.** b	
2. b		**7.** h		**12.** c	
3. a		**8.** d		**13.** a	
4. b		**9.** e		**14.** c	
5. c		**10.** f		**15.** b	

26

Biologically Active Nitrogen Compounds

Objectives

- To review the synthesis of physiologically important metabolites that are derived from amino acids.
- To identify the chemical and structural characteristics of intact heme.
- To identify the source of all the atoms of ALA.
- To identify the subcellular location of each enzyme and intermediate on the pathway from glycine and succinylCoA to porphobilinogen.
- To identify the main regulation on heme biosynthesis.

- To identify the difference between "type I" and "type III" porphyrinogens.
- To define the role of uroporphyrinogen III cosynthase.
- To distinguish the porphyrias caused by genetic defects of enzymes early in the heme pathway from those of enzymes late in the pathway.
- To identify the compounds on the pathway from heme to bilirubin; to identify the solubilized form of bilirubin that is excreted in the bile.
- To identify the basis for the difference in color of bilirubin and biliverdin.
- To identify the association that exists between ornithine decarboxylase (ODC) activity, cell replication, and the polyamines.
- To identify the intermediates on the pathway from tyrosine to epinephrine and the main physiological activities of all the intermediates, as well as the catecholamines.
- To identify the pathway to serotonin and melatonin and the physiological role of these compounds.

Concepts

Although the two most prominent fates of amino acids are incorporation into protein and catabolite degradation, the biosynthesis of numerous other physiologically important compounds, such as porphyrin and nitric oxide (NO), require the availability of amino acids and their constituent nitrogen. The syntheses of the most prominent of these compounds are reviewed in this chapter.

The structure of heme was presented in Figure 18–5. Aside from its importance as the prosthetic group of hemoglobin and a small number of enzymes (eg, cytochromes and P450) heme is important because a number of genetic disease states are associated with deficiencies of the enzymes used in its biosynthesis. Some are readily diagnosed because they cause δ-aminolevulinic acid (ALA) and abnormally colored heme intermediates to appear in the circulation, the urine, and in other tissues such as teeth and bones. Some disorders of heme biosynthesis are more insidious, however. The disease known as **acute intermittent porphyria**, although easily diagnosed, often leads to death because of an associated neuropathology that may remain unrecognized.

Table 26–1 lists the key characteristics of the heme biosynthetic pathway. These and other features of the pathway will be reviewed more completely in the following material.

Synthesis of Porphobilinogen

The first reaction in heme biosynthesis takes place in the mitochondrion. As illustrated in Figure 26–1, it involves the condensation of 1 glycine and 1 succinylCoA by the pyridoxal phosphate-containing enzyme, **δ-aminolevulinic acid synthase (ALA synthase)**. This reaction is both the rate-limiting reaction of heme biosynthesis, and the most highly regulated reaction. Regulation takes place mainly at the gene level, where it has been demonstrated that diminished heme levels lead to increased transcription of ALA synthase mRNA with proportionate increases in heme production. Likewise, raising heme levels reduces cellular levels of ALA synthase. Mitochondrial δ-aminolevulinic acid (ALA) is transported to the cytosol, where ALA dehydratase dimerizes 2 molecules of ALA (Figure 26–1) to produce the pyrrole known as porphobilinogen.

The next step in the pathway involves the head-to-tail condensation of 4 molecules of porphobilinogen to produce hydroxymethylbilane, as illustrated in Figure 26–2. Note that the acetate and propionate residues (noted in Figure 26–2 as **A** and **P,** respectively) remain on the pryrrole rings during condensation. The enzyme for this condensation is **uropophyrinogen I synthase**, a cytosolic enzyme. Hydroxymethylbilane has two main fates. The most important is regulated, enzymatic conversion to uroporphyrinogen III, the next intermediate on the path to heme. This step is mediated by a holoen-

Table 26–1. Characteristics of heme biosynthesis.

1.	The enzymes catalyzing the first and the last three reactions of the pathway are mitochondrial and the intervening reactions are catalyzed by cytosolically localized enzymes.
2.	The heme ring system and all substituents on the ring are formed from eight residues of glycine and 8 residues of succinyl-CoA.
3.	Reactions that modify the side chain residues, associated with the four pyrrole ring systems, take place before the molecule becomes fully conjugated and colored. Thus accumulation of metabolic pathway intermediates up to this point is not necessarily signaled by accumulation of colored products.

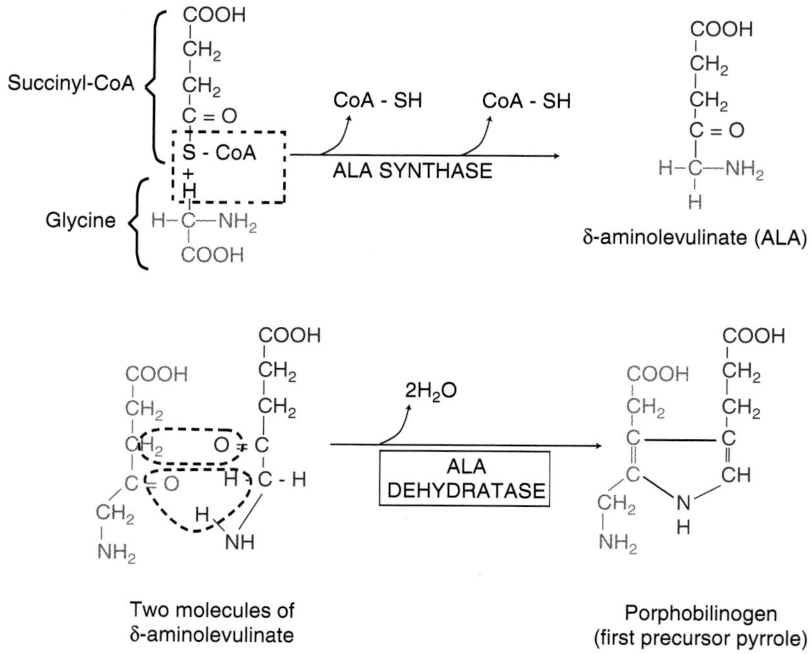

Figure 26–1. Boiosynthesis of porphobilinogen. ALA synthase occurs in the mitochondria, whereas ALA dehydratase is present in the cytosol.

zyme comprised of **uroporphyrinogen synthase** plus a protein known as **uroporphyrinogen III cosynthase**. The key feature of enzymatic ring-closure is that the last pyrrole (marked with arrows in Figure 26–2) is flipped in its orientation with respect to the other 3 pyrroles during closure. Consequently, in uroporphyrinogen III, the substituents on this ring are inverted relative to the orientation expected from the structure of the linear tetrapyrrole.

Genetic defects that cause increased ALA synthase activity or decreased uroporphyrinogen I synthase activity lead to the disease known as acute intermittent porphyria, which is diagnosed by the excretion of excess porphobilinogen (a condition that is not obvious from the color of the urine).

Genetic defects in enzymes beyond hydroxymethylbilane in the biosynthetic pathway result in the accumulation of excess hydroxymethylbilane, leading to an alternative non-enzymatic cyclization. In **erythropoietic uroporphyria**, bone marrow uroporphyrinogen III cosynthase is only present at about 30% of the normal level; the result is that massive amounts of Type I uroporphyrinogen and its highly colored oxidation products are found in the urine and deposited in a variety of tissues, including teeth and bones. It is noteworthy that the spontaneously cyclized compound, uroporphyrinogen I, retains the original, expected arrangement of substituents on the 4 pyrrole rings.

In the cytosol, the acetate substituents of uroporphyrinogen (normal uroporphyrinogen III or abnormal uroporphyrinogen I) are all decarboxylated by the enzyme uroporphyrinogen decarboxylase. The resultant products have methyl/groups in place of acetate and are known as coproporphyrinogens, with **coproporphyrinogen III** being the important, normal intermediate in heme synthesis.

Coproporphyrinogen III is transported to the interior of the mitochondrion, where 2 propionate residues are decarboxylated, yielding vinyl (V) substituents on the 2 pyrrole rings. The colorless product is protoporphyrinogen IX. In the mitochondrion, protoporphyrinogen IX is converted to protoporphyrin IX by protoporphyrinogen IX oxidase. The oxidase reaction requires molecular oxygen and results in the loss of 6 protons and 6 electrons, yielding a completely conjugated ring system (Figure 26–2), which is responsible for the characteristic red color of hemes. The final reaction in heme synthesis also takes place in the mitochondrion and involves the insertion of the iron atom into the ring system. The enzyme catalyzing this reaction is known as ferrochelatase.

The enzymes ferrochelatase and ALA synthase are particularly sensitive to lead poisoning, which is often brought about by direct ingestion (eg, by eating leaded paints). Affected individuals become severely anemic and excrete large quantities of coproporphyrinogen and ALA in the urine.

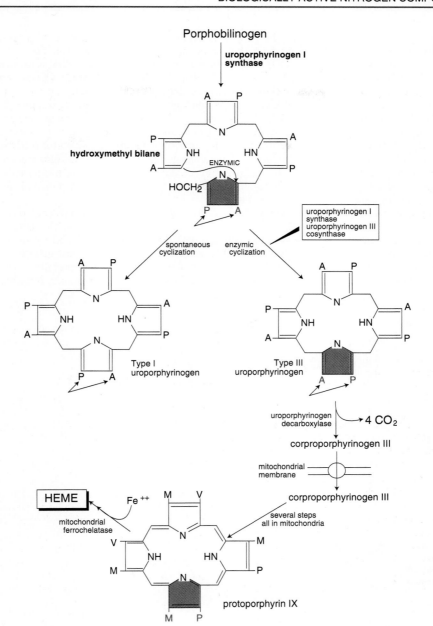

Figure 26–2. Key elements in the biosynthesis of heme.

HEME CATABOLISM

The largest repository of heme in the human body is in red blood cells, which have a life span of about 120 days. There is thus a turnover of about 6 g/day of hemoglobin, which presents two problems. First, the porphyrin ring is hydrophobic and must be solubilized to be excreted. Second, iron must be conserved for new heme synthesis.

Normally, senescent red blood cells and heme from other sources are engulfed by cells of the reticuloendothelial system. The globin is recycled or converted into amino acids, which in turn are recycled or catabolized as required. Heme is oxidized, with the heme ring being opened by the endoplasmic reticulum enzyme, **heme oxygenase**. The oxidation step requires heme as a substrate, and any hemin (Fe^{+3}) is reduced to heme (Fe^{+2}) prior to oxidation by heme oxygenase. The oxidation, shown in Figure 26–3, occurs on a specific carbon, producing the linear tetrapyrrole biliverdin, ferric iron ($Fe+3$), and carbon monoxide (CO). **This is the only reaction in the body that is known to produce CO. Most of the CO is excreted through the lungs, with the result that the CO content of expired air is a direct measure of the activity of heme oxidase in an individual.**

In the next reaction a second bridging methylene (between rings III and IV) is reduced by biliverdin reductase, producing bilirubin. Bilirubin is significantly less extensively conjugated than biliverdin causing a change in the color of the molecule from blue-green (biliverdin) to yellow-red (bilirubin). The latter catabolic changes in the structure of tetrapyrroles are responsible for the progressive changes in color of a hematoma, or bruise, in which the damaged tissue changes its color from an initial dark blue to a red-yellow and finally to a yellow color before all the pigment is transported out of the affected tissue. Peripherally arising bilirubin is transported to the liver in association with albumin, where the remaining catabolic reactions take place.

In hepatocytes, UDP glucuronyl transferase adds 2 equivalents of glucuronic acid to bilirubin to produce the more water soluble, diglucuronide derivative shown in Figure 26–4. The increased water solubility of the tetrapyrrole facilitates its excretion with the remainder of the bile as the bile pigments. In individuals with abnormally high red cell lysis, or liver damage with obstruction of the bile duct, the bilirubin and its precursors accumulate in the circulation; the result is hyperbilirubinemia, the cause of the abnormal body pigmentation known as **jaundice**. In normal individuals, intestinal bilirubin is

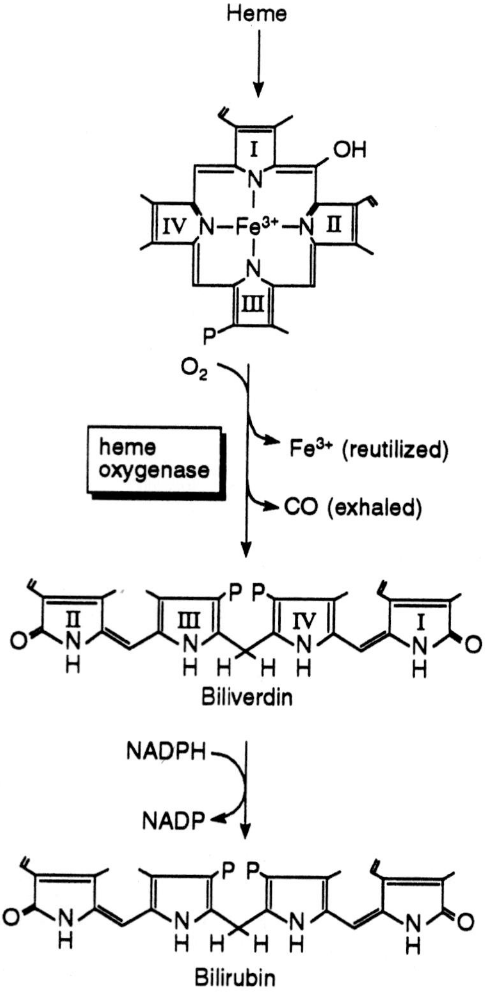

Figure 26–3. Heme oxygenase catalyzed catabolism of heme to bilirubin.

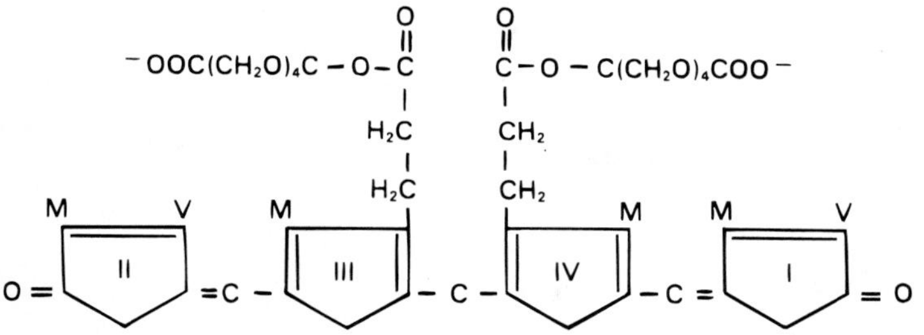

Figure 26–4. Structure of bilirubin diglucuronide. Bilirubin is conjugated in the liver with two equivalents of glucuronic acid which is attached in ester linkage to the propionate residues of bilirubin. (Reproduced, with permission, from Murray, RK: *Harper's Biochemistry*, 23rd ed. Appleton & Lange, 1993.)

acted on by bacteria to produce the final porphyrin products, urobilinogens and urobilins, that are found in the feces. Bilirubin and its catabolic products are collectively known as the bile pigments.

Polyamines: Spermine and Spermidine

One of the earliest signals that cells have entered their replication cycle is the appearance of elevated levels of mRNA for ornithine decarboxylase (ODC), and then increased levels of the enzyme, which is the first enzyme in the pathway to synthesis of the polyamines. Because of the latter, and because the polyamines are highly cationic and tend to bind nucleic acids with high affinity, it is believed that the polyamines are important participants in DNA synthesis, or in its regulation.

The pathway to the polyamines is illustrated in Figure 26–5. The key features of the pathway are that it involves putrescine, an ornithine catabolite not previously encountered, and S-AdoMet as a donor of 2 propylamine residues that combine with putrescine to form spermidine.

The function of ODC is to produce the 4-carbon saturated diamine, putrescine. At the same time, S-AdoMet decarboxylase cleaves the S-AdoMet carboxyl residue, producing decarboxylated S-AdoMet (S-adenosymethylthiopropylamine), which retains the methyl group usually involved in S-AdoMet methyltransferase activity. S-AdoMet decarboxylase activity is regulated by product inhibition and allosterically stimulated by **putrescine**. Spermidine synthase catalyzes the condensation reaction, producing spermidine and methylthioadenosine. Spermidine adds a second propylamine residue as illustrated, producing spermine.

Catabolism of the polyamines is carried out by the enzyme **polyamine oxidase**, which cleaves the two propylamines to β-aminopropionaldehyde and putrescine as shown in Figure 26–6.

The signal for regulating ornithine decarboxylase is unknown, but since the product of its activity, putrescine, regulates S-AdoMet decarboxylase activity, it appears that polyamine production is principally regulated by ODC concentration.

Active Metabolites of Tyrosine

The pathway from tyrosine to epinephrine is illustrated in Figure 26–7. Tyrosine hydroxylase is catalytically similar to phenylalanine hydroxylase and also employs biopterin as a cofactor. The product, DOPA, is decaroxylated in specific regions of the brain to produce the neurotransmitter **dopamine**. Decreased production of dopamine is responsible for **Parkinson's disease**. In neurons, dopamine is metabolized to norepinephrine and stored in vesicles of axon terminals, where it acts in synaptic neural transmission. Dopamine is also made in the adrenal medulla, where it is further metabolized to the catecholamines, endocrine hormones, **norepinephrine**, and **epinephrine**, and stored until secreted into the circulation.

Serotonin and Melatonin Synthesis

Serotonin and melatonin are produced from tryptophan. Serotonin is a neurotransmitter and a peripheral vasodilator, while melatonin's principal function is as an endocrine regulator of reproductive functions.

$$\text{tryptophan} \xrightarrow[\text{bioptern}]{\substack{\text{tryptophan}\\\text{hydroxylase}}} 5 - \text{hydroxy tryptophan} \xrightarrow{\text{decarboxylase}} CO_2 + \text{serotonin} \qquad \textbf{26.1}$$

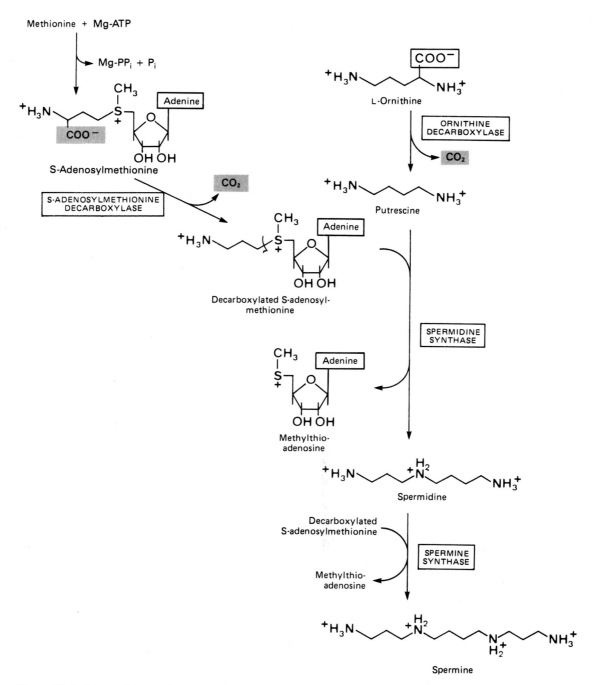

Figure 26–5. Intermediates and enzymes that participate in the biosynthesis of spermidine and spermine. Methylene groups are abbreviated to facilitate visualization of the overall process. (Reproduced, with permission, from Murray, RK: *Harper's Biochemistry*, 23rd ed. Appleton & Lange, 1993.)

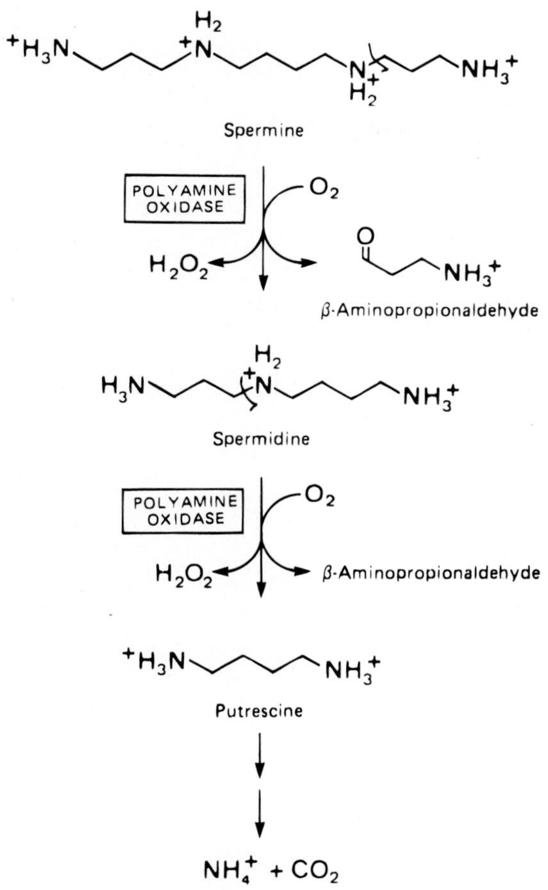

Figure 26–6. Catabolism of polyamines. Structures are abbreviated to facilitate presentation. (Reproduced, with permission, from Murray, RK: *Harper's Biochemistry*, 23rd ed. Appleton & Lange, 1993.)

The pathway to serotonin is shown in Equation 26.1; the structures of 5-hydroxytryptophan, serotonin (5-hydroxytryptamine), and melatonin are illustrated in Figure 26–8. Serotonin is produced in a number of tissues, but melatonin is produced from serotonin only by the pineal gland, in a process that is regulated by the light/dark cycle. In the dark, epinephrine stimulates the conversion of serotonin to melatonin by an acetylation using acetyl-CoA and a methylation using S-AdoMet.

A number of other physiologically important compounds are derived from tyrosine. Melanin, the skin-darkening pigment, is a polymeric oxidation product of tyrosine and is also found in retinal pigments and in the substantia nigra of the brain.

Nitric Oxide Formation from Arginine

Recently it was discovered that the active product made from glyceryl trinitrate (nitroglycerin) and used to treat angina pectoris is nitric oxide (NO). Further, the function of NO from glyceryl trinitrate is to replace or supplement the naturally occurring NO produced from the action of the enzyme **NO synthase** on arginine, as illustrated in Equation 26.2.

$$\text{arginine} + \text{NADPH} + \text{H}^+ + \text{O}_2 \xrightarrow{\text{NO synthase}} \text{citrulline} + \text{NO} + \text{NADP}^+ \qquad \textbf{26.2}$$

In the case of angina the NO acts as a relaxant of vascular smooth muscle cells, relieving painful vascular spasms. It is now recognized that the physiological role of NO is much more widespread than previously suspected, with important functions in the vascular system, the brain, and the immune system. The roles of NO are so diverse and widespread because it activates the production of intracellular cGMP, which is a potent second messenger.

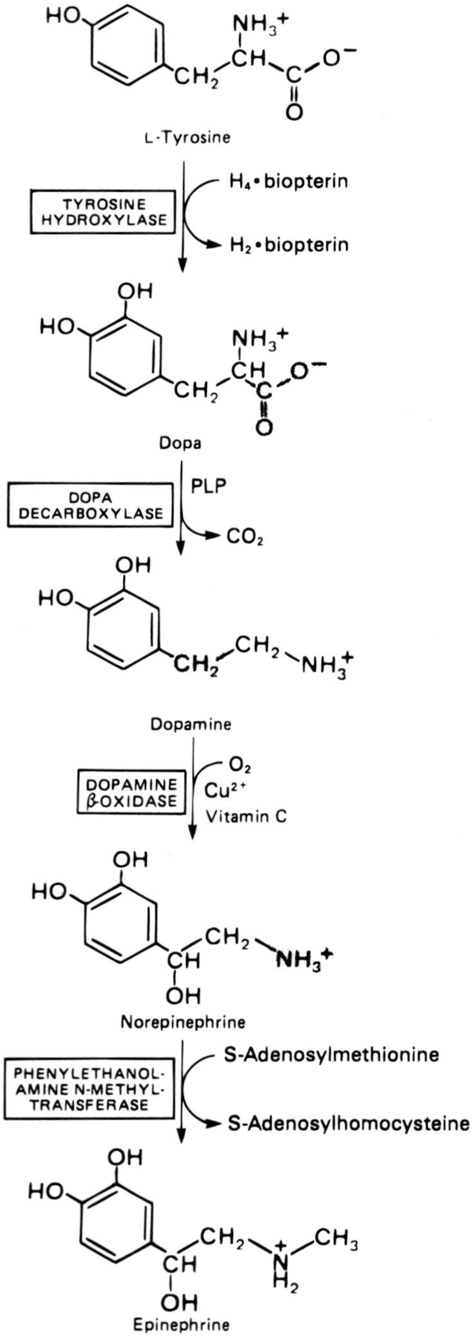

Figure 26-7. Conversion of tyrosine to epinephrine and norepinephrine in neuronal and adrenal cells. PLP, pyridoxal phosphate. (Reproduced, with permission, from Murray, RK: *Harper's Biochemistry*, 23rd ed. Appleton & Lange, 1993.)

Questions

DIRECTIONS (items 1–11): Each numbered item or incomplete statement is followed by answers or by completions of the statement. Select the ONE lettered answer or completion that is BEST in each case.

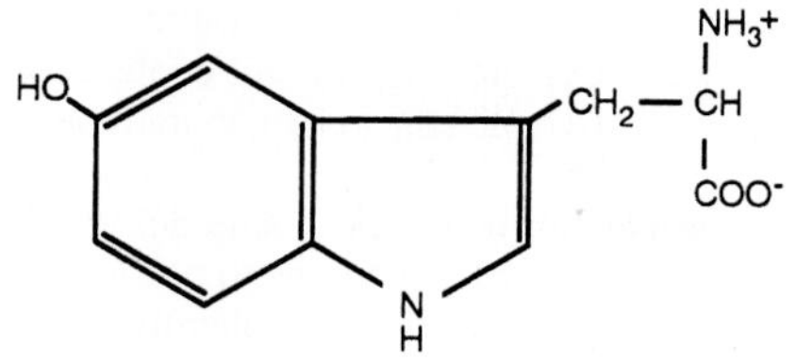

5-Hydroxytryptophan

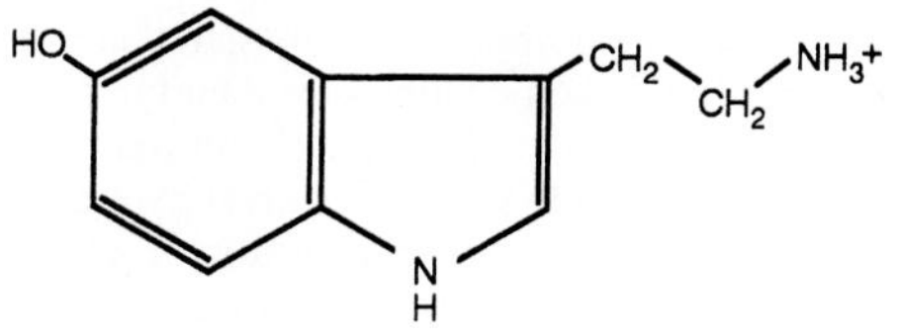

5-Hydroxytryptamine (serotonin)

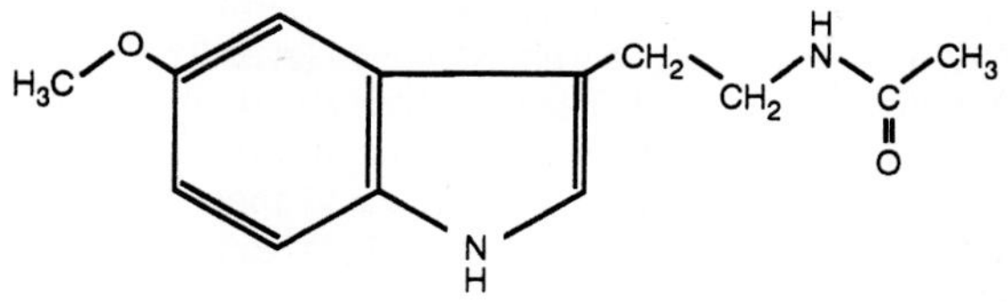

Melanotonin
(N-acetyl-5-methoxyserotonin)

Figure 26–8. Structures of tryptophan metabolites 5-hydroxy tryptophan, serotonin and melatonin.

1. The number of glycine equivalents required to form δ-aminolevulinic acid (ALA), porphobilinogen, and heme is:
   a. 1, 2, and 8 respectively.
   b. 1, 4, and 16 respectively.
   c. 2, 8, and 32 respectively.
   d. 4, 8, and 16 respectively.

2. ALA synthase is an enzyme of:
   a. The endoplasmic reticulum.
   b. The cytosol (soluble).
   c. Peroxisomes.
   d. Mitochondria.

3. Coproporphyrinogens are uncolored because:
   a. The individual pyrrole rings do not contain unsaturated bonds.

    b. The individual pyrrole rings contain 2 unsaturated bonds each.

    c. The bridges between pyrrole rings do not contain unsaturated bonds.

    d. The bridges between pyrrole rings each contain an unsaturated bond.

4. The reaction catalyzed by aminolevulinic acid dehydratase:

    a. Splits water from 2 different molecules to produce δ-aminolevulinic acid.

    b. Splits water from 4 molecules of -aminolevulinic acid to produce a tetrapyrrole.

    c. Splits water from 2 molecules to produce porphobilinogen.

    d. Catabolizes -aminolevulinic acid to glycine and succinate.

5. The difference between uroporphyrinogen type I and uroporphyrinogen type III is:

    a. Type I molecules are completely conjugated and thus colored; type III are not conjugated.

    b. Type III molecules are completely conjugated and thus colored; type I are not conjugated.

    c. Type III uroporphyrinogen occurs later than type I uroporphyrinogen in the pathway to heme and is colored; type I molecules are not colored.

    d. Type III uroporphyrinogen molecules arise by enzymatic ring closure of a tetrapyrrole; type I molecules arise by spontaneous cyclization of a tetrapyrrole.

6. A 52-year-old woman presented to the emergency room with severe generalized abdominal pains. She had been vomiting, was constipated, and had a distended abdomen. Her history showed she had recently been prescribed griseofulvin and sufonamide for a persistent infection. Initial laboratory results were generally only modestly different from normal. Because the abdominal difficulty persisted, a laparotomy was considered. Before surgery it was discovered that the patient had a markedly elevated urinary porphobilinogen level. The drugs were discontinued and the patient's symptoms slowly improved. The most likely cause of the patient's difficulty was:

    a. A deficiency in a heme biosynthesis enzyme located late in the pathway, after synthesis of porphyrin, so that porphyrin accumulated and appeared in the urine.

    b. A deficiency in uroporphyrinogen III cosynthase, leading to the appearance of hydroxymethylbilane in the urine.

    c. Abnormally high δ-aminolevulinate synthase activity, leading to elevated steady state levels of porphobilinogen which was excreted in the urine.

    d. Abnormal low activity of δ-aminolevulinate dehydratase, leading to accumulation of δ-aminolevulinate and its excretion in the urine.

7. An 18-month-old boy was referred to the pediatrics clinic because of persistent anemia and associated failure to thrive. Laboratory analysis confirmed a microcytic anemia and revealed blood lead-levels of 50 μg/dL (2 times normal levels) and high levels of corproporphyrinogen III in the urine. The child was put on chelation therapy and recovered uneventfully. The cause of the child's difficulty was most likely:

    a. Inhibition of ALA dehydratase activity by lead.

    b. Lead inhibition of mitochondrial ferrochelatase.

    c. Lead inhibition of an enzyme located in the heme biosynthetic pathway between corproporphyrinogen III and protoporphyrin IX.

    d. Lead inhibition of uroporphyrinogen III cosynthase.

8. A 12-year-old girl playing Little League baseball slid vigorously into home plate, sustaining a large bruise that was initially dark red. The girl had difficulty walking and was transported to the hospital, where X-rays were negative; meanwhile, the hematoma had become dark blue in color. Hydrotherapy was recommended and the patient was discharged. In a few days the parents and child returned to the physician, concerned that the affected area of the child's thigh had dramatically changed to a vivid yellow and red color and was almost twice as large as shortly after the trauma. The physician calmed the parent's concerns by explaining the events which were transpiring in simple lay terms. In biochemical terms, the explanation of the events is best described as:

    a. The early change from red to blue was due to conversion of heme to bilirubin by heme oxygenase.

    b. The delayed change to the yellow/red color was due to reduction of a bridging methylene group of bilirubin by an NADPH-dependent reductase, thus markedly decreasing the extent of conjugation in the molecule and changing its color.

    c. Biliverdin was formed shortly after the accident by heme oxygenase and then to bilirubin by reduction of a methylene bridge of biliverdin, with attendant decrease in bond conjugation and color change.

    d. The black and blue color due to biliverdin was converted to the yellow/red of bilirubin which was then conjugated with glucuronic acid, increasing the solubility of the tetrapyrrole and allowing it to diffuse away from the injury site thus increasing the affected area.

9. Spermine synthesis is associated with the following:
    a. Addition of a carbon and nitrogen group to ornithine directly from S-adenosyl methionine.
    b. Addition of a carbon and nitrogen group to putrescine directly from S-adenosyl methionine.
    c. Addition of a carbon and nitrogen group to spermidine directly from S-adenosyl methionine.
    d. Entry of cells into the replicative cell cycle.

10. Choose the correct precursor-product relationship from among the following:
    a. Tyrosine is a precursor to dihydroxyphenylalanine.
    b. Tyrosine is a precursor to melatonin.
    c. Melatonin is a precursor of serotonin.
    d. Dopamine is a precursor to norepinephrine.
    e. 5-hydroxy tryptamine is a precursor of serotonin.

11. Nitric oxide is produced by:
    a. An oxygenase acting on arginine.
    b. An oxygenase acting on glutamine.
    c. An oxygenase acting on asparagine.
    d. An oxygenase acting on urea.

Answers

1. a	**5.** d	**9.** a
2. d	**6.** c	
3. c	**7.** b	
4. c	**8.** a	

Nucleotide Metabolism

27

Objectives

- To outline the synthesis and catabolism of the nucleotides.
- To outline the biosynthesis of the purine nucleotides, adenine, and guanine.
- To outline the control of purine nucleotide biosynthesis.
- To describe how purine nucleotides are catabolized and salvaged.
- To describe the physiological significance of disorders in purine nucleotide metabolism.
- To outline the biosynthesis of the pyrimidine nucleotides, cytosine and uracil.
- To outline the synthesis of the thymine nucleotides.
- To describe several clinically relevant aspects to thymidine nucleotide synthesis.
- To outline the control of pyrimidine biosynthesis.

- To describe how pyrimidine nucleotides are catabolized and salvaged.
- To describe the physiological significance of disorders in pyrimidine nucleotide metabolism.
- To outline the pathway for the conversion of ribonucleotides to deoxyribonucleotides.
- To outline the regulation of ribonucleotide reductase.
- To outline nucleoside phosphate interconversions.
- To outline the synthesis of the nucleotide containing coenzymes.

Concepts

The nucleotides utilized by the cell can be synthesized de novo, or else from preformed purine and pyrimidine bases by the highly efficient salvage pathways. The ability to salvage nucleotides from sources within the body obviates any nutritional requirement for nucleotides. Nucleotide synthesis is maximal just before DNA replication in dividing cells; however, the synthesis, turnover, and salvage of nucleotides continue throughout the life cycle of a cell. Owing to the continual salvage of the purine and pyrimidine bases, they are found primarily in the form of the nucleotides; the absolute level of free bases is extremely small.

Inasmuch as there is no requirement of nucleotides in the diet, nucleic acids that are ingested are for the most part enzymatically degraded and excreted. The hydrolysis of ingested nucleic acids occurs through the concerted actions of endonucleases, phosphodiesterases, and nucleoside phosphorylases. **Endonucleases** degrade DNA and RNA at internal sites, leading to the production of oligonucleotides. Oligonucleotides are further digested by **phosphodiesterases**, which act inward from both ends to yield free nucleosides. The bases are hydrolyzed from nucleosides by the action of **phosphorylases**, which yield ribose-1-P and free bases. The purine bases are further oxidized to uric acid, absorbed, and then excreted in the urine.

PURINE NUCLEOTIDE BIOSYNTHESIS

The salvage pathway and the de novo pathway of synthesis both produce nucleoside-5′-phosphates by using an activated sugar intermediate and a class of enzymes called **phosphoribosyltransferases**. The activated sugar used is 5-phospho-α-D-ribosyl-1-pyrophosphate, **PRPP**. This compound is generated by the action of **PRPP synthetase** and requires energy in the form of ATP (Figure 27–1).

The major site of purine synthesis is in the liver (Figure 27–1). Synthesis of the purine nucleotides begins with PRPP and leads to the first fully formed nucleotide, inosine 5′-monophosphate (**IMP**). The nucleotide base is built upon the ribose by several **amidotransferase** and **transformylation** reactions (Figure 27–1).

The synthesis of IMP requires five moles of ATP, two moles of glutamine, one mole of glycine, one mole of CO_2, one mole of aspartate and two moles of formate (Figure 27–2). The **formyl** moieties are carried on tetrahydrofolate (**THF**) in the form of N^5,N^{10}-methenyl-THF and N^{10}-formyl-THF (Figure 27–1).

IMP represents a branch point for purine biosynthesis, because it can be converted into either AMP or GMP through two distinct reaction pathways (Figure 27–3). The pathway leading to AMP requires energy in the form of GTP; that leading to GMP requires energy in the form of ATP. The utilization of GTP in the pathway to AMP synthesis allows the cell to control the proportions of AMP and GMP to near-equivalence. The accumulation of excess GTP will lead to accelerated AMP synthesis from IMP instead, at the expense of GMP synthesis. Conversely, since the conversion of IMP to GMP requires ATP, the accumulation of excess ATP leads to accelerated synthesis of GMP over that of AMP.

Regulation of Purine Biosynthesis

The essential rate-limiting steps in purine biosynthesis occur at the first two reactions of the pathway (Figure 27–4). The synthesis of PRPP by PRPP synthetase is feedback-inhibited by purine-5′-nucleotides (predominantly AMP and GMP). Combinatorial effects are greatest—eg, inhibition is maximal—when the correct concentration of both adenine and guanine nucleotides is achieved.

The amidotransferase reaction catalyzed by **PRPP amidotransferase** is also allosterically fee-back-inhibited by ATP, ADP, and AMP at one inhibitory site and GTP, GDP, and GMP at another. Conversely, the activity of the enzyme is stimulated by PRPP.

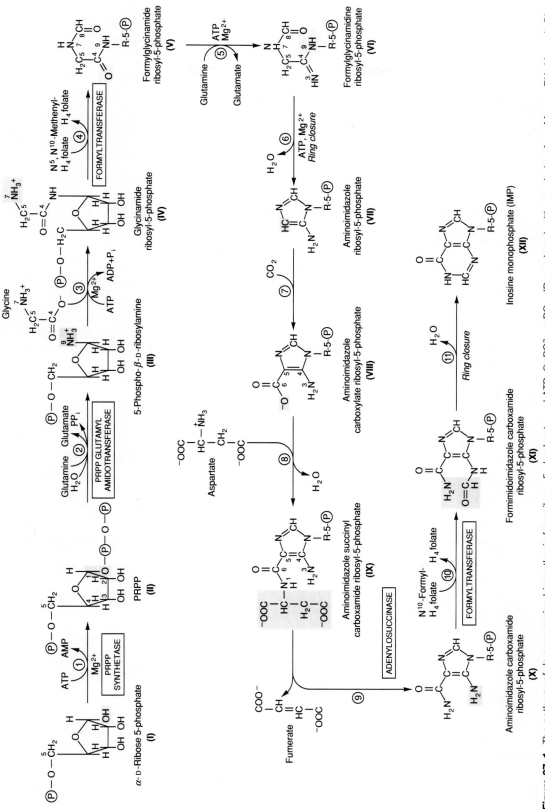

Figure 27–1. The pathway of de novo purine biosynthesis from ribose 5-phosphate and ATP. ⓟ, PO_3^{2-} or PO_2^-. (Reproduced, with permission, from Murray, RK: *Harper's Biochemistry*, 23rd ed. Appleton & Lange, 1993.)

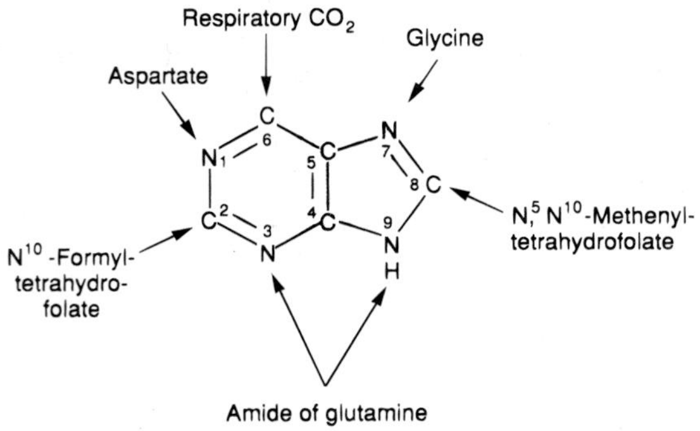

Figure 27–2. The sources of the nitrogen and carbon atoms of the purine ring. Atoms 4, 5, and 7 (shaded) derive from glycine. (Reproduced, with permission, from Murray, RK: *Harper's Biochemistry*, 23rd ed. Appleton & Lange, 1993.)

Additionally, purine biosynthesis is regulated in the branch pathways from IMP to AMP and GMP. As discussed earlier, the accumulation of excess ATP leads to the accelerated synthesis of GMP, and excess GTP leads to accelerated production of AMP.

Salvage and Catabolism of Purine Nucleotides

Catabolism of the purine nucleotides leads ultimately to the production of **uric acid**, (Figure 27–5) which is insoluble and is excreted in the urine as sodium urate crystals.

The formation of nucleotides from the purine (and pyrimidine, see below) bases and nucleosides takes place in a series of steps known as the **salvage pathways**. The free purine bases—adenine, guanine, and hypoxanthine—can be reconverted to their corresponding nucleotides by **phosphoribosylation**. Two key transferase enzymes are involved in the salvage of purines: adenosine phosphoribosyltransferase (**APRT**), which catalyzes reaction 27.1:

$$\text{adenine} + \text{PRPP} \Leftrightarrow \text{AMP} + \text{PP}_i \qquad\qquad 27.1$$

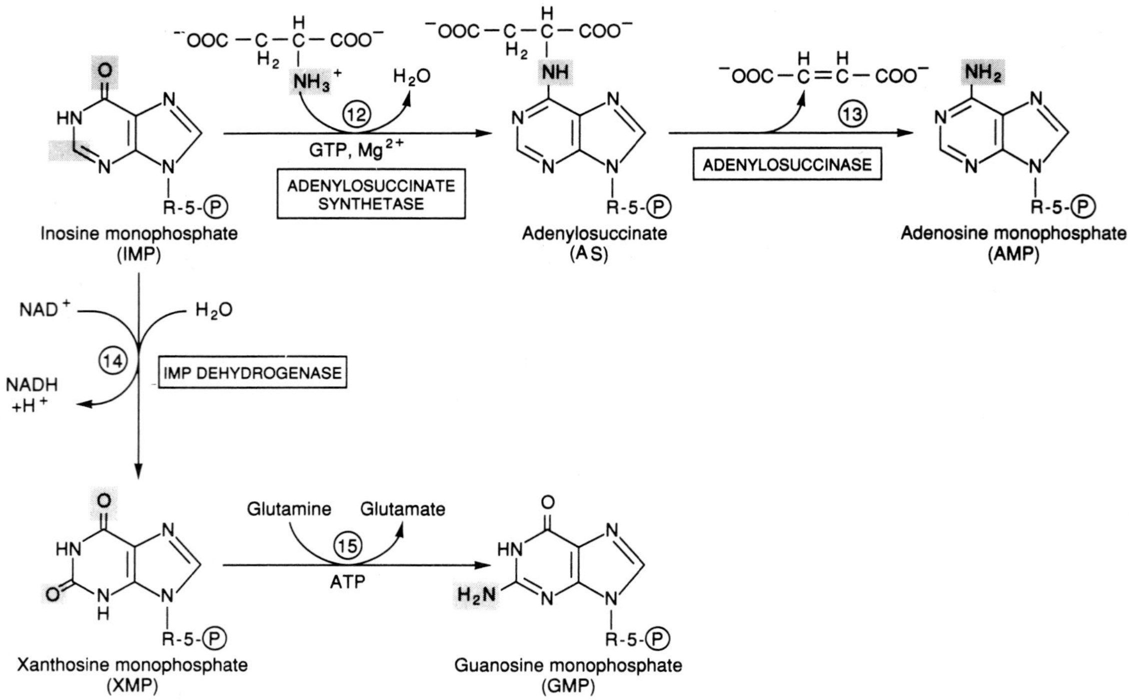

Figure 27–3. Conversion of IMP to AMP and GMP. (Reproduced, with permission, from Murray, RK: *Harper's Biochemistry*, 23rd ed. Appleton & Lange, 1993.)

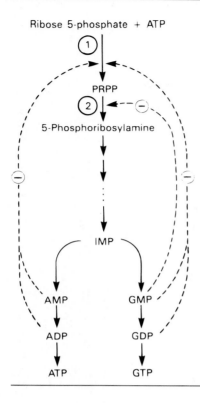

Ribose 5-phosphate + ATP

PRPP

5-Phosphoribosylamine

IMP

AMP   GMP

ADP   GDP

ATP   GTP

Figure 27–4. Control of the rate of de novo purine nucleotide synthesis. Solid lines represent chemical flow, and broken lines represent feedback inhibition ($\ominus$) by end products of the pathway. Reactions ① and ② are catalyzed by PRPP synthetase and by PRPP glutamyl amidotransferase, respectively. (see Figure 25–1) (Reproduced, with permission, from Murray, RK: *Harper's Biochemistry*, 23rd ed. Appleton & Lange, 1993.)

and hypoxanthine-guanine phosphoribosyltransferase **(HGPRT)**, which catalyzes reactions 27.2 and 27.3:

$$\text{hypoxanthine} + \text{PRPP} \Leftrightarrow \text{IMP} + \text{PP}_i \qquad \textbf{27.2}$$

$$\text{guanine} + \text{PRPP} \Leftrightarrow \text{GMP} + \text{PP}_i \qquad \textbf{27.3}$$

Purine nucleotide phosphorylases can also contribute to the salvage of the bases through a reversal of the catabolism pathways (see Figure 27–5). However, this pathway is less significant than those catalyzed by the phosphoribosyltransferases.

The synthesis of AMP from IMP and the salvage of IMP via AMP catabolism have the net effect of deaminating aspartate to fumarate. This process has been termed the **purine nucleotide cycle** (Figure 27–6). This cycle is very important in muscle cells. Increases in muscle activity create a demand for an increase in TCA cycle activity, in order to generate more NADH for the production of ATP. However, muscle lacks most of the enzymes of the major anaplerotic reactions (see Chapter 17). Muscle replenishes TCA-cycle intermediates in the form of fumarate generated by the purine nucleotide cycle.

CLINICAL SIGNIFICANCE

Clinical problems associated with nucleotide metabolism in humans are predominantly the result of abnormal catabolism of the purines (Table 27–1). The clinical consequences of abnormal purine metabolism range from mild to severe and even fatal disorders. Clinical manifestations of abnormal purine catabolism arise from the insolubility of the degradation byproduct, uric acid. Excess accumulation of uric acid leads to **hyperuricemia**, more commonly known as **gout**. This condition results from the precipitation of sodium urate crystals in the synovial fluid of the joints, leading to severe inflammation and arthritis.

Most forms of gout are the result of excess purine or of a partial deficiency in the salvage enzyme, **HGPRT**. Most forms of gout can be treated by administering the **antimetabolite**, allopurinol. This

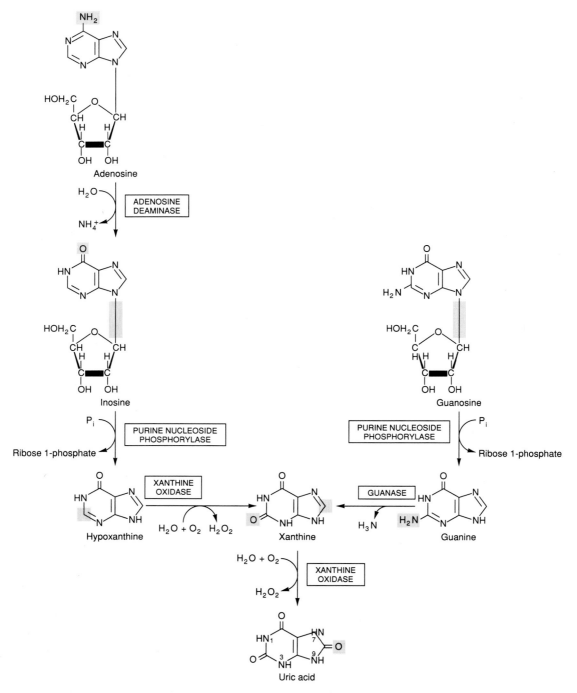

Figure 27–5. Formation of uric acid from purine nucleosides by way of the purine bases hypoxanthine, xanthine, and guanine. Purine deoxyribonucleosides are degraded by the same catabolic pathway and enzymes, all of which exist in the mucosa of the mammalian gastrointestinal tract. (Reproduced, with permission, from Murray, RK: *Harper's Biochemistry*, 23rd ed. Appleton & Lange, 1993.)

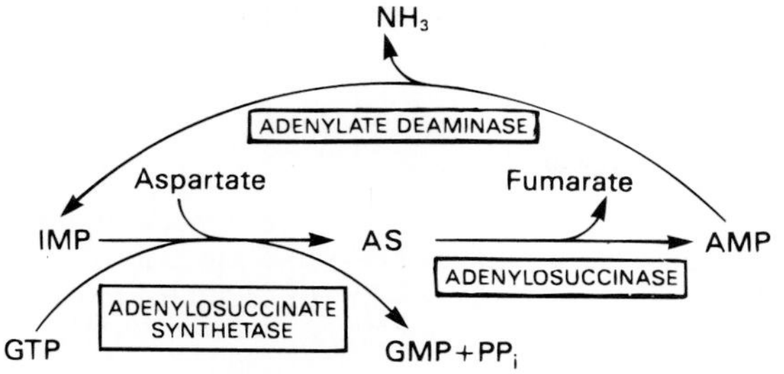

Figure 27–6. The purine nucleotide cycle. (Reproduced, with permission, from Murray, RK: *Harper's Biochemistry*, 23rd ed. Appleton & Lange, 1993.)

compound is a structural analog of hypoxanthine that strongly inhibits **xanthine oxidase** (see Figure 27–5).

Two severe disorders, both quite well described, are associated with defects in purine metabolism: Lesch-Nyhan syndrome and severe combined immunodeficiency disease (SCID). **Lesch-Nyhan syndrome** results from the loss of a functional HGPRT gene. The disorder is inherited as a sex-linked trait, with the HGPRT gene on the X chromosome. Patients with this defect exhibit not only severe symptoms of gout but also a severe malfunction of the nervous system. In the most serious cases, patients resort to self-mutilation. Death usually occurs before patients reach their 20th year.

SCID is caused by a deficiency in the enzyme adenosine deaminase (**ADA**). This is the enzyme responsible for converting adenosine to inosine in the catabolism of the purines (see Figure 27–5). This deficiency selectively leads to a destruction of B- and T-lymphocytes, the cells that mount immune responses to invasion by any "non-self" particles. In the absence of ADA, deoxyadenosine is phosphorylated to yield levels of dATP that are 50-fold higher than normal. The levels are especially high in lymphocytes, which have abundant amounts of the salvage enzymes, including **nucleoside kinases**. High concentrations of dATP inhibit **ribonucleotide reductase** (see below), thereby preventing other dNTPs from being produced. The net effect is to inhibit DNA synthesis. Since lymphocytes must be able to proliferate dramatically in response to antigenic challenge, the inability to synthesize DNA seriously impairs the immune responses, and the disease is usually fatal in infancy unless special protective measures are taken. A less severe immunodeficiency results when there is a lack of purine nucleoside phosphorylase (**PNP**), another purine-degradative enzyme (see Figure 27–5).

One of the many glycogen storage diseases **von Gierke's disease** (see Chapter 15) also leads to excessive uric acid production. This disorder results from a deficiency in **glucose 6-phosphatase** activity. The increased availability of G6P increases the rate of flux through the pentose phosphate path-

Table 27–1. Disorders of purine metabolism.

Disorder	Defect	Nature of Defect	Clinical Sign
Gout	PRPP synthetase	Increased enzyme activity due to elevated V_{max}	Hyperuricemia
Gout	PRPP synthetase	Enzyme is resistant to feedback inhibition	Hyperuricemia
Gout	PRPP synthetase	Enzyme has increased affinity for ribose-5-phosphate (lowered K_M)	Hyperuricemia
Gout	PRPP amidotransferase	Loss of feedback inhibition of enzyme	Hyperuricemia
Gout	HGPRT[a]	Partially defective enzyme	Hyperuricemia
Lesch-Nyhan syndrome	HGPRT	Lack of enzyme	See text
SCID	ADA[b]	Lack of enzyme	See text
Immunodeficiency	PNP[c]	Lack of enzyme	See text
Renal lithiasis	APRT[d]	Lack of enzyme	2,8-Dihydroxyadenine renal lithiasis
Xanthinuria	Xanthine oxidase	Lack of enzyme	Hypouricemia and xanthine renal lithiasis
von Gierke's disease	Glucose-6-phosphatase	Enzyme deficiency	See text

[a]Hypoxanthine-guanine phosphoribosyltransferase; [b]adenosine deaminase; [c]purine nucleotide phosphorylase; [d]adenosine phosphoribosyltransferase

way, yielding an elevation in the level of R5P and consequently PRPP. The increases in PRPP then result in excess purine biosynthesis.

PYRIMIDINE NUCLEOTIDE BIOSYNTHESIS

Synthesis of the pyrimidines is less complex than that of the purines, since the base is much simpler. The first completed base is derived from 1 mole of glutamine, one mole of ATP and one mole of CO_2 (which form carbamoyl phosphate) and one mole of aspartate (Figure 27–7). An additional mole of glutamine and ATP are required in the conversion of UTP to CTP (Figure 27-7).

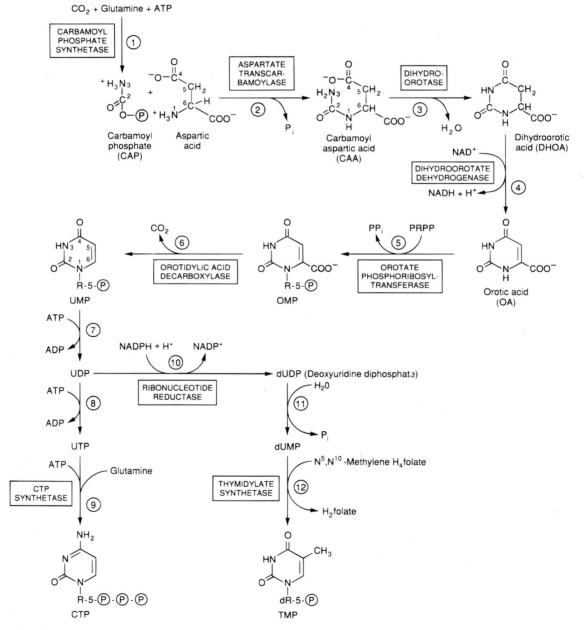

Figure 27–7. The biosynthetic pathway for pyrimidine nucleotides. (Reproduced, with permission, from Murray, RK: *Harper's Biochemistry*, 23rd ed. Appleton & Lange, 1993.)

The synthesis of pyrimidines differs in two significant ways from that of purines. First, the ring structure is assembled as a free base, not built upon PRPP (Figure 27–7). PRPP is added to the first fully formed pyrimidine base (orotic acid), forming orotate monophosphate (OMP), which is subsequently decarboxylated to UMP (Figure 27–7). Second, there is no branch in the pyrimidine synthesis pathway. UMP is phosphorylated twice to yield UTP (ATP is the phosphate donor). The first phosphorylation is catalyzed by **uridylate kinase** and the second by ubiquitous **nucleoside diphosphate kinase** (see below). Finally UTP is aminated by the action of **CTP synthase**, generating CTP (Figure 27–7). The thymine nucleotides are in turn derived by de novo synthesis from dUMP or by salvage pathways from deoxyuridine or deoxythymidine (Figure 27–7).

Synthesis of the Thymine Nucleotides

The de novo pathway to dTTP synthesis first requires the use of dUMP from the metabolism of either UDP or CDP (Figure 27–7). The dUMP is converted to dTMP by the action of **thymidylate synthase** (Figure 27–7). The methyl group (recall that thymine is 5-methyl uracil) is donated by **tetrahydrofolate**, similarly to the donation of methyl groups during the biosynthesis of the purines.

The salvage pathway to dTTP synthesis involves the enzyme **thymidine kinase** which can use either thymidine or deoxyuridine as substrate:

$$\text{deoxythymidine} + \text{ATP} \Leftrightarrow \text{dTMP} + \text{ADP} \qquad 27.4$$

$$\text{deoxyuridine} + \text{ATP} \Leftrightarrow \text{dUMP} + \text{ADP} \qquad 27.5$$

The activity of thymidine kinase (one of the various deoxyribonucleotide kinases) is unique in that it fluctuates with the cell cycle, rising to peak activity during the phase of DNA synthesis (see Chapter 29); it is inhibited by dTTP.

CLINICAL SIGNIFICANCE

Tetrahydrofolate (THF) is regenerated from the dihydrofolate (DHF) product of the thymidylate synthase reaction by the action of dihydrofolate reductase **(DHFR)**, an enzyme that requires NADPH. Cells that are unable to regenerate THF suffer defective DNA synthesis and eventual death. For this reason, as well as the fact that dTTP is utilized only in DNA, it is possible therapeutically to target rapidly proliferating cells over nonproliferative cells through the inhibition of **thymidylate synthase**. Many anticancer drugs act directly to inhibit thymidylate synthase, or indirectly, by inhibiting DHFR.

The class of molecules used to inhibit thymidylate synthase is called the "suicide substrates," because they irreversibly inhibit the enzyme. Molecules of this class include **5-fluorouracil** and **5-fluorodeoxyuridine**. Both are converted within cells to **5-fluorodeoxyuridylate, FdUMP**. It is this drug metabolite that inhibits thymidylate synthase. Many DHFR inhibitors have been synthesized, including **methotrexate, aminopterin**, and **trimethoprim**. Each of these is an analog of folic acid.

Regulation of Pyrimidine Biosynthesis

The regulation of pyrimidine synthesis occurs mainly at the first step (Figure 27–8) which is catalyzed by aspartate transcarbamoylase, **ATCase** (see Figure 27–7), Inhibited by CTP and activated by ATP, ATCase is a multifunctional protein in mammalian cells. It is capable of catalyzing the formation of **carbamoyl phosphate, carbamoyl aspartate**, and **dihydroorotate**. The carbamoyl synthetase activity of this complex is termed carbamoyl phosphate synthetase II **(CPS-II)** as opposed to CPS-I, which is involved in the urea cycle (see Chapter 23). ATCase, and therefore the activity of CPS-II, is localized to the cytoplasm and prefers glutamine as a substrate. CPS-I of the urea cycle is localized in the mitochondria and utilizes ammonia. The **CPS-II** domain is activated by ATP and inhibited by UDP, UTP, dUTP, and CTP.

The role of glycine in ATCase regulation is to act as a competitive inhibitor of the glutamine binding site. As in the regulation of purine synthesis, ATP levels also regulate pyrimidine biosynthesis at the level of PRPP formation. An increase in the level of PRPP results in an activation of pyrimidine synthesis.

There is also regulation of **OMP decarboxylase**: this enzyme is competitively inhibited by UMP and, to a lesser degree, by CMP. Finally, **CTP synthase** is feedback-inhibited by CTP and activated by GTP.

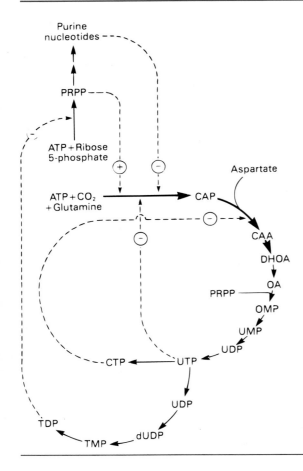

Figure 27–8. Control of pyrimidine nucleotide synthesis. Solid lines represent chemical flow. Broken lines represent positive ($\oplus$) and negative ($\ominus$) feedback regulation. (Reproduced, with permission, from Murray, RK: *Harper's Biochemistry*, 23rd ed. Appleton & Lange, 1993.)

Salvage and Cabolism of Pyrimidine Nucleotides

Catabolism of the pyrimidine nucleotides leads ultimately to β-**alanine** (when CMP and UMP are degraded) or β-**aminoisobutyrate** (when dTMP is degraded) and NH_3 and CO_2 (Figure 27–9). The β-alanine and β-aminoisobutyrate serve as NH_2 donors in transamination of α-ketoglutarate to glutamate. A subsequent reaction converts the products to **malonyl-CoA** (which can be diverted to fatty acid synthesis) or **methylmalonyl-CoA** (which is converted to succinyl-CoA and can be shunted to the TCA cycle).

The salvage of pyrimidine bases has less clinical significance than that of the purines, owing to the solubility of the by-products of pyrimidine catabolism. However, as indicated above, the salvage pathway to thymidine nucleotide synthesis is especially important in the preparation for cell division. Uracil can be salvaged to form UMP through the concerted action of **uridine phosphorylase** and **uridine kinase**, as indicated:

$$\text{uracil + ribose-1-phosphate} \Leftrightarrow \text{uridine} + P_i \qquad \textbf{27.6}$$

$$\text{uridine + ATP} \rightarrow \text{UMP + ADP} \qquad \textbf{27.7}$$

Deoxyuridine is also a substrate for uridine phosphorylase. Formation of dTMP, by salvage of dTMP requires **thymine phosphorylase** and the previously encountered thymidine kinase:

$$\text{thymine + deoxyribose-1-phosphate} \Leftrightarrow \text{thymidine} + P_i \qquad \textbf{27.8}$$

$$\text{thymidine + ATP} \rightarrow \text{dTMP + ADP} \qquad \textbf{27.9}$$

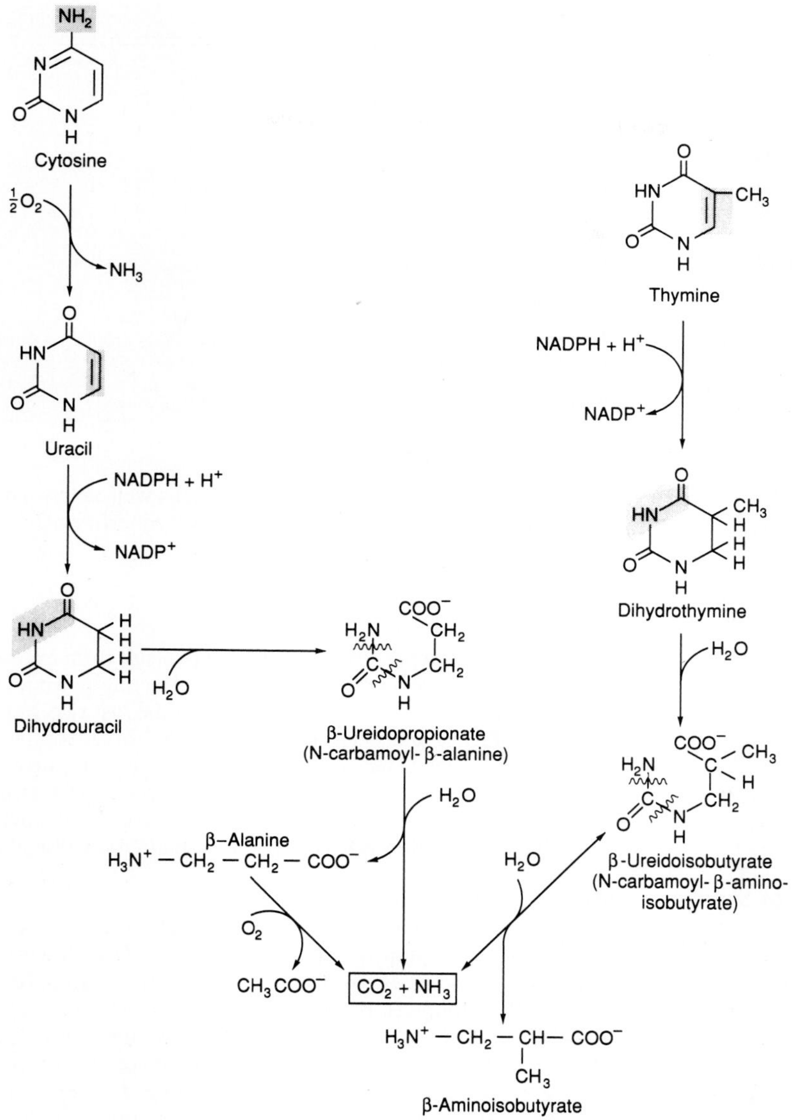

Figure 27–9. Catabolism of pyrimidines. (Reproduced, with permission, from Murray, RK: *Harper's Biochemistry*, 23rd ed. Appleton & Lange, 1993.)

The salvage of deoxycytidine is catalyzed by **deoxycytidine kinase:**

$$\text{deoxycytidine} + \text{ATP} \Leftrightarrow \text{dCMP} + \text{ADP} \qquad \textbf{27.10}$$

Deoxyadenosine and deoxyguanosine are also substrates for deoxycytidine kinase, although the K_M for these substrates is much higher than for deoxycytidine.

The major function of the pyrimidine nucleoside kinases is to maintain a cellular balance between the level of pyrimidine nucleosides and pyrimidine nucleoside monophosphates. However, since the

Table 27–2. Disorders of pyrimidine metabolism.

Disorder	Defective enzyme	Comments
Orotic aciduria, Type I	Orotate phosphoribosyl transferase and OMP decarboxylase	See text
Orotic aciduria, Type II	OMP decarboxylase	See text
Orotic aciduria (mild, no hematologic component)	The urea cycle enzyme, ornithine transcarbamoylase, is deficient	Increased mitochondrial carbamoyl phosphate exits and augments pyrimidine biosynthesis; hepatic encephalopathy
β-aminoisobutyric aciduria	Transaminase affects urea cycle function during deamination of α-amino acids to of α-keto acids	Benign, frequent in Asians
Drug-induced orotic aciduria	OMP decarboxylase	Allopurinol and 6-azauridine treatments cause orotic acidurias without a hematologic component; their catabolic by-products inhibit OMP decarboxylase

overall cellular and plasma concentrations of the pyrimidine nucleosides, as well as those of ribose-1-phosphate, are low, the salvage of pyrimidines by these kinases is relatively inefficient.

CLINICAL SIGNIFICANCE

Because the products of pyrimidine catabolism are soluble, few disorders result from excess levels of their synthesis or catabolism (see Table 27–2). Two inherited disorders affecting pyrimidine biosynthesis are the result of deficiencies in the bifunctional enzyme catalyzing the last two steps of UMP synthesis, **orotate phosphoribosyl transferase** and **OMP decarboxylase**. These deficiencies result in orotic aciduria that causes retarded growth, and severe anemia caused by hypochromic erythrocytes and megaloblastic bone marrow. Leukopenia is also common in orotic acidurias. The disorders can be treated with uridine and/or cytidine, which leads to increased UMP production via the action of nucleoside kinases. The UMP then inhibits CPS-II, thus attenuating orotic acid production.

Formation of the Deoxyribonucleotides

The typical cell contains 5–10 times as much RNA (mRNAs, rRNAs, and tRNAs) as DNA. Therefore, the majority of nucleotide biosynthesis has as its purpose the production of rNTPs. However, because proliferating cells need to replicate their genomes, the production of dNTPs is also necessary. This process begins with the reduction of rNDPs, followed by phosphorylation to yield the dNTPs (Figure 27–10). The phosphorylation of dNDPs to dNTPs is catalyzed by the same **nucleoside diphosphate kinases** that phosphorylate rNDPs to rNTPs, using ATP as the phosphate donor (see below).

Ribonucleotide reductase (**RR**) is a multifunctional enzyme that contains redox-active thiol groups for the transfer of electrons during the reduction reactions. In the process of reducing the rNDP to a

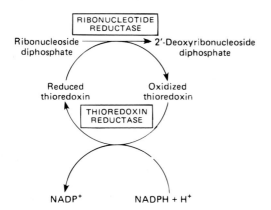

Figure 27–10. Reduction of ribonucleosides diphosphates to 2'-deoxyribonucleoside diphosphates. (Reproduced, with permission, from Murray, RK: *Harper's Biochemistry*, 23rd ed. Appleton & Lange, 1993.)

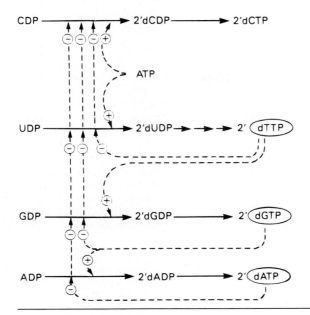

Figure 27–11. Regulation of the reduction of purine and pyrimidine ribonucleotides to their respective 2'-deoxyribonucleotides. Solid line represents chemical flow, and broken lines represent negative ($\ominus$) or positive ($\oplus$) feedback regulation. (Reproduced, with permission, from Murray, RK: *Harper's Biochemistry*, 23rd ed. Appleton & Lange, 1993.)

dNDP, RR becomes oxidized. RR is reduced in turn, by either **thioredoxin** or **glutaredoxin**. The ultimate source of the electrons is NADPH. The electrons are shuttled through a complex series of steps involving enzymes that regenerate the reduced forms of thioredoxin or glutaredoxin. These enzymes are **thioredoxin reductase** and **glutathione reductase** respectively (Figure 27–10).

Regulation of dNTP Formation

Ribonucleotide reductase is the only enzyme used in the generation of all the deoxyribonucleotides. Therefore, its activity and substrate specificity must be tightly regulated to ensure balanced production of all four of the dNTPs required for DNA replication. Such regulation occurs by binding of nucleoside triphosphate effectors to either the activity sites or the specificity sites of the enzyme complex. The activity sites bind either ATP or dATP with low affinity, whereas the specificity sites bind ATP, dATP, dGTP, or dTTP with high affinity. The binding of ATP at activity sites leads to increased enzyme activity, while the binding of dATP inhibits the enzyme. The binding of nucleotides at specificity sites effectively allows the enzyme to detect the relative abundance of the four dNTPs and to adjust its affinity for the less abundant dNTPs, in order to achieve a balance of production (Figure 27–11).

Interconversion of the Nucleoside Mono-, Di- and Triphosphates

During the catabolism of nucleic acids, nucleoside mono- and diphosphates are released. The nucleosides do not accumulate to any significant degree, owing to the action of **nucleoside kinases**. These include both nucleoside monophosphate **(NMP) kinases** and nucleoside diphosphate **(NDP) kinases**. The NMP kinases catalyze ATP-dependent reactions of the type:

$$\text{(d)NMP} + \text{ATP} \Leftrightarrow \text{(d)NDP} + \text{ADP} \qquad \textbf{27.11}$$

There are four classes of NMP kinases that catalyze, respectively, the phosphorylation of:

1. AMP and dAMP; this kinase is known as **adenylate kinase**
2. GMP and dGMP
3. CMP, UMP, and dCMP
4. dTMP

The enzyme adenylate kinase is important for ensuring adequate levels of energy in cells such as liver and muscle. The predominant reaction catalyzed by adenylate kinase is:

$$\text{2ADP} \Leftrightarrow \text{AMP} + \text{ATP} \qquad \textbf{27.12}$$

The NDP kinases catalyze reaction of the type:

$$N_1TP + N_2DP \Leftrightarrow N_1DP + N_2TP \qquad\qquad 27.13$$

N_1 can represent a purine ribo- or deoxyribonucleotide; N_2 a pyrimidine ribo- or deoxyribonucleotide. The activity of the NDP kinases can range from 10 to 100 times higher than that of the NMP kinases. This difference in activity maintains a relatively high intracellular level of (d)NTPs relative to that of (d)NDPs. Unlike the substrate specificity seen for the NMP kinases, the NDP kinases recognize a wide spectrum of (d)NDPs and (d)NTPs.

Nucleotide Co-factors

The nucleotide coenzymes, FMN, FAD, NAD$^+$, NADP$^+$ and coenzyme A, are involved in numerous metabolic reactions. The flavin nucleotide coenzymes, FMN and FAD, are derived from the vitamin **riboflavin** (see Chapter 36). In addition, the ATP-dependent enzyme, **flavokinase**, converts riboflavin to riboflavin-5-phosphate, or FMN (Figure 27–12A). The latter is interconvertible with FAD in a reaction catalyzed by the ATP-dependent enzyme, **flavin nucleotide pyrophosphorylase** (Figure 27–12B).

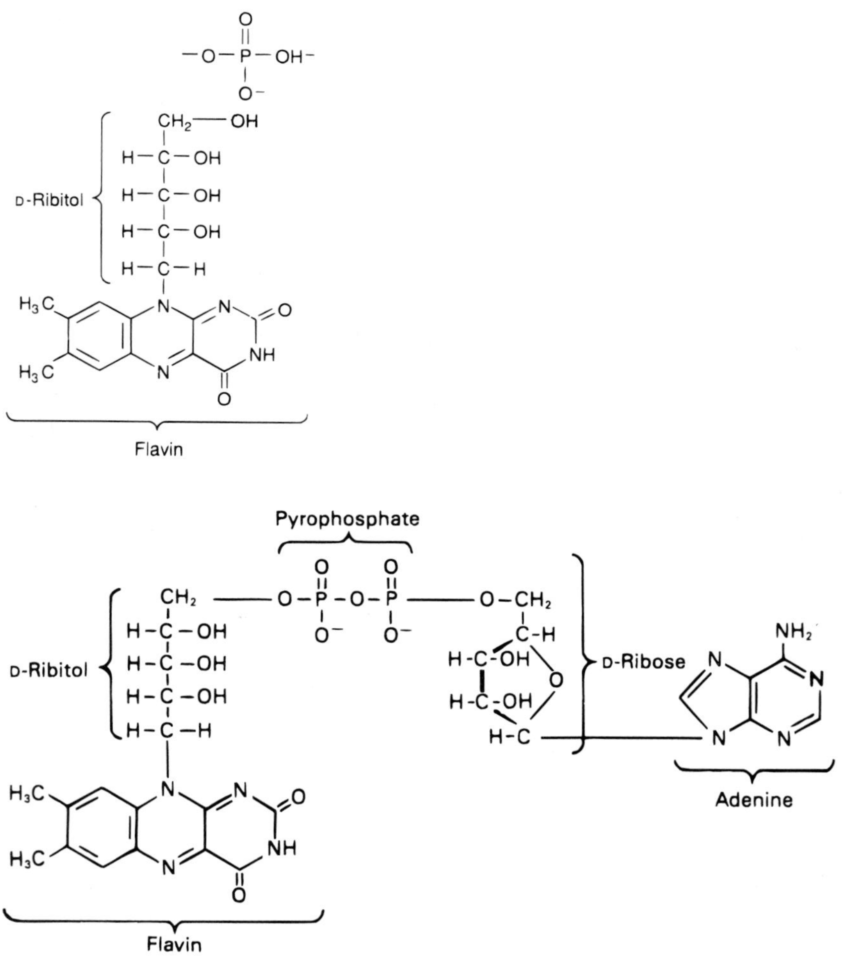

Figure 27–12. A. Riboflavin, In riboflavin phosphate (flavin mononucleotide, FMN), the −OH is replaced by phosphate. B. Flavin adenine dinucleotide (FAD). (Reproduced, with permission, from Murray, RK: *Harper's Biochemistry*, 23rd ed. Appleton & Lange, 1993.)

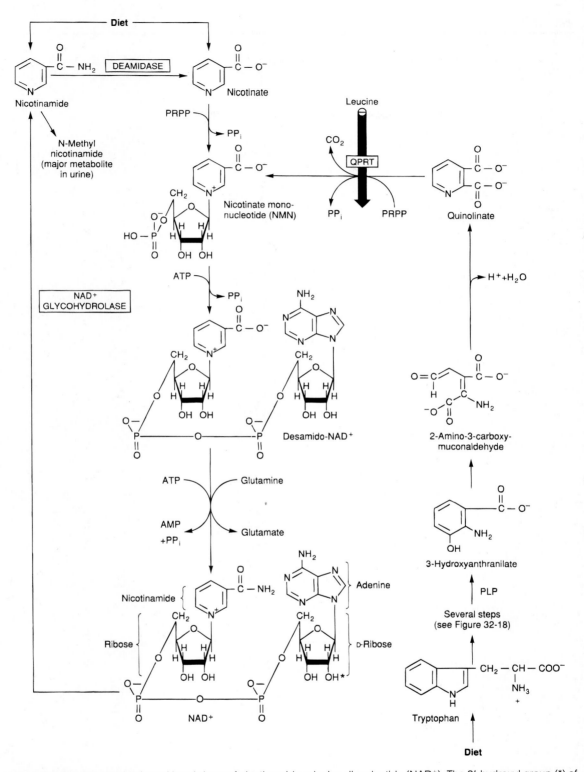

Figure 27–13. The synthesis and breakdown of nicotinamide adenine dinucleotide (NAD^+). The 2'-hydroxyl group (*) of the adenosine moiety is phosphorylated in nicotinamide dinucleotide phosphate ($NADP^+$). Humans, but not cats, can provide all of their niacin requirement from tryptophan if there is a sufficient amount in the diet. Normally, about two-thirds comes from the source PRPP, 5-phosphoribosyl-1-pyrophosphate; QPRT, quinolinate phosphoribosyl transferase; PLP pyridoxal phosphate. (Reproduced, with permission, from Murray, RK: *Harper's Biochemistry*, 23rd ed. Appleton & Lange, 1993.)

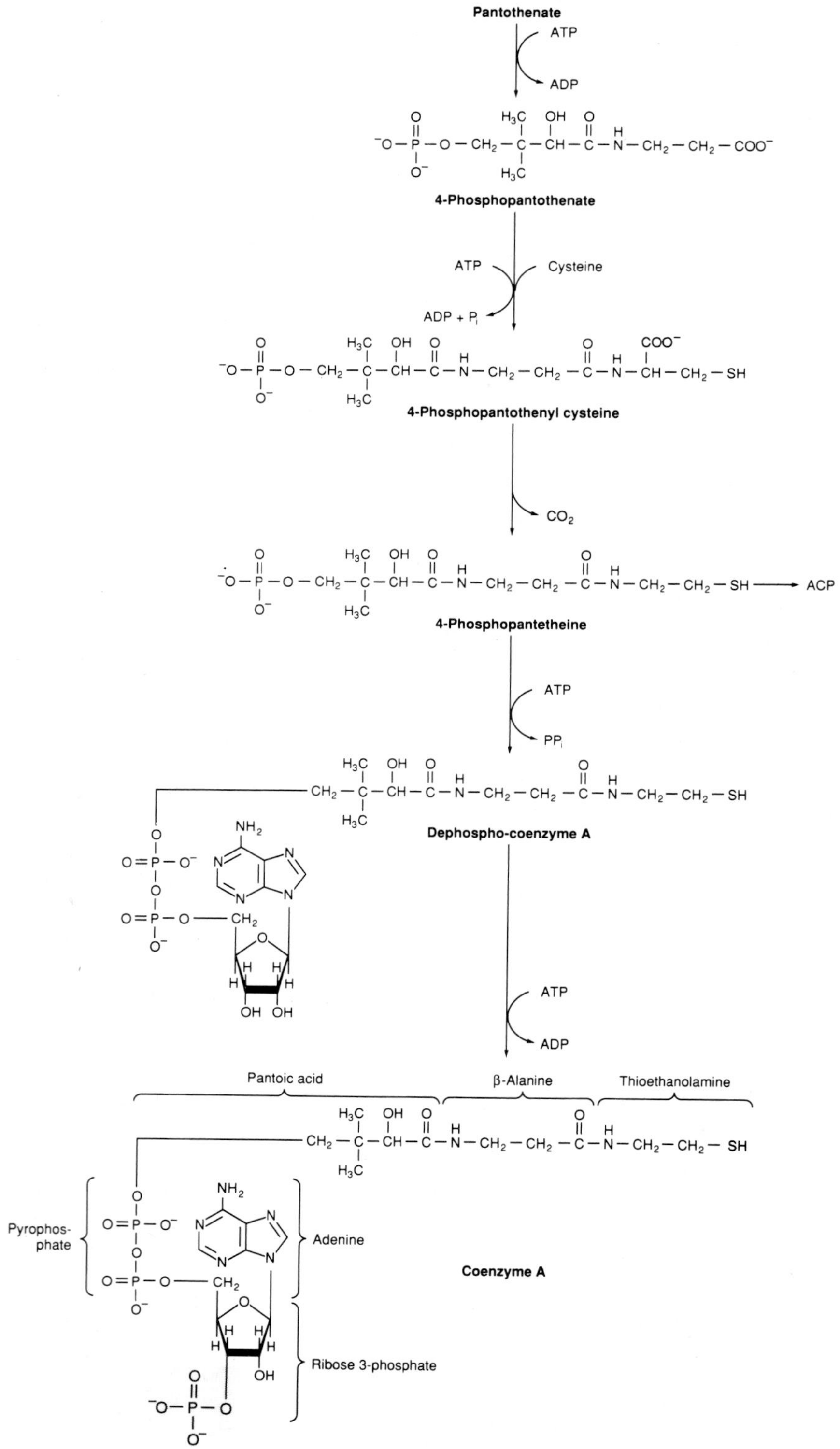

Figure 27–14. The synthesis of coenzyme A from pantothenic acid. ACP, acyl carrier protein. (Reproduced, with permission, from Murray, RK: *Harper's Biochemistry*, 23rd ed. Appleton & Lange, 1993.)

The synthesis of the nicotinamide-based coenzymes can be initiated with either the vitamin **nicotinic acid** (see Chapter 36) or **tryptophan**. Both pathways derive the ribose portion of the molecule from PRPP (Figure 27–13). Conversion of NAD^+ to $NADP^+$ is catalyzed by a specific NAD^+ **kinase**. NADH is not a substrate for the enzyme. It competitively inhibits the enzyme.

Coenzyme A is synthesized in the liver from the vitamin **pantothenic acid** (see Chapter 36) in a series of five steps (Figure 27–14).

Questions

DIRECTIONS (items 1–15): Each numbered item or incomplete statement is followed by answers or by completions of the statement. Select the ONE lettered answer or completion that is BEST in each case.

1. The purine nucleotide cycle is the mechanism by which skeletal muscle acquires fumarate, the TCA cycle intermediate, during:
   a. The interconversion of GMP to AMP.
   b. The interconversion of GMP to IMP.
   c. The interconversion of ATP to GTPP.
   d. The interconversion of IMP to AMP.
   e. The interconversion of ADP to GDPP.

2. The de novo synthesis of a completely formed AMP requires how many moles of high energy phosphate?
   a. 6
   b. 7
   c. 5
   d. 8
   e. 9

3. Reduction of ribonucleotides requires which electron donor?
   a. NADH.
   b. NADPH.
   c. $FADH_2$.
   d. $FADH^+$.
   e. NAD^+.

4. Allopurinol is used to treat gout because:
   a. It activates thymidylate synthase.
   b. It inhibits PRPP synthase.
   c. It inhibits xanthine oxidase.
   d. It inhibits hypoxanthine-guanine phosphoribosyltransferase.
   e. It activates PRPP synthase.

5. 5-fluorodeoxyuridine monophosphate (FdUMP) is useful as an anti-tumor agent because it inhibits:
   a. Dihydrofolate reductase.
   b. Thymidine kinase.
   c. Thymidylate synthase.
   d. Nucleoside diphosphate kinase.
   e. Ribonucleotide reductase.

6. The enzyme deficiency seen in patients with Lesch-Nyhan syndrome is:
   a. Hypoxanthine-guanine phosphoribosyltransferase.
   b. Xanthine oxidase.
   c. PRPP synthetase.

    d. Aspartate transcarbamoylase.
    e. Adenosine phosphoribosyltransferase.

7. Catabolism of deoxythymidine can feed the TCA cycle in the form of:
    a. Malonyl-CoA.
    b. Succinyl-CoA.
    c. Acetyl-CoA.
    d. Fumarate.
    e. Aspartate.

8. Folate analogs are useful anticancer drugs because:
    a. They inhibit thymidine kinase.
    b. They inhibit ribonucleotide reductase by increasing the production of dATP.
    c. They inhibit thymidylate synthase.
    d. They activate purine nucleoside phosphorylase, thereby inhibiting purine salvage.
    e. They inhibit dihydrofolate reductase.

9. Aside from obvious structural differences, purine synthesis differs from that of pyrimidine synthesis in which way?
    a. Purine nucleotides are preformed prior to ribosylation.
    b. Purine synthesis does not require folate analogs.
    c. Purine synthesis requires carbamoyl phosphate.
    d. Pyrimidine nucleotides are "constructed" upon PRPP.
    e. Pyrimidine biosynthesis requires carbamoyl phosphate.

10. The disorder identified as severe combined immune deficiency (SCID) results from a defect in:
    a. Adenosine deaminase.
    b. Thymidylate synthase.
    c. Purine nucleotide phosphorylase.
    d. PRPP synthase.
    e. Dihydrofolate reductase.

11. The following clinical findings in an 18-month-old boy suggest which disease? Severe mental retardation, extrapyramidal signs, spasticity, and hyperuricemia.
    a. von Gierke's disease.
    b. PNP deficiency syndrome.
    c. Lesch-Nyhan syndrome.
    d. Type I orotic aciduria.
    e. SCID.

12. Which of the following enzymes is the target of the anticancer drug methotrexate?
    a. Thymidylate synthase.
    b. DHFR.
    c. Thymidine kinase.
    d. ATCase.
    e. PRPP synthetase.

13. Which of the following describes the function of NMP kinases?
    a. Uridine + ATP ⇔ UMP + ADP.
    b. dGMP + ATP ⇔ dGDP + ADP.
    c. Deoxycytidine + ATP ⇔ dCMP + ADP.
    d. Thymidine + ATP ⇔ dTMP + ADP.
    e. 2ADP ⇔ AMP + ATP.

14. Synthesis of PRPP is predominantly controlled by which nucleotide?
    a. TTP.
    b. GTP.
    c. CTP.

    d. ATP.

    e. UTP.

15. A 37-year-old male is presented with tophaceous deposits within the articular cartilage, synovium, tendons, tendon sheaths, pinnae, and the soft tissue on the extensor surface of the forearms. These clinical observations suggest the patient is suffering from:

    a. Gout.

    b. Lesch-Nyhan syndrome.

    c. von Gierke's disease.

    d. PNP deficiency.

    e. ADA deficiency.

Answers

1. d	**6.** a	**11.** c
2. b	**7.** b	**12.** b
3. b	**8.** e	**13.** b
4. c	**9.** e	**14.** d
5. c	**10.** a	**15.** a

28   Interrelationships of the Major Organs

Objectives

- To describe metabolism in the well-fed versus fasting states.
- To describe the responses of the liver, muscle, adipose tissue, brain, and heart to changing nutritional status.
- To describe the metabolic effects of diabetes.
- To describe the hormonal control of metabolism.

Concepts

With respect to energy interrelationships, the important organs and tissues are the **liver, brain, heart, skeletal muscle**, and **adipose tissue** (Table 28–1).

The major metabolic pathways influenced by the nutritional and hormonal status of the body are **glycolysis, gluconeogenesis, glycogen synthesis and degradation, fatty acid and cholesterol synthesis and degradation, the TCA cycle**, and **oxidative phosphorylation and amino acid synthesis and degradation**. The flux through each of these pathways is influenced by the level of substrates and products as well as by specific hormones (Table 28–2).

In all the pathways listed above, two compounds are the most important biological products or precursors. These are **acetyl-CoA** and **pyruvate**. Acetyl-CoA is the primary oxidizable fuel of the TCA cycle; it generates large amounts of reduced electron carriers, which are in turn re-oxidized with concomitant synthesis of the cellular energy molecule **ATP**. Acetyl-CoA is a degradation product obtained from lipids, carbohydrates, and proteins. Pyruvate is obtained from the oxidation of carbohydrates in glycolysis and the oxidation of lactate as well as that of several amino acids. The fate of pyruvate depends upon the metabolic state of the organism. It can be oxidized to acetyl-CoA for entry into the TCA cycle or incorporation into fatty acids and cholesterol. Pyruvate also can be converted to OAA and further oxidized in the TCA cycle. Alternatively, the OAA can be shunted into the gluconeogenesis pathway and ultimately into the carbohydrate and amino acid pools.

The major fuel depots from which acetyl-CoA and pyruvate are derived consist of three types: **triacylglycerols** stored in adipose tissue, **protein** primarily from skeletal muscle, and **glycogen** primarily stored in liver but also in muscle. The "decision" of a particular organ to store food byproducts for later use or to oxidize them in order to produce ATP energy depends on the overall needs of the body and the major function of the particular organ. Factors such as hormonal status and disease state can have profound effects on the metabolic nature of a given organ.

METABOLISM IN THE WELL-FED VERSUS THE FASTING STATE

The consumption of a well-balanced diet allows the intestine to absorb the primary energy-supplying molecules: glucose, amino acids, and fat. Glucose and amino acids enter the circulation through the portal vein after their direct uptake by intestinal epithelial cells (Figure 28–1). Absorbed fats are "packaged" into chylomicrons and delivered to the circulation via the lymphatic system that drains the intestine. At the thoracic duct, chylomicrons enter the circulation by way of the subclavian vein. The delivery of metabolized food to the circulation allows all tissues to absorb the necessary molecules for their specific requirements.

Table 28–1. Summary of the major and unique features of metabolism of the principal organs.

Organ	Major Function	Major Pathways	Main Substrates	Major Products	Specialist Enzymes
Liver	Service for the other organs and tissues	Most represented, including gluco-neogenesis; β-oxidation; keto-genesis; lipopro-tein formation; urea, uric acid & bile acid forma-tion; cholesterol synthesis	Free fatty acids, glucose (well fed), lactate, glycerol, fructose, amino acids (Ethanol)	Glucose, VLDL (triacylglycerol), HDL, ketone bodies, urea, uric acid, bile acids, plasma proteins (Acetate)	Glucokinase, glu-cose-6-phospha-tase, glycerol kinase, phospho-enolpyruvate carboxykinase, fructokinase, arginase, HMG-CoA synthase and lyase, 7 α-hydroxylase
Brain	Coordination of the nervous system	Glycolysis, amino acid metabolism	Glucose, amino acids, ketone bodies (in starva-tion) Polyunsaturated fatty acids in neonate	Lactate	
Heart	Pumping of blood	Aerobic pathways, eg, β-oxidation and citric acid cycle	Free fatty acids, lactate, ketone bodies, VLDL and chylomicron tria-cylglycerol, some glucose		Lipoprotein lipase. Respiratory chain well developed
Adipose tissue	Storage and breakdown of triacylglycerol	Esterification of fatty acids and lipolysis	Glucose, lipopro-tein triacylglycerol	Free fatty acids, glycerol	Lipoprotein lipase, hormone-sensitive lipas
Muscle Fast twitch Slow twitch	Rapid movement Sustained move-ment	Glycolysis Aerobic pathways, eg, β-oxidation and citric acid cycle	Glucose Ketone bodies, triacylglycerol in VLDL and chylo-microns, free fatty acids	Lactate	Lipoprotein lipase. Respiratory chain well developed

In the fasting state, no nutrients enter the circulation and the liver is rapidly depleted of glycogen stores. As the level of blood glucose falls, the level of glucagon secretion increases. One of the major effects of glucagon is to increase triacylglycerol hydrolysis in adipocytes (Figure 28–2). This process supplies the body with fatty acids as a source of energy. However, several tissues, in particular the brain, require a continuous supply of glucose. This supply is met by hepatic gluconeogenesis (Figure 28–2).

Liver

The portal circulation from the intestine drains into the liver, providing this major organ with primary access to all incoming nutrients, excluding fats (Figure 28–1). It is the major function of the liver to process these nutrients and from them to supply the rest of the body with energy-rich compounds such as glucose, fatty acids, and ketone bodies. In the well-fed state the liver is the first major organ to receive glucose. In fact the liver acts as a blood buffer for glucose, since low levels lead to increased synthesis and release by the liver whereas high levels lead to uptake and conversion to glycogen or to fatty acids (via acetylCoA generated from pyruvate oxidation). However, in the well-fed state the liver does not generate glucose via gluconeogenesis. When glucose is taken up by the liver in excess of its needs or capacity to store it, the unused glucose is oxidized via the pentose phosphate pathway for the generation of NADPH used in reductive biosynthetic pathways. Oxidation of excess glucose also leads to the generation of acetylCoA which is subsequently converted to lipid for storage.

Although the liver has the first opportunity to utilize dietary amino acids, most of them pass through the liver and are not absorbed. This allows other tissues access to essential amino acids necessary for protein synthesis. Since the K_M values for hepatic amino acid–catabolizing enzymes are high relative

Table 28–2. Summary of the major regulators of metabolic pathways.

Pathway	Major Regulatory Enzyme(s)	Activator	Inhibitor	Effector Hormone	Remarks
Citric acid cycle	Citrate synthase		ATP, long-chain acyl-CoA		Regulated mainly by the need for ATP and therefore by the supply of NAD⁺
Glycolysis and pyruvate oxidation	Phosphofructokinase	AMP, fructose 2,6-bisphosphate in liver, fructose 1,6-bisphosphate in muscle	Citrate (fatty acids, ketone bodies), ATP, cAMP	Glucagon ↓	Induced by insulin
	Pyruvate dehydrogenase	CoA, NAD, ADP, pyruvate	Acetyl-CoA, NADH, ATP (fatty acids, ketone bodies)	Insulin ↑ (in adipose tissue)	Also important in regulating the citric acid cycle
Gluconeogenesis	Pyruvate carboxylase Phosphoenolpyruvate carboxykinase Fructose-1,6-bisphosphatase	Acetyl-CoA cAMP? cAMP	ADP AMP, fructose 2,6-bisphosphate	Glucagon? Glucagon	Induced by glucocorticoids, glucagon, cAMP Repressed by insulin
Glycogenesis	Glycogen synthase		Phosphorylase (in liver) cAMP, Ca²⁺ (muscle)	Insulin ↑ Glucagon ↓ (liver) Epinephrine ↓	Induced by insulin
Glycogenolysis	Phosphorylase	cAMP, Ca²⁺ (muscle)		Insulin ↓ Glucagon ↑ (liver) Epinephrine ↑	
Pentose phosphate pathway	Glucose-6-phosphate dehydrogenase	NADP⁺	NADPH		Induced by insulin
Lipogenesis	Acetyl-CoA carboxylase	Citrate	Long-chain acyl-CoA, cAMP	Insulin ↑ Glucagon ↓ (liver)	Induced by insulin
Cholesterol synthesis	HMG-CoA reductase		Cholesterol, cAMP	Insulin ↑ Glucagon ↓ (liver)	Inhibited by certain drugs, eg, lovastatin

to the aminoacyl-tRNA synthetases, the use of dietary amino acids for protein synthesis tends to prevail over their use for energy production. Catabolism of amino acids occurs only after sufficiently high intracellular levels are attained. Within the liver, excess amino acids can be oxidized and the intermediates utilized for lipid synthesis. The released nitrogen is converted to urea and excreted.

The liver is the major site for fatty acid synthesis. The level of malonyl-CoA is the major regulator of hepatic fatty acid synthesis. When other fuels are abundant the level of malonyl-CoA is elevated because of the increased production of acetyl-CoA by various oxidative pathways. The increased malonyl-CoA, in turn, inhibits mitochondrial carnitine acyl-transferase I, preventing the entry of fatty acids into the mitochondria for oxidation and ketogenesis. The converse is true when energy sources are low: malonyl-CoA levels fall, and fatty acyl-CoAs enter the mitochondria for oxidation and ketogenesis. During prolonged fasting, levels of hepatic oxaloacetate become depleted to the point where the liver can no longer oxidize fatty acid-derived acetyl-CoA in the TCA cycle. Under these conditions the level of hepatic ketogenesis increases.

The brain is dependent on glucose as a fuel, and skeletal muscle prefers glucose during exertion. To meet these demands during the fasting state, the liver becomes the primary source of glucose production via gluconeogenesis. The carbon atoms required for this process derive primarily from **lactate, glycerol,** and **alanine.** Additional sources of carbons for hepatic glucose synthesis are those of acetyl-CoA produced during fatty acid oxidation, and other amino acids derived from muscle proteolysis. Both the **Cori** and **alanine cycles** are important for the maintenance of hepatic gluconeogenesis. However, these pathways do not support net glucose production (see Chapter 14).

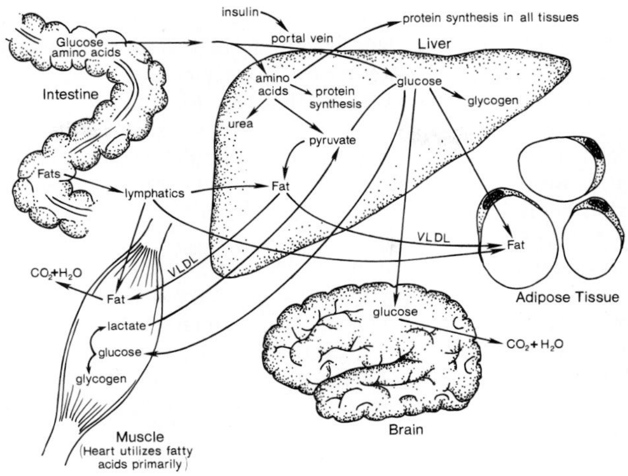

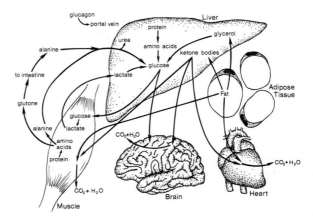

Figure 28–1. Interrelationships of the major organs following a meal in the well-fed state. Glucose absorption from the intestine stimulates the secretion of insulin leading to increased glucose uptake by the tissues, primarily the liver. Fat synthesized by the liver is utilized by other tissues for energy and any excess fat is stored as triacylglycerols in adipose tissue.

During fasting, the oxidation of fatty acids within hepatocytes provides the ATP energy necessary to support gluconeogenesis. However, like the Cori and alanine cycles, glucose production from the acetyl-CoA generated during fatty acid oxidation does not contribute to net glucose production. Since the brain oxidizes glucose completely, net glucose synthesis is required during fasting. The glycerol backbone of hydrolyzed adipocyte triacylglycerols, as well as amino acids derived from skeletal muscle proteolysis, are the sources of carbons for this process.

Muscle

Skeletal muscle can utilize glucose, fatty acids, and ketone bodies for energy production. Since the energy demands of skeletal muscle can vary dramatically, its fuel sources must also be able to change rapidly and efficiently to meet continuously changing needs. During rest, skeletal muscle derives the majority of its energy from the oxidation of fatty acids. In the early stages of exertion, glucose becomes the major energy molecule utilized by the muscle. This glucose is derived primarily from the breakdown of muscle glycogen stores. Muscle cells, like most other cells excluding hepatocytes, lack G6Pase (see Chapter 14); they therefore oxidize all glucose derived from glycogen and do not supply the blood with free glucose. Also, skeletal muscle cells do not contain glucagon receptors and there-

Figure 28–2. Interrelationships of the major organs during the fasting state. The liver supplies ketone bodies to the blood and adipose tissue supplies fatty acids. The glucose-alanine cycle between muscle and liver is very active.

fore do not respond to low blood glucose. However, muscle cells do contain epinephrine receptors, which stimulate the cAMP cascade leading to increased glycogenolysis. In muscle, PKA-mediated phosphorylation of PFK-2 activates its kinase activity (as opposed to its phosphatase activity as occurs in the liver), leading to increased F-2,6-BP levels and thus increased PFK-1 activity.

The degradation of muscle protein also serves as a readily mobilizable source of energy during exertion. Indeed, during periods of fasting, skeletal muscle protein is metabolized to amino acids. Many of the amino acids are converted to pyruvate, which in turn is transaminated to alanine. Alanine is then transported to the liver for reconversion to pyruvate and eventually glucose, in a process termed the **alanine cycle** (see Chapter 14).

Muscle cells do not participate in gluconeogenesis; therefore, any pyruvate that is not oxidized in the TCA cycle is reduced to lactate. (The production of lactate is significant during periods of anaerobic exercise.) The lactate is then transported to the liver for input into the gluconeogenic pathway, in a phenomenon known as the **Cori cycle** (see Chapter 14). The Cori cycle has the effect of shifting the respiratory burden of the muscle to that of the liver. Following intense exercise, elevated consumption of oxygen continues for some time in order to fuel the production of ATP in the liver. This ATP is necessary to convert the muscle-derived lactate into glucose via gluconeogenesis.

During periods of anaerobic exercise, skeletal muscle becomes effectively isolated from the rest of the body because of the constriction of blood vessels within this tissue. Under these conditions the muscle derives needed ATP energy from the hydrolysis of **phosphocreatine**, according to the formula:

$$\text{phosphocreatine} + \text{ADP} \Leftrightarrow \text{creatine} + \text{ATP} \qquad \textbf{28.1}$$

The supply of phosphocreatine is extremely limited and can be exhausted in as little as 5 seconds. Therefore, stimulation of muscle glycogenolysis must, and does, occur rapidly to support the production of ATP via glycolysis. The rate of glycolysis, however, exceeds the capacity of the muscle cell to completely oxidize all the pyruvate generated. The majority of the pyruvate is reduced to lactate, which is released to the blood for reconversion to glucose in the liver. All these mechanisms are included within the phenomenon of the Cori cycle.

Adipose Tissue

Adipose tissue is the major fuel depot of an animal. Energy is stored in the form of triacyglycerols within adipose cells; adipose cells, in turn, are designed for the continuous synthesis of triacylglycerols. There is no direct mechanism to inhibit fat storage in adipose tissue. Fat metabolism, on the other hand, is handled primarily by hormonal control of triacylglycerol mobilization. Indeed, the higher the level of excess energy consumed, the more adipose cells are generated to store this excess energy in the form of triacylglycerols (Figure 28–3). Upon metabolic demand, the fatty acids of the triacylglycerols are released via the action of hormone sensitive lipase. The free fatty acids diffuse into the blood, are bound by albumin, and are transported to other organs for oxidation and energy production.

Adipose cells require a certain basal level of glycolysis to take place as a mechanism to generate DHAP for conversion to glycerol-3-phosphate, the backbone of triacylglycerols. Because adipose cells lack glycerol kinase, they require glycolysis as a source of the building block for triacylglycerols. The glycerol that results from the hydrolysis of triacylglycerols in adipose tissue is transported in the blood to the liver for conversion to glucose via gluconeogenesis. **Glucose, therefore, effectively controls the rate of triacylglycerol synthesis in adipose cells.** When levels of glucose are adequate, sufficient oxidation occurs to yield the DHAP necessary for storage of fats. As the level of glucose falls, so too does the level of DHAP. Under these conditions, fatty acids are released for export to other tissues for use as sources of fuel.

Brain

As an organ, the brain is the single most voracious consumer of energy in the body. The brain is an extremely aerobic organ, utilizing approximately 20% of the total oxygen intake. It also accounts for approximately 60% of the glucose oxidized by the body at rest. A large percentage of the energy utilized by the brain goes toward powering the Na^+/K^+-ATPase that is necessary for the maintenance of nerve membrane electrochemical potentials.

The brain has no significant energy stores of lipid, carbohydrate, or protein; it must therefore have an uninterrupted supply of glucose and oxygen. During periods of starvation the brain utilizes ketone bodies, generated in the liver, as a source of energy.

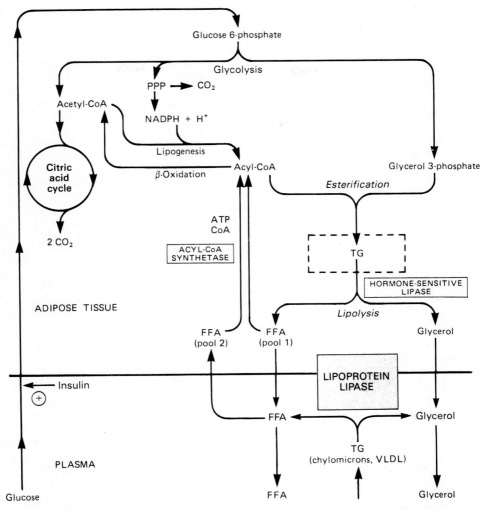

Figure 28–3. Metabolism of adipose tissue. Hormone-sensitive lipase is activated by ACTH, TSH, glucagon, epinephrine, norepinephrine, and vasopressin and inhibited by insulin, prostaglandin E_1, and nicotinic acid. PPP, pentose phosphate pathway; TG, triacylglycerol; FFA, free fatty acids; VLDL, very low density lipoprotein. (Reproduced, with permission, from Murray, RK: *Harper's Biochemistry*, 23rd ed. Appleton & Lange, 1993.)

Heart

Like the brain, the heart is a completely aerobic organ (except for very brief periods of extreme exertion) and contains negligible energy reserves in the form of glycogen or lipid. Therefore, the heart also must have a continuous supply of energy. However, unlike the brain, the heart muscle can utilize fatty acids and lactate as energy sources in addition to glucose and ketone bodies.

Unlike the metabolism of skeletal muscle, metabolism within the heart muscle cell does not vary as dramatically and cannot operate anaerobically. Fatty acids are the fuel of choice for heart muscle, but if the work load increases dramatically the heart begins to use glucose, derived mostly from its own limited glycogen stores.

Metabolic Effects of Diabetes

Insulin-dependent diabetes (IDDM or Type I diabetes) results from a defect in the production of insulin by the β-cells of the pancreas. This defect produces an abnormal ratio of insulin to glucagon, even in the well-fed state in which the level of blood glucose is high. The primary effect of the failure of the diabetic to produce insulin is that all the major tissues remain in a state of catabolism irrespective of nutritional status (Figure 28–4).

In diabetes, the liver remains gluconeogenic in spite of the elevation in blood glucose that typically follows a meal. The result is severe hyperglycemia. In the absence of insulin, skeletal muscle fails to absorb glucose, a condition that exacerbates the hyperglycemia. During periods of fasting, the hyperglycemia is maintained because of increased skeletal muscle proteolysis. The net effect then is an increase in the availability of gluconeogenic substrates within the liver.

The altered regulation of the insulin-to-glucagon ratio also leads to hyperlipoproteinemia. This condition results from uncontrolled lipolysis in adipose tissue, in response to the altered ratio of insulin to glucagon. The increased release of fatty acids by adipose tissue also contributes to an elevation in the rate of ketone body production. If the ketone bodies are not consumed, the potentially fatal condition termed **ketoacidosis** can develop. Ketone bodies are acids, and an elevation in their production disrupts normal blood pH buffering. In an attempt to control the pH of the blood, the diabetic system excretes excess H^+ ions in the urine. Accompanying the H^+ ions are Na^+, K^+, P_i, and H_2O—a combined lack that produces severe dehydration. Indeed, excessive thirst is a classic symptom of diabetes.

In contrast to Type I diabetes, **non-insulin dependent diabetes (NIDDM or Type II diabetes) results primarily from a reduction in the ability of cells to *respond* to insulin** rather than from a lack of insulin secretion. Type II diabetes is characterized by its late onset and is most frequently associated with obesity; the primary site of deficiency is in the insulin receptors. Persons suffering from Type II diabetes usually have lower numbers of peripheral tissue insulin receptors, a reduced affinity of the receptors for insulin, or defects in insulin receptor signaling pathways. When Type II diabetes sets in at an early age (*maturity-onset type diabetes of the young: MODY*) it is associated, at high frequency, with defects in the **glukokinase** gene. Because Type II diabetics have a relatively normal insulin/glucagon ratio in the blood, there is not an associated increase in adipose tissue lipolysis. Therefore, these individuals do not develop ketoacidosis.

HORMONAL REGULATION OF METABOLISM

Rapid control over the direction of flux through the various metabolic pathways, in response to nutritional status, is primarily effected by two major hormones **insulin** and **glucagon**. Additionally, important rapid metabolic regulation is exerted by the **catecholamines (primarily epinephrine)**. The predominant mechanism by which these hormones act on the enzymes involved in metabolism is the regulation of cAMP levels. Both glucagon and epinephrine stimulate an increase in the intracellular level of cAMP, which in turn activates PKA. Conversely, insulin opposes the actions of the other two hormones by a mechanism that likely involves the activation of an inhibitory G-protein.

The major short-term effect of PKA activation is an increase in the phosphorylation of key metabolic enzymes as well as in the phosphorylation of other regulatory proteins, in particular **phosphoprotein phosphatase inhibitor-1, PPI-1** (Table 28–3). As the level of cAMP falls, so does the level of active PKA. The overall effect is a reduction in protein phosphorylation, principally due to an increase in the activity of specific phosphatases. In the well-fed organism, most of the important enzymes involved in catabolism are in the dephosphorylated and active state. However, three key enzymes—

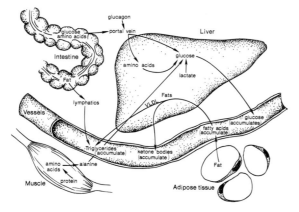

Figure 28–4. Interrelationships of the major organs in Type I diabetes. The continued secretion of glucagon leads to a maintenance of the catabolic state of the major organs.

Table 28–3. Rapid hormonal influences on several key metabolic enzymes.

Enzyme[a]	Dephosphorylated in response to insulin	Phosphorylated in response to glucagon and epinephrine	Enzyme Function
Glycogen synthase	More active	Less active	Incorporation of UDP-glucose into glycogen
Phosphorylase kinase	Less active	More active	Phosphorylation of glycogen synthase and phosphorylase
Phosphorylase	Less active	More active	Release of glucose-1-phosphate from glycogen
PPI-1	More active	Less active	Binds to and inhibits phosphoprotein phosphatase-1
ACC	More active	Less active	Synthesis of malonyl-CoA
Hormone sensitive lipase	Less active	More active	Release of fatty acids from adipocyte triacylglycerols
HMG-CoA reductase	More active	Less active	Conversion of HMG-CoA to mevalonate
Reductase kinase	Less active	More active	Phosphorylation of HMG-CoA reductase
Pyruvate kinase	Active	Less active, liver and adipose tissue only	Conversion of PEP to pyruvate
PFK-2	More active as PFK-2 kinase	Less active as PFK-2 kinase	Synthesis of F-2,6-BP from F-6-P
F-2,6-BPase	Less active as PFK-2 phosphatase	More active as PFK-2 phosphatase	Conversion of F-2,6-BP to F-6-P

This list of enzymes regulated by PKA-mediated phosphorylation is not intended to be complete. These are, however, the major metabolic regulatory enzymes whose activities are rapidly influenced by hormonal stimulation of a variety of cell types.

glycogen phosphorylase, phosphorylase kinase, and **fructose-2,6-bisphosphatase** are inactive in the dephosphorylated state. In periods of fasting, the opposite state of activity of the key enzymes is induced in response to glucagon-induced activation of PKA.

It should be remembered that additional control of key metabolic enzymes is exerted through the action of allosteric effector molecules and by substrate availability. Longer term control over pathway fluxes is effected by the hormonal regulation of enzyme synthesis, which elevates or decreases the absolute level of individual enzymes.

Questions

DIRECTIONS (items 1–10): Each numbered question or incomplete statement is followed by answers or by completions of the statement. Select the ONE lettered answer or completion that is BEST in each case.

1. Which of the following statements is true for the indicated hepatic enzymes during starvation?
   a. PFK-2 is phosphorylated and active as a kinase.
   b. Carnitine acyltransferase I is active.
   c. PFK-1 is allosterically activated.
   d. Pyruvate kinase is phosphorylated and its activity increased.
   e. Fatty acid synthase is phosphorylated and its activity increased.

2. In response to epinephrine, which of the following occurs?
   a. Muscle ACC is activated by phosphorylation, whereas liver enzyme is inhibited by phosphorylation.
   b. Muscle and liver PFK-2 is active as a kinase.
   c. Muscle glycogen phosphorylase is active, liver enzyme inhibited.

    d. Muscle pyruvate kinase is active, liver enzyme is inhibited.
    e. Muscle adenylate cyclase is active, liver enzyme is inhibited.

3. Which of the following statements is true for insulin-induced responses?
    a. Cardiac muscle lipoprotein lipase is activated.
    b. Adipose lipoprotein lipase is inhibited.
    c. Liver ACC is phosphorylated and activated.
    d. Liver PFK-2 is active as the F-2,6-BPase.
    e. Muscle pyruvate kinase is phosphorylated and inhibited.

4. Which of the following describes an organ response to starvation?
    a. Liver ACC is phosphorylated and activated.
    b. Hormone-sensitive lipase in adipose tissue hormone sensitive lipase is inhibited.
    c. Cardiac lipoprotein lipase is inhibited.
    d. Adipose tissue glycerol kinase is activated.
    e. Liver pyruvate kinase is phosphorylated and activated.

5. Which of the following best describes the cause of hyperlipoproteinemia associated with insulin-dependent diabetes?
    a. Synthesis of fatty acids in the liver is greatly accelerated.
    b. The brain cannot oxidize ketone bodies; therefore they are converted to fatty acids by the liver.
    c. The lack of insulin alters the ratio of insulin to glucagon in the blood, bringing about accelerated lipolysis in adipose tissue.
    d. Liver cells exhibit unregulated synthesis of VLDL-associated apoproteins, which then form a complex with serum fatty acid.
    e. Adipose cells lack the ability to absorb free fatty acids.

6. Liver cirrhosis results predominantly in which metabolic defect?
    a. Hyperlipoproteinemia due to increased hepatic VLDL production.
    b. Ketoacidosis due to increased hepatic ketone body production.
    c. Elevated serum ammonia due to decreased hepatic urea cycle function.
    d. Hypotriglyceridemia due to defective hepatic fatty acid synthesis.
    e. Hyperglycemia due to uncontrolled hepatic gluconeogenesis.

7. During the early phase of anaerobic exercise the muscle relies upon which of the following for the generation of ATP?
    a. Adipose tissue fatty acids.
    b. Blood glucose.
    c. Muscle protein.
    d. Muscle glycogen.
    e. Muscle phosphocreatine.

8. Excess consumption of alcohol inhibits gluconeogenesis and fatty acid oxidation due to which of the following?
    a. Metabolism of ethanol yields large quantities of NADH, thereby disrupting the $NAD^+/NADH$ ratio with its concomitant effects on NAD^+-requiring reactions.
    b. Ethanol is an allosteric inhibitor of key dehydrogenases of the fat oxidation and gluconeogenic pathways.
    c. Metabolism of ethanol yields large quantities of NAD^+, thereby disrupting the $NAD^+/NADH$ ratio with its concomitant effects on NADH-requiring reactions.
    d. Metabolism of ethanol yields large quantities of acetaldehyde, which inhibits central nervous system function, thereby disrupting the ability of catecholamines to induce hepatic gluconeogenesis and fatty acid oxidation.
    e. Ethanol allosterically activates PFK-1 and ACC, increasing the rate of glycolysis and fat synthesis respectively.

9. Which of the following responses to glucagon is true?
    a. Hepatic phosphorylase kinase is inactivated by phosphorylation.
    b. Adipose tissue pyruvate kinase is inactivated by phosphorylation.

    c. Adipose tissue hormone sensitive lipase is phosphorylated and inactive.
    d. Hepatic PKF-2 is active as F-2,6-BPase.
    e. Hepatic pyruvate kinase is activated by phosphorylation.

10. Type II diabetes is most often characterized by:
    a. Impaired glucagon-dependent adipose tissue lipolysis.
    b. Elevated insulin secretion.
    c. Decreased glucagon secretion.
    d. Decreased insulin secretion.
    e. Impaired of insulin-dependent glucose uptake.

Answers

1. b	**5.** c	**9.** b
2. d	**6.** c	**10.** e
3. a	**7.** e	
4. b	**8.** a	

Part VII: Biological Information Processing

29

DNA Synthesis

Objectives

- To describe the nucleotide composition of eukaryotic genomes.
- To describe the structure of the nucleosome.
- To describe a typical eukaryotic cell cycle.
- To describe the process of DNA replication.
- To describe the major postreplicative modification of eukaryotic DNA.
- To describe the processes of DNA restructuring.
- To describe the processes of DNA repair.

Concepts

All cells undergo a division cycle during their life span. Some cells are continually dividing (eg, stem cells), others divide a specific number of times until cell death (**apoptosis**) occurs, and still others divide a few times before entering a **terminally differentiated or quiescent** state. Most cells of the body fall into the latter category of cells. During the process of cell division everything within the cell must be duplicated in order to ensure the survival of the two resulting daughter cells. Of particular importance for cell survival is the accurate, efficient, and rapid duplication of the cellular genome. This process is termed **DNA replication**.

Eukaryotic Genomes

The size of eukaryotic genomes is vastly larger than those of prokaryotes. This is partly due to the complexity of eukaryotic organisms compared to prokaryotes. However, the size of a particular eukaryotic genome is not directly correlated to the organism's complexity. This is the result of the presence of a large amount of noncoding DNA. The functions of these noncoding nucleic acid sequences are only partly understood. Some sequences are involved in the control of gene expression while others may simply be present in the genome to act as an evolutionary buffer able to withstand nucleotide mutation without disrupting the integrity of the organism.

One abundant class of DNA is termed **repetitive DNA**. There are 2 different subclasses of repetitive DNA, **highly repetitive and moderately repetitive**. Highly repetitive DNA consists of short sequences 6–10 bp long reiterated from 10^5–10^6 times. The DNA of the genome consisting of the genes (coding sequences) is identified as non-repetitive DNA since most genes occur but once in an organism's haploid genome. However, it should be pointed out that several genes exist as tandem clusters of multiple copies of the same gene ranging from 50 to 10,000 copies such as is the case for the rRNA genes and the histone genes.

Another characteristic feature that distinguishes eukaryotic from prokaryotic genes is the presence of **introns**. Introns are stretches of nucleic acid sequences that separate the coding exons of a gene. The existence of introns in prokaryotes is extremely rare. Essentially all humans genes contain introns. A notable exception is the histone genes which are intronless. In many genes the presence of introns separates exons into coding regions exhibiting distinct functional domains.

Chromatin Structure

Chromatin is a term designating the structure in which DNA exists within cells. The structure of chromatin is determined and stabilized through the interaction of the DNA with DNA-binding proteins.

There are 2 classes of DNA-binding proteins. The histones are the major class of DNA-binding proteins involved in maintaining the compacted structure of chromatin. There are 5 different histone proteins identified as **H1, H2A, H2B, H3, and H4**.

The other class of DNA-binding proteins is a diverse group of proteins called simply, **non-histone proteins**. This class of proteins includes the various transcription factors, polymerases, hormone receptors and other nuclear enzymes. In any given cell there are greater than 1000 different types of non-histone proteins bound to the DNA.

The binding of DNA by the histones generates a structure called the nucleosome (Figure 29–1). The nucleosome core contains an octamer protein structure consisting of 2 subunits each of H2A, H2B, H3, and H4. Histone H1 occupies the internucleosomal DNA and is identified as the linker histone. The nucleosome core contains approximately 150 bp of DNA. The linker DNA between each nucleosome can vary from 20 to more than 200 bp. These nucleosomal core structures would appear as beads on a string if the DNA were pulled into a linear structure (Figure 29–2).

The nucleosome cores themselves coil into a solenoid shape which itself coils to further compact the DNA. These final coils are compacted further into the characteristic chromatin seen in a karyotyping spread (Figure 29–2). The protein-DNA structure of chromatin is stabilized by attachment to a non-histone protein scaffold called the nuclear matrix.

Eukaryotic Cell Cycle

The cell cycle is defined as the sequence of events that occurs during the lifetime of a cell. The eukaryotic cell cycle is divided into 4 major periods (Figure 29–3). During each period a specific sequence of events occurs. The ultimate conclusion of one cell cycle is **cytokinesis** resulting in two identical daughter cells.

The 4 phases of a typical cell cycle and the events occurring during each phase are outlined:

1. **M phase** is the period when cells prepare for and then undergo cytokinesis. M phase stands for **mitotic phase or mitosis**. During mitosis the chromosomes are paired and then divided prior to cell division. The events in this stage of the cell cycle leading to cell division are **prophase, metaphase, anaphase, and telophase**.
2. **G₁ phase** corresponds to the **gap** in the cell cycle that occurs following cytokinesis. During this phase cells make a decision to either exit the cell cycle and become **quiescent or terminally differentiated** or to continue dividing. Terminal differentiation is identified as a nondividing state for a cell. Quiescent and terminally differentiated cells are identified as being in **G₀ phase**. Cells in G₀ can remain in this state for extended periods of time. Specific stimuli may induce the G₀ cell to re-enter the cell cycle at the G₁ phase or alternatively may induce permanent terminal differentiation.

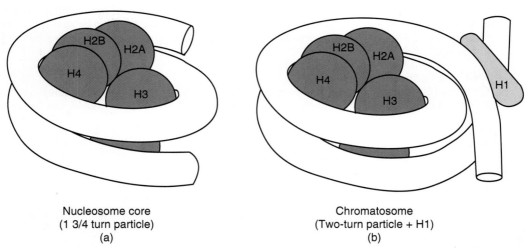

Nucleosome core
(1 3/4 turn particle)
(a)

Chromatosome
(Two-turn particle + H1)
(b)

Figure 29–1. *(a)* Structure of a nucleosome core and *(b)* a structure refered to as the chromatosome containing the linker histone H1. (Reproduced, with permission from Devlin, T: *Textbook of Biochemistry with Clinical Correlations*, 3rd ed. Wiley-Liss, Inc. 1992.)

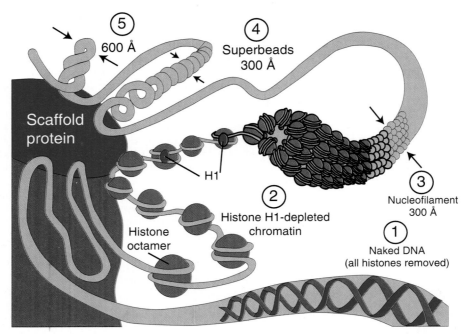

Figure 29–2. Various levels of organization of chromatin within the cell. Section #5 represents the 600-A knob-like structure of terminal packaging of chromatin as it is attached to the scaffold. (Reproduced, with permission from Devlin, T: *Textbook of Biochemistry with Clinical Correlations*, 3rd ed. Wiley-Liss, Inc. 1992.)

During G_1 cells begin synthesizing all the cellular constituents needed in order to generate two identically complemented daughter cells. As a result the size of cells begins to increase during G_1.

3. **S phase** is the phase of the cell cycle during which the DNA is replicated. This is the DNA synthesis phase. Additionally, some specialized proteins are synthesized during S phase, particularly the histones.

4. Following completion of DNA replication the cell enters the **G_2 phase**. During G_2 the chromosomes begin condensing, the nucleoli disappear and two microtubule organizing centers begin polymerizing tubulins for eventual production of the spindle poles.

Typical eukaryotic cell cycles occupy approximately 16–24 hrs when grown in culture. However, in the context of the multicellular organization of organisms the cell cycles can be as short as 6–8 hrs to greater than 100 days. The high variability of cell cycle times is due to the variability of the G_1 phase of the cycle.

DNA Replication

Replication of DNA occurs during the process of normal cell division cycles. Because the genetic complement of the resultant daughter cells must be the same as the parental cell, DNA replication must

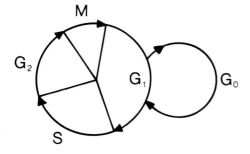

Figure 29–3. Typical eukaryotic cell-cycle indicating the four major phases, M = mitosis, G_1 = gap 1 prior to DNA synthesis, S = DNA synthesis, G_2 = gap 2 prior to mitosis. Phase G_0 represents a stage of quiescence from which a cell may or may not (terminal differentiation) return to the cycle.

possess a very high degree of fidelity. The entire process of DNA replication is complex and involves multiple enzymatic activities.

The mechanics of DNA replication was originally characterized in the bacterium, *E. coli*. *E. coli* contain 3 distinct enzymes capable of catalyzing the replication of DNA. These have been identified as DNA polymerase (pol) I, II, and III. Pol I is the most abundant replicating activity in *E. coli* but has as its primary role to ensure the fidelity of replication through the repair of damaged and mismatched DNA. Replication of the *E. coli* genome is the job of pol III. This enzyme is much less abundant than pol I, however, its activity is nearly 100 times that of pol I.

There have been 5 distinct eukaryotic DNA polymerases identified, α, β, γ, δ, and ε. The identity of the individual enzymes relates to its subcellular localization as well as its primary replicative activity. The polymerase of eukaryotic cells that is the equivalent of *E. coli* pol III is pol-α. The pol I equivalent in eukaryotes is pol-β. Polymerase-γ is responsible for replication of mitochondrial DNA. The precise functional roles of eukaryotic polymerases δ and ε are not yet fully defined.

The ability of DNA polymerases to replicate DNA requires a number of additional accessory proteins.

These include (not ordered with respect to importance):

1. Primase
2. Processivity accessory proteins
3. Single strand binding proteins
4. Helicase
5. DNA ligase
6. Topoisomerases
7. Uracil-DNA N-glycosylase

The combination of polymerases with several of the accessory proteins yields an enzyme complex identified as **DNA polymerase holoenzyme**.

The process of DNA replication begins at specific sites in the chromosomes termed **origins of replication**, requires a **primer** bearing a free 3'-OH, proceeds specifically in the **5' → 3' direction** on both strands of DNA concurrently and results in the copying of the template strands in a **semiconservative** manner. The semiconservative nature of DNA replication means that the newly synthesized daughter strands remain associated with their respective parental template strands.

The large size of eukaryotic chromosomes and the limits of nucleotide incorporation during DNA synthesis, make it necessary for multiple origins of replication to exist in order to complete replication in a reasonable period of time. The precise nature of origins of replication in higher eukaryotic organisms is unclear. However, it is clear that at a replication origin the strands of DNA must dissociate and unwind in order to allow access to DNA polymerase. Unwinding of the duplex at the origin as well as along the strands as the replication process proceeds is carried out by **helicases**. The resultant regions of single-stranded DNA are stabilized by the binding of **single-strand binding proteins**. The stabilized single-stranded regions are then accessible to the enzymatic activities required for replication to proceed. The site of the unwound template strands is termed the **replication fork** (Figure 29-7).

In order for DNA polymerases to synthesize DNA they must encounter a free 3'-OH which is the substrate for attachment of the 5'-phosphate of the incoming nucleotide. During repair of damaged DNA the 3'-OH can arise from the hydrolysis of the backbone of one of the two strands. During replication the 3'-OH is supplied through the use of an **RNA primer**, synthesized by the **primase** activity. The primase utilizes the DNA strands as templates and synthesizes a short stretch of RNA generating a primer for DNA polymerase (Figure 29–4).

Synthesis of DNA proceeds in the 5' → 3' direction through the attachment of the 5'-phosphate of an incoming dNTP to the existing 3'-OH in the elongating DNA strands with the concomitant release of pyrophosphate. Initiation of synthesis, at origins of replication, occurs simultaneously on both strands of DNA (Figure 29–5). Synthesis then proceeds **bidirectionally**, with one strand in each direction being copied **continuously** and one strand in each direction being copied **discontinuously**. During the process of DNA polymerases incorporating dNTPs into DNA in the 5' → 3' direction they are moving in the 3' → 5' direction with respect to the template strand. In order for DNA synthesis to occur simultaneously on both template strands as well as bidirectionally one strand appears to be synthesized in the 3' → 5' direction. In actuality one strand of newly synthesized DNA is produced discontinuously.

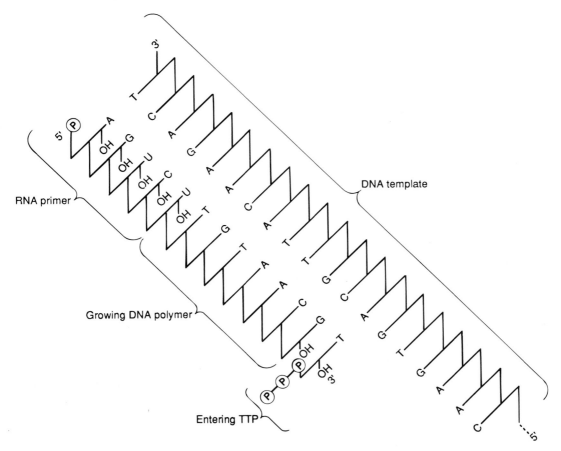

Figure 29–4. The RNA-primed synthesis of DNA demonstrating the template function of the complementary strand of parental DNA. (Reproduced, with permission, from Murray, RK: *Harper's Biochemistry*, 23rd ed. Appleton & Lange, 1993.)

The strand of DNA synthesized continuously is termed the **leading strand** and the discontinuous strand is termed the **lagging strand**. The lagging strand of DNA is composed of short stretches of RNA primer plus newly synthesized DNA approximately 100–200 bases long (the approximate distance between adjacent nucleosomes). The lagging strands of DNA are also called **Okasaki fragments** (Figure 29–6). The concept of continuous strand synthesis is somewhat of a misnomer since DNA polymerases do not remain associated with a template strand indefinitely. The ability of a particular

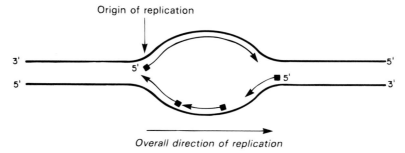

Overall direction of replication

Figure 29–5. The process of semidiscontinuous, simultaneous replication of both strands of double-stranded DNA. The leading strand (top) is replicated continuously in a 5′ to 3′ direction. The lagging strand (bottom) is replicated discontinuously, and in shorter segments, in a 3′ to 5′ direction. (Reproduced, with permission, from Murray, RK: *Harper's Biochemistry*, 23rd ed. Appleton & Lange, 1993.)

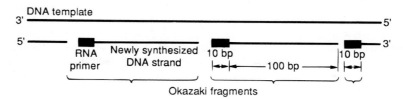

Figure 29–6. The discontinuous plymerization of deoxyribonucleotides and formation of Okazaki fragments. (Reproduced, with permission, from Murray, RK: *Harper's Biochemistry*, 23rd ed. Appleton & Lange, 1993.)

polymerase to remain associated with the template strand is termed its' **processivity**. The longer it associates the higher the processivity of the enzyme. DNA polymerase processivity is enhanced by additional protein activities identified as **processivity accessory proteins**.

How is it that DNA polymerase can copy both strands of DNA in the 5' → 3' direction simultaneously? A model has been proposed where DNA polymerases exist as dimers associated with the other necessary proteins at the replication fork and identified as the **replisome**. The template for the lagging strand is temporarily looped through the replisome such that the DNA polymerases are moving along both strands in the 3' → 5' direction simultaneously for short distances, the distance of an Okazaki fragment. As the replication forks progress along the template strands the newly synthesized daughter strands and parental template strands reform a DNA double helix. The means that only a small stretch of the template duplex is single-stranded at any given time (Figure 29–7).

The progression of the replication fork requires that the DNA ahead of the fork be continuously unwound. Due to the fact that eukaryotic chromosomal DNA is attached to a protein scaffold the progressive movement of the replication fork introduces severe torsional stress into the duplex ahead of the fork. This torsional stress is relieved by **DNA topoisomerases**. These special enzymes relieve torsional stresses in duplexes of DNA by introducing either double- (topoisomerases II) or single-stranded (topoisomerases I) breaks into the backbone of the DNA. These breaks allow unwinding of the duplex

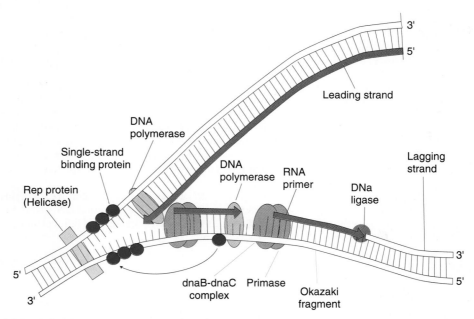

Figure 29–7. Model for the replication of DNA based upon studies in *E. coli*. The initial stages of replication at one side of a replication bubble are depicted. Replication at the other side of the bubble (not shown) would proceed identically. Rep protein is the helicase, and dnaB and dnaC are components of the RNA primase activity of *E. coli*. (Reproduced, with permission from Devlin, T: *Textbook of Biochemistry with Clinical Correlations*, 3rd ed. Wiley-Liss, Inc. 1992.)

and removal of the replication-induced torsional strain. The nicks are then resealed by the topoisomerases.

The RNA primers of the leading strands and Okasaki fragments are removed by the repair DNA polymerases simultaneously replacing the ribonucleotides with deoxyribonucleotides. The gaps that exist between the 3'-OH of one leading strand and the 5'-phosphate of another as well as between one Okasaki fragment and another are repaired by **DNA ligases** (Figure 29–7), thereby, completing the process of replication.

Additional Enzymatic Activities of DNA Polymerases

The main enzymatic activity of DNA polymerases is the 5' → 3' synthetic activity. However, DNA polymerases possess two additional activities of importance for both replication and repair (see below). These additional activities include a 5' → 3' exonuclease function and a 3' → 5' exonuclease function. The 5' → 3' exonuclease activity allows the removal of ribonucleotides of the RNA primer, utilized to initiate DNA synthesis, along with their simultaneous replacement with deoxyribonucleotides by the 5' → 3' polymerase activity. The 5' → 3' exonuclease activity is also utilized during the repair of damaged DNA. The 3' → 5' exonuclease function is utilized during replication to allow DNA polymerase to remove mismatched bases. It is possible (but rare) for DNA polymerases to incorporate an incorrect base during replication. These mismatched bases are recognized by the polymerase immediately due to the lack of Watson-Crick base-pairing. The mismatched base is then removed by the 3' → 5' exonuclease activity and the correct base inserted prior to progression of replication.

Post-Replicative Modification

One of the major post-Replicative reactions that modifies the DNA is **methylation**. The sites of natural methylation (ie, not chemically induced) of eukaryotic DNA is always on cytosine residues that are present in CpG dinucleotides. However, it should be noted that not all CpG dinucleotides are methylated at the C residue. The cytosine is methylated at the 5 position of the pyrimidine ring generating 5-methylcytosine (Figure 29–8).

Methylation of DNA in prokaryotic cells also occurs. The function of this methylation is to prevent degradation of host DNA in the presence of enzymatic activities synthesized by bacteria called **restriction endonucleases** (see Chapter 38). These enzymes recognize specific nucleotide sequences of DNA. The role of this system in prokaryotic cells (called the restriction-modification system) is to degrade invading viral DNAs. Since the viral DNAs are not modified by methylation they are degraded by the host restriction enzymes. The methylated host genome is resistant to the action of these enzymes.

The precise role of methylation in eukaryotic DNA is unclear. It was originally thought that methylated DNA would be less transcriptionally active. Indeed, experiments have been carried out to demonstrate that this is true for certain genes. For example, under-methylation of the MyoD gene (a master control gene regulating the differentiation of muscle cells through the control of the expression of muscle-specific genes) results in the conversion of fibroblasts to myoblasts. The experiments were carried out by allowing replicating fibroblasts to incorporate 5-azacytidine into their newly synthesized DNA. This analog of cytidine prevents methylation. The net result is that the maternal pattern of methylation is lost and numerous genes become under methylated. However, lack of methylation nor the presence of methylation is a clear indicator of whether a gene will be transcriptionally active or silent.

The pattern of methylation is copied post-Replicatively by the **maintenance methylase** system. This activity recognizes the pattern of methylated C residues in the maternal DNA strand following replication and methylates the C residue present in the corresponding CpG dinucleotide of the daughter strand.

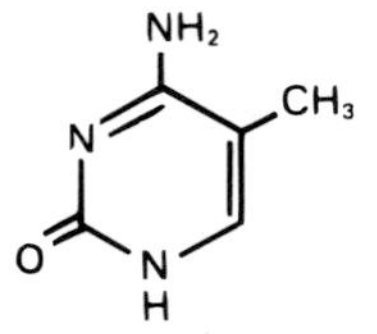

5-Methylcytosine

Figure 29–8. The uncommon yet naturally occuring pyrimidine, 5-methylcytosine. Cytosine residues residing with the dinucleotide sequence, CG, are frequently methylated on carbon 5 in eukaryotic DNA. (Reproduced, with permission, from Murray, RK: *Harper's Biochemistry*, 23rd ed. Appleton & Lange, 1993.)

DNA RESTRUCTURING

Recombination

DNA recombination refers to the phenomenon whereby two parental strands of DNA are spliced together resulting in an exchange of portions of their respective strands. This process leads to new molecules of DNA that contain a mix of genetic information from each parental strand. There are 3 main forms of genetic recombination. These are **homologous recombination, site-specific recombination and transposition**.

Homologous recombination is the process of genetic exchange that occurs between any two molecules of DNA that share a region (or regions) of homologous DNA sequences. This form of recombination occurs frequently while sister chromatids are paired during meiosis (Figure 29–9). Indeed, it is the process of homologous recombination between the maternal and paternal chromosomes that imparts genetic diversity to an organism. Homologous recombination generally involves exchange of large regions of the chromosomes.

Site-specific recombination involves exchange between much smaller regions of DNA sequence (approximately 20–200 base pairs) and requires the recognition of specific sequences by the proteins involved in the recombination process. Site-specific recombination events occur primarily as a mechanism to alter the program of genes expressed at specific stages of development. The most significant site-specific recombinational events in humans are the somatic cell gene rearrangements that take place in the immunoglobulin genes during B-cell differentiation in response to antigen presentation. These gene rearrangements in the immunoglobulin genes result in an extremely diverse potential for antibody production. A typical antibody molecule is composed of both heavy and light chains. The genes for both these peptide chains undergo somatic cell rearrangement, yielding the potential for approximately 3000 different light chain combinations and approximately 5000 heavy chain combinations (Figure 29–10). Then because any given heavy chain can combine with any given light chain the potential diversity exceeds 10^7 possible different antibody molecules.

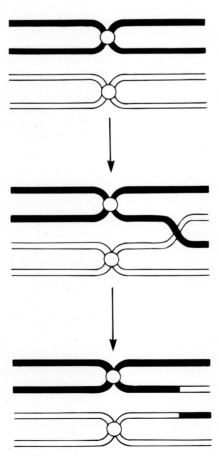

Figure 29–9. The process of crossing-over between homologous chromosomes to generate recombinant chromosomes. (Reproduced, with permission, from Murray, RK: *Harper's Biochemistry*, 23rd ed. Appleton & Lange, 1993.)

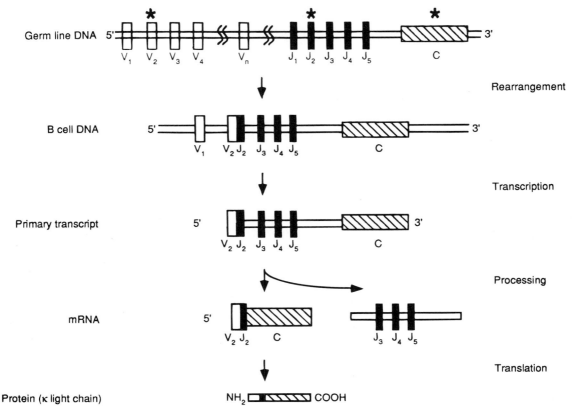

Figure 29–10. Recombination events leading to a V_2J_2 κ light chain. The 500 V_L (variable), 5–6 V_j (joining), and single V_C (constant) segments exist in alinear array in the chromosome. These components span thousands of kilobase pairs. Upon receipt of a signal to differentiate, the DNA in the B cell undergoes rearrangement so that the 3′ end of V_2 is ligated to the 5′ end of J_2. Transcription initiates through a promoter at the 5′ end of V_2 and proceeds through C. In order to produce a V_2J_2 C mRNA, the sequence from the 3′ end of J_2 to the 5′ end of C is excised, then V_2J_2 is spliced to C. This mRNA can then be translated into a κ light chain that contains the amino acid sequence encoded by V_2J_2 and C. Lthe same general procedure can result in κ light chains consisting of any combination of V and J segments with C. (Reproduced, with permission, from Murray, RK: *Harper's Biochemistry*, 23rd ed. Appleton & Lange, 1993.)

Transposition

Transposition is a unique form of recombination where **mobile genetic elements** can virtually move from one region to another within one chromosome or to another chromosome entirely. There is no requirement for sequence homology for a transpositional event to occur. Because the potential exists for the disruption of a vitally important gene by a transposition event this process must be tightly regulated. The exact nature of how transpositional events are controlled is unclear.

Transposition occurs with a higher frequency in bacteria and yeasts than it does in humans. The identification of the occurrence of transposition in the human genome resulted when it was found that certain processed genes were present in the genome. These processed genes are nearly identical to the mRNA encoded by the normal gene. The processed genes contain the poly(A) tail that would have been present in the RNA and they lack the introns of the normal gene. These particular forms of genes must have arisen through a reverse transcription event, similar to the life cycle of retroviral genomes, and then been incorporated into the genome by a transpositional event. Since most of the processed genes that have been identified are non-functional they have been termed **pseudogenes**.

Repair of Damaged DNA

Cancer, in most non-viral induced cases, is the most severe medically relevant consequence of the inability to repair damaged DNA. It is clear that multiple somatic cell mutations in DNA can lead to the genesis of the transformed phenotype. Therefore, it should be obvious that complete understanding of DNA repair mechanisms would be invaluable in the design of potential therapeutic agents in the treatment of cancer.

DNA damage can occur as the result of exposure to environmental stimuli such as alkylating chemicals or ultraviolet or radioactive irradiation and free radicals generated spontaneously in the oxidizing environment of the cell. These phenomena can, and do, lead to the introduction of mutations in the coding capacity of the DNA. Mutations in DNA can also, but rarely, arise from the spontaneous tautomerization of the bases.

Modification of the DNA bases by alkylation (predominately the incorporation of -CH_3 groups) predominately occurs on purine residues. Methylation of G residues allows them to base pair with T instead of C. A unique activity called **O^6-alkylguanine transferase** removes the alkyl group from G residues. The protein itself becomes alkylated and is no longer active, thus, a single protein molecule can remove only one alkyl group.

Mutations in DNA are of two types. **Transition** mutations result from the exchange of one purine, or pyrimidine, for another purine, or pyrimidine. **Transversion** mutations result from the exchange of a purine for a pyrimidine or visa versa.

The prominent byproduct from *uv* irradiation of DNA is the formation of thymine dimers. These form from two adjacent T residues in the DNA. Repair of thymine dimers is most understood from consideration of the mechanisms used in *E coli*. However, several mechanism are common to both prokaryotes and eukaryotes.

Thymine dimers are removed by several mechanisms. Specific glycohydrolases recognize the dimer as abnormal and cleave the N-glycosidic bond of the bases in the dimer. This results in the base leaving and generates an apyrimidinic site in the DNA. This is repaired by DNA polymerase and ligase. Glycohydrolases are also responsible for the removal of other abnormal bases, not just thymine dimers.

Another, widely distributed activity, is **DNA photolyase** or photoreactivating enzyme. This protein binds to thymine dimers in the dark. In response to visible light stimulation the enzyme cleaves the pyrimidine rings. The chromophore associated with this enzyme that allows visible light activation is $FADH_2$.

Humans defective in DNA repair, (in particular the repair of thymine dimers), due to autosomal recessive genetic defects suffer from the disease **Xeroderma pigmentosum**. There are at least nine distinct genetic defects associated with this disease. One of these is due to a defect in the gene coding for the glycohydrolase that cleaves the N-glycosidic bond of the thymine dimers.

Questions

DIRECTIONS (items 1–10): Each numbered item or incomplete statement is followed by answers or by completions of the statement. Select the ONE lettered answer or completion that is BEST in each case.

1. Which of the following statements concerning nucleosomes is true?
   a. The core structure is composed solely of a piece of DNA.
   b. The core structure contains 4 histone proteins along with DNA.
   c. DNA is believed to be coated by the histone core.
   d. The histones of the core are comprise a total of 9 proteins.
   e. The histones of the core include H2A, H2B, H3, and H4.

2. DNA replication requires all of the following **EXCEPT**:
   a. $5' \rightarrow 3'$ exonuclease activity.
   b. An RNA primer.
   c. Energy in the form of ATP.
   d. A free 5'-OH.
   e. DNA ligase.

3. The enzyme DNA photolyase catalyzes:
   a. Insertion of uracil for cytidine in DNA following *uv* irradiation.
   b. Removal of O^6-alkylguanine from DNA following *uv* irradiation.

    c. Removal of apyrimidinic sites in DNA following *uv* irradiation.
    d. The light activated removal of 5-methylcytidine.
    e. Removal of thymine dimers in DNA.

4. DNA replication:
    a. Is bidirectional proceeding from a single initiation point on each chromosome in eukaryotic cells.
    b. Is unidirectional in prokaryotic cells.
    c. Does not require ATP.
    d. Is initiated on RNA primers.
    e. Involves the formation, and not the cleavage, of phosphodiester bonds.

5. Ultraviolet light produces thymine dimers in DNA strands. Removal of the dimer and repair of the DNA requires all of the enzymes listed **EXCEPT**:
    a. $5' \rightarrow 3'$ exonuclease.
    b. DNA polymerase I.
    c. DNA ligase.
    d. Topoisomerase.
    e. Two of the above.

6. Xeroderma pigmentosum is a genetic disease characterized by:
    a. Loss of melanin pigment in the skin.
    b. Hyper-proliferation of dermal fibroblasts.
    c. A defect in DNA repair-replication.
    d. Two of the above.
    e. All of the above.

7. The role of topoisomerases in DNA replication:
    a. Is to stabilize double-stranded regions in the DNA.
    b. Is to relieve induced strain in the DNA.
    c. Is to unwind the replication fork.
    d. Is to ligate the nicks between Okasaki fragments following removal of the RNA primer.
    e. Is to prevent the strands from reassociating until after the replisome has sufficiently moved down the template.

8. Incorporation of 5-bromouracil into DNA will result in which conversion:
    a. AT to GC.
    b. GC to AT.
    c. TA to AT.
    d. CG to GC.

9. The term semi-conservative replication refers to:
    a. The fact that the newly synthesized strands of DNA associate with each other.
    b. The fact that the newly synthesized strands of DNA associate with their respective template strands.
    c. The fact that some of the enzymatic machinery is utilized simultaneously to synthesize both strands of the duplex.
    d. The fact that two distinct DNA polymerases are required to complete the replication process.
    e. The fact that all of the enzymatic machinery is utilized simultaneously to synthesize both strands of the duplex.

10. What is meant by the statement that DNA synthesis is discontinuous?
    a. Replication of different regions of the DNA occurs at different times.
    b. One direction of synthesis from an origin of replication is carried out following completion of synthesis in the opposing direction.
    c. One strand of the DNA duplex is replicated in short stretches as opposed to continuously as for the opposite strand of the duplex.
    d. One chromosome is replicated at a time before the enzymatic machinery can begin the replication of additional chromosomes.

e. The order of events requires that the helix must first be unwound, then single stranded regions stabilized by protein binding then an RNA primer synthesized prior to actual DNA synthesis.

DIRECTIONS (items 11–15): Match the lettered list of proteins/activities to identify the corresponding numbered arrows in the diagram. Letters can only be used once.

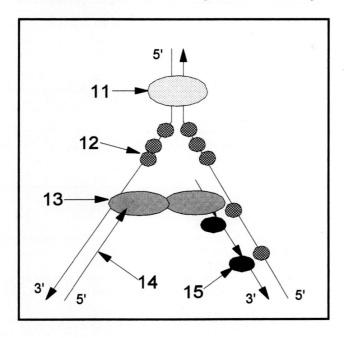

a. Helicase, duplex unwinding activity.
b. Single strand binding proteins.
c. Primase, RNA primer synthesizing activity.
d. Leading strand.
e. Lagging strand.
f. DNA polymerase.
g. Topoisomerase.
h. DNA ligase.

Answers

1. e	**6.** c	**11.** a
2. d	**7.** b	**12.** b
3. e	**8.** a	**13.** f
4. d	**9.** b	**14.** d
5. d	**10.** c	**15.** h

30

RNA Synthesis

Objectives

- To describe the 3 classifications of RNA.
- To describe the 3 classes of RNA polymerases.
- To describe the process of transcription, including the steps of initiation, elongation, and termination.
- To describe the posttranscriptional processing of RNAs.
- To describe the different mechanisms for the splicing of RNAs.
- To describe the clinical significances of alternative or aberrant RNA splicing.

Concepts

Transcription is the mechanism by which a template strand of DNA is used by specific RNA polymerases to generate 1 of the 3 different classes of RNA. These classes are:

1. **Messenger RNAs (mRNAs).** These RNAs are the genetic "coding" templates used by the translational machinery to determine the order of amino acids incorporated into an elongating polypeptide (see Chapter 31).
2. **Transfer RNAs (tRNAs).** These small RNAs form covalent attachments to individual amino acids and recognize the encoded sequences of the mRNAs, to allow the correct insertion of amino acids into the elongating polypeptide chain (see Chapter 31).
3. **Ribosomal RNAs (rRNAs).** This class of RNAs is assembled, together with numerous ribosomal proteins, to form the **ribosomes**. Ribosomes engage the mRNAs and form a catalytic domain into which the tRNAs enter with their attached amino acids. The proteins of the ribosomes catalyze several functions of polypeptide synthesis (see Chapter 31).

All RNA polymerases are dependent upon a DNA template in order to synthesize RNA. The resultant RNA is, therefore, complimentary to the template strand of the DNA duplex and identical to the non-template strand. The non-template strand is called the **coding strand** because its sequences are identical to those of the RNA. However, in RNA, the nucleotide U is substituted for T.

RNA POLYMERASE CLASSES

In prokaryotic cells, all 3 classes of RNA are synthesized by a single polymerase. In eukaryotic cells there are 3 distinct classes of RNA polymerase, **RNA polymerase (pol) I, II, and III.** Each polymerase is responsible for the synthesis of a different class of RNA.

The capacity of the various polymerases to synthesize different RNAs was shown with the toxin **α-amanitin.** At low concentrations of α-amanitin, the synthesis of mRNAs is affected but not that of rRNAs nor tRNAs. At high concentrations, synthesis of both mRNAs and tRNAs is affected. These observations have allowed the identification of which polymerase synthesizes which class of RNAs. RNA pol I is responsible for rRNA synthesis (excluding the 5S rRNA). There are 4 major rRNAs in eukaryotic cells, designated by their sedimentation size. The 28S, 5S, and 5.8S rRNAs are associated with the large ribosomal subunit; the 18S rRNA is associated with the small ribosomal subunit (see Chapter 31). RNA pol II synthesizes the mRNAs and some of the small nuclear RNAs (snRNAs) involved in RNA splicing. RNA pol III synthesizes the tRNAs, the 5S rRNA, and some snRNAs.

The synthesis of RNA exhibits several features that are synonymous with DNA replication. RNA synthesis requires accurate and efficient initiation; elongation proceeds in the $5' \rightarrow 3'$ direction (ie, the polymerase moves relative to the template strand of DNA in the $3' \rightarrow 5'$ direction); and RNA synthesis requires distinct and accurate termination.

Transcription, however, exhibits several features that are distinct from replication:

1. Transcription initiates, both in prokaryotes and eukaryotes, from many more sites than replication.
2. There are many more molecules of RNA polymerase per cell than of DNA polymerase.
3. RNA polymerase proceeds at a rate much slower than DNA polymerase (approximately 50–100 bases/sec for RNA, as compared with near 1000 bases/sec for DNA).
4. The fidelity of RNA polymerization is much lower than that of DNA. This is allowable since the aberrant RNA molecules can simply be turned over and new, correct molecules made.

PROKARYOTIC AND EUKARYOTIC TRANSCRIPTION

Initiation

Signals within the DNA template act in *cis* to stimulate the initiation of transcription. These sequence elements are termed **promoters**. Promoter sequences assist in the ability of RNA polymerases to recognize the nucleotide at which initiation begins. Additional sequence elements are present within genes that act in *cis* to enhance polymerase activity even further. These sequence elements are termed **enhancers**. The role of transcriptional promoter and enhancer elements in the regulation of transcriptional activity will be discussed in Chapter 32.

E. coli RNA polymerase is composed of 5 distinct polypeptide chains. The association of several of these generates the **RNA polymerase holoenzyme** (Figure 30–1). The sigma (σ) subunit is only transiently associated with the holoenzyme. This subunit is required for the accurate initiation of transcription: it provides polymerase with the proper cues that a start site has been encountered. Following initiation, the σ subunit dissociates from the holoenzyme.

In both prokaryotic and eukaryotic transcription, the first incorporated ribonucleotide is a purine, and it is incorporated as a triphosphate. In *E. coli,* several additional nucleotides are added before the σ subunit dissociates (Figure 30–2).

Elongation involves the addition of the 5'-phosphate of ribonucleotides to the 3'-OH of the elongating RNA, with the concomitant release of pyrophosphate. Nucleotides continue to be added until specific termination signals are encountered. After termination, the core polymerase dissociates from the template. The core and σ subunit can then reassociate, forming a holoenzyme that is once again ready to initiate a round of transcription (Figure 30–2).

In *E. coli,* transcriptional termination occurs by both factor-dependent and factor-independent means. There have been 2 structural features identified as common to all *E. coli* factor-independently terminating genes. The first of these is the presence of 2 symmetrical GC-rich segments that are capable of forming a stem-loop structure in the RNA, and the second is a downstream A-rich sequence in the template. The formation of the stem-loop in the RNA destabilizes the association between polymerase and the DNA template. This is further destabilized by the weaker nature of the AU base pairs

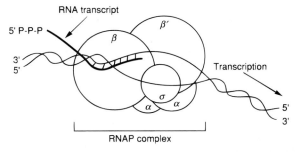

Figure 30–1. RNA polymerase catalyzes the polymerization of ribonucleotides into an RNA sequence that is complementary to the template strand of the gene. The RNA transcript has the same polarity (5' to 3') as the coding strand but contains U rather than T. *E. Coli* RNAP consists of a core complex of two α subunits and two β subunits (β and β'). The holoenzyme contains the o subunit when near the transcription start site (within ~10 bp) but then transcription proceeds with just the core complex. The transcription "bubble" is s 17-bp area of melted DNA, and the entire complex covers 30–75 bp, depending on the conformation of RNAP. (Reproduced, with permission, from Murray, RK: *Harper's Biochemistry*, 23rd ed. Appleton & Lange, 1993.)

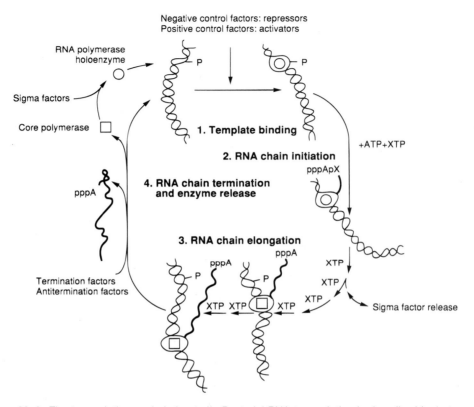

Figure 30–2. The transcription cycle in bacteria. Bacterial RNA transcription is described in 4 steps: **(1.) Template Binding:** RNA polymerase (RNAP) binds to BNA and locates a promoter. **(2.) Chain initiation:** RNAP holoenzyme (core + sigma factors) catalyzes the coupling of the first base (usually ATP or GTP) to a second ribonucleoside triphophate to form a dinucleotide. **(3.) Chain elongation:** successive residues are added to the 3′-OH teminus of the nascent RNA molecule; sigma factor dissociates from the holoenzyme after the chain length of about 10 is achieved. **(4.) Chain termination and releases:** the complete RNA chain and RNAP are released from the template. The RNAP holoenzyme reforms, finds a promoter, and the cycle is repeated. (Slightly modified and reproduced, with permission, from Chamerlain M: Bacterial DNA-dependent RNA polymerases. *The Enzymes.* Academic Press, 1982.)

that are formed between the template and the RNA following the stem-loop. This leads to dissociation of the polymerase and the termination of transcription (Figure 30–3). Most genes in *E. coli* terminate by this method.

Factor-dependent termination requires the recognition of termination sequences by the termination protein, rho (ρ). The ρ factor recognizes and binds to sequences in the 3′ portion of the RNA. This binding destabilizes the polymerase-template interaction, leading to dissociation of the polymerase and the termination of transcription.

POSTTRANSCRIPTIONAL PROCESSING

When the transcription of bacterial rRNAs and tRNAs is completed, they are immediately ready for use in translation. No additional processing takes place. The translation of bacterial mRNAs can begin even before transcription is completed, thanks to the lack of a nuclear-cytoplasmic separation that exists in eukaryotes. The ability to initiate the translation of prokaryotic RNAs while transcription is still in progress affords a unique opportunity for regulating the transcription of certain genes (see Chapter 32). An additional feature of bacterial mRNAs is that most are **polycistronic**, meaning that multiple polypeptides can be synthesized from a single primary transcript. This does not occur in eukaryotic mRNAs.

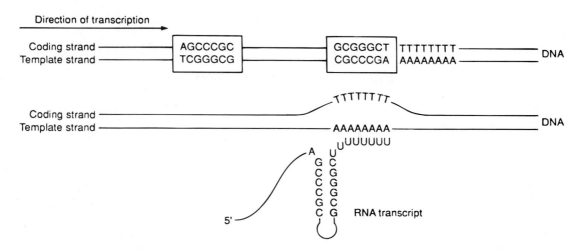

Figure 30–3. The bacterial transcription termination signal in the gene contains an inverted, hyphenated repeat (the 2 boxed areas) followed by a stretch of AT base pairs (top figure). The inverted repeat, when transcribed into RNA, can generate the secondary structure in the RNA transcript shown in the bottom figure. (Reproduced, with permission, from Murray, RK: *Harper's Biochemistry*, 23rd ed. Appleton & Lange, 1993.)

In contrast to bacterial transcripts, eukaryotic RNAs (all 3 classes) undergo significant post-transcriptional processing such as that for mRNAs shown in Figure 30–4. All 3 classes of RNA are transcribed from genes that contain introns. The sequences encoded by the intronic DNA must be removed from the primary transcript before the RNAs can be biologically active. The process of intron removal is called **RNA splicing** (see below). Additional processing of mRNAs takes place as well: The 5′ end of all eukaryotic mRNAs is **capped** with a unique 5′ → 5′ linkage to a 7-methylguanosine residue (Figure 30–5). The capped end of the mRNA is thus protected from exonucleases; more importantly, it is recognized by specific proteins of the translational machinery (see Chapter 31).

Messenger RNAs also are polyadenylated at the 3′ end. A specific sequence, **AAUAAA**, is recognized by the endonuclease activity of **polyadenylate polymerase**, which cleaves the primary transcript approximately 11–30 bases 3′ of the sequence element (Figure 30–6). A stretch of 20–250 A residues is then added to the 3′ end by the polyadenylate polymerase activity.

In addition to intron removal in tRNAs, extra nucleotides at both the 5′ and 3′ ends are cleaved, the sequence -CCA is added to the 3′ end of all tRNAs (Figure 30–7), and several nucleotides undergo modification. More than 60 different modified bases have been identified in tRNAs.

Both prokaryotic and eukaryotic rRNAs are synthesized as long precursors termed **preribosomal RNAs**. In eukaryotes, a 45S preribosomal RNA serves as the precursor for the 18S, 28S, and 5.8S rRNAs (Figure 30–8).

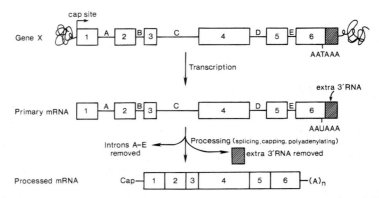

Figure 30–4. Processing of a typical eukaryotic messenger RNA gene. Boxes represent exons and lines, introns.

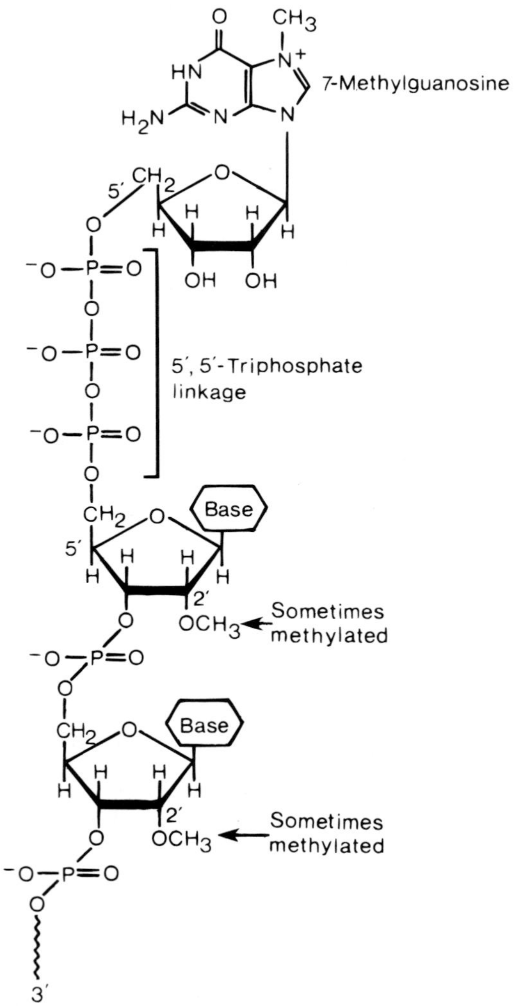

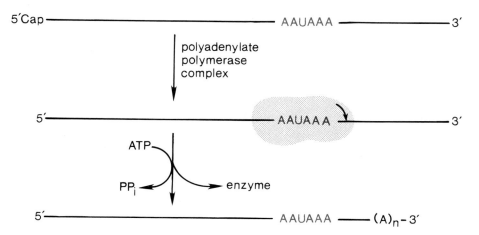

Figure 30–5. Structure of the 5'-cap of a typical eukaryotic mRNA. The 2'-OH position of the up to three 5'-terminal nucleotides is occasionally methylated.

Figure 30–6. Enzymatic process by which the 3'-polyadenylate tail is added post-transcriptionally to mRNAs. The enzyme complex recognizes the AAUAAA consensus sequences, cleaves the primary transcript some 20–30 bases 3' of this signal and then adds from 50–250 adenosine residues to the 3'-end.

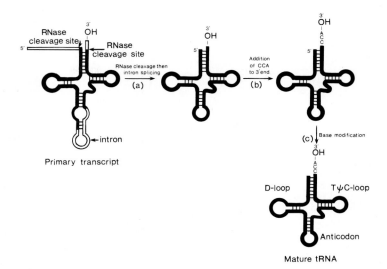

Figure 30–7. Processing events for a typical tRNA. The extra RNA at the 5' and 3' ends is removed by specific ribonucleases (RNases), the intron (blue) is removed, the consensus sequence, CCA, is added to the 3' end of all tRNAs and numerous bases are modified.

RNA SPLICING

There are 4 different classes of introns, with the 2 most common being the **group I** and **group II** introns. Group I introns are found in nuclear, mitochondrial, and chloroplast rRNA genes; group II, in mitochondrial and chloroplast mRNA genes. Many of the group I and group II introns are **self-splicing**: no additional protein factors are necessary in order for the intron to be accurately and efficiently spliced out.

Group I introns require an external guanosine nucleotide as a cofactor. The 3'-OH of the guanosine nucleotide acts as a nucleophile to attack the 5'-phosphate of the 5' nucleotide of the intron. The resultant 3'-OH at the 3' end of the 5' exon then attacks the 5' nucleotide of the 3' exon, releasing the intron and covalently attaching the 2 exons together (Figure 30–9). The 3' end of the 5' exon is termed the **splice donor site**, and the 5' end of the 3' exon is termed the **splice acceptor site**.

Group II introns are spliced similarly, except that instead of an external nucleophile, the 2'-OH of an adenine residue within the intron serves as the nucleophile. This residue attacks the 3' nucleotide of the 5' exon, forming an internal loop called a **lariat structure**. The 3' end of the 5' exon then attacks the 5' end of the 3' exon as in group I splicing, releasing the intron and covalently attaching the 2 exons together (Figure 30–10).

The third class of introns is the largest class found in nuclear mRNAs. This class undergoes a splicing reaction similar to that of group II introns, forming an internal lariat structure is formed. However, the splicing is catalyzed by specialized RNA-protein complexes called **small nuclear ribonucleoprotein particles** (snRNPs, pronounced "snurps"). The RNAs found in snRNPs are identified as **U1, U2, U4, U5, and U6.** The genes encoding these snRNAs are highly conserved in vertebrates and insects and are also found in organisms as divergent as yeasts and slime molds—an indication of their importance.

Analysis of a large number of mRNA genes has led to the identification of highly conserved consensus sequences at the 5' and 3' ends of essentially all mRNA introns (Figure 30–11). The U1 RNA contains sequences that are complementary to those near the 5' end of the intron. The binding of U1 RNA distinguishes the GU at the 5' end of the intron from other randomly placed GU sequences in mRNAs. The U2 RNA also recognizes sequences in the intron, in this case near the 3' end. The addition of U4, U5, and U6 RNAs forms a complex, the **spliceosome**, that then removes the intron and joins the 2 exons together (Figure 30–12).

The fourth class of introns is composed of those found in certain tRNAs. These introns are spliced by a specific splicing endonuclease that utilizes the energy of ATP hydrolysis to catalyze intron removal and ligation of the 2 exons together.

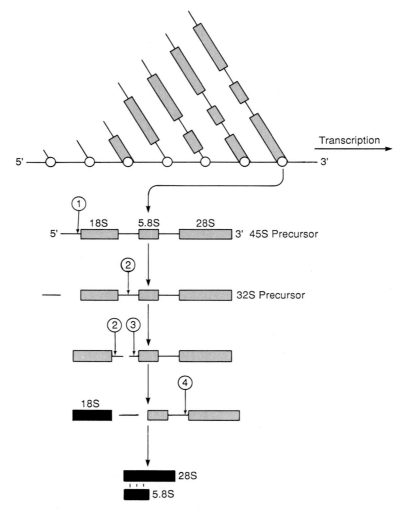

Figure 30–8. Diagrammatic representation of the processing of ribosomal RNA from large RNA precursor molecules. The primary transcript has a size of 45S and is about 13,000 nucleotides in length. It is synthesized from 5' to 3' by RNA polymerase I (solid circles). The densely clustered genes are transcribed at a rapid rate, so the elongating RNA molecules, easily seen in chromatin preparations, resemble the "trees" in the electron micrograph. The final products are formed through a series of endo- and exo-nuclease reactions involving rRNA-specific ribonucleases. A possible sequence of these reactions is shown by the numbered circles and arrows. The sizes and nucleotide length of the 18S, 5.8S and 28S products are shown. The 18S rRNA is incorporated into the small ribosomal subunit. The 28S and 5.8S rRNAs are incorporated into the large subunit. (Reproduced, with permission, from Murray, RK: *Harper's Biochemistry*, 23rd ed. Appleton & Lange, 1993.)

CLINICAL SIGNIFICANCE OF ALTERNATIVE AND ABERRANT SPLICING

The presence of introns in eukaryotic genes might appear to be an extreme waste of cellular energy, given the number of nucleotides incorporated into the primary transcript only to be removed later as well as the energy expended in the synthesis of the splicing machinery. However, the presence of introns can protect the genetic makeup of an organism from genetic damage by outside influences such as chemical or radiation. Another important function of introns is to allow alternative splicing to occur, thereby increasing the genetic diversity of the genome without increasing the overall number of genes. By altering the pattern of exons from a single primary transcript that are spliced together, different pro-

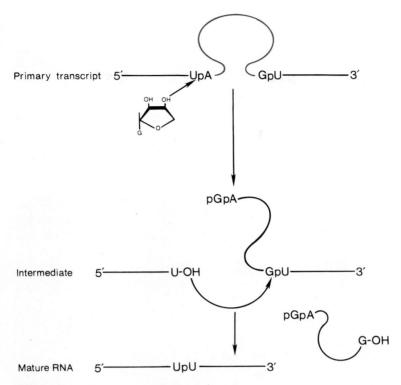

Figure 30–9. Mechanism of splicing group I self-splicing introns. The guanosine nucleotide that acts as the nucleophile in attacking the 5′ end of the intron can be GMP, GDP or GTP.

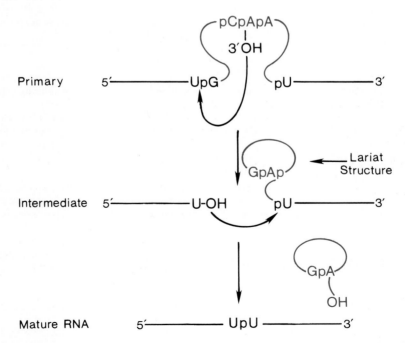

Figure 30–10. Mechanism of splicing group II self-splicing introns. The chemistry of this reaction is smilar to that of group I self-splicing, however the nucleophile in this reaction is a nucleotide within the intron itself. The lariat structure that is formed contains a novel 2′,5′-phosphodiester bond.

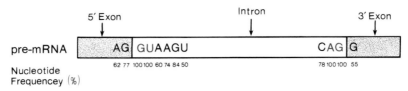

Figure 30–11. Nucleotide frequencies at the boundaries of eukaryotic pre-mRNAs. The GU at the 5′ end of the intron and the AG at the 3′ end are invariant.

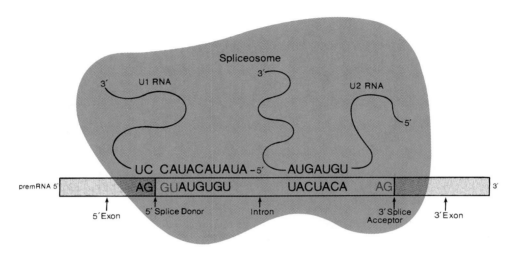

Figure 30–12. Mechanism of splice site recognition in eukaryotic pre-mRNAs. The splicosome contains multiple small nuclear ribonucleoproteins (snRNPs, "snurps") two of which contain small RNAs (U1 and U2) that contain sequences complimentary to regions of the intron. These complimentary sequences allow the splisosome to engage the pre-mRNA and accurately recognize the 5′ and 3′ splice sites.

teins can arise from the processed mRNA from a single gene. Alternative splicing can occur either at specific developmental stages or in different cell types.

This process of alternative splicing has been found to occur in the primary transcripts from at least 40 different genes. As an example, depending upon the site of transcription, the calcitonin gene yields an RNA that synthesizes calcitonin (thyroid) or calcitonin-gene related peptide (CGRP, brain). Even more complex is the alternative splicing that occurs in the α-tropomyosin transcript. At least 8 different alternatively spliced α-tropomyosin mRNAs have been identified.

Abnormalities in the splicing process can lead to various disease states. Many defects in the β-globin genes are known to exist; these lead to β-thalassemias. Some of these defects are caused by mutations in the sequences of the gene required for intron recognition; they therefore result in abnormal processing of the β-globin primary transcript.

Patients suffering from a number of different connective tissue diseases exhibit humoral autoantibodies that recognize cellular RNA-protein complexes. Patients suffering from **systemic lupus erythematosus** have autoantibodies that recognize the U1 RNA of the spliceosome.

Questions

DIRECTIONS (items 1–9): Each numbered item or incomplete statement is followed by answers or by completions of the statement. Select the ONE lettered response that is BEST in each case.

1. Most mRNA molecules formed in the nucleus of an animal cell undergo which of the following?
   a. The formation of a large precursor RNA is followed by exon rearrangement to yield the mature mRNA.
   b. Nuclear processing mechanisms modify both the 5′ and 3′ ends.
   c. Transport out of the nucleus for intron removal.
   d. Cytoplasmic processing of the 5′ and 3′ ends.
   e. Removal of the poly(A) tail following completion of translation.

2. Which of the following statements is true of *E. coli* RNA polymerase?
   a. The σ subunit dissociates from the core enzyme shortly after the elongation process begins.
   b. The ρ protein is part of the core enzyme.
   c. The σ subunit is involved in the selection of termination sites on a natural DNA template.
   d. If the σ subunit is missing, the core polymerase cannot bind to the DNA template.
   e. The σ subunit allows RNA polymerase to recognize the "CAT-box," thereby increasing specific initiation at the correct nucleotide.

3. The sequence of an mRNA molecule synthesized from a DNA template strand having the sequence 5′ CTA TTT AAA AGC CAT 3′ is:
   a. 5′ CUA UUU AAA AGC CAU 3′.
   b. 5′ CAU AGG AAA UUU CUA 3′.
   c. 5′ ATG GCT TTT AAA CTA 3′.
   d. 5′ AUG GCU UUU AAA UAG 3′.
   e. 5′ UAC CGA AAA UUU AUC 3′.

4. The CAPPED 5′ end of eukaryotic RNA:
   a. Consists of poly(A) sequences.
   b. Contains a 5′ → 5′ link between ribose residues mediated by phosphate ester bridges.
   c. Is transcribed directly from DNA sequences.
   d. Is of NO significance for the activity of mRNA in translation.
   e. Is directly adjacent to the AUG initiator codon in mRNA.

5. All of the following are true for RNA polymerase **EXCEPT**:
   a. It requires a template.
   b. It reaction produces PP_i as a product.
   c. Synthesis proceeds in the 5′ → 3′ direction.
   d. It utilizes 5′-nucleoside triphosphates as substrates.
   e. It requires a primer.

6. All of the following processing events occur upon completion of transcription of tRNAs **EXCEPT**:
   a. Removal of the 5′ and 3′ terminal nucleotides.
   b. Addition of the trinucleotide sequence, CCA, to the 3′ end.
   c. Addition of the poly(A) tail.
   d. Removal of intron sequences.
   e. Modification of nucleotides within the anticodon loop.

7. Which of the following statements is most correct in describing class I introns?
   a. They are removed by complex small nuclear ribonucleoprotein particles termed spliceosomes.
   b. They contain the consensus sequences GT and AG at their 5′ and 3′ ends, respectively.
   c. They are found predominantly in mRNAs.
   d. They utilize a nucleotide cofactor to catalyze the splicing reaction.
   e. They form a lariat structure during the initial stages of the splicing process.

8. Systemic lupus erythematosus is a disorder characterized by which biochemical involvement?
   a. The production of autoantibodies against components of the splicing machinery.
   b. The production of autoantibodies against components of the transcriptional machinery.
   c. The production of autoantibodies against the small 5S rRNAs.

    d. The production of autoantibodies against several tRNAs.
    e. The production of autoantibodies against components of the tRNA modifying enzymes.

9. One form of β-thalassemia has been shown to result from which defect in the β-globin gene?
    a. Loss of the transcriptional initiation sequences.
    b. Loss of the transcriptional termination sequences.
    c. Loss the poly(A) addition signal.
    d. Loss of the ribosome recognition sequences.
    e. Loss of the splicing consensus sequences within 1 intron.

    DIRECTIONS (items 10–15): Match the following lettered descriptions to the numbered arrows with the diagram of a nuclear precursor mRNA. Each letter can be used only once.

    a. Splice donor site.
    b. Poly(A) recognition sequence location.
    c. Cap site.
    d. Splice acceptor site.
    e. Intron.
    f. Exon.

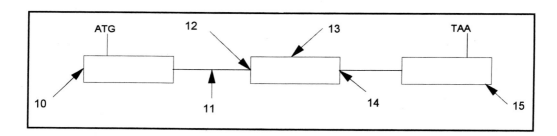

Answers

1. b	**6.** c	**11.** e
2. a	**7.** d	**12.** d
3. d	**8.** a	**13.** f
4. b	**9.** e	**14.** a
5. e	**10.** c	**15.** b

Protein Synthesis

31

Objectives

- To describe the genetic code.
- To describe amino acid activation.
- To describe the events of translational initiation.
- To describe the events of translational elongation.
- To describe the events of translational termination.
- To describe the major mechanisms that regulate translation.
- To define the major protein synthesis inhibitors.
- To describe the processes of posttranslational modification of proteins.

Concepts

Translation is the RNA-directed synthesis of polypeptides. This process requires all 3 classes of RNA. Although the chemistry of peptide bond formation is relatively simple, the processes leading to the ability to form a peptide bond are exceedingly complex. The "template" for the correct addition of individual amino acids is the mRNA, yet both tRNAs and rRNAs are involved in the process. The tRNAs carry activated amino acids into the ribosome which is composed of rRNA and ribosomal proteins. The ribosome is associated with the mRNA, ensuring correct access of activated tRNAs and containing the necessary enzymatic activities to catalyze peptide bond formation.

THE GENETIC CODE

Early genetic experiments demonstrated the existence of colinearity between the sequences of genes and those of the encoded protein—ie, mutations in a gene corresponded to the same changes observable in the protein. The **genetic code** is thus the sequences of a gene that direct the incorporation of amino acids into proteins. Evidence from a number of genetic mutations has demonstrated that:

- The genetic code is read in a sequential manner starting near the 5′ end of the mRNA. This means that translation proceeds along the mRNA in the 5′ → 3′ direction, which corresponds to the N-terminal to C-terminal direction of the amino acid sequences within proteins.
- The code is composed of a triplet of nucleotides.
- All 64 possible combinations of the 4 nucleotides code for amino acids. In other words, since there are only 20 amino acids, the code is **degenerate**.
- The precise dictionary of the genetic code was determined with the use of in-vitro translation systems and polyribonucleotides. The results of these experiments confirmed that some amino acids are encoded by more than 1 triplet codon; hence the degeneracy of the genetic code. These experiments also established the identity of translational termination codons (Table 31–1).

AMINO ACID ACTIVATION

During the course of in-vitro protein synthesis and labeling experiments it was shown that the amino acids became transiently bound to a low-molecular-weight mass fraction of RNA. This fraction has been termed the transfer RNAs **(tRNAs)**, since they transfer amino acids to the elongating polypeptide. These results indicate that accurate translation requires 2 equally important steps of recognition:

Table 31–1. The genetic code (codon assignments in messenger RNA).[1]

First Nucleotide	Second Nucleotide				Third Nucleotide
	U	C	A	G	
U	Phe	Ser	Tyr	Cys	U
	Phe	Ser	Tyr	Cys	C
	Leu	Ser	Term	Term[2]	A
	Leu	Ser	Term	Trp	G
C	Leu	Pro	His	Arg	U
	Leu	Pro	His	Arg	C
	Leu	Pro	Gln	Arg	A
	Leu	Pro	Gln	Arg	G
A	Ile	Thr	Asn	Ser	U
	Ile	Thr	Asn	Ser	C
	Ile[2]	Thr	Lys	Arg[2]	A
	Met	Thr	Lys	Arg[2]	G
G	Val	Ala	Asp	Gly	U
	Val	Ala	Asp	Gly	C
	Val	Ala	Glu	Gly	A
	Val	Ala	Glu	Gly	G

[1]The terms first, second, and third nucleotide refer to the individual nucleotides of a triplet codon. U, uridine nucleotide; C, cytosine nucleotide; A, adenine nucleotide; G, guanine nucleotide; Met, chain initiator codon; Term, chain terminator codon. AUG, which codes for Met, serves as the initiator codon in mammalian cells. (Abbreviations of amino acids are explained in Chapter 4.)

[2]In mammalian mitochondria, AUA codes for Met and UGA for Trp, and AGA and AGG serve as chain terminators.

1. The correct choice of amino acid must be made for attachment to the correspondingly correct tRNA.
2. The correct amino acid-charged tRNA must be selected by the mRNA. This second process is facilitated by the ribosomes.

More than 300 different tRNAs have been sequenced, either directly or from their corresponding DNA sequences. The tRNAs vary in length from 60 to 95 nucleotides (18–28 kDa). The majority contain 76 nucleotides. As illustrated in Figure 31–1, all tRNAs:

- Exhibit a cloverleaf-like secondary structure.
- Have a 5'-terminal phosphate.
- Have a 7-bp stem that includes the 5'-terminal nucleotide and may contain non-Watson-Crick base pairs (eg, GU). This portion of the tRNA is called the **acceptor**, since the amino acid is carried by the tRNA while attached to the 3'-terminal OH group.
- Have a D loop and a T + C loop.
- Have an anticodon loop.
- Terminate at the 3'-end with the sequence 5'-CCA-3'.
- Contain 13 invariant positions and 8 semi-variant positions.
- Contain numerous modified nucleotide bases.

The role of tRNAs in translation is to "carry" activated amino acids to the elongating polypeptide chain. The activation of amino acids is carried out by a 2-step process catalyzed by **aminoacyl-tRNA synthetases**. Each tRNA, and the amino acid it carries, are recognized by individual aminoacyl-tRNA synthetases. There are at least 21 aminoacyl-tRNA synthetases, since the initiator met-tRNA of both prokaryotes and eukaryotes is distinct from the normal met-tRNA.

The activation of amino acids requires energy in the form of ATP and occurs in a 2-step reaction catalyzed by the aminoacyl-tRNA synthetases (Figure 31–2). First the enzyme attaches the amino acid to the γ-phosphate of ATP, with the concomitant release of pyrophosphate. This is termed an **aminoacyl-adenylate** intermediate. Then the enzyme catalyzes transfer of the amino acid to either the 2'- or

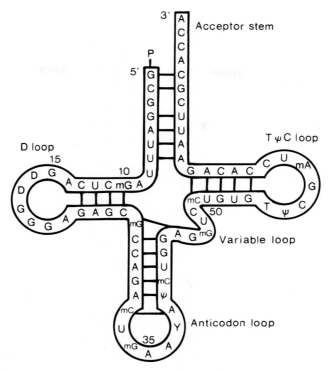

Figure 31–1. Two-dimensional structure of a typical tRNA molecule. Transfer RNAs contain numerous modified nucleotides. D, dihydrouridine; mG, either N⁷-methylguanosine or N²,N²-dimethylguanosine; mA, 1-methyladenosine, mC, 3-methylcytidine; Ψ, pseudouridine (5-azauracil).

3′-OH of the ribose portion of the 3′-terminal A residue of the tRNA, generating the activated aminoacyl-tRNA. Although these reactions are freely reversible, the forward reaction is favored by the coupled hydrolysis of PP_i.

Accurate recognition of the correct amino acid as well as the correct tRNA is different for each aminoacyl-tRNA synthetase. Since the different amino acids have different R-groups, the enzyme for each amino acid has a different binding pocket for its specific amino acid. It is not the anticodon that determines the tRNA utilized by the synthetases. Although the exact mechanism is not known for all synthetases, it is likely to be a combination of the presence of specific modified bases and the secondary structure of the tRNA that is correctly recognized by the synthetases.

It is absolutely necessary that the discrimination of correct amino acid and correct tRNA be made by a given synthetase prior to release of the aminoacyl-tRNA from the enzyme. Once the product is

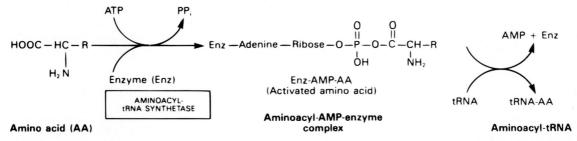

Figure 31–2. Formation of aminoacyl-tRNA. A 2-step reaction, involving the enzyme animoacyl-tRNA synthetase, results in the formation of aminoacyl-tRNA. The first reaction involves the formation of an AMP-amino acid-enzyme complex. This activated amino acid is next transferred to the corresponding tRNA molecule. The AMP and enzyme are released, and the latter can be reutilized. (Reproduced, with permission, from Murray, RK: *Harper's Biochemistry*, 23rd ed. Appleton & Lange, 1993.)

released there is no further way to "proofread" whether a given tRNA is coupled to its corresponding tRNA. Erroneous coupling would lead to the wrong amino acid being incorporated into the polypeptide, since the discrimination of amino acids during protein synthesis comes from the recognition of the anticodon of a tRNA by the codon of the mRNA and not by recognition of the amino acid itself. This was demonstrated by reductive desulfuration of cys-tRNA^{cys} with Raney nickel generating ala-tRNA^{cys}. Alanine was then incorporated into an elongating polypeptide where cysteine should have been.

As discussed above, 3 of the possible 64 triplet codons are recognized as translational termination codons. The remaining 61 codons might be considered as being recognized by individual tRNAs. Most cells contain **isoaccepting tRNAs**—different tRNAs that are specific for the same amino acid—however, many tRNAs bind to 2 or 3 codons specifying their cognate amino acids. As an example, yeast tRNA^{phe} has the anticodon 5'-GmAA-3' ("Gm" indicates a methylated G) and can recognize the codons 5'-UUC-3' and 5'-UUU-3'. It is possible, therefore, for non-Watson-Crick base-pairing to occur at the third codon position, ie, the 3' nucleotide of the mRNA codon and the 5' nucleotide of the tRNA anticodon. This phenomenon is described by the **wobble hypothesis**.

Initiation

Initiation of translation in both prokaryotes and eukaryotes requires a specific initiator tRNA, **tRNA$_i^{met}$**, that is used to incorporate the initial methionine residue into all proteins. In *E. coli* a specific version of tRNA$_i^{met}$ is required to initiate translation, tRNA$_i^{fmet}$. The methionine attached to this initiator tRNA is formylated. Formylation requires N^{10}-formyl-THF and is carried out after the methionine is attached to the tRNA. The fmet-tRNA$_i^{fmet}$ still recognizes the same codon, AUG, as regular tRNA$_i^{met}$. Although tRNA$_i^{met}$ is specific for initiation in eukaryotes it is not a tRNA$_i^{fmet}$.

The initiation of translation requires recognition of an AUG codon. In the polycistronic prokaryotic RNAs, this AUG codon is located adjacent to a **Shine-Delgarno** element in the mRNA. The Shine-Delgarno element is recognized by complementary sequences in the small subunit rRNA (16S in *E. coli*). In eukaryotes, initiator AUGs are generally (although not always) the first encountered by the ribosome. A specific sequence context surrounding the initiator AUG aids ribosomal discrimination. This context is A/GCCA/GCCAUGG/G in most mRNAs.

The specific non-ribosomally associated proteins required for accurate translational initiation are termed **initiation factors**. In *E. coli* they are **IFs**; in eukaryotes they are **eIFs**. Numerous eIFs have been identified (Table 31–2).

As shown in Figure 31–3, the initiation of translation requires 4 specific steps:

1. A ribosome must dissociate into its 40S and 60S subunits.
2. A ternary complex termed the **preinitiation complex** is formed, consisting of the initiator tRNA$_i^{met}$, GTP, eIF-2, and the 40S subunit.

Table 31–2. Eukaryotic initiation factors and their function in translation initiation.

Initiation Factor	Subunit mass (kDa)	Activity
eIF-1	15	Repositioning of met-tRNA to facilitate mRNA binding
eIF-2	35, 50, 55	Ternary complex formation
eIF-2B	34, 40, 55, 65, 82	eIF-2 recycling
eIF-3	~10 subunits, 28–160	Ribosome subunit antiassociation, binding to 40S subunit
eIF-4A	50	mRNA binding to 40S subunit, ATPase
eIF-4B	80	mRNA binding to 40S subunit
eIF-4C	17	Ribosome subunit antiassociation, 60S subunit joining
eIF-4E (24kDa CBP[a]; CBP I)	24	5' cap recognition (is a subunit of eIF-4F)
eIF-4F (CBP II, CBP complex)	24, 50, 220	mRNA binding to 40S subunit, ATPase, mRNA melting activity
eIF-5	150	Release of eIF-2 and eIF-3, ribosome-dependent GTPase
eIF-6	25.5	Ribosome subunit antiassociation

[a]CBP = Cap binding protein

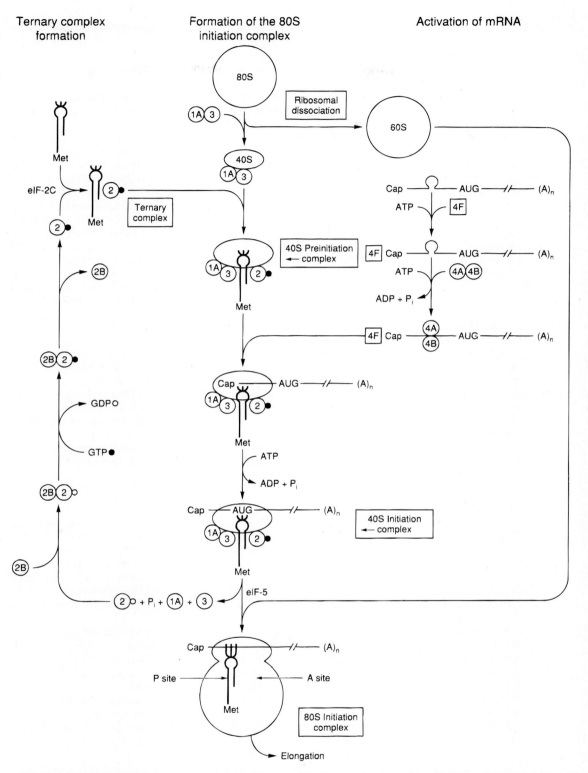

Figure 31–3. Diagrammatic representation of the initiation of protein synthesis on the mRNA template containing a 5' cap (G^mTP-5') and 3' poly(A) terminus [3' (A)$_n$]. This process proceeds in 3 steps: (1) activation of mRNA; (2) formation of the ternary complex consisting of tRNAmet$_i$, initiation factor eIF-2, and GTP; and (3) formation of the active 80S initiation complex. (See text for details.) GTP, ·; GDP, ○. The various intiation factors appear in abbreviated form as circles or squares, eg, eIF-3 (③), eIF-4 (④F). (Reproduced, with permission, from Murray, RK: *Harper's Biochemistry*, 23rd ed. Appleton & Lange, 1993.)

3. The mRNA is bound to the preinitiation complex.

4. The 60S subunit associates with the preinitiation complex to form the 80S **initiation complex**.

The initiation factors eIF-1 and eIF-3 bind to the 40S ribosomal subunit favoring antiassociation to the 60S subunit. The prevention of subunit reassociation allows the preinitiation complex to form (Figure 31–3).

The first step in the formation of the preinitiation complex is the binding of GTP to eIF-2 to form a binary complex. eIF-2 is composed of 3 subunits, α, β, and γ. The binary complex then binds to the activated initiator tRNA, tRNA$_i^{met}$, forming a ternary complex that then binds to the 40S subunit forming the preinitiation complex. The preinitiation complex is stabilized by the earlier association of eIF-3 and eIF-1 to the 40S subunit.

The cap structure of eukaryotic mRNAs is bound by specific eIFs prior to association with the preinitiation complex. Cap binding is accomplished by the initiation factor **eIF-4F** (Figure 31–3). This factor is actually a complex of 3 proteins. The protein **eIF-4E** is a 24-kDa protein which physically recognizes and binds to the cap structure; **eIF-4A** is a 46-kDa protein which binds and hydrolyzes ATP and exhibits RNA helicase activity. Unwinding of mRNA secondary structure is necessary to allow the ribosomal subunits access. Once the eIF-4A subunit is bound, the initiation factor eIF-4B binds. It is the combination of eIF-4A and eIF-4B that is believed to unwind mRNA secondary structure. An additional component of eIF-4F is a 220-kDa protein that may be involved in stabilizing eIF-4A and eIF-4E.

Once the mRNA is properly aligned onto the preinitiation complex and the initiator met-tRNA is bound to the initiator AUG codon (a process facilitated by eIF-1), the 60S subunit associates with the complex. The association of the 60S subunit requires the activity of eIF-5 which has first bound to the preinitiation complex. The energy needed to stimulate the formation of the 80S initiation complex comes from the hydrolysis of the GTP bound to eIF-2. The GDP-bound form of eIF-2 then binds to eIF-2B, which stimulates the exchange of GTP for GDP on eIF-2. When GTP is exchanged, eIF-2B dissociates from eIF-2. This is termed the **eIF-2 cycle** (Figure 31-3). This cycle is absolutely required in order for eukaryotic translational initiation to occur. The GTP exchange reaction can be affected by phosphorylation of the α-subunit of eIF-2 (see below).

At this stage the initiator met-tRNA is bound to the mRNA within a site of the ribosome termed the **P- (for "peptide") site**. The other site within the ribosome to which incoming charged tRNAs bind is termed the **A-site**, (for "amino acid site").

Elongation

The process of elongation, like that of initiation, requires specific non-ribosomal proteins. In *E. coli* these are **EFs**, and in eukaryotes **eEFs**. The elongation of polypeptides occurs in a cyclic manner: at the end of one complete round of amino acid addition, the A site is empty and ready to accept the incoming aminoacyl-tRNA dictated by the next codon of the mRNA (Figure 31–4). This means that not only must the incoming amino acid be attached to the peptide chain but the ribosome must move down the mRNA to the next codon. Each incoming aminoacyl-tRNA is brought to the ribosome by an **eEF-1α-GTP** complex. When the correct tRNA is deposited into the A site, the GTP is hydrolyzed and the eEF-1α-GDP complex dissociates.

The peptide attached to the tRNA in the P site is transferred to the amino group at the aminoacyl-tRNA in the A site. This reaction is catalyzed by **peptidyltransferase**. This process is termed **transpeptidation**. The elongated peptide now resides on a tRNA in the A site. The A site needs to be freed in order to accept the next aminoacyl-tRNA. The process of moving the peptidyl-tRNA from the A site to the P site is termed **translocation**. Translocation is catalyzed by **eEF-2** coupled to GTP hydrolysis. In the process of translocation, the ribosome is moved along the mRNA in such a way that the next codon of the mRNA resides under the A site. The cycle can now begin again.

Following translocation, eEF-2-GDP is released from the ribosome. In order for additional translocation events to take place, the GDP must be exchanged for GTP. This is carried out by **eEF-1βγ** similarly to the GTP exchange that occurs with eIF-2 catalyzed by eIF-2B.

Termination

Like initiation and elongation, translational termination requires specific protein factors that have been identified as **releasing factors, RFs**, in *E. coli* and **eRFs** in eukaryotes. There are 2 RFs in *E. coli* and 1 in eukaryotes; The signals for termination are the same in each case and consist of termination codons present in the mRNA. There are 3 such codons: UAG, UAA and UGA.

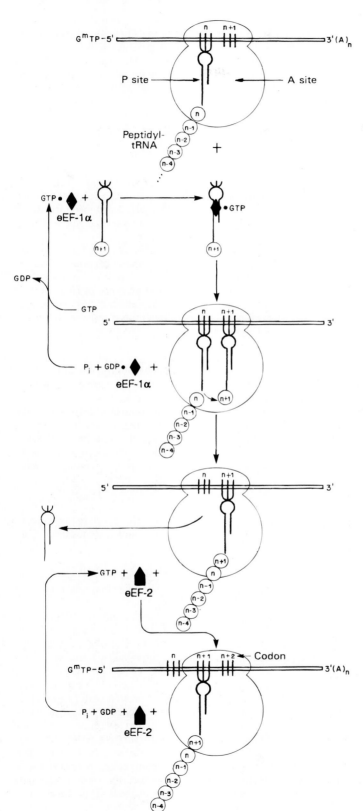

Figure 31–4. Diagrammatic representation of the peptide elongation process of protein synthesis. The small circles labeled n − 1, n, n + 1, etc, represent the amino acid residues of the newly formed protein molecule. eEF-1α and eEF-2 represent elongation factors 1 and 2, respectively. The peptidyl-tRNA and aminoacyl-tRNA sites on the ribosome are represented by P site and A site, respectively. (Reproduced, with permission, from Murray, RK: *Harper's Biochemistry*, 23rd ed. Appleton & Lange, 1993.)

In *E. coli* the termination codons UAA and UAG are recognized by RF-1, whereas RF-2 recognizes the termination codons UAA and UGA. The eRF binds to the A site of the ribosome in conjunction with GTP. The binding of eRF to the ribosome stimulates the peptidyltransferase activity to transfer the peptidyl group to water instead of an aminoacyl-tRNA (Figure 31–5). The resulting uncharged tRNA left in the P site is expelled with concomitant hydrolysis of GTP. The inactive ribosome then releases its mRNA and the 80S complex dissociates into the 40S and 60S subunits, ready for another round of translation.

REGULATION OF PROTEIN SYNTHESIS

Heme Control of Translation

Initiation in eukaryotes is regulated by phosphorylation of a ser residue in the α subunit of eIF-2 (Figure 31–6). When eIF-2 is phosphorylated, the GDP-bound complex is stabilized and exchange for GTP is inhibited. The exchange of GDP for GTP is mediated by eIF-2B. When eIF-2 is phosphorylated it binds eIF-2B more tightly, thus slowing the rate of exchange. It is this inhibited exchange that affects the rate of initiation.

Phosphorylated eIF-2 in the absence of eIF-2B is just as active an initiator as non-phosphorylated eIF-2. The phosphorylation of eIF-2 is the result of an enzyme called **heme-controlled inhibitor (HCI)**. HCI is generated in the absence of heme, a mitochondrial product. The removal of phosphate is catalyzed by a specific eIF-2 phosphatase that is unaffected by heme. The presence of HCI was first seen in an in-vitro translation system derived from lysates of reticulocytes. Reticulocytes tend to synthesize hemoglobin almost exclusively, and at an extremely high rate. In an intact reticulocyte, eIF-2 is protected from phosphorylation by a specific 67-kDa protein.

Interferon Control of Translation

Regulation of translation can also be induced in virally infected cells. This capability may have its origin in the cell's defense system: it would benefit a virally infected cell to turn off protein synthesis to prevent propagation of the viruses. The regulation is accomplished by the induced synthesis of **interferons (IFNs)**. There are 3 classes of IFs: the **leukocyte** or **α-IFNs**, the **fibroblast** or **β-IFNs**, and the **lymphocyte** or **γ-IFNs**. The IFs are induced by dsRNAs and themselves induce a specific kinase that phosphorylates eIF-2, thereby shutting of translation (Figure 31–7). Additionally, IFNs induce the synthesis of **2′5′-oligoadenylate, pppA(2′p5′A)$_n$**, which activates a pre-existing ribonuclease, RNase L. In turn, RNase L degrades all classes of mRNAs and thereby shuts off translation by another means) (Figure 31–7).

Protein Synthesis Inhibitors

Many of the antibiotics utilized for the treatment of bacterial infections as well as certain toxins function through the inhibition of translation. Inhibition can be effected at all stages of translation, from initiation to elongation to termination (Table 31–3).

Protein Modification

The post-translational modification of proteins is generally of 2 types: proteolytic cleavage/processing or covalent modification.

Most proteins undergo proteolytic cleavage. The simplest form of this is the removal of the initiation methionine. Many proteins are synthesized as inactive precursors that are activated under proper physiological conditions by limited proteolysis; examples include pancreatic digestive enzymes and enzymes involved in blood clotting. Inactive precursor proteins that are activated by removal of peptides are termed **proproteins**.

Proteins that are membrane-bound or are destined for excretion are synthesized by ribosomes associated with the membranes of the endoplasmic reticulum (ER). The ER associated with ribosomes is termed rough ER (RER). This class of proteins all contain an N-terminus termed a **signal sequence** or **signal peptide**. The signal peptide is usually composed of 13–36 predominantly hydrophobic residues and is recognized by a multi-protein complex termed the **signal recognition particle (SRP)**. This signal peptide is removed after passage through the endoplasmic reticulum membrane, in a step catalyzed by **signal peptidase** (Figure 31–8). Proteins that contain a signal peptide are called **preproteins**, to distinguish them from **proproteins**. However, some proteins that are destined for secretion are also further proteolyzed following secretion, and therefore they contain 'pro' sequences. This class of pro-

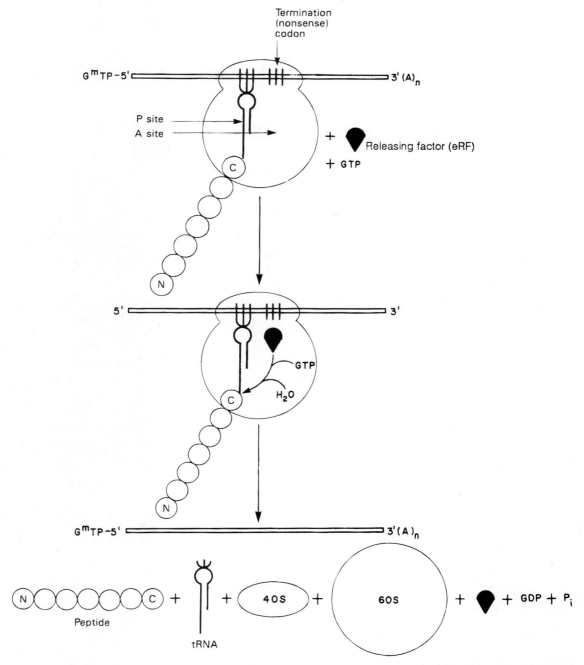

Figure 31–5. Diagrammatic representation of the termination process of protein synthesis. The peptidyl-tRNA and aminoacyl-tRNA sites are indicated as P site and A site, respectively. The hydrolysis of the peptidyl-tRNA complex is shown by the entry of H_2O. N and C indicate the amino- and carboxy-terminal amino acids, respectively, and illustrates the polarity of protein synthesis. (Reproduced, with permission, from Murray, RK: *Harper's Biochemistry*, 23rd ed. Appleton & Lange, 1993.)

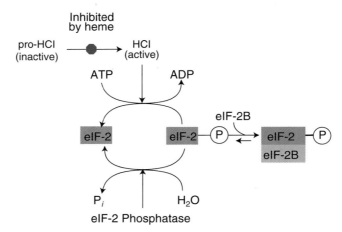

Figure 31–6. Model for the control of translation in reticulocytes by the heme-controlled repressor. (Reproduced, with permission from Voet, D. and Voet, JG: *Biochemistry*, 1st ed. John Wiley and Sons, Inc. 1990.)

teins is termed **preproproteins**. A complex example of posttranslational processing of a prepoprotein is the cleavage of **prepro-opiomelanocortin** synthesized in the pituitary. This preproprotein undergoes complex cleavage pathways that differ depending upon the cellular location of its synthesis. Another example of a preproprotein is insulin, which is secreted from the pancreas and therefore has a prepeptide. Following cleavage of the 24–amino acid signal peptide, the protein folds into proinsulin. Proinsulin is further cleaved, yielding active insulin.

The covalent modification of proteins may be of many types. Chemical modifications can occur at the N- and C-termini of proteins as well as at the functional groups of individual amino acid side chains. Over 150 different types of side-chain modifications are known, involving all amino acids except A, G, I, L, M, and V. These modifications include acetylations, hydroxylations, methylations, nucleotidylations, phosphorylations, ADP-ribosylations and glycosylations. Glycosylations occur in almost infinite variety and are necessary for altering protein structure, targeting intracellular or membrane location, and interactions with extracellular constituents as well as for cell-cell communications. Glycosylation is the most common form of protein modification (see Chapter 19 for the processes of carbohydrate modification of proteins).

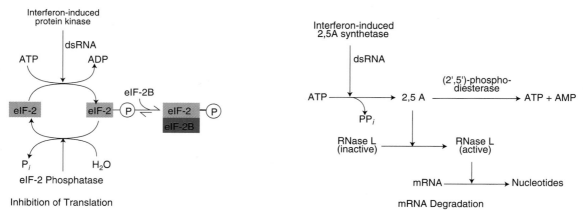

Figure 31–7. Model for interferon-regulated translation. The viral dsRNA causes *(a)* the inhibition of translation and *(b)* the degradation of mRNA. (Reproduced, with permission from Voet, D. and Voet, JG: *Biochemistry*, 1st ed. John Wiley and Sons, Inc. 1990.)

Table 31–3. Several antibiotic and toxin inhibitors of translation.

Inhibitor	Comments
Chloramphenicol	Inhibits prokaryotic peptidyl transferase
Streptomycin	Inhibits prokaryotic peptide chain initiation, also induces mRNA misreading
Tetracycline	Inhibits prokaryotic aminoacyl-tRNA binding to the ribosome small subunit
Neomycin	Similar in activity to streptomycin
Erythromycin	Inhibits prokaryotic translocation through the ribosome large subunit
Fusidic acid	Similar to erythromycin only by preventing EF-G from dissociating from the large subunit
Puromycin	Resembles an aminoacyl-tRNA, interferes with peptide transfer resulting in premature termination in both prokaryotes and eukaryotes
Diptheria toxin	Catalyzes ADP-ribosylation of and inactivation of eEF-2
Ricin	Found in castor beans, catalyzes cleavage of the eukaryotic large subunit rRNA
Cycloheximide	Inhibits eukaryotic peptidyl transferase

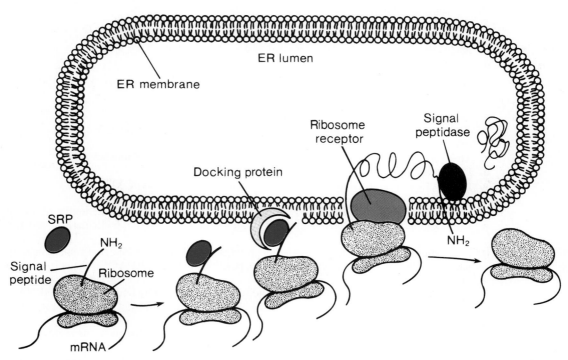

Figure 31–8. Model for the translation of secreted and membrane bound proteins. The signal sequence at the N-terminus of the newly synthesizing protein is recognized by the signal recognition particle (SRP). The complex of SRP-signal sequence binds to the ER membrane bound docking protein. The newly synthesized protein is extruded through the membrane into the lumen of the ER where the signal sequence is removed by signal peptidase.

Questions

DIRECTIONS (items 1–7): Each question or incomplete statement is followed by answers or by completions of the statement. Select the ONE response that is BEST in each case.

1. Translation in eukaryotes differs from that in prokaryotes with respect to:
   a. The use of UGG as a termination codon instead of UAG.
   b. The use of AUA as the initial codon for methionine.
   c. The presence of many more releasing factors at termination.
   d. The larger size of ribosomes found in the cytoplasm.
   e. The fact that eukaryotic translation utilizes the same proteins for initiation and elongation.

2. Proteins that are destined for secretion travel from the ER to the Golgi complex. Then:
   a. Their subsequent processing includes packaging in vesicles that move to the phospholipid bilayer and fuse with it, releasing the proteins to the exterior.
   b. An invariable intermediate step in this preparation for secretion is ubiquitinylation of lysines in the peptide sequence.
   c. After completion of translation, such proteins first move to the cytoplasm.
   d. In order to complete the processing necessary for secretion, this class of proteins is synthesized on the smooth ER.
   e. After completion of translation the proteins are carried to the plasma membrane by the signal recognition particle protein coupled to the docking protein.

3. Heme controlled inhibitor is activated:
   a. When heme is in relatively high concentration.
   b. And is then able to phosphorylate one of the subunits of eIF-2.
   c. And is then able to dephosphorylate one of the subunits of eIF-2.
   d. And is then able to phosphorylate one of the subunits of eEF-2.
   e. And is then able to ADP-ribosylate one of the subunits of eRF.

4. Eukaryotic translation differs from that of prokaryotes in several ways:
   a. The initiating codon frequently differs from AUG.
   b. There are many more releasing factors, a number of which dissociate into subunits that participate in control of the process.
   c. There are many more initiating factors, several of which dissociate into subunits.
   d. Prokaryotic translation requires an RNA cap binding factor.
   e. Eukaryotic translation requires GTP to charge the tRNAs instead of ATP.

5. Placement of mRNA on the small ribosomal subunit during initiation of a functioning ribosome:
   a. Is random because the tRNA^{met} invariably aligns with the first AUG, particularly in prokaryotes.
   b. Is determined by the fact that cap-binding proteins and mRNA caps interact in eukaryotes, apparently followed by a scanning process that positions the fmet-tRNA^{fmet} at the correct initiation AUG.
   c. Depends upon the presence of all 20 amino acids, each covalently bound to its own tRNA.
   d. Depends upon the interaction of cap-binding factor with met-tRNA^{met} after it first recognizes the cap structure of the mRNA.
   e. Requires the interaction of eIF-2C with the met-tRNA^{met} and subsequent interaction of this complex with the small subunit prior to engagement with the mRNA.

6. Which of the following best describes the reason for the virulence of *Corynebacterium diphtheriae?*
   a. A subunit of a membrane-binding toxic protein of *C. diphtheriae* origin acts as an enzyme in cytoplasm of human cells and ADP-ribosylates a specific amino-acid residue in eEF-2.
   b. An analogue of adenosine to which methyltyrosine is covalently bound is secreted by *C. diphtheriae*. This binds to the A site and accepts the amino acid or peptide from the tRNA on the P site, in a reaction catalyzed by peptidyl transferase.
   c. An amino acid analog of bacterial origin that contains a peptide bond binds to the A site and inhibits peptidyl transferase.

    d. An analogue of adenosine to which methyltyrosine is covalently bound is secreted by *C. diphtheriae*. This binds to the P site prior to translocation, effectively terminating further translation.

    e. A subunit of a membrane-binding toxic protein of *C. diphtheriae* origin acts as an enzyme in the cytoplasm of human cells and ADP-ribosylates a specific amino-acid residue in eIF-2.

7. Proteins that are destined to be secreted from the cell:
    a. Are synthesized on the rough endoplasmic reticulum.
    b. Are synthesized on the smooth endoplasmic reticulum.
    c. Are synthesized on the Golgi apparatus.
    d. Contain a leader region rich in acidic amino acids.
    e. Contain a leader region rich in basic amino acids.

DIRECTIONS (items 8–12): Match each of the following lettered key phrases with the numbered description that BEST fits. Each key phrase is used once:

    a. eEF-2 and GTP.
    b. aminoacyl-tRNA and GTP.
    c. eRF and GTP.
    d. Peptidyl transferase.
    e. P site.

8. Identified by a met-tRNA^{met} on an initiated ribosome.

9. External protein that alters the reaction usually catalyzed by peptidyl transferase.

10. Unable to function without eEF-1α.

11. Necessary for translocation.

12. Of ribosomal origin.

DIRECTIONS (items 13–15): Each numbered item or incomplete statement is followed by answers or by completions of the statement. Select the ONE response that is BEST in each case.

13. Aminoacyl-tRNAs control the amino acids incorporated into peptides primarily by:
    a. Their interaction with the aminoacyl-tRNA synthetases.
    b. The nature of the bond between the amino acid and the acceptor end of the tRNA.
    c. The interaction between their anticodons and the codons in mRNA.
    d. Correctly interacting with the peptidyltransferase of the ribosomes.
    e. Their ability to discriminate the correct amino acid covalently attached to their 3′-adenosine residue during the interaction of the anticodon and codon sequences.

14. Puromycin inhibits translation by:
    a. Resembling an aminoacyl-tRNA and interfering with peptide transfer, which results in premature termination in both prokaryotes and eukaryotes.
    b. Catalyzing ADP-ribosylation of and inactivation of eEF-2.
    c. Catalyzing cleavage of the eukaryotic large subunit rRNA.
    d. Inhibiting prokaryotic aminoacyl-tRNA binding to the ribosome small subunit.
    e. Inhibiting prokaryotic peptidyl transferase.

15. Ricin inhibits translation by:
    a. Resembling an aminoacyl-tRNA and interfering with peptide transfer, which results in premature termination in both prokaryotes and eukaryotes.
    b. Catalyzing ADP-ribosylation of and inactivation of eEF-2.
    c. Catalyzing cleavage of the eukaryotic large subunit rRNA.
    d. Inhibiting prokaryotic aminoacyl-tRNA binding to the ribosome small subunit.
    e. Inhibiting prokaryotic peptidyl transferase.

Answers

1. d	**6.** a	**11.** a
2. a	**7.** a	**12.** d
3. b	**8.** e	**13.** c
4. c	**9.** c	**14.** a
5. e	**10.** b	**15.** c

32

Control of Gene Expression

Objectives

- To describe the major mechanism for control of gene expression n prokaryotes.
- To describe the control of the *lac* operon.
- To describe the control of the *trp* operon.
- To describe the mechanisms available to control gene expression in eukaryotes.
- To describe the mechanisms controlling eukaryotic transcriptional initiation.
- To describe the major structural motifs, and their functions, in eukaryotic transcription regulating DNA-binding proteins.
- To describe the control of eukaryotic RNA polymerase III transcription.

Concepts

The controls that act on gene expression (ie, the ability of a gene to produce a biologically active protein) are much more complex in eukaryotes than in prokaryotes. A major difference is the presence in eukaryotes of a nuclear membrane, which prevents the simultaneous transcription and translation that occurs in prokaryotes. Whereas, in prokaryotes, control of transcriptional initiation is the major point of regulation, in eukaryotes the regulation of gene expression is controlled nearly equivalently from many different points.

Gene Control in Prokaryotes

In bacteria, genes are clustered into **operons**: gene clusters that encode the proteins necessary to perform coordinated function, such as biosynthesis of a given amino acid. The RNA that is transcribed from an operon is **polycistronic**—ie, multiple proteins are encoded in a single transcript.

In bacteria, control of the rate of transcriptional initiation is the predominant site for control of gene expression. As with the majority of prokaryotic genes, initiation is controlled by two DNA sequences that are approximately 35 bases and 10 bases, respectively, to the 5′ side of the site of transcriptional initiation. These are termed **promoter** sequences, because they "promote" recognition of transcriptional start sites by RNA polymerase. The consensus sequence for the −35 position is TTGACA, and for the −10 position, TATAAT. (The −10 position is also known as the "Pribnow-box.") These promoter sequences are recognized and contacted by RNA polymerase.

The activity of RNA polymerase at a given promoter is in turn regulated by interaction with accessory proteins, which affect its ability to recognize start sites. These regulatory proteins can act both positively (activators) and negatively (repressors). The accessibility of promoter regions of prokaryotic DNA is in many cases regulated by the interaction of proteins with sequences termed **operators**. The operator region is adjacent to the promoter elements in most operons and binds activators or repressors.

Two major modes of transcriptional regulation operate in bacteria, *E. coli*. Both involve repressor proteins. One mode of regulation is exerted upon operons that produce gene products necessary for the utilization of energy; these are **catabolite**-regulated operons. The other mode regulates operons that produce gene products necessary for the synthesis of small biomolecules, eg, amino acids. Expression from the latter operons is **attenuated** by sequences within the RNA. A classic example of a catabolite-regulated operon is the *lac* **operon**, responsible for obtaining energy from β-galactosides such as lactose. A classic example of an attenuated operon is the *trp* **operon**, responsible for the biosynthesis of tryptophan.

A. The *lac* Operon: The *lac* operon consists of one regulatory gene (the *i* gene) and three structural genes (*z*, *y*, and *a*). The *i* gene codes for the repressor of the *lac* operon. The *z* gene codes for β-galactosidase (β-gal), which is primarily responsible for conversion of lactose to galactose and glucose. The *y* gene codes for permease, which increases permeability of the cell to β-galactosides. The *a* gene codes for transacetylase.

During normal growth on a glucose-based medium, the *lac* repressor is bound to the operator region of the *lac* operon, preventing transcription (Figure 32–1). However, in the presence of an inducer of the *lac* operon, the repressor protein binds the inducer and is therefore incapable of interacting with the operator region of the operon. RNA polymerase is thus able to bind at the promoter region, and transcription of the operon ensues (Figure 32–1).

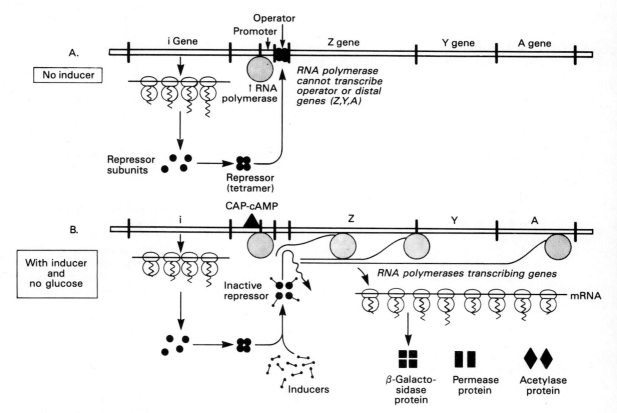

Figure 32–1. The mechanism of repression and derepression of the lactose operon. When no inducer is present **(A)**, the I gene products that are synthesized constitutively form a repressor molecule which binds at the operator locus to prevent the binding of RNA polymerase at the promoter locus and thus to prevent the subsequent transcription of the Z, Y, and A structural genes. When inducer is present **(B)**, the constitutively expressed I gene forms repressor molecules that are inactivated by the inducer and cannot bind to the operator locus. In the presence of cAMP and its binding protein (CAP), the RNA polymerase can transcribe the structural genes Z, Y, and A, and the polycistronic mRNA molecule formed can be translated into the corresponding protein molecules β-galactosidase, permease, and acetylase, allowing for the catabolism of lactose. (Reproduced, with permission, from Murray, RK: *Harper's Biochemistry*, 23rd ed. Appleton & Lange, 1993.)

The *lac* operon is repressed, even in the presence of lactose, if glucose is also present. This repression is maintained until the glucose supply is exhausted. The repression of *lac* under these conditions is termed **catabolite repression** (Figure 30–1) and is a result of the low levels of cAMP that result from an adequate glucose supply. The repression of *lac* is relieved in the presence of glucose if excess cAMP is added. As the glucose levels in the medium fall, the level of cAMP increases. Simultaneously there is an increase in inducer binding to the *lac* repressor. The net result is an increase in transcription from the *lac* operon (Figure 32–1).

The ability of cAMP to activate expression from the *lac* operon results from an interaction of cAMP with a protein termed **CAP** (for catabolite activator protein, also termed CRP for cAMP receptor protein). The cAMP-CAP complex binds to a region of the *lac* operon just upstream of the region bound by RNA polymerase. The binding of the cAMP-CAP complex to the *lac* operon stimulates RNA polymerase activity 20-to-50-fold.

B. The *trp* Operon: The *trp* operon encodes the genes for the synthesis of tryptophan. This cluster of genes, like the *lac* operon, is regulated by a repressor that binds to the operator sequences (Figure 32–2). The activity of the *trp* repressor for binding the operator region is enhanced when it binds tryptophan; in this capacity, tryptophan is known as a **corepressor**. Since the activity of the *trp* repressor is enhanced in the presence of tryptophan, the rate of expression of the *trp* operon is graded in response to the level of tryptophan in the cell.

Expression of the *trp* operon is also regulated by **attenuation**. The attenuator region is involved in controlling transcription from the operon after the RNA polymerase has initiated synthesis. The attenuator of polymerase consists of sequences in the leader region of the RNA (which is located prior to the start of the coding region for the *trpE* protein). The attenuator region contains codons for a small "leader" polypeptide, that contains tandem tryptophan codons. This region of the RNA is also capable of forming several different stable stem-loop structures (Figure 32–3).

Depending on the level of tryptophan in the cell—and hence the level of charged trp-tRNAs—the position of ribosomes on the leader polypeptide and the rate at which they are translating allows different stem-loops to form (Figure 32–3). If tryptophan is abundant, the ribosome prevents stem-loop 1-2 from forming and thereby favors stem-loop 3-4. The latter is found near a region rich in uracil and acts as the transcriptional terminator loop. Consequently, RNA polymerase is dislodged from the template (Figure 32–3). The operons coding for genes necessary for the synthesis of a number of other amino acids are also regulated by this attenuation mechanism. It should be clear, however, that this type of transcriptional regulation is not feasible for eukaryotic cells.

Gene Control in Eukaryotes

In eukaryotic cells, the ability to express biologically active proteins comes under regulation at several points.

A. Chromatin Structure: The physical structure of the DNA, as it exists compacted into chromatin (see Chapter 29), can affect the ability of transcriptional regulatory proteins (termed **transcription factors**) and RNA polymerases to find access to specific genes and to activate transcription from them. The presence of the histones and CpG methylation (see Chapter 29) most affect accessibility of the chromatin to RNA polymerases and transcription factors.

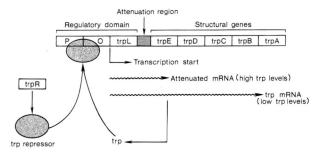

Figure 32–2. Transcription control exerted upon the *trp* operon in *E. coli*. When the operon is completely active trp will be synthesized. As the level of trp increases some binds to the repressor, thereby activating it which in turn leads to inhibition of transcription. The attenuator region controls the rate of RNA polymerase movement across the operon (see Figure 32–3).

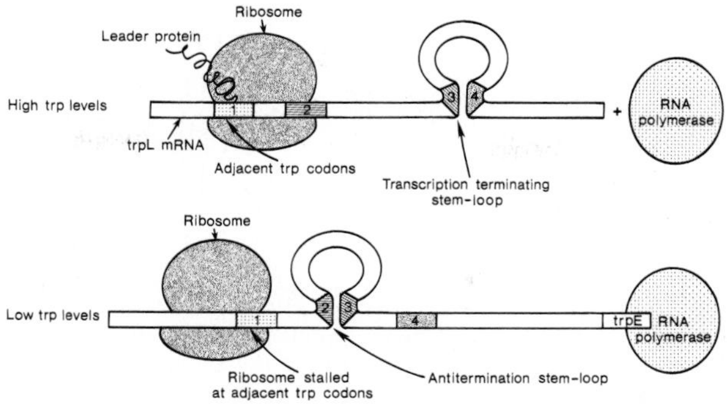

Figure 32–3. Two modes of transcription in the trp operon. Under conditions of low trp-tRNAs (low trp levels) the ribosomes stall in the leader peptide region which allows stem-loop 2-3 to form with the subsequent complete transcription of the operon. With abundant trp-tRNAs (high trp levels) the ribosome moves rapidly over the leader peptide allowing stem-loop 3-4 to form which functions as a transcriptional terminator structure.

B. Transcriptional Initiation: This is the most important mode for control of eukaryotic gene expression. Specific factors that exert control include the strength of promoters, the presence or absence of **enhancer sequences** (which enhance the activity of RNA polymerase at a given promoter by binding specific transcription factors), and the interaction between multiple activator proteins and inhibitor proteins.

C. Transcript Processing and Modification: Eukaryotic mRNAs must be capped and polyadenylated, and the introns must be accurately removed (Chapter 30). Several genes have been identified that undergo tissue-specific patterns of alternative splicing, which generate biologically different proteins from the same gene (Chapter 30).

D. RNA Transport: A fully processed mRNA must leave the nucleus in order to be translated into protein.

E. Transcript Stability: Unlike prokaryotic mRNAs, whose half-lives are all in the range of 1–5 minutes, eukaryotic mRNAs can vary greatly in their stability. Certain unstable transcripts have sequences (predominately, but not exclusively, in the 3′-non-translated regions) that are signals for rapid degradation.

F. Translational Initiation, Elongation and Termination: Since many mRNAs have multiple methionine codons, the ability of ribosomes to recognize and initiate synthesis from the correct AUG codon can affect the expression of a gene product. Several examples have emerged demonstrating that some eukaryotic proteins initiate at non-AUG codons. This phenomenon has been known to occur in *E. coli* for quite some time, but only recently has it been observed in eukaryotic mRNAs.

G. Post-Translational Modification: Common modifications include glycosylation, acetylation, fatty acylation, disulfide bond formations, etc.

H. Protein Transport: In order for proteins to be biologically active following translation and processing, they must be transported to their site of action.

I. Control of Protein Stability: Many proteins are rapidly degraded, whereas others are highly stable. Specific amino acid sequences in some proteins have been shown to bring about rapid degradation.

Control of Transcription Initiation

Transcription of the different classes of RNAs in eukaryotes is carried out by three different polymerases (see Chapter 30). RNA pol I synthesizes the rRNAs, except for the 5S species. RNA pol II synthesizes the mRNAs and some small nuclear RNAs involved in RNA splicing. RNA pol III synthesizes the 5S rRNA and the tRNAs. Almost all eukaryotic RNAs undergo post-transcriptional processing (see Chapter 30).

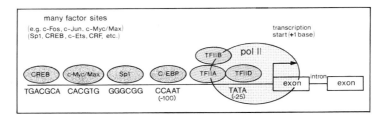

Figure 32–4. Structure of the upstream region of a typical eukaryotic mRNA gene that hypothetically contains 2 exons and a single intron. The diagram indicates the TATA-box and CCAAT-box basal elements and, upstream of them and of the transcriptional start site, several enhancer elements. (See Table 32–1 for a description of several transcription factors.) The location of the transcription factors is schematic here, and is not necessarily indicative of their relative location and order in all mRNA genes.

The most complex controls observed in eukaryotic genes are those that regulate the expression of RNA pol II-transcribed genes, the mRNA genes. Almost all eukaryotic mRNA genes contain basal promoters of two types and any number of different transcriptional regulatory domains (Figure 32–4). The basal promoter elements are termed **CCAAT** (pronounced "cat") boxes and **TATA** ("tata") boxes because of their sequence motifs. The TATA box resides 20–30 bases upstream of the transcriptional start site and is similar in sequence to the prokaryotic Pribnow-box (consensus TATA$^T/_A$A$^T/_A$, where $^T/_A$ indicates that either base may be found at that position). Numerous proteins identified as **TFIIA, B, C,** etc. (for transcription factors regulating RNA pol II), have been observed to interact with the TATA-box. The CCAAT box (consensus GG$^T/_C$CAATCT) resides 50–130 bases upstream of the transcriptional start site. The protein identified as **C/EBP** (for *C*CAAT-box/*e*nhancer *b*inding *p*rotein) binds to the CCAATbox element.

The are many other regulatory sequences in mRNA genes, as well, that bind various transcription factors. Theses regulatory sequences are predominantly located upstream (5′) of the transcription initiation site, although some elements occur downstream (3′) or even within the genes themselves. The number and type of regulatory elements to be found varies with each mRNA gene. Different combinations of transcription factors also can exert differential regulatory effects upon transcriptional initiation. The various cell types each express characteristic combinations of transcription factors; this is the major mechanism for cell-type specificity in the regulation of mRNA gene expression.

Structural Motifs Common to DNA-Binding Proteins

A. Helix-Loop-Helix *(HLH)* (identified as helix-turn-helix in prokaryotes): A motif composed of two regions of α-helix separated by a region of variable length which forms a loop between the 2 α-helices. The α-helical domains are structurally similar and are necessary for protein interaction with sequence elements that exhibit a twofold axis of symmetry (Figure 32–5). This class of transcription factor most often contains a region of basic amino acids located on the N-terminal side of the HLH domain (termed bHLH proteins) that is necessary in order for the protein to bind DNA at specific sequences. The HLH domain is necessary for homo- and heterodimerization. Examples of bHLH proteins include MyoD, c-Myc, and Max (see Table 32–1). Several HLH proteins that do not contain the basic region act as repressors because of this lack. These HLH proteins repress the activity of other bHLH proteins by forming heterodimers with them and preventing DNA binding.

B. Zinc Fingers: This motif consists of specific spacings of C and H residues that allow the protein to bind zinc atoms. The metal atom coordinates the sequences around the C and H residues into a fingerlike domain. The finger domains can "interdigitate" into the major groove of the DNA helix. The spacing of the zinc finger domain in this class of transcription factor coincides with a half-turn of the double helix. The classic example is the RNA pol III transcription factor, TFIIIA (see below).

C. Leucine Zipper: A motif generated by a repeating distribution of L residues spaced 7 amino acids apart within α-helical regions of the protein (Figure 32–6). These L residues end up with their R-groups protruding from the α-helical domain in which the L residues reside. The protruding R-groups are thought to "interdigitate" with L-R groups of another leucine zipper domain, thus stabilizing homo- or heterodimerization. The leucine zipper domain is present in many DNA-binding proteins, such as c-Myc, c-Fos, c-Jun and C/EBP.

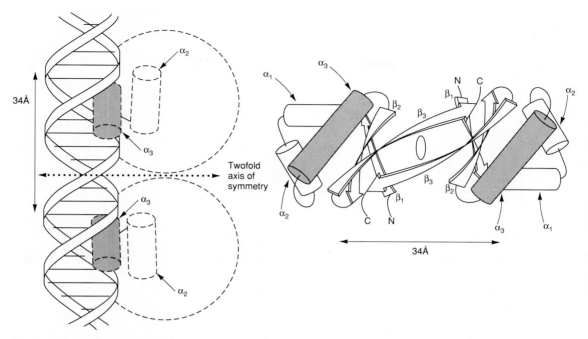

Figure 32–5. A schematic representation of the 3-dimensional structure of cor protein and its binding to DNA by the helix-turn-helix motifl The cro monomer consists of 3 antiparallel β sheets (β$_1$-β$_3$) and 3 α helices (α$_1$-α$_3$). The helix-turn-helix motif is formed becuase the α$_3$ and α$_2$ helices are held at 90 degrees to each other by a turn of 4 amino acids. The α$_3$ helix of cro is the DNA recognition surface (shaded). Two monomers associate through the antiparallel β$_3$ sheets to form a dimer that has a 2-fold axis of symmetry (right). A cro dimer binds to DNA through its α$_3$ helices, each of which contacts about 5 bp on the same surface of the major groove. The distance between comparable points on the two DNA α helices is 34Å, which is the distance required for one complete turn of the double helix. (Courtesy of B Mathews.) (Reproduced, with permission, from Murray, RK: *Harper's Biochemistry*, 23rd ed. Appleton & Lange, 1993.)

CLINICAL SIGNIFICANCE

There is no doubt that transcriptional factors are of major importance in controlling cell growth and differentiation. Their potency is exemplified by findings, for example, of several transcriptional factors encoded in mammalian and avian genes that, under certain circumstances, are involved in the development of neoplasia (see Table 32–1). Several oncogenic retroviruses have been shown to harbor mutant versions of normal cellular genes that encode transcription factors. In some cases, these potentially oncogenic proteins (termed **proto-oncogenes** because of their potential to become oncogenic under the proper circumstances, see Chapter 35) have also been shown to be involved in non-viral mediated oncogenesis as a result of mutations and/or loss of control of the expression of these genes.

Transcription by RNA Pol III

The major transcripts generated by pol III are the tRNAs and the 5S rRNA. Regulation of 5S rRNA transcription has been effectively studied in the clawed toad, *Xenopus laevis*. This organism contains an oocyte-specific 5S rRNA that is distinct from the somatic cell version of the 5S rRNA. Control over the pattern of expression of these two different 5S rRNA genes is the result of proteins that regulate RNA pol III activity: TFIIIA, TFIIIB, etc.

The major regulatory protein is TFIIIA, a zinc-finger protein containing 9 finger domains. This protein binds to two regulatory domains that reside within the 5S rRNA gene, as opposed to the 5' end of the gene (Figure 32–7). The binding of TFIIIA to the central regulatory region of the 5S rRNA genes stimulates the binding and subsequent transcriptional activity of RNA pol III (Figure 32–7). The presence of the zinc fingers allows TFIIIA to remain associated with the 5S genes as RNA pol III progresses along the genes with only temporary, local displacement.

Table 32–1. Description of several well-characterized mRNA gene transcription factors.*

Factor	Sequence Motif	Comments
c-Myc and Max	CACGTG	c-Myc first identified as retroviral oncogene; Max specifically associates with C-Myc in cells
c-Fos and c-Jun	TGA$^C/_G$T$^C/_A$A	Both first identified as retroviral oncogenes; associated in cells, also known as the factor AP-1
CREB	TGACG$^C/_T$$^C/_A$$^G/_A$	Binds to the cAMP response element; family of at least 10 factors resulting from different genes or alternative splicing; can form dimers with c-Jun
c-ErbA; also TR (thyroid hormone receptor)	GTGTCAAAGGTCA	First identified as retroviral oncogene; member of the steroid/thyroid hormone receptor superfamily; binds thyroid hormone
c-Ets	$^G/_C$$^A/_CGGA^A/_TG^T/_C$	First identified as retroviral oncogene; predominates in B- and T-cells
GATA	$^T/_C$GATA	Family of erythroid cell-specific factors, GATA-1 to -4
c-Myb	$^T/_C$AAC$^G/_T$$_G$	First identified as retroviral oncogene; hematopoietic cell-specific factor
MyoD	CAACTGAC	Controls muscle differentiation
NF-κB and c-Rel	GGGA$^A/_C$TN$^T/_C$CC[1]	Both factors identified independently; c-Rel first identified as retroviral oncogene; predominant in B- and T-cells
RAR (retinoic acid receptor)	ACGTCATGACCT	Also binds to c-Jun/c-Fos site
SRF (serum response factor)	GGATGTCCATATTAGGACATCT	Exists in many genes that are inducible by the growth factors present in serum

*This list is only representative of the hundreds of identified factors, some emphasis is placed on several factors that exhibit oncogenic potential. [1]N signifies that any base can occupy that position.

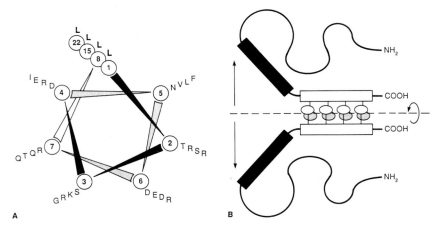

Figure 32–6. The leucine zipper motif. *A* shows a helical wheel analysis of a carboxy-terminal portion of the DNA binding protein C/EBP. The amino acid sequence is displayed end-to-end down the axis of a schematic α helix. The helical wheel consists of 7 spokes which correspond to the 7 amino acids that comprise every 2 turns of the α helix. Note that leucine residues (L) occur at every seventh position. Other proteins with "leucine zippers" have a similar helical wheel pattern. *B* is a schematic model of the DNA-binding domain of C/EBP. Two identical C/EBP polypeptide chains are held in dimer formation by the leucine zipper domain of each polypeptide (denoted by the rectangles and attached ovals). This association is apparently required to hold the DNA-binding domains of each polypeptide (the shaded rectangles) in the proper conformation for DNA binding. (Courtesy of S. McKnight.) (Reproduced, with permission, from Murray, RK: *Harper's Biochemistry*, 23rd ed. Appleton & Lange, 1993.)

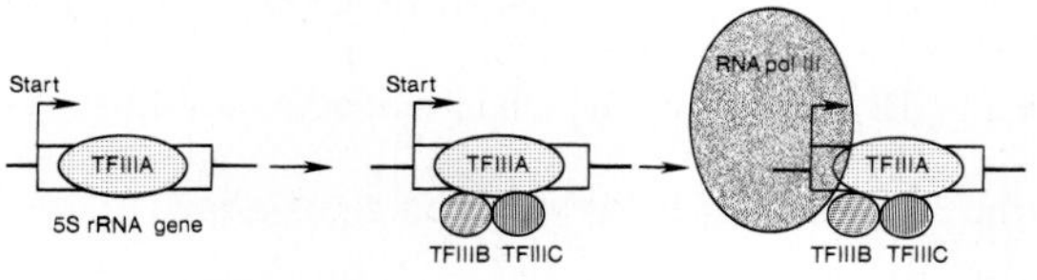

Figure 32–7. Model of TFIIIA and other associated TFIII proteins in regulating expression of the 5S rRNA genes. TFIIIA binds to the control region first, then TFIIIB and TFI-IIC; finally RNA pol III binds and transcription is initiated.

Questions

DIRECTIONS (items 1–10): Each numbered item or incomplete statement is followed by answers or by completions of the statement. Select the ONE lettered answer or completion that is BEST in each case.

1. The cAMP receptor protein (CRP):
   a. Stimulates replication.
   b. Inhibits transcription.
   c. Binds the promoter region of the appropriate operon in the presence of cAMP.
   d. Binds the operator gene in the absence of cAMP.
   e. Binds the regulator gene in the presence of cAMP and stimulates the synthesis of repressor proteins.

2. All of the following statements about the repressor of the lac operon of *E. coli* are true **except** which one?
   a. The repressor is the product of a regulatory gene, *i*.
   b. The repressor binds to the operator region of the DNA of the lac operon.
   c. The repressor is a protein.
   d. The repressor interacts with the operator in such a way as to directly interfere with the synthesis of β-galactosidase.
   e. The repressor can combine with lactose to form a complex that will no longer bind to the operator region.

3. Transcriptional regulatory sequences identified as "enhancers" function by:
   a. Enhancing the activity of RNA polymerase at a single promoter site.
   b. Enhancing the activity of RNA polymerase at multiple promoter sites simultaneously.
   c. Enhancing the activity of polymerase antiterminators at transcriptional termination sites in the gene.
   d. Enhancing the activity of polymerase to allow it to transcribe through terminator regions of the gene.

4. Synthesis of tryptophan in *E. coli* is inhibited by:
   a. Repressor binding of tryptophan.
   b. Reduced levels of charged trp-tRNAs.
   c. Decreased levels of glucose leading to an increase in cAMP levels.
   d. Repressor release of tryptophan.
   e. Binding of CRP-cAMP to the trp operon.

5. *E. coli* cells grown on a mixture of glucose and lactose selectively use glucose because:
   a. Lactose provides less energy than glucose.
   b. CAP is not present when glucose is available.
   c. Glucose lowers cAMP levels, which reduces the level of the cAMP-CAP complex.
   d. Lactose cannot enter the cells, owing to glucose-induced inhibition of the lactose transporter.
   e. The synthesis of the lac repressor is elevated in the presence of glucose.

6. Transcription factor TFIIIA regulates transcription of the 5S rRNA genes by:
   a. Binding to RNA pol III and increasing its activity.
   b. Binding to RNA pol III and inhibiting its activity.

   c. Binding to regulatory sequences at the 5′ end of the 5S rRNA genes, thereby being positioned to prevent initiation of transcription.
   d. Increasing the affinity of RNA pol III for TFIIIB and TFIIIC, which in turn activates the polymerase.
   e. Binding to regulatory sequences within the 5S rRNA genes, thereby being positioned to activate RNA pol III.

7. Which of the following is most responsible for regulating the level of eukaryotic mRNA gene products?
   a. Chromatin structure.
   b. Transcript transport.
   c. Translational initiation.
   d. Transcriptional initiation.
   e. Post-translational processing and transport.

8. The characteristic that most closely resembles the zinc finger domain is:
   a. Selectively positioned M and H residues, enabling metal ion coordination.
   b. Selectively positioned C and H residues, enabling metal ion coordination.
   c. Selectively positioned L residues, enabling metal ion coordination.
   d. Selectively positioned C and H, residues, enabling protein-protein interactions.
   e. Selectively positioned L residues, enabling protein-protein interactions.

9. Which of the following best describes the observation that the insulin gene is specifically expressed in pancreatic islet cells as opposed to all cells of the body?
   a. The insulin gene is found only in the β-islet cells of the pancreas.
   b. The β-islet cells contain the proper processing enzymes to process insulin correctly and to transport it.
   c. All other cells of the body contain specific proteases that digest insulin instead of activating it.
   d. The β-islet cells express specific factors that activate the insulin gene.
   e. The β-islet cells do not synthesize the repressor of the insulin gene that is expressed in all other cells of the body.

10. Which of the following would most likely result from an alteration in the pattern of CpG methylation following the replication of DNA?
   a. A variety of genes would likely be less transcriptionally active.
   b. The daughter cells would die.
   c. A variety of genes would likely be more transcriptionally active.
   d. The daughter cells would become muscle cells.
   e. The daughter cells would develop into tumorigenic cells.

Answers

1. c	**5.** c	**9.** d
2. d	**6.** e	**10.** c
3. a	**7.** d	
4. a	**8.** b	

Part VIII: Cellular Regulation

Growth Factors and Cytokines

33

Objectives

- To describe the major role of the following growth factors: PDGF, EGF, TGF-α, TGF-β, FGF, NGF, Epo, IGF-I, and IGF-II.
- To define the terms: cytokine, lymphokine, monokine.
- To describe the major roles of the cytokines; IL-1, IL-2, IL-6, IL-8, TNF-α, TNF-β, and IFN-γ.
- To describe the term: colony stimulating factors (CSFs).

Concepts

Growth factors are proteins that bind to receptors on the cell surface, with the primary result of activating cellular proliferation and/or differentiation. Many growth factors are quite versatile, stimulating cellular division in numerous different cell types; while others are specific to a particular cell-type (Table 33–1). Among the majority of growth factors, which stimulate proliferation, there are two noted exceptions. Both **tumor necrosis factor-α (TNF-α)** and some forms of the **transforming growth factor-β (TGF-β)**, can inhibit the proliferation of some cells types, while stimulating proliferation in other cells.

Cytokines are a unique family of growth factors. Secreted primarily from leukocytes, cytokines stimulate both the humoral and cellular immune responses, as well as the activation of phagocytic cells. Cytokines that are secreted from lymphocytes are termed **lymphokines**, whereas those secreted by monocytes or macrophages are termed **monokines**. A large family of cytokines are produced by various cells of the body (see Table 33–2). Many of the lymphokines are also known as **interleukins (ILs)**, since they are not only secreted by leukocytes but also able to affect the cellular responses of leukocytes. Specifically, interleukins are growth factors targeted to cells of hematopoietic origin. The list of interleukin molecules currently numbers 13 and is still growing.

Epidermal Growth Factor (EGF)

EGF, like all growth factors, binds to specific high-affinity, low-capacity receptors on the surface of responsive cells. Intrinsic to the EGF receptor is tyrosine kinase activity, which is activated in response to EGF binding. The kinase domain of the EGF receptor phosphorylates the EGF receptor itself ('autophosphorylation') as well as other proteins, in signal transduction cascades, that associate with the receptor following activation (see Chapter 34). Experimental evidence has shown that the Neu proto-oncogene is a homolog of the EGF receptor (see Chapter 35).

EGF has proliferative effects on cells of both mesodermal and ectodermal origin, particularly keratinocytes and fibroblasts. EGF exhibits negative growth effects on certain carcinomas as well as hair follicle cells. Growth-related responses to EGF include the induction of nuclear proto-oncogene expression, such as Fos, Jun and Myc (see Chapter 35). EGF also has the effect of decreasing gastric acid secretion.

Platelet-Derived Growth Factor (PDGF)

PDGF is composed of two distinct polypeptide chains, A and B, that form homodimers (AA or BB) or heterodimers (AB). The c-Sis proto-oncogene has been shown to be homologous to the PDGF B chain.

Table 33–1. Characteristic properties of the major growth factors.

Factor	Principal Source	Primary Activity	Comments
PDGF	Platelets, endothelial cells, placenta	Promotes proliferation of connective tissue, glial and smooth muscle cells	Two different protein chains form 3 distinct dimer forms; AA, AB and BB
EGF	Submaxillary gland, Brunners gland	Promotes proliferation of mesenchymal, glial and epithelial cells	
TGF-α	Common in transformed cells	May be important for normal wound healing	Related to EGF
FGF	Wide range of cells; protein is associated with the ECM	Promotes proliferation of many cells; inhibits some stem cells; induces mesoderm to form in early embryos	At least 7 family members
NGF		Promotes neurite outgrowth and neural cell survival	
Erythropoietin	Kidney	Promotes proliferation and differentiation of erythrocytes	
TGF-β	See Table 33–2	See Table 33–2	At least 60 different family members including BMPs and activins
IGF-I	Primarily liver	Promotes proliferation of many cell types	Related to IGF-II and proinsulin, also called Somatomedin C
IGF-II	Variety of cells	Promotes proliferation of many cell types primarily of fetal origin	Related to IGF-I and proinsulin

Only the dimeric forms of PDGF interact with the PDGF receptor. Two distinct classes of PDGF receptor have been cloned, one specific for AA homodimers and another that binds BB and AB type dimers. Like the EGF receptor, the PDGF receptors have intrinsic tyrosine kinase activity. Following autophosphorylation of the PDGF receptor, numerous signal-transducing proteins associate with the receptor and are subsequently tyrosine phosphorylated (see Chapter 34).

Proliferative responses to PDGF action are exerted on many mesenchymal cell types. Other growth-related responses to PDGF include cytoskeletal rearrangement and increased polyphosphoinositol turnover. Again, like EGF, PDGF induces the expression of a number of nuclear localized proto-oncogenes, such as Fos, Myc and Jun (see Chapter 35). The primary effects of TGF-β are due to the induction, by TGF-β, of PDGF expression.

Fibroblast Growth Factors (FGFs)

There are at least 7 distinct members of the FGF family of growth factors. The two originally characterized FGFs were identified by biological assay and are termed aFGF (acidic) and bFGF (basic). Kaposi's sarcoma cells (prevalent in patients with AIDS) secrete a homolog of FGF called the K-FGF (KGF) proto-oncogene. In mice the mammary tumor virus integrates at two predominant sites in the mouse genome identified as Int-1 and Int-2. The protein encoded by the Int-2 locus is a homolog of the FGF family of growth factors.

The predominant effect of FGF is to increase the growth of endothelial cells; however, proliferative effects are observed on a number of other cells of mesodermal and neuroectodermal origin. FGFs also are neurotrophic for cells of both the peripheral and central nervous system. Additionally, several members of the FGF family are potent inducers of mesodermal differentiation in early embryos. Non-proliferative effects include regulation of pituitary and ovarian cell function.

The FGF receptor has intrinsic tyrosine kinase activity like both the EGF and PDGF receptors. As with all transmembrane receptors that have tyrosine kinase activity, autophosphorylation of the receptor is the immediate response to FGF binding. Following activation of the FGF receptor, numerous signal-transducing proteins associate with the receptor and become tyrosine-phosphorylated. The Flg proto-oncogene is a homolog of the FGF receptor. The FGF receptor also has been shown to be the portal of entry into cells for herpes viruses.

Table 33–2. Major properties of human interleukins and other immunoregulatory cytokines.

	Earlier Terms	Principal Cell Source	Principal Effects[1]
Interleukins IL-1 α and β	Lymphocyte-activating factor, B cell activating factor, hematopoietin	Macrophages, other APCs, other somatic cells	Costimulation of APCs and T cells B cell proliferation and Ig production Acute-phase response of liver Phagocyte activation Inflammation and fever Hematopoiesis
IL-2	T cell growth factor	Activated T$_H$1 cells, Tc cells, NK cells	Proliferation of activated T cells NK and Tc cell functions B cell proliferation and IgG2 expression
IL-3	Multi-colony-stimulating factor	T lymphocytes	Growth of early hematopoietic progenitors
IL-4	B cell growth factor I, B cell stimulatory factor I	T$_H$2 cells, mast cells	B cell proliferation, IgE expression, and class II MHC expression T$_H$2- and Tc-cell proliferation and functions Eosinophil and mast cell growth and function Inhibition of monokine production
IL-5		T$_H$2 cells, mast cells	Eosinophil growth and function
IL-6	IFN-β2, hepatocyte-stimulating factor, hybridoma growth factor	Activated T$_H$2 cells, APCs, other somatic cells	Synergistic effects with IL-1 or TNF to costimulate T cells Acute-phase response of liver B-cell proliferation and Ig production Thrombopoiesis
IL-7		Thymic and marrow stromal cells	T and B lymphopoiesis Tc cell functions
IL-8		Macrophages, other somatic cells	Chemoattractant for neutrophils and T cells
IL-9		Cultured T cells	Some hematopoietic and thymopoietic effects
IL-10	Cytokine synthesis inhibitory factor	Activated T$_H$2, CD8 T, and B lymphocytes, macrophages	Inhibition of cytokine production by T$_H$1 cells, NK cells, and APCs Promotion of B cell proliferation and antibody responses Suppression of cellular immunity Mast cell growth
IL-11		Stromal cells	Synergistic effects on hematopoiesis and thrombopoiesis
IL-12	Cytotoxic lymphocyte maturation factor, NK cell stimulatory factor	B cells, macrophages	Proliferation and function of activated Tc and NK cells IFN γ production T$_H$1 cell induction; suppresses T$_H$2 cell functions Promotion of cell-mediated immune responses
IL-13		T$_H$2 cells	IL-4-like effects
Other cytokines TNFα	Cachectin	Activated macrophages, other somatic cells	IL-1-like effects Vascular thrombosis and tumor necrosis
TNFβ	Lymphotoxin	Activated T$_H$1 cells	IL-1-like effects Vascular thrombosis and tumor necrosis

(*continued*)

Table 33–2. Major properties of human interleukins and other immunoregulatory cytokines (continued).

	Earlier Terms	Principal Cell Source	Principal Effects[1]
Other cytokines (cont'd.)			
INF α and β	Leukocyte interferons, type I interferons	Macrophages; neutrophils, other somatic cells	Antiviral effects Induction of class I MHC on all somatic cells Activation of macrophages and NK cells
INF γ	Immune interferon, type II interferon	Activated TH1 and NK cells	Induction of class I MHC on all somatic cells Induction of class II MHC on APCs and somatic cells Activation of macrophages, neutro-phils, and NK cells Promotion of cell-mediated immu-nity (inhibits TH2 cells) Induction of high endothelial ven-ules Antiviral effects
TGFβ		Activated T lymphocytes, platelets, macrophages, other somatic cells	Anti-inflammatory (suppression of cytokine production and class II MHC expression) Anti-proliferative for macrophages and lymphocytes Promotion of B-cell expression of IgA Promotion of fibroblast proliferation and wound healing

[1]All of the listed processes are enhanced unless otherwise indicated.

Transforming Growth Factors-β (TGFs-β)

TGF-β was originally characterized as a protein (secreted from a tumor cell line) that was capable of inducing a transformed phenotype in non-neoplastic cells in culture. This effect was reversible, as demonstrated by the reversion of the cells to a normal phenotype following removal of the TGF-β. Subsequently, many proteins homologous to TGF-β have been identified. The four closest relatives are TGF-β1 (the original TGF-β) through TGF-β5 (TGF-β1 = TGF-β4). All four of these proteins share extensive regions of similarity in their amino acids. Many other proteins, possessing distinct biological functions, have stretches of amino-acid homology to the TGF-β family of proteins, particularly the C-terminal region of these proteins.

The TGF-β-related family of proteins includes the activin and inhibin proteins. There are activin A, B and AB proteins, as well as an inhibin A and inhibin B protein. The Mullerian inhibiting substance (MIS) is also a TGF-β-related protein, as are members of the bone morphogenetic protein (BMP) family of bone growth-regulatory factors. Indeed, the TGF-β family may comprise as many as 60 distinct proteins, all with at least one region of amino-acid sequence homology.

There are several classes of cell-surface receptors that bind different TGFs-β with differing affinities. There also are cell-type specific differences in receptor sub-types. Unlike the EGF, PDGF and FGF receptors, the TGF-β family of receptors all have intrinsic serine/threonine kinase activity and, therefore, induce distinct cascades of signal transduction.

TGFs-β have proliferative effects on many mesenchymal and epithelial cell types. Under certain conditions TGFs-β will demonstrate anti-proliferative effects on endothelial cells, macrophages, and T- and B-lymphocytes. Such effects include decreasing the secretion of immunoglobulin and suppressing hematopoiesis, myogenesis, adipogenesis and adrenal steroidogenesis. Several members of the TGF-β family are potent inducers of mesodermal differentiation in early embryos, in particular TGF-β2 and activin A.

Transforming Growth Factor-α (TGF-α)

TGF-α, like the β form, was first identified as a substance secreted from certain tumor cells that, in conjunction with TGF-β1, could reversibly transform certain types of normal cells in culture. TGF-α binds to the EGF receptor, as well as its own distinct receptor, and it is this interaction that is thought to be responsible for the growth factor's effect. The predominant sources of TGF-α are carcinomas, but activated macrophages and keratinocytes (and possibly other epithelial cells) also secrete TGF-α. In normal cell populations, TGF-α is a potent keratinocyte growth factor; forming an autocrine growth loop by virtue of the protein activating the very cells that produce it.

Erythropoietin (Epo)

Epo is synthesized by the kidney and is the primary regulator of erythropoiesis. Epo stimulates the proliferation and differentiation of immature erythrocytes; it also stimulates the growth of erythoid progenitor cells (eg, erythrocyte burst-forming and colony-forming units) and induces the differentiation of erythrocyte colony-forming units into proerythroblasts. When patients suffering from anemia due to kidney failure are given Epo, the result is a rapid and significant increase in red blood cell count.

Insulin-Like Growth Factor-I (IGF-I):

IGF-I (originally called **somatomedin C**) is a growth factor structurally related to insulin. IGF-I is the primary protein involved in responses of cells to growth hormone (GH): that is, IGF-I is produced in response to GH and then induces subsequent cellular activities, particularly on bone growth. It is the activity of IGF-I in response to GH that gave rise to the term "somatomedin." Subsequent studies have demonstrated, however, that IGF-I has autocrine and paracrine activities in addition to the initially observed endocrine activities on bone. The IGF-I receptor, like the insulin receptor, has intrinsic tyrosine kinase activity. Owing to their structural similarities IGF-I can bind to the insulin receptor but does so at a much lower affinity than does insulin itself.

Insulin-Like Growth Factor-II (IGF-II)

IGF-II is almost exclusively expressed in embryonic and neonatal tissues. Following birth, the level of detectable IGF-II protein falls significantly. For this reason IGF-II is thought to be a fetal growth factor. The IGF-II receptor is identical to the mannose-6-phosphate receptor that is responsible for the integration of lysosomal enzymes (which contain mannose-6-phosphate residues) to the lysosomes (see Chapter 17).

Cytokines

A. IL-1: IL-1 is one of the most important immune response-modifying interleukins (Figure 33–1). The predominant function of IL-1 is to enhance the activation of T-cells in response to antigen. The activation of T-cells, by IL-1, leads to increased T-cell production of IL-2 and of the IL-2 receptor, which in turn augments the activation of the T-cells in an autocrine loop. IL-1 also induces expression of interferon-γ (IFN-γ) by T-cells. This effect of T-cell activation by IL-1 is mimicked by TNF-α (see below) which is another cytokine secreted by activated macrophages. There are 2 distinct IL-1 proteins, termed IL-1α and -1β, that are 26% homologous at the amino acid level. The IL-1s are secreted primarily by macrophages but also from neutrophils, endothelial cells, smooth muscle cells, glial cells, astrocytes, B- and T-cells, fibroblasts and keratinocytes. Production of IL-1 by these different cell types occurs only in response to cellular stimulation. In addition to its effects on T-cells, IL-1 can induce proliferation in non-lymphoid cells.

B. IL-2: IL-2, produced and secreted by activated T-cells, is the major interleukin responsible for clonal T-cell proliferation. IL-2 also exerts effects on B-cells, macrophages, and natural killer (NK) cells. The production of IL-2 occurs primarily by CD4$^+$ T-helper cells. As indicated above, the expression of both IL-2 and the IL-2 receptor by T-cells is induced by IL-1. Indeed, the IL-2 receptor is not expressed on the surface of resting T-cells and is present only transiently on the surface of T-cells, disappearing within 6–10 days of antigen presentation. In contrast to T-helper cells, NK cells constitutively express IL-2 receptors and will secrete TNF-α, IFN-γ and GM-CSF (see below) in response to IL-2, which in turn activate macrophages.

C. IL-6: IL-6 is produced by macrophages, fibroblasts, endothelial cells and activated T-helper cells. IL-6 acts in synergy with IL-1 and TNF-α in many immune responses, including T-cell activation. In particular, IL-6 is the primary inducer of the **acute-phase response** in liver. IL-6 also enhances the differentiation of B-cells and their consequent production of immunoglobulin. Glucocorticoid synthesis is also enhanced by IL-6. Unlike IL-1, IL-2 and TNF-α, IL-6 does not induce cytokine expression; its main effects, therefore, are to augment the responses of immune cells to other cytokines.

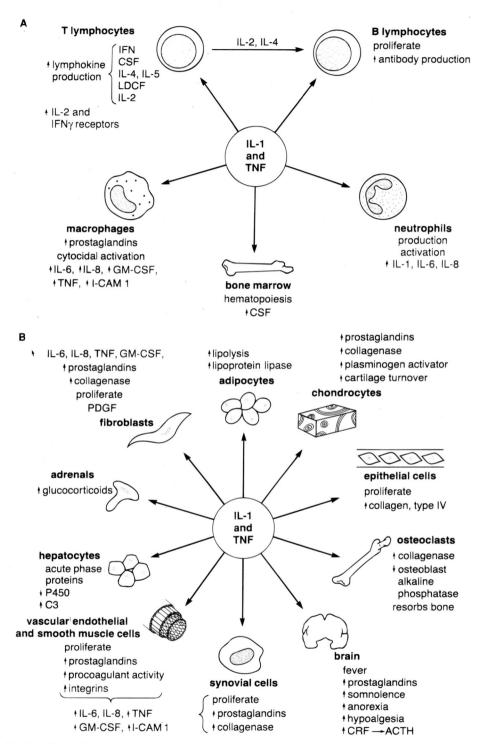

Figure 33–1. Actions of IL-1 and TNF on hematopoietic and lymphoid tissues *(A)* and nonlymphoid cells and tissues. *(B)* Activities of the two individual cytokines differ in some respects. (Reproduced, with permission, from Stites, DP: *Basic & Clinical Immunology*, 8th ed. Appleton & Lange, 1994.)

D. IL-8: IL-8 is an interleukin that belongs to an ever-expanding family of proteins that exert chemoattractant activity to leukocytes and fibroblasts. This family of proteins is termed the **chemokines**. IL-8 is produced by monocytes, neutrophils, and NK cells and is chemoattractant for neutrophils, basophils and T-cells. In addition, IL-8 activates neutrophils to degranulate.

Tumor Necrosis Factor-α (TNF-α)

TNF-α (also called **cachectin**), like IL-1 is a major immune response–modifying cytokine produced primarily by activated macrophages (Figure 33–1). Like IL-1, TNF-α induces the expression of other autocrine growth factors, increases cellular responsiveness to growth factors and induces signaling pathways that lead to proliferation. TNF-α acts synergistically with EGF and PDGF on some cell types. Like other growth factors, TNF-α induces expression of a number of nuclear proto-oncogenes as well as of several interleukins.

Tumor Necrosis Factor-β (TNF-β)

TNF-β (also called **lymphotoxin**) is characterized by its ability to kill a number of different cell types, as well as the ability to induce terminal differentiation in others. One significant non-proliferative response to TNF-β is an inhibition of lipoprotein lipase present on the surface of vascular endothelial cells (see Chapter 20). The predominant site of TNF-β synthesis is T-lymphocytes, in particular the special class of T-cells called cytotoxic T-lymphocytes (CTL cells). The induction of TNF-β expression results from elevations in IL-2 as well as the interaction of antigen with T-cell receptors.

Interferon-γ (IFN-γ)

IFN-α, IFN-β and IFN-ω are known as **type I interferons**: they are predominantly responsible for the antiviral activities of the interferons. In contrast, IFN-γ is a **type II** or **immune interferon**. Although IFN-γ, has antiviral activity it is significantly less active at this function than the type I IFNs. Unlike the type I IFNs, IFN-γ is not induced by infection nor by double-stranded RNAs. IFN-γ is secreted primarily by CD8$^+$ T-cells. Nearly all cells express receptors for IFN-γ and respond to IFN-γ binding by increasing the surface expression of class I MHC proteins, thereby promoting the presentation of antigen to T-helper (CD4$^+$) cells. IFN-γ also increases the presentation of class II MHC proteins on class II cells further enhancing the ability of cells to present antigen to T-cells.

Colony Stimulating Factors (CSFs)

CSFs are cytokines that stimulate the proliferation of specific pluripotent stem cells of the bone marrow in adults. Granulocyte-CSF (**G-CSF**) is specific for proliferative effects on cells of the granulocyte lineage. Macrophage-CSF (**M-CSF**) is specific for cells of the macrophage lineage. Granulocyte-macrophage-CSF (**GM-CSF**) has proliferative effects on both classes of lymphoid cells. **Epo** is also considered a CSF as well as a growth factor (see above), since it stimulates the proliferation of erythrocyte colony-forming units. **IL-3** (secreted primarily from T-cells) is also known as **multi-CSF**, since it stimulates stem cells to produce all forms of hematopoietic cells.

Questions

DIRECTIONS (item 1): Select the ONE lettered answer.

1. Of the following cytokines, which is considered a lymphokine as opposed to a monokine?
   a. IL-1.
   b. IL-2.
   c. TNF-β.
   d. IFNγ.
   e. IL-3.

DIRECTIONS (items 2–10): Match the following growth factors and cytokines with the correct description of their function.

   a. Fetal growth factor.
   b. Induction of antigen presentation CD4$^+$ T-cells.
   c. Major T-cell mitogen.

    d. Acute phase response of hepatocytes.
    e. Inhibits the growth of hair follicle cells.
    f. Potent inducer of early mesodermal differentiation.
    g. Induction of immunoglobulin production.
    h. Primary regulator of erythropoiesis.
    i. Induces proliferation of smooth muscle cells.

  2. TGF-β.

  3. IL-1.

  4. EGF.

  5. PDGF.

  6. IL-2.

  7. Epo.

  8. IFNγ.

  9. IGF-II.

10. IL-6.

Answers

1. a	**5.** i	**9.** a
2. f	**6.** c	**10.** d
3. g	**7.** h	
4. e	**8.** b	

34   Mechanisms of Signal Transduction

Objectives

- To describe the three major classifications of signal transducing receptors.
- To describe the activity of the receptor and non-receptor tyrosine kinases.
- To describe the activity of the receptor and non-receptor serine/threonine kinases.
- To describe the role of protein kinase C (PKC).
- To describe the role of phospholipids and phospholipases in signal transduction.
- To describe the role of phosphatidylinositol-3-kinase in signal transduction.
- To describe the role of mitogen activated protein kinases (MAP kinases) in signal transduction.
- To describe the role of G-protein coupled receptors.
- To describe the role of G-protein regulators.
- To describe the role of intracellular hormone receptors.
- To describe the role of phosphatases in signal transduction.

Concepts

Signal transduction at the cellular level refers to the movement of signals from outside the cell to inside (Figure 34–1). The movement of signals can be simple, like that associated with receptor molecules of the acetylcholine class. These receptors constitute channels that, upon ligand interaction, allow signals to be passed by the movement of small ions into or out of the cell. These ion movements result in changes in the electrical potential of the cells that, in turn, propagate the signal along the cell. A more complex level of signal transduction involves the coupling of ligand-receptor interactions to many intracellular events (Figure 34–1). These events include phosphorylations by tyrosine kinases and/or by serine/threonine kinases. Protein phosphorylations change enzyme activities and protein conformations. The eventual outcome is an alteration in the program of genes expressed.

There are three major classifications of signal-transducing receptors:

1. Receptors that penetrate the plasma membrane and have **intrinsic enzymatic activity**. Receptors of this class include the **tyrosine kinases** (PDGF, insulin, EGF, and FGF receptors), **tyrosine phosphatases** (eg, the CD45 ["cluster determinant-45"] protein of T-cells and macrophages), **guanylate cyclases** (eg, natriuretic peptide receptors) and **serine/threonine kinases** (eg, activin and TGF-β receptors). Receptors with intrinsic tyrosine kinase activity are capable of autophosphorylation as well as phosphorylation of other substrates.

    Additionally, some receptors lack intrinsic enzyme activity, yet are coupled to intracellular tyrosine kinases by direct protein-protein interactions. For example, the CD4 and CD8 proteins of T-cells interact with the proto-oncogenic protein, lck ("lick"), a tyrosine kinase of the src ("sarc") family (see Chapter 35 for definitions of the proto-oncogenes mentioned in this Chapter).
2. Receptors that are coupled to GTP-binding and hydrolyzing proteins inside the cell **(G-proteins)**. All receptors of the class that interact with G-proteins have a structure characterized by 7 membrane-spanning domains (Figure 34–1). These receptors are termed **serpentine receptors**. Examples of this class are the adrenergic receptors, odorant receptors, and certain hormone receptors (eg, glucagon, angiotensin, vasopressin, and bradykinin).
3. Receptors that are found within the cell and that, upon ligand binding, migrate to the nucleus where the ligand-receptor complex directly affects gene transcription.

Tyrosine Kinases

A.   Receptor Tyrosine Kinases (RTKs):   The proteins encoding RTKs contain four major domains: an extracellular ligand-binding domain, an intracellular tyrosine kinase domain, an intracellular regulatory domain, and a transmembrane domain. The amino-acid sequences of the tyrosine kinase domains of RTKs are highly related to those of cAMP-dependent protein kinase (PKA) within the ATP-binding and substrate-binding regions. Within the family of RTK proteins there are at least 9 subfamilies, each distinguished by its own structural features (Figure 34–2).

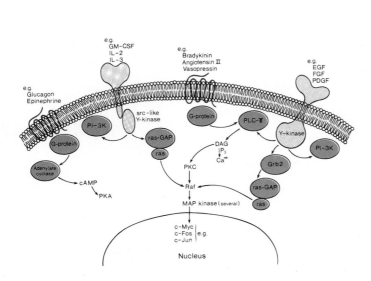

Figure 34–1. Cartoon of various transmembrane receptors and associated intracellular signaling molecules. Two classes of "serpentine" receptors are depicted; one couples to G-proteins that activate adenylate cyclase and one to G-proteins that activate PLCγ. Receptors with intrinsic tyrosine kinase activity couple to several intracellular effector molecules, including PLCγ, PI-3K and Grb2. Other receptors that have no intrinsic enzymatic activity can be coupled to intracellular tyrosine kinases of the src family that, in turn, activate proteins such as PI-3K or rasGAP. Many of the different types of receptors activate signal transduction pathways that converge upon equivalent kinases (for example MAP kinases) and eventually simi-lar nuclear transcriptional effector proteins. Receptors that couple to guanylate cyclases, those with intrinsic serine/threonine kinase activity and those with intrinsic tyrosine phosphatase activity are not represented. Only the more well-characterized intracellular signal transduction molecules are shown, many others with various activities, are known to exist within cells.

Class I is exemplified by the EGF receptor. This class of receptors contains cysteine-rich sequences in the extracellular domain.

Class II is exemplified by the insulin receptor and includes the IGF-I receptor. This class also contains cysteine-rich sequences in the extracellular. An additional feature of the class II receptors is that they are generally heterotetrameric with the individual polypeptides linked by disulfide bonding.

Class III is exemplified by the PDGF receptors. This class of receptors is identified by the presence of 5 immunoglobulin-like domains in the extracellular portion, as well as by the insertion of a short stretch of amino acids into the center of the kinase domain (the latter is termed the **kinase insert**).

Class IV is exemplified by the FGF receptors. This group also has immunoglobulin-like domains (only 3, however), as well as the kinase insert.

Class V is exemplified by the vascular endothelial cell growth factor (VEGF) receptor. Similar to class III and IV receptors, class V receptors contain 7 immunoglobulin-like domains as well as the kinase insert.

Class VI is exemplified by the hepatocyte growth factor (HGF) and scatter factor (SC) receptors. The receptors in this class are heterodimeric like class II receptors, except that one of the two protein subunits is completely extracellular. The HGF receptor is a proto-oncogene that was originally identified as the **met oncogene** and subsequently shown to bind HGF.

Class VII constitutes the neurotrophin receptor family, with nerve growth factor (NGF) being the most notable. This class of receptors, like class VI, contains no cysteine-rich or immunoglobulin domains in the extracellular sequences. A potential accessory protein has been identified that may be required in order for this class of receptors to exhibit high-affinity ligand binding. The NGF receptor is also a proto-oncogene that was originally identified as the **trk ("track") oncogene** and subsequently shown to bind NGF. Additional trk-related proteins (trkA and trkB) are the receptors for other neurotrophins.

Since **class VIII** and **class IX** proteins are either orphan receptors (that is, they have no known ligands) or their ligands are not well known, they will not be described here.

Many receptors that have intrinsic tyrosine kinase activity, and the tyrosine kinases that are associated with cell-surface receptors, contain tyrosine residues; upon becoming phosphorylated, the residues interact with other proteins of the signaling cascade. These other proteins contain a domain of amino-acid sequences that is homologous to one first identified in the src proto-oncogene, and for this reason they are termed **SH2 (src homology 2) domains** (Figure 34–3). When proteins containing SH2 domains interact with RTKs or receptor-associated tyrosine kinases, the result is tyrosine phosphorylation of the SH2 proteins. This phosphorylation, in SH2-containing proteins that have enzymatic activity, brings about an alteration (either positively or negatively) in that activity. Several proteins that come under this heading are phospholipase C-γ (PLCγ), the proto-oncogene ras-associated GTPase activating protein (ras-GAP), phosphatidylinositol-3-kinase (PI-3K), and protein tyrosine phosphatase-1C (PTP1C), as well as members of the src family of protein tyrosine kinases (PTKs). (For descriptions, see below.)

B. Non-Receptor Tyrosine Kinase: There are numerous intracellular PTKs that are responsible for phosphorylating intracellular proteins on tyrosine residues after the activation of signals for cellu-

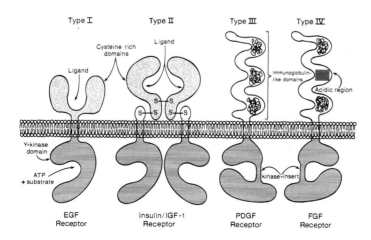

Figure 34–2. Cartoons of the four most common forms of receptors with intrinsic tyrosine kinase activity, types I, II, III, and IV.

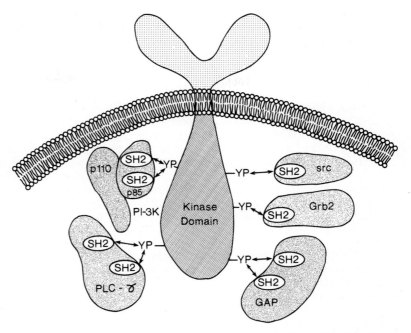

Figure 34–3. Typical interactions of SH2 containing intracellular effector proteins with phosphorylated tyrosines present in a transmembrane receptor tyrosine kinase.

lar growth and proliferation. The archetypal PTK is the src protein. Additionally, there is a class of receptors that lack intrinsic tyrosine kinase activity but are able to couple through protein-protein interaction with proteins of the src family of tyrosine kinases. This class of receptors includes the IL-2 receptor and the CD4 and CD8 cell-surface glycoproteins of T-cells. In T-cells the N-terminus of the proto-oncogene, lck, interacts with the cytoplasmic tails of the CD4 and CD8 proteins. This mode of coupling receptors to intracellular PTKs suggests a split form of RTK. An additional example of the split RTK is the T-cell antigen receptor (TCR), a complex of at least 7 distinct proteins. The ζ chain is tyrosine-phosphorylated in response to the presentation of an antigen. The TCR is not itself a tyrosine kinase but is associated with the non-receptor tyrosine kinase, fyn ("fin," an src family PTK).

Another example of receptor signaling through protein interaction involves the insulin receptor (IR). This receptor has intrinsic tyrosine kinase activity but does not directly interact, after autophosphorylation, with enzymatically active proteins containing SH2 domains. Instead, it uses as its principal substrate a protein termed IRS-1, which contains several motifs that resemble SH2 binding consensus sites for the catalytically active subunit of PI-3K (see below). The binding thus allows complexes to form between IRS-1 and PI-3K. This model suggests that IRS-1 acts as a **docking** or **adapter** protein to couple the IR to SH2 containing signaling proteins. Additional adapter proteins have been identified, the most common being a protein termed **growth factor receptor-binding protein 2, Grb2** (Figure 34–3).

An example of altered receptor activity in response to association with an intracellular PTK is found in the nicotinic acetylcholine receptor (AChR). This receptor comprises an ion channel consisting of 4 distinct subunits (α, β, γ, and δ). In response to the binding of acetylcholine the β, γ, and δ subunits are tyrosine-phosphorylated, which leads to an increase in the rate of desensitization to acetylcholine.

Serine/Threonine Kinases

A. Receptor Serine/Threonine Kinases: The receptors for the activins and TGFs-β have intrinsic serine/threonine kinase activity. Although both these families of proteins can induce and/or inhibit cellular proliferation or differentiation, the signaling pathways utilized are different from those of receptors with intrinsic tyrosine kinase activity or those associated with intracellular tyrosine kinases. The precise mechanisms by which the receptor serine/threonine kinases induce cellular responses are not as clearly understood as those of RTKs. One nuclear protein involved in the responses of cells to TGF-β is the proto-oncogene myc ("mick"), which directly affects the expression of genes harboring myc-binding elements.

B. Non-Receptor Serine/Threonine Kinases: Several serine/threonine kinases play a role in the transmission of signals from the outside of cells to the inside. The two more commonly known are cAMP-dependent protein kinase (PKA) and protein kinase C (PKC). PKA will not be reviewed here; its activities are described in the chapters that discuss intermediary metabolism. Additional serine/threonine kinases important for signal transduction are the **mitogen activated protein kinases (MAP kinases)**.

Protein Kinase C (PKC)

PKC was originally identified as a serine/threonine kinase that was maximally active in the presence of diacylglycerols (DAG) and calcium ion. Subsequent to its original characterization, at least 10 highly related proteins with PKC activity have been identified. Each of these enzymes exhibits specific patterns of tissue expression and activation by lipid and calcium. PKCs are involved in the signal transduction pathways initiated by certain hormones, growth factors, and neurotransmitters. The phosphorylation of various proteins by PKC can lead to either increased or decreased activity. Of particular importance is the phosphorylation of the EGF receptor by PKC, which down-regulates the tyrosine kinase activity of the receptor. This effectively limits the length of the cellular responses initiated through the EGF receptor.

Phospholipases and Phospholipids

Phospholipases and phospholipids are involved in the processes of transmitting ligand-receptor—induced signals from the plasma membrane to intracellular proteins. The primary protein affected by the activation of phospholipases is PKC, which is maximally active in the presence of calcium ion and DAG; this is because the generation of DAG occurs in response to agonist activation of various phospholipases. The principal mediators of PKC activity are receptors coupled to the activation of phospholipase C-γ (PLCγ). PLCγ contains SH2 domains that allow it to interact with tyrosine-phosphorylated RTKs (Figure 34–3). This allows PLCγ to be intimately associated with the signal transduction complexes of the membrane as well as with the membrane phospholipids that are its substrates. Activation of PLCγ leads primarily to the hydrolysis of membrane phosphatidylinositol bisphosphate (PIP$_2$), which in turn leads to an increase in intracellular DAG and inositol trisphosphate (IP$_3$). The released IP$_3$ interacts with intracellular membrane receptors, causing an increased release of stored calcium ions. Together, the increased DAG and intracellular concentrations of free calcium ions increase the activity of PKC.

Recent evidence indicates that phospholipases D and A$_2$ (PLD and PLA$_2$) also are involved in the sustained activation of PKC, through their hydrolysis of membrane phosphatidylcholine (PC). PLD action on PC leads to the release of phosphatidic acid, which in turn is converted to DAG by a specific phosphatidic acid phosphomonoesterase. PLA$_2$ hydrolyzes PC to yield free fatty acids and lysoPC, both of which have been shown to potentiate the DAG-mediated activation of PKC. Of medical significance is the ability of phorbol ester tumor promoters to activate PKC directly. This leads to elevated and unregulated activation of PKC and a consequent loss of control over normal cellular growth and proliferation, resulting ultimately in neoplasia.

A. Phosphatidylinositol-3-Kinase: PI-3K is tyrosine-phosphorylated, and subsequently activated, by various RTKs and receptor-associated and non-receptor tyrosine kinases. PI-3K is a heterodimeric protein containing 2 subunits, one 85 kDa and the other 110 kDa. The p85 subunit contains SH2 domains that interact with activated receptors or other receptor-associated and non-receptor tyrosine kinases (Figure 34–3); the subunit itself is then tyrosine-phosphorylated and activated. The 85-kDa subunit is non-catalytic but does contain a domain homologous to GAP proteins. It is the 110-kDa subunit that is enzymatically active. PI-3K associates with and is activated by the PDGF, EGF, insulin, IGF-I, HGF and NGF receptors. PI-3K phosphorylates various phosphatidylinositols at the 3 position of the inositol ring. This activity generates additional substrates for PLCγ, allowing a cascade of DAG and IP$_3$ to be generated by a single activated RTK or by other protein tyrosine kinases.

MAP Kinases

MAP kinases were identified by virtue of their activation in response to growth factor–stimulation of cells in culture; hence the name **mitogen-activated protein kinases**. MAP kinases are also called ERKs, for **extracellular-signal regulated kinases**. On the basis of in vitro substrates the MAP kinases have been variously called microtubule associated protein-2 kinase (MAP-2 kinase), myelin basic protein kinase (MBP kinase), ribosomal S6 protein kinase (RSK)-kinase (ie, a kinase that phosphorylates a kinase), and EGF receptor threonine kinase (ERT kinase). All these proteins have similar biochemical properties, immunological cross-reactivities, amino-acid sequences, and ability in vitro to phosphorylate ribosomal S6 protein kinase (RSK).

Maximal MAP kinase activity requires that both tyrosine and threonine residues are phosphorylated. This indicates that MAP kinases act as switch kinases that transduce information of increased intracellular tyrosine phosphorylation to that of serine/threonine phosphorylation. Although MAP kinase activation was first observed in response to activation of the EGF, PDGF, NGF, and insulin RTKs, other cellular stimuli such as T-cell activation (which signals through the lck tyrosine kinase), phorbol esters (which function through activation of PKC), thrombin, bombesin, and bradykinin (which function through G-proteins) as well as N-methyl-D-aspartate (NMDA) receptor activation and electrical stimulation, rapidly induce tyrosine phosphorylation of MAP kinases.

MAP kinases are, however, not the direct substrates for RTKs nor for receptor associated tyrosine kinases, but are in fact activated by additional enzymes termed the MAP kinase kinases (MAPK kinases) and even the MAPK kinase kinases (MAPKK kinases). One of the MAPK kinases has been identified as the proto-oncogenic serine/threonine kinase, Raf1. The ultimate targets of the MAP kinases are several transcriptional regulators—for instance, serum response factor (SRF) and the proto-oncogenes Fos, Myc and Jun—as well as members of the steroid/thyroid hormone receptor superfamily of proteins (Figure 34–1).

G-Protein Coupled Receptors

There are several different classifications of receptors that couple signal transduction to G-proteins (Figure 34–1). Three different classes are reviewed here:

1. G-protein coupled receptors that **modulate adenylate cyclase activity**. One class of adenylate cyclase–modulating receptors activates the enzyme leading to the production of cAMP as the second messenger. Receptors of this class include the β-adrenergic, glucagon, and odorant molecule receptors. Increases in the production of cAMP lead to an increase in the activity of PKA in the case of β-adrenergic and glucagon receptors. In the case of odorant molecule receptors, the increase in cAMP leads to the activation of ion channels. In contrast to increased adenylate cyclase activity, the α-type adrenergic receptors are coupled to inhibitory G-proteins that repress adenylate cyclase activity upon receptor activation.
2. G-protein coupled receptors that activate PLCγ, leading to the **hydrolysis of polyphosphoinositides** (eg, PIP$_2$), which in turn generates the second messengers diacylglycerol (DAG) and inositoltrisphosphate (IP$_3$). This class of receptors includes the angiotensin, bradykinin, and vasopressin receptors.
3. A novel class of G-protein coupled receptors are the **photoreceptors**. This class is coupled to a G-protein termed transducin, which activates a phosphodiesterase that leads to a decrease in the level of cGMP. The drop in cGMP then results in the closing of a Na^+/Ca^{2+} channel, leading to hyperpolarization of the cell (see Chapter 36).

G-Protein Regulators

The activity of G-proteins with respect to GTP hydrolysis is in turn regulated by a family termed **GTPase activating proteins (GAPs)**. The proto-oncogenic protein, ras, is a G-protein involved in the genesis of numerous forms of cancer when it sustains specific mutations. Of particular clinical significance is the involvement of ras in the development of colorectal cancers, where it is found to be implicated more frequently than any other gene. The regulation of ras GTPase activity is controlled by rasGAP.

Several GAP proteins besides rasGAP are important in signal transduction; two have clinical significance. One is the gene product of the **neurofibromatosis type-1 (NF1)** susceptibility locus. The NF1 gene is a tumor suppressor gene (see Chapter 35), and the protein encoded is called **neurofibromin**. The second is the protein encoded by the **BCR locus (break point cluster region gene)**. The BCR locus is rearranged in the so-called **Philadelphia$^+$ chromosome** (Ph$^+$), which is observed with high frequency in **chronic myelogenous leukemias (CMLs)** and **acute lymphocytic leukemias (ALLs)**.

Intracellular Hormone Receptors

Hormone receptors are proteins that, by residing in the cytoplasm, effectively bypass all the signal transduction pathways described thus far. Additionally, all hormone receptors are bi-functional: they are capable of binding hormone as well as directly activating gene transcription.

The steroid/thyroid hormone receptor superfamily (which includes, for instance, **glucocorticoid, vitamin D, retinoic acid, and thyroid hormone receptors**), is a class of proteins that reside in the cytoplasm and bind the lipophilic steroid/thyroid hormones. These hormones are capable of freely penetrating the hydrophobic plasma membrane. Upon binding ligand, the hormone-receptor complex translocates to the nucleus and binds to specific DNA sequences termed "hormone response elements" (HREs). The binding of the complex to an HRE alters the transcription rates of the associated gene.

Phosphatases

Substantial evidence links both tyrosine and serine/threonine phosphorylation with increased cellular growth, proliferation, and differentiation. Removal of the incorporated phosphates is necessary in order to turn off proliferative signals; this observation suggests that phosphatases may function as anti-oncogenes or growth suppressor genes. The loss of a functional phosphatase involved in regulating growth-promoting signals could lead to neoplasia. However, instances are known in which dephosphorylation is required for the promotion of cell growth. This is particularly true of specialized kinases that are directly involved in regulating cell cycle progression. Therefore, it is difficult to envision all phosphatases as being tumor suppressor genes.

There are two broad classes of protein tyrosine phosphatases (PTPs). One class is the transmembrane enzymes, which contain the phosphatase activity domain in the intracellular portion of the protein. The other are intracellularly localized enzymes.

The first transmembrane PTP characterized was the **leukocyte common antigen protein, CD45**. This protein was shown to have homology to the intracellular PTP, PTP1B. There are now six subclasses of the transmembrane PTPs known. As yet no clearly defined ligands have been assigned to any transmembrane PTP.

The clearest studies of a role for transmembrane PTPs in signal transduction have involved the CD45 protein. These studies have shown that CD45 is involved in the regulation of the tyrosine kinase activity of lck in T cells. As indicated above, lck is associated with T-cell antigens CD4 and CD8, generating a split RTK that is involved in T-cell activation. It is suspected that CD45 dephosphorylates a regulatory tyrosine phosphorylation site in the C-terminus of lck, thereby increasing the activity of lck towards its substrate(s).

The second class of PTPs is the intracellular proteins. The C-terminal residues of most if not all intracellular PTPs are very hydrophobic, which suggests that these sites serve as membrane-attachment domains. One role of intracellular PTPs is in the maturation of *Xenopus* oocytes in response to hormone. It has been found that the over-expression of PTP1B in oocytes markedly retards the rate of insulin- and progesterone-induced maturation. These results suggest a role for PTP1B in countering the signals leading to cellular activation. As with the transmembrane PTPs, little is known about the regulation of the activity of the intracellular PTPs.

Two recently identified intracellular PTPs (PTP1C and PTP1D) have been shown to contain SH2 domains. These SH2 domains allow the PTPs to interact directly with tyrosine-phosphorylated RTKs and PTKs, thereby dephosphorylating the tyrosines in these proteins. Following receptor stimulation of signal-transduction events, the PTPs that contain SH2 domains are directed to several of the RTKs and/or PTKs; the net effect is a termination of the signaling events by tyrosine dephosphorylation.

Other phosphatases that recognize serine and/or threonine phosphorylated proteins also exist in cells. These are referred to as **protein serine phosphatases (PSPs)**. At least 15 distinct PSPs have been identified. The type 2A PSPs exhibit selective substrate specificity towards PKC-phosphorylated proteins, particularly serine and threonine phosphorylated receptors. Type 2A PSPs are more effective than other PSPs in dephosphorylating RSKs, proteins that are involved in signaling cascades by phosphorylating ribosomal S6 protein (see above). However, a type 1 PSP is required for dephosphorylating S6 itself.

The type 2A PSPs have 2 subunits (one regulatory and one catalytic), both of which can associate with one of the tumor antigens of the DNA tumor virus, polyoma. Transformation by DNA tumor viruses such as polyoma appears to be mediated by the formation of a signal-transduction unit consisting of a virally encoded T-antigen and several host-encoded proteins. Several host proteins are tyrosine kinases of the src family. Polyoma middle-T antigen (the T antigens of DNA tumor viruses are the major viral proteins detected in virally transformed cells) also can bind to PI-3K. The association of type 2A PSPs in these complexes may lead to dephosphorylation of regulatory serine/threonine phosphorylated sites, resulting in increased signal transduction and subsequent cellular proliferation.

Questions

DIRECTIONS (items 1–3): Each numbered item or incomplete statement is followed by answers or by completions of the statement. Select the ONE lettered answer or completion that is BEST in each case.

1. Which of the following characteristics defines the Class I RTKs?
   a. Seven immunoglobulin-like domains in the extracellular portion.
   b. Three immunoglobulin-like domains in the extracellular portion.
   c. Cysteine-rich extracellular domain.
   d. Contains an insert in the kinase domain.
   e. Composed of heterodimers.

2. The role of PLA_2 in the activation of PKC is:
   a. To hydrolyze phospholipids generating IP_3, which activates PKC.
   b. To hydrolyze phospholipids generating DAG, which activates PKC.
   c. To yield phosphatidic acid upon phospholipid hydrolysis, which is then converted to DAG, which then activates PKC.
   d. To hydrolyze an inhibitory lipid from the N-terminus of PKC, allowing the enzyme to leave its inhibitory membrane-bound location.
   e. To generate free fatty acids from phospholipids, which can then potentiate the DAG-mediated activation of PKC.

3. Receptors that are coupled to G-proteins are defined by a structure that includes:
   a. Cysteine-rich extracellular domains.
   b. Seven transmembrane spanning domains.
   c. Immunoglobulin-like extracellular domains.
   d. The formation of heterodimers.
   e. A domain for binding lipophilic hormones.

DIRECTIONS (items 4–10): Match the following numbered statements and protein names to the lettered protein classifications or descriptive statements of function in signal transduction. Each letter is used once.

   a. Binds ligand intracellularly, then interacts with DNA to regulate transcription.
   b. Composed of 2 subunits, 1 of which contains SH2 domains.
   c. Dephosphorylates and thereby inhibits lck activity.
   d. Possesses tyrosine kinase activity.
   e. Coupled to cAMP production via a specific G-protein.
   f. Hydrolyzes PIP_2, yielding DAG and IP_3.
   g. Possesses serine/threonine kinase activity.

4. EGF receptors.

5. PI-3K.

6. PLCγ.

7. TGF-β receptors.

8. β-adrenergic receptors.

9. Leukocyte common antigen.

10. Glucocorticoid receptors.

Answers

1. c	**5.** b	**9.** c	
2. e	**6.** f	**10.** a	
3. b	**7.** g		
4. d	**8.** e		

Growth Factors and Proto-Oncogenes in Cancer

Objectives

- To define the terms proto-oncogene and oncogene.
- To describe the differences between dominant and negative oncogenes.
- To develop an appreciation for the numerous growth regulatory molecules that have been shown to be able to induce a transformed phenotype in cells.
- To describe the characteristics of the tumor suppressor genes, Rb, WT1, NF-1, FAP (or APC), DCC, and p53.

Concepts

Most, if not all, cancer cells contain genetic damage that appears to be the factor leading to tumorigenesis. The genetic damage that exists in a "parental" tumorigenic cell is maintained (in other words, it does not correct itself) and thus it is a heritable trait of all cells of subsequent generations. The genetic damage found in cancer cells falls into two categories:

1. **Dominant: the genes are proto-oncogenes.** The distinction between the terms **proto-oncogene** and **oncogene** relates to the activity of the protein product of the gene. A proto-oncogene is a gene whose protein product has the capacity to induce cellular transformation if it sustains some genetic insult. An oncogene is a gene that has sustained some genetic damage leading to over-expression or production of an abnormal protein; either event being capable of cellular transformation.
2. **Recessive: termed tumor suppressors, growth suppressors, recessive oncogenes or anti-oncogenes.**

Given the complexity of inducing and regulating cellular growth, proliferation and differentiation (as indicated in Chapter 34), researchers suspected for many years that genetic damage to genes encoding growth factors, growth factor receptors and/or the proteins of the various signal-transduction cascades would lead to cellular transformation. This suspicion has been borne out with the identification of numerous genes whose products function in cellular signaling, that are involved in some way in the genesis of the tumorigenic state.

Tumor cells also can arise by nongenetic means through the actions of specific tumor viruses. Such tumor viruses are of two distinct types: those with DNA genomes (eg, papilloma and adenoviruses) and those with RNA genomes (termed **retroviruses**).

RNA tumor viruses are common in chickens, mice, and cats, but rare in humans. The only currently known human retroviruses are the **human T-cell leukemia viruses (HTLVs)** and the related retrovirus, **human immunodeficiency virus (HIV)**.

Retroviruses can induce the transformed state within cells by means of two mechanisms, both of which are related to the viral life cycle. When a retrovirus infects a cell its RNA genome is converted into DNA by the viral encoded RNA-dependent DNA polymerase (**reverse transcriptase**). The DNA then integrates into the genome of the host cell, where it can remain and be copied as the host genome is duplicated during the process of cellular division. Contained within the sequences at the ends of the retroviral genome are powerful transcriptional promoter sequences, termed **long terminal repeats (LTRs)**. The LTRs promote the transcription of viral DNA, leading to the production of new virus particles.

At some frequency the integration process leads to rearrangement of the viral genome and the consequent incorporation of a portion of the host genome into the viral genome. This process, termed

transduction, occasionally leads to the virus acquiring a gene from the host that is normally involved in cellular growth control. Because of the alteration of the host gene during transduction, as well as the gene being transcribed at a higher rate because of its association with retroviral LTRs, the transduced gene confers a growth advantage to the infected cell. The end result is unrestricted cellular proliferation, leading to tumorigenesis. The transduced genes are termed **oncogenes**; the normal cellular gene in its unmodified, non-transduced form is termed a **proto-oncogene**, since it has the capacity to transform cells if altered in some way or expressed in an uncontrolled manner. At least 30 different oncogenes have been discovered in the genomes of transforming retroviruses.

The second mechanism by which retroviruses can transform cells relates to the powerful transcription-promoting effect of the LTRs. When a retrovirus genome integrates into a host genome, it does so randomly. At some frequency this integration process leads to the placement of LTRs close to a gene that encodes a growth regulating protein. If the protein is expressed at an abnormally elevated level it can result in cellular transformation. This mechanism is termed **retroviral integration–induced transformation**. It has recently been shown that HIV induces certain forms of cancers in infected individuals by this process.

Cellular transformation by DNA tumor viruses, in most cases, has been found to be the result of protein-protein interaction. Proteins encoded by the DNA tumor viruses, termed **tumor antigens or T antigens**, can interact with cellular proteins. This interaction effectively sequesters the cellular proteins away from their normal functional locations within the cell. The predominant types of proteins that are sequestered by viral T antigens have been shown to be of the tumor-suppressor type. It is the loss of their normal suppressor functions that results in cellular transformation (see below for several descriptions).

Proto-Oncogenes and Oncogenes

The process of activating proto-oncogenes to oncogenes can include retroviral transduction or retroviral integration (see above), point mutations, insertion mutations, gene amplification, chromosomal translocation, and/or protein-protein interactions.

Proto-oncogenes can be classified into many different groups on the basis of their normal function within cells or of their sequence homology to other known proteins. As predicted, proto-oncogenes have been identified at all levels of the various signal-transduction cascades that control cell growth, proliferation, and differentiation. The list of proto-oncogenes identified to date is too lengthy to include here; only those genes that have been highly characterized are described. Proto-oncogenes that were originally identified as resident in transforming retroviruses are designated with the prefix **"c-"** to indicate their cellular origin, as opposed to **"v-"** to signify original identification in a retrovirus.

A.   Growth Factors: The c-sis gene (simian, sarcoma) encodes the PDGF B chain, simian sarcoma. The int-2 gene encodes an FGF-related growth factor. The K-FGF (also KGF and hst) gene also encodes an FGF-related growth factor and was identified in gastric carcinoma and Kaposi's sarcoma cells.

B.   Receptor Tyrosine Kinases: The flg ("flag") gene encodes the FGF receptor. The neu ("new") gene was identified as an EGF receptor-related gene in an ethylnitrosourea-induced *neu*roblastoma. The conversion of proto-oncogenic to oncogenic neu requires a change in only a single amino acid in the transmembrane domain. The c-fms gene encodes the colony stimulating factor-1 (CSF-1) receptor. The trk gene encodes the NGF receptor and was first found in a pancreatic cancer. The met gene encodes the hepatocyte growth factor(HGF)/scatter factor (SF) receptor. The c-kit gene encodes the mast cell growth factor receptor.

C.   Membrane Associated Non-Receptor Protein Tyrosine Kinases: The c-src gene was the first oncogene to be identified. It is the archetypal protein tyrosine kinase. The c-lck gene was isolated from a T-cell tumor line (*L*YST*RA c*ell *k*inase) and has been shown to be associated with the CD4 and CD8 antigens of T-cells.

D.   G-Protein Coupled Receptors: The mas gene was identified in a mammary carcinoma and has been shown to be the angiotensin receptor.

E.   Membrane Associated G-Proteins: There are 3 different homologs of the ras gene, each of which was identified in a different type of tumor cell. The ras gene is one of the genes most frequently involved in colorectal carcinomas.

F. Serine/Threonine Kinases: c-Raf1 gene is involved in the signaling pathway of RTKs. One role of the c-Raf1 kinase is to threonine phosphorylate MAP kinase following receptor activation.

G. Nuclear DNA-Binding/Transcription Factors: The Myc gene was originally identified in the avian myelocytomatosis virus. A disrupted human c-Myc gene has been found to be involved in numerous hematopoietic neoplasias. The disruption of c-Myc has been shown to result from retroviral integration and transduction, as well as from chromosomal rearrangements. The c-Myc protein binds to the DNA sequence motif CACGTG.

The Fos gene was identified in the feline osteosarcoma virus. The c-Fos protein interacts with a second proto-oncogenic protein, Jun, to form a transcriptional regulator.

The p53 gene was originally identified as a major nuclear antigen in transformed cells. The p53 gene is the single most commonly found mutant protein in human tumors. Mutant forms of the p53 protein interfere with cell growth suppressor effects of wild-type p53 indicating that the p53 gene product is actually a tumor suppressor (see below).

Tumor Suppressor Genes

Since the vast majority of cells in the human body are not actively dividing at any given time, DNA tumor viruses normally infect non-dividing cells. Therefore, in order to replicate, the infecting viruses must alter the program of the host cell and induce synthesis of the proteins required for DNA replication. Recent evidence has shown that the T-antigens of DNA tumor viruses are capable of binding to cellular proteins encoded by tumor suppressor genes. The complexing of T-antigens with cellular tumor suppressor proteins effectively inactivates the suppressive function and inductes cellular proliferation.

Several tumor suppressor genes have been isolated. These include genes identified through the study of familial cancers, such as the **retinoblastoma susceptibility gene (RB), Wilms' tumors (WT1), neurofibromatosis type-1 (NF1), and familial adenomatosis polyposis coli (FAP or APC),** as well as those identified through loss of heterozygosity, such as **colorectal carcinoma** (called DCC for detected in colon carcinoma) **and p53**.

A. Retinoblastoma (RB): In the familial form of this disease, individuals inherit a mutant allele from an affected parent that causes loss of function. Later, an additional somatic mutation takes place, inactivating the remaining normal allele and leading to the development of retinoblastoma. The result is an apparently dominant mode of inheritance. However, the fact that an additional somatic mutation must occur at the unaffected allele in order for the disease to manifest itself means that penetration of the defect is not always complete.

In sporadic forms of tumors involving the RB locus *two* somatic mutational events must occur; the second one affects the descendants of the cell undergoing the first mutation. Understandably, this combination of mutational events is extremely rare. The germline mutations at RB occur predominantly during spermatogenesis as opposed to oogenesis; however, the somatic mutations occur with equal frequency at the paternal or maternal locus. In contrast, somatic mutations at RB in sporadic osteosarcomas occur preferentially at the paternal locus. This may be the result of genomic imprinting.

The RB gene encodes a 110-kDa nuclear localized phosphoprotein (pRB). Detectable levels of pRB can be found in most proliferating cells, although only a limited number of tissues are actually affected by mutations in the RB gene (ie retina, bone, and connective tissue). In contrast, pRB is not detectable in any retinoblastoma cells.

Many different types of mutations can result in the loss of RB function. The highest percentage (30%) of retinoblastomas contain large-scale deletions. RNA splicing errors, point mutations, and small deletions in the promoter region have also been observed in some retinoblastomas.

The regulation of cell cycle progression also involves pRB. The ability of this protein to regulate the cell cycle correlates to its state of phosphorylation. Phosphorylation is maximal at the start of S phase and lowest after mitosis and entry into G_1. Stimulation of quiescent cells with mitogen induces phosphorylation of pRB, whereas differentiation induces hypophosphorylation. It is, therefore, the hypophosphorylated form of pRB that suppresses cell proliferation. Transformation by the DNA tumor viruses—adeno, polyoma and papilloma—is accomplished by binding of the T-antigens of these viruses to pRB. This protein-protein interaction occurs when pRB is in the hypophosphorylated (inhibitory) state.

B. Wilms' Tumor (WT1): Genetic evidence indicates that at least three distinct loci may be involved in the development of Wilms' tumor, a cancer of the kidney found in children. Either one (unilateral) or both (bilateral) kidneys can be involved. Sporadic evolution of Wilms' tumors is associated with two different deletions from the short arm of chromosome 11. There are also familial forms of Wilms' tumors that do not involve deletions of chromosome 11. These studies suggest that lesions at three different loci could lead to Wilms' tumors.

Only a single candidate locus has thus far been characterized. The potential gene for Wilms' tumor was found on chromosome 11 in a deleted region of about 345 kbp. This region contains a single transcription unit identified as WT1 that spans 50–60 kb. The expression of WT1 is very restrictive. Wilms' tumors that are homozygous for the chromosome 11 deletions do not contain WT1 mRNA. Most other Wilms' tumors show high expression of WT1 mRNA but this mRNA is likely to be produced from mutated WT1 genes. The WT1 gene codes for a protein containing 4 zinc finger domains (see Chapter 32), which may be a transcription factor.

C. Neurofibromatosis Type 1 (NF1): All cases of neurofibromatosis arise by inheritance of a mutant allele. Roughly 50% of all affected individuals carry new mutations; these appear to arise paternally, possibly reflecting genomic imprinting. Germline mutations at the NF1 locus result in multiple abnormal melanocytes (**café-au-lait spots**) and benign neurofibromas. Some patients also develop benign pheochromocytomas and tumors of the central nervous system. A small percentage of patients develop neurofibrosarcomas, which are likely to be derived from the Schwann cells.

The assignment of the NF1 locus to the short arm of chromosome 17 was carried out by linkage studies of affected pedigrees. The NF1 gene is extremely large: it encodes an mRNA of 11–13 kb, including a 7.5 kb protein coding region. The protein encoded therein consists of 2485 amino acids and shares striking homology to rasGAP. Expression of NF1 is observed in all tissues that have been examined.

The development of benign neurofibromas or of malignant neurofibrosarcomas may reflect the difference between inactivation of one NF1 allele or of both. However, changes elsewhere, besides the NF1 locus, are clearly indicated in the genesis of neurofibrosarcomas. A consistent loss of genetic material on the short arm of chromosome 17 is seen in neurofibrosarcomas but not neurofibromas. The chromosome 17 deletions also affect the wild-type p53 locus and may be associated with a mutant p53 allele on the other chromosome.

D. Familial Adenomatosis Polyposis (FAP or APC): FAP exhibits a dominant pattern of inheritance. The development of multiple colonic polyps characterizes the disease. These polyps arise during the second and third decades of life and become malignant carcinomas and adenomas later in life. Genetic linkage analysis assigned the FAP locus to the long arm of chromosome 5, a region that is also involved in nonfamilial forms of colon cancer. FAP adenomas appear to be loss-of-function mutants. One allele can be normal while the other is affected.

Identification of the FAP gene was aided by the observation in two patients of deletions at a particular locus spanning 100 kb. Three candidate genes in this region were examined for mutations that could be involved in FAP. Of these, one coding region (the DP2.5 gene) had sustained 4 distinct mutations specific to FAP patients, indicating this to be the FAP gene. The FAP gene contains 15 exons spanning approximately 125 kb of DNA encoding an 8.5 kb coding region in the mRNA. The protein coding region of the FAP gene is also extremely large, encompassing 2844 amino acids. No similarities to known proteins have been identified except for several stretches of sequences related to intermediate filament proteins.

E. p53: Loss of heterozygosity at the short arm of chromosome 17 has been associated with tumors of the lung, colon and breast. This region of chromosome 17 includes the p53 gene, which was originally discovered because the protein product complexed with the large T antigen of the SV40 DNA tumor virus. It was first thought that p53 was a dominant oncogene because cDNA clones isolated from tumor lines were able to cooperate with the oncogenic form of the ras gene in cellular transformation assays. This proved to be misleading, since the cDNA clones used in all these studies were mutated forms of wild-type p53; cDNAs from normal tissue were later shown to be incapable of ras co-transformation.

The mutant p53 proteins were shown to be altered in stability and conformation as well as their ability to bind the 70-kDa heat shock protein, hsp70. Subsequent analysis of several murine leukemia cell lines showed that the p53 locus was lost by either insertions or deletions on both alleles. This suggested that wild-type p53 may be a tumor suppressor rather than a dominant proto-oncogene. Direct confirmation came when it was shown that wild-type p53 could suppress the transformed phenotype in certain tumor cell lines generated by mutant p53 and oncogenic ras. It has now been demonstrated that mutation at the p53 locus occurs in cancers of the colon, breast, liver, and lung. Indeed, p53 is involved more frequently in neoplasia than is any other known tumor suppressor or dominant proto-oncogene.

The protein encoded by p53 is a nuclear localized phosphoprotein. A domain near the N-terminus of the p53 protein is highly acidic, as are similar domains found in various transcription factors. This suggests that p53 may be involved in transcriptional regulation. Indeed, p53 has been shown to bind DNA, *in vitro*, that contains at least two copies of the motif 5′-PuPuC(A/T)(A/T)GPyPyPy-3′ (Pu = purine; Py = pyrimidine). This motif suggests that p53 may bind to DNA as a tetramer—a form of binding that would explain the fact that mutant p53 proteins act in a dominant manner. They are present in complexes with wild type p53 and alter the function of the normal tetramer.

Like pRB, p53 forms a complex with SV40 large T antigen, as well as the E1B transforming protein of adenovirus and the E6 protein of human papilloma viruses. Complexing with these tumor antigens increases the stability of the p53 protein. The complexes between T-antigens and p53 renders p53 incapable of binding to DNA. A cellular protein, termed MDM2, has been shown to bind to p53. Complexes of p53 and MDM2 result in loss of p53-mediated activation of gene expression. An observation with clinical significance is the finding of chromosomal amplification of the MDM2 gene in a large fraction of human sarcomas.

F. Deleted in Colon Carcinoma (DCC): Loss of heterozygosity on the long arm of chromosome 18 is a frequent occurrence in colorectal carcinomas but not colorectal adenomas. A large transcription unit within this region has been identified and termed the DCC locus. The DCC RNA is transcribed at a low level in colonic mucosa and other tissues but is absent in a large proportion of colorectal carcinoma cell lines. The protein encoded by DCC appears to be a member of the immunoglobulin superfamily of proteins, with homology to neural cell adhesion molecules (NCAMs).

Questions

DIRECTIONS (items 1–10): Match the following lettered predicted functions with the known genes. Each answer can be used only once.

    a. Growth factor related to FGF.
    b. Mast cell growth factor.
    c. Tyrosine kinase associated with T cell CD4 and CD8 proteins.
    d. Transcription factor-related tumor suppressor containing zinc-finger domains.
    e. Tumor suppressor transcription factor that functions as a tetramer.
    f. Associates with tumor suppressor proteins.
    g. Proto-oncogene encoding the PDGF B chain.
    h. Serine/threonine kinase involved in signal transduction.
    i. Cell cycle regulating tumor suppressor.
    j. Transcription factor proto-oncogene that contains a helix-loop-helix domain.

1. p53.

2. c-sis.

3. int-2.

4. c-kit.

5. Raf1.

6. Myc.

7. Lck.

8. RB.

   9. WT1.

   10. SV40 T antigen.

Answers

1. e	**5.** h	**9.** d
2. g	**6.** j	**10.** f
3. a	**7.** c	
4. b	**8.** i	

36

Vitamins and Minerals

Objectives

- To describe the role of thiamine (vitamin B_1).
- To describe the clinical significance of thiamine deficiency.
- To describe the role of riboflavin (vitamin B_2).
- To describe the clinical significance of flavin deficiency.
- To describe the role of niacin.
- To describe the clinical significance of niacin and nicotinic acid.
- To describe the role of pantothenic acid.
- To describe the role of the B_6 vitamins (pyridoxal, pyridoxamine, and pyridoxine).
- To describe the role of biotin.
- To describe the role of cobalamin (vitamin B_{12}).
- To describe the clinical significance of cobalamin deficiency.
- To describe the role of folic acid.
- To describe the clinical significance of folate deficiency.
- To describe the role of ascorbic acid (vitamin C).
- To describe the clinical significance of vitamin C deficiency.
- To describe the roles of vitamin A, including in gene control and vision.
- To describe the clinical significance of vitamin A deficiency.
- To describe the role of vitamin D.
- To describe the clinical significance of vitamin D deficiency.
- To describe the role of vitamin E.
- To describe the clinical significance of vitamin E deficiency.
- To describe the role of vitamin K.
- To describe the clinical significance of vitamin K deficiency.
- To describe the role of minerals in the diet.

Concepts

Vitamins are organic molecules that function in a wide variety of capacities within the body. The most prominent function is that of cofactor in various enzymatic reactions. A distinguishing feature of the vitamins is that they generally cannot be synthesized by mammalian cells and, therefore, must be supplied in the diet. Vitamins fall into two categories: those that are soluble in water (**thiamin, riboflavin, niacin, pantothenic acid, B_6, biotin, cobalamin, folic acid and ascorbic acid**), and those that are soluble in lipid (**vitamin A, vitamin D, vitamin E, and vitamin K**). The minerals used by the body also can be divided into two broad groups, defined as the **macrominerals** and the **microminerals** (also termed the **trace minerals**). The macrominerals are those that the adult human diet must include at a level greater than 100 mg/day; trace minerals are required in smaller daily amounts.

WATER-SOLUBLE VITAMINS

Thiamine (Vitamin B_1)

Thiamine is also known as **vitamin B_1**. This vitamin is derived from a substituted pyrimidine and a thiazole, which are coupled by a methylene bridge (Figure 36–1).

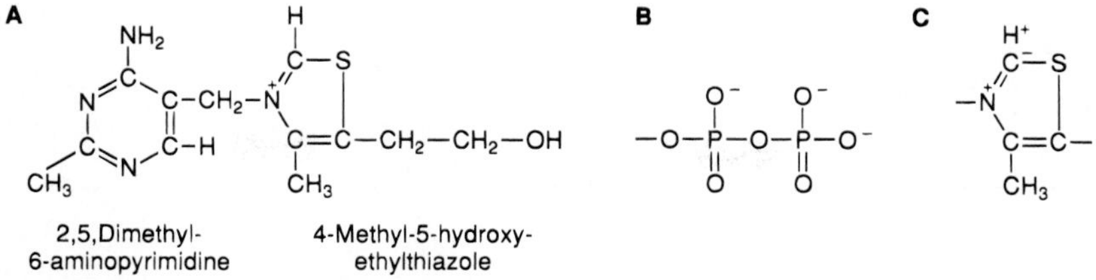

A 2,5,Dimethyl-6-aminopyrimidine   4-Methyl-5-hydroxy-ethylthiazole

Figure 36–1. Thiamine **A:** The free vitamin **B:** In thiamin diphosphate, the −OH group is replaced by pyrophosphate. **C:** Carbanion form. (Reproduced, with permission, from Murray, RK: *Harper's Biochemistry*, 23rd ed. Appleton & Lange, 1993.)

Thiamine is rapidly converted to its active form, thiamin pyrophosphate, **TPP** (Figure 36–1), in the brain and liver by a specific enzyme, **thiamine diphosphotransferase**. TPP is necessary as a cofactor for reactions that are catalyzed by **pyruvate dehydrogenase** and α-**ketoglutarate dehydrogenase**, as well as for the **transketolase**-catalyzed reactions of the **pentose phosphate pathway**. A deficiency in thiamine intake severely reduces the capacity of cells to generate energy, since it serves a vital role in these reactions.

The dietary requirement for thiamine is proportional to the caloric intake of the diet and ranges from 1.0 to 1.5 mg/day for normal adults. If the carbohydrate content of the diet is excessive, additional thiamine will be required.

Riboflavin (Vitamin B₂)

Riboflavin is also known as **vitamin B$_2$**. This vitamin, whose structure is shown in Figure 36–2, is the precursor for the coenzymes flavin mononucleotide (**FMN**) and flavin adenine dinucleotide (**FAD**). (See Chapter 27 for synthesis of FMN and FAD).

The enzymes that require FMN or FAD as cofactors are termed **flavoproteins**. Several flavoproteins also contain metal ions and are termed **metalloflavoproteins**. Both classes of enzymes are involved in a wide range of redox reactions; **succinate dehydrogenase** and **xanthine oxidase** (Chapter 27) are two examples. During the course of the enzymatic reactions involving the flavoproteins, the reduced forms of FMN and FAD are generated, yielding FMNH$_2$ and FADH$_2$, respectively.

The normal requirement for riboflavin is 1.2–1.7 mg/day for normal adults.

Niacin (Vitamin B₃)

Niacin (nicotinic acid and nicotinamide) is also known as **vitamin B$_3$**. Both nicotinic acid and nicotinamide can serve as dietary sources of vitamin B$_3$. Niacin is required for the synthesis of the active forms of vitamin B$_3$, nicotinamide adenine dinucleotide (**NAD$^+$**) and nicotinamide adenine dinucleotide phosphate (**NADP$^+$**). (See Chapter 27 for the structure and synthesis of NAD$^+$ and NADP$^+$) Both NAD$^+$ and NADP$^+$ function as cofactors for numerous dehydrogenases, such as lactate and malate dehydrogenases.

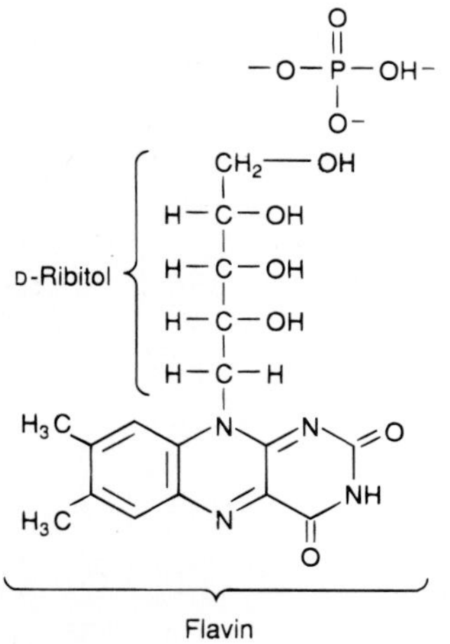

Figure 36–2. Riboflavin. In riboflavin phosphate (flavin mononucleotide, FMN), the −OH is replaced by phosphate. (Reproduced, with permission, from Murray, RK: *Harper's Biochemistry*, 23rd ed. Appleton & Lange, 1993.)

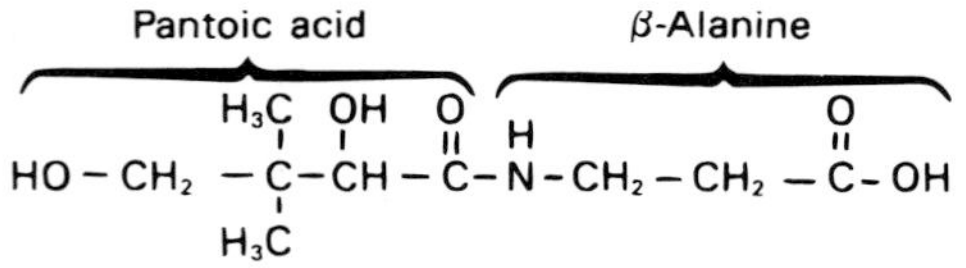

Figure 36–3. Pantothenic acid. (Reproduced, with permission, from Murray, RK: *Harper's Biochemistry*, 23rd ed. Appleton & Lange, 1993.)

Niacin is not a true vitamin according to the strict definition, since it can be derived from the amino acid tryptophan. However, the ability to utilize tryptophan for niacin synthesis is inefficient (60 mg of tryptophan are required to synthesize 1 mg of niacin). Also, the synthesis of niacin from tryptophan requires vitamins B_1, B_2, and B_6, which would pose a limitation themselves in the case of a marginal diet. The recommended requirement for niacin is 13–19 niacin equivalents (NE) per day for a normal adult. One NE is equivalent to 1 mg of free niacin.

Pantothenic Acid (Vitamin B₅)

Pantothenic acid is also known as **vitamin B_5**. This vitamin is formed from β-alanine and pantoic acid (Figure 36–3). Pantothenate is required for the synthesis of coenzyme A, **CoA** (see Chapter 27 for structure and synthesis of CoA) and is a component of the acyl carrier protein (**ACP**) domain of fatty acid synthase. Pantothenate is, therefore, required for the metabolism of carbohydrate via the TCA cycle and of all fats and proteins. At least 70 enzymes have been identified as requiring CoA or ACP derivatives for their function.

Vitamin B₆

Pyridoxal, pyridoxamine, and **pyridoxine** are collectively known as **vitamin B_6** (Figure 36–4). All three compounds are efficiently converted to the biologically active form of vitamin B_6, **pyridoxal phosphate** (Figure 36–5). This conversion is catalyzed by the enzyme **pyridoxal kinase**, which requires ATP.

Pyridoxal phosphate functions as a cofactor in enzymes involved in transamination reactions—those required for the synthesis and catabolism of the amino acids—as well as in glycogenesis as a cofactor for **glycogen phosphorylase**. The requirement for vitamin B_6 in the diet is proportional to the level of protein consumed and ranges from 1.4 to 2.0 mg/day for a normal adult. During pregnancy and lactation, the requirement for vitamin B_6 increases approximately 0.6 mg/day.

Biotin

Biotin (Figure 36–6) is the cofactor required of enzymes such as **acetyl-CoA carboxylase** and **pyruvate carboxylase**, which are involved in carboxylation reactions. Biotin is found in numerous foods and also is synthesized by intestinal bacteria; deficiencies of the vitamin are therefore rare.

Cobalamin (Vitamin B₁₂)

Cobalamin is more commonly known as **vitamin B_{12}**. Vitamin B_{12} is composed of a complex tetrapyrrol ring structure (corrin ring) and a cobalt ion in the center (Figure 36–7).

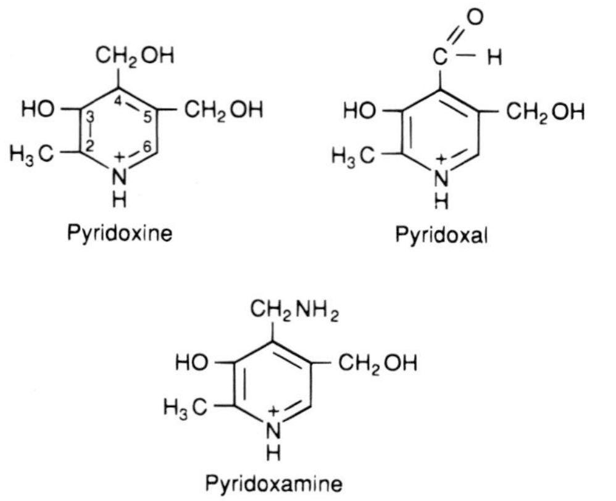

Figure 36–4. Naturally occurring forms of vitamin B_6. (Reproduced, with permission, from Murray, RK: *Harper's Biochemistry*, 23rd ed. Appleton & Lange, 1993.)

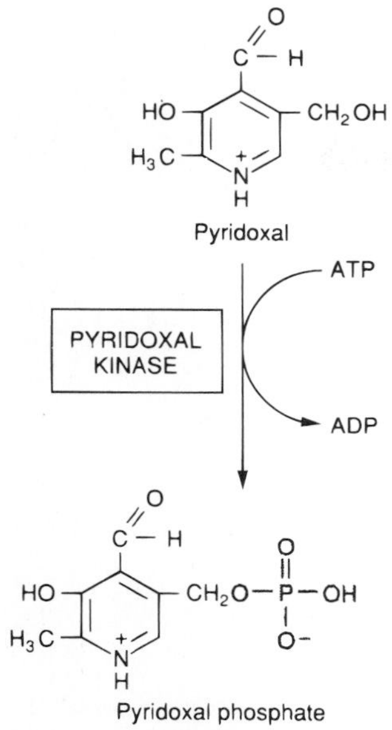

Figure 36–5. The phosphorylation of pyridoxal by pyridoxal kinase to form pyridoxal phosphate. (Reproduced, with permission, from Murray, RK: *Harper's Biochemistry*, 23rd ed. Appleton & Lange, 1993.)

Vitamin B_{12} is synthesized exclusively by microorganisms and is found in the liver of animals bound to protein as methycobalamin or 5′-deoxyadenosylcobalamin. The vitamin must be hydrolyzed from protein in order to be active. Hydrolysis is carried out in the stomach by gastric acids, or in the intestines by trypsin digestion, after the consumption of meat. The vitamin is then bound by **intrinsic factor**, a protein secreted by parietal cells of the stomach, and carried to the ileum where it is absorbed. Following absorption, the vitamin is transported to the liver in the blood bound to **transcobalamin II**.

There are only two reactions in the body that require vitamin B_{12} as a cofactor. During the catabolism of fatty acids with an odd number of carbon atoms and the amino acids valine, isoleucine, and threonine, the resultant propionyl-CoA is converted to succinyl-CoA for oxidation in the TCA cycle. One of the enzymes in this pathway, **methylmalonyl-CoA mutase**, requires vitamin B_{12} as a cofactor in the conversion of methylmalonyl-CoA to succinyl-CoA (Figure 36–8). The 5′-deoxyadenosine derivative of cobalamin is required for this reaction.

The second reaction requiring vitamin B_{12} catalyzes the conversion of homocysteine to methionine and is catalyzed by **methionine synthase**. This reaction results in the transfer of the methyl group from N^5-methyltetrahydrofolate to hydroxycobalamin, generating tetrahydrofolate and methylcobalamin during the process of the conversion (Figure 36–8).

Folic Acid

Folic acid is a conjugated molecule consisting of a pteridine ring structure linked to ρ-aminobenzoic acid (**PABA**) that forms pteroic acid (Figure 36–9). Folic acid itself is then generated through the conjugation of glutamic-acid residues to pteroic acid (Figure 36–9). Folic acid is obtained primarily from yeasts, leafy vegetables, and animal liver. Animals cannot synthesize PABA nor attach glutamate residues to pteroic acid, and therefore require folate in their diet.

When ingested or stored in the liver, folic acid exists in a polyglutamate form. Intestinal mucosal cells remove the glutamate residues through the action of the lysosomal enzyme, **conjugase**. The free folic acid is then reduced to tetrahydrofolate (**THF**, also **H_4 folate**) through the action of dihydrofolate reductase (**DHFR**), an enzyme that requires NADPH (see Chapter 27).

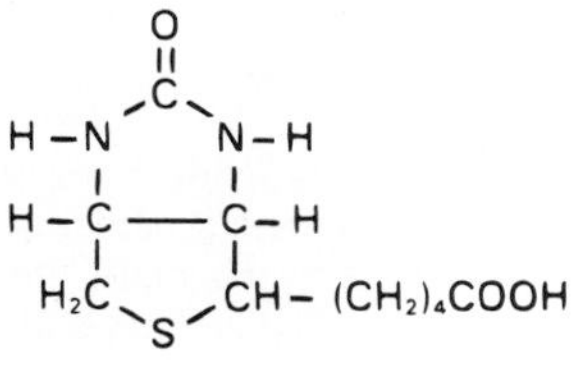

Figure 36–6. Biotin. (Reproduced, with permission, from Murray, RK: *Harper's Biochemistry*, 23rd ed. Appleton & Lange, 1993.)

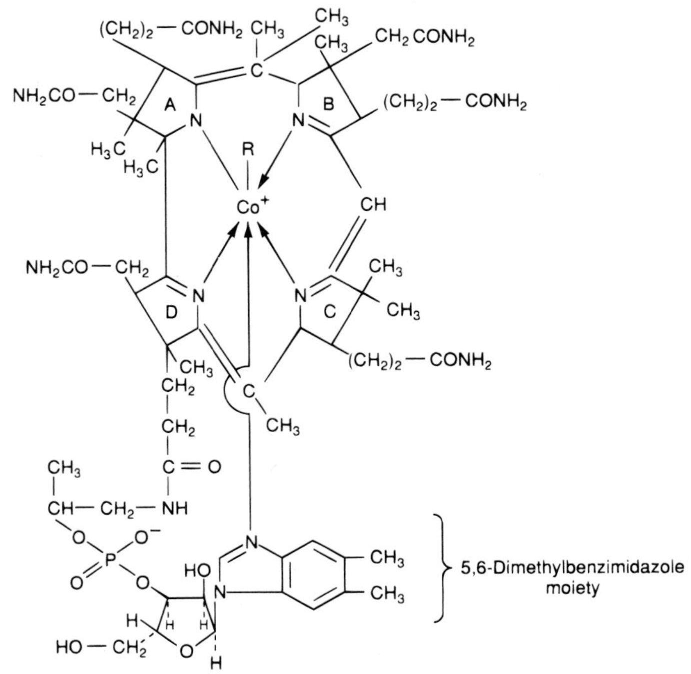

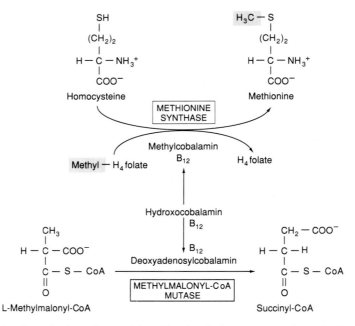

Figure 36–7. Vitamin B_{12} (cobalamin). R may be varied to give the various forms of the vitamin, eg, R = CN in cyanocobalamin; R = OH in hydroxocobalamin; R = 5'-deoxyadenosyl I 5'-deoxyadenosylcobalamin; and R = CH_3 in methyl-cobalamin. (Reproduced, with permission, from Murray, RK: *Harper's Biochemistry*, 23rd ed. Appleton & Lange, 1993.)

Figure 36–8. The two important reactions catalyzed by vitamin B_{12} coenzyme-dependent enzymes. Vitamin B_{12} deficiency leads to inhibition of both methylmalonyl-CoA mutase and methionine synthase activity, leading to methylmalonic aciduria, homocystinuria, and the trapping of folate as methyl-H_4 folate (the folate trap). (Reproduced, with permission, from Murray, RK: *Harper's Biochemistry*, 23rd ed. Appleton & Lange, 1993.)

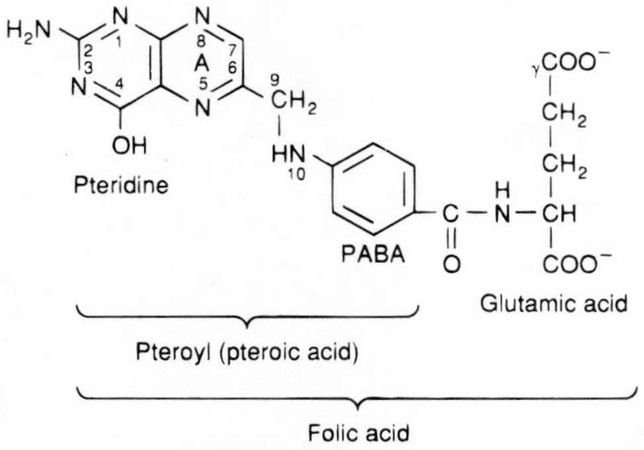

Figure 36–9. The structure and numbering of atoms of folic acid. (Reproduced, with permission, from Murray, RK: *Harper's Biochemistry*, 23rd ed. Appleton & Lange, 1993.)

The function of THF derivatives is to carry and transfer various forms of one one-carbon units during biosynthetic reactions. The units are either **methyl, methylene, methenyl, formyl,** or **formimino** groups (Figure 36–10). These one-carbon transfer reactions are required in the biosynthesis of **serine, methionine, glycine, choline,** and the **purine nucleotides** and **dTMP**.

The ability to acquire choline and amino acids from the diet and to salvage the purine nucleotides makes the role of N^5, N^{10}-methylene-THF in dTMP synthesis (see Chapter 27) the most metabolically significant function for this vitamin. The role of vitamin B_{12} and N^5-methyl-THF in the conversion of homocysteine to methionine (see above) also can have a significant impact on the ability of cells to regenerate needed THF.

Ascorbic Acid (Vitamin C)

Ascorbic acid is more commonly known as **vitamin C**. Ascorbic acid is derived from glucose via the uronic acid pathway (see Chapter 19). The enzyme **L-gulonolactone oxidase**, responsible for the conversion of gulonolactone to ascorbic acid, is absent in primates; hence ascorbic acid is required in the diet (Figure 36–11).

The active form of vitamin C is ascorbic acid itself. The main function of ascorbate is to serve as a reducing agent in a number of different reactions. Vitamin C has the potential to reduce cytochromes a and c of the respiratory chain, as well as molecular oxygen. The most important reaction requiring ascorbate as a cofactor is the hydroxylation of proline residues in collagen. Vitamin C is, therefore, required for the maintenance of normal connective tissue as well as for the healing of wounds, since the synthesis of connective tissue is the first event in wound tissue remodeling. Vitamin C also is necessary for bone remodeling, owing to the presence of collagen in the organic matrix of bones.

Several other metabolic reactions require vitamin C as a cofactor. These include the catabolism of tyrosine, the synthesis of epinephrine from tyrosine, and the synthesis of the bile acids. It is thought that vitamin C is also involved in the process of steroidogenesis, because the adrenal cortex contains high levels of vitamin C that become depleted when adrenocorticotropic hormone (ACTH) stimulates the gland.

CLINICAL SIGNIFICANCE

Thiamine (Vitamin B₁)

The earliest symptoms of thiamine deficiency include constipation, appetite suppression, and nausea, as well as mental depression, peripheral neuropathy, and fatigue. Chronic thiamine deficiency leads to more severe neurological symptoms, including ataxia, mental confusion, and a loss of eye coordination. Other clinical signs of prolonged thiamine deficiency come in the form of cardiovascular and muscular defects. The severe disease known as **beriberi** is the result of a diet that is rich in carbohydrates and deficient in thiamine. Also related to thiamine deficiency is **Wernicke-Korsakoff syndrome**, a disease most commonly found in chronic alcoholics as a result of poor diet.

Riboflavin (Vitamin B₂)

Riboflavin deficiencies are rare in the United States, thanks to adequate amounts of the vitamin in eggs, milk, meat, and cereals. Cases are, however, often seen in chronic alcoholics as a result of poor diet.

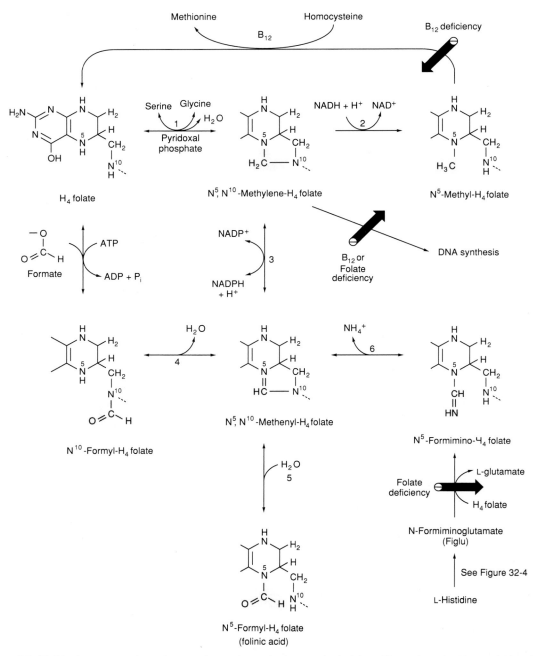

Figure 36–10. The interconversions of one-carbon unit attached to tetrahydrofolate. (Reproduced, with permission, from Murray, RK: *Harper's Biochemistry*, 23rd ed. Appleton & Lange, 1993.)

Symptoms associated with riboflavin deficiency include glossitis, seborrhea, angular stomatitis, cheilosis, and photophobia. Because riboflavin decomposes when exposed to visible light, riboflavin deficiencies can develop in newborns who are treated for hyperbilirubinemia by phototherapy.

Niacin (Vitamin B₃)

A diet deficient in niacin (as well as tryptophan) leads to glossitis of the tongue, dermatitis, weight loss, diarrhea, depression, and dementia. Depression, dermatitis, and diarrhea are associated with the con-

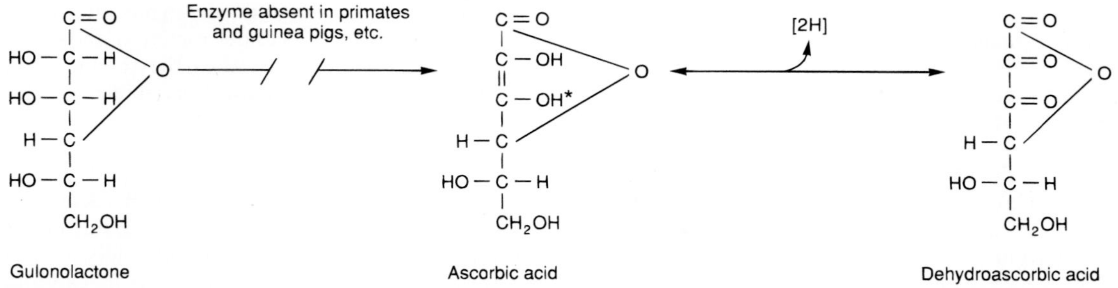

Figure 36–11. Ascorbic acid, its source in nonprimates, and its oxidation to dehydroascorbic acid. *(ionizes in ascorbate). (Reproduced, with permission, from Murray, RK: *Harper's Biochemistry*, 23rd ed. Appleton & Lange, 1993.)

dition known as **pellagra**. Several physiological conditions (eg, **Hartnup disease** and **malignant carcinoid syndrome**), as well as certain drug therapies (eg, isoniazid) can lead to niacin deficiency. In Hartnup disease tryptophan absorption is impaired, and in malignant carcinoid syndrome the metabolism of tryptophan is altered, resulting in excess synthesis of serotonin. Isoniazid (the hydrazide derivative of isonicotinic acid) is the primary drug for the chemotherapy of tuberculosis.

When administered in pharmacological doses of 2–4 g/day, nicotinic acid (but not nicotinamide) lowers plasma cholesterol levels and has been shown to be a useful therapeutic for hypercholesterolemia. The major action of nicotinic acid in this capacity is to reduce the mobilization of fatty acids from adipose tissue. Although nicotinic-acid therapy lowers blood cholesterol, it also depletes glycogen stores and fat reserves in skeletal and cardiac muscle. It also causes an elevation in blood glucose and uric acid production. For these reasons, nicotinic-acid therapy is not recommended for people with diabetes or gout.

Pantothenic Acid (Vitamin B$_5$)
Deficiency of pantothenic acid is extremely rare, owing to its widespread distribution in whole-grain cereals, legumes, and meat. Symptoms of pantothenate deficiency are difficult to assess, since they are subtle and resemble those of other B-vitamin deficiencies.

Vitamin B$_6$
Deficiencies of vitamin B$_6$ are rare and usually are related to an overall deficiency of all the B-complex vitamins. Isoniazid (see the discussion of niacin deficiencies above) and penicillamine (used to treat rheumatoid arthritis and cystinurias) are two drugs that associate with pyridoxal and pyridoxal phosphate, resulting in a deficiency of vitamin B$_6$.

Biotin
Deficiencies are generally seen only after long antibiotic therapies, which deplete the intestinal fauna, or as the result of excessive consumption of raw eggs. The latter is due to the affinity of the egg white protein, **avidin**, for biotin, which prevents intestinal absorption of the vitamin.

Cobalamin (Vitamin B$_{12}$)
The liver can store up to 6 years worth of vitamin B$_{12}$; hence, deficiencies of this vitamin are rare. **Pernicious anemia** is a megaloblastic anemia resulting from vitamin B$_{12}$ deficiency; this condition develops because of a lack of intrinsic factor in the stomach, leading to malabsorption of the vitamin. The anemia results from impaired DNA synthesis, which in turn is due to a block in purine and thymidine biosynthesis. The latter is a consequence of the effect of vitamin B$_{12}$ on folate metabolism. When vitamin B$_{12}$ is deficient, essentially all of the folate becomes "trapped" as the N^5-methyltetrahydrofolate derivative, as a result of the loss of functional methionine synthase. This trapping prevents the synthesis of other tetrahydrofolate derivatives required for the purine and thymidine nucleotide biosynthesis pathways (see Chapter 27).

Neurological complications also are associated with vitamin B$_{12}$ deficiency; these result from a progressive demyelination of nerve cells. The demyelination is thought to arise from the increase in methylmalonyl-CoA that comes with vitamin B$_{12}$ deficiency. Methylmalonyl-CoA is a competitive inhibitor of malonyl-CoA in fatty acid biosynthesis, as well as being able to substitute for malonyl-CoA in any fatty acid biosynthesis that may occur. Since the myelin sheath is continually in flux, the methyl-malonyl-CoA-induced inhibition of fatty acid synthesis eventually leads to destruction of the sheath. The incorporation of methylmalonyl-CoA into fatty acid biosynthesis causes the production of branched-chain fatty acids that may severely alter the architecture of the normal membrane structure of nerve cells.

Folic Acid

Folate deficiency results in complications nearly identical to those described for vitamin B_{12} deficiency. The most pronounced effect on cellular processes is found in DNA synthesis, owing to an impairment in dTMP synthesis (see Chapter 27). This leads to cell cycle arrest in the S-phase of rapidly proliferating cells, particularly hematopoietic cells. The result, as for vitamin B_{12} deficiency, is **megaloblastic leukemia**. The inability to synthesize DNA during erythrocyte maturation leads to abnormally large erythrocytes, a condition called **macrocytic anemia**.

Folate deficiencies are rare, thanks to adequate folate in most foods. Poor diet, (as in chronic alcoholism) can lead to folate deficiency. The principal causes of folate deficiency in non-alcoholics are impaired absorption or metabolism, or an increased demand for the vitamin. The predominant condition requiring an increase in the daily intake of folate is pregnancy. This is due to the increased number of rapidly proliferating cells in the blood. By the third trimester of pregnancy, the need for folate is nearly double that of an average adult. Certain drugs such as anticonvulsants and oral contraceptives can impair the absorption of folate. Anticonvulsants also increase the rate of folate metabolism.

The ability to target the metabolism of the folate pool (through inhibition of DHFR) with chemotherapeutic drugs for cancer such as **methotrexate** and **aminopterin** has been discussed in Chapter 27.

Ascorbic Acid (Vitamin C)

Because ascorbic acid is required for the synthesis of collagen, a deficiency of vitamin C leads to **scurvy**. This disease is characterized by easily bruised skin, muscle fatigue, soft swollen gums, decreased wound healing and hemorrhaging, osteoporosis, and anemia. Vitamin C is readily absorbed, and therefore the primary cause of vitamin C deficiency is poor diet and/or increased requirements. The primary physiological state leading to an increased requirement for vitamin C is severe stress (or trauma), which rapidly depletes the adrenal stores of the vitamin. The reason for the depletion is unclear, but may be due either to redistribution of the vitamin to areas that need it or to an overall increase in utilization.

LIPID-SOLUBLE VITAMINS

Vitamin A

Vitamin A consists of three biologically active molecules: **retinol, retinal** (retinaldehyde), and **retinoic acid** (Figure 36–12). Each of these compounds is derived from the plant precursor molecule β-**carotene**, (a member of a family of molecules known as **carotenoids**). Beta-carotene, which consists of two molecules of retinal linked at their aldehyde ends, is also referred to as the 'provitamin' form of vitamin A.

Ingested β-carotene is cleaved in the lumen of the intestine by β-**carotene dioxygenase** to yield retinal. Retinal is reduced to retinol by **retinaldehyde reductase**, an enzyme within the intestines that requires NADPH (Figure 36–12). Retinol is then esterified to palmitic acid and delivered to the blood via chylomicrons. The uptake of chylomicron remnants by the liver leads to the delivery of retinol to this organ for storage as a lipid ester within **lipocytes**. Transport of retinol from the liver to extrahepatic tissues occurs by binding of hydrolyzed retinol to aporetinol binding protein (**RBP**); the retinol-RBP complex is then transported to the cell surface within the Golgi and secreted. Within extrahepatic tissues, retinol is bound to cellular retinol binding protein (**CRBP**). The plasma transport of retinoic acid is accomplished by means of binding to albumin.

A. Gene Control Exerted by Retinol and Retinoic Acid: Within cells, both retinol and retinoic acid bind to specific receptor proteins. The receptor-vitamin complex then interacts with specific sequences in several genes involved in growth and differentiation, thereby affecting the expression of these genes. In this capacity retinol and retinoic acid are considered hormones of the steroid/thyroid hormone superfamily of proteins. Vitamin D also acts in a similar capacity (see below). Several genes whose patterns of expression are altered by retinoic acid are involved in the earliest processes of embryogenesis, including differentiation of the three germ layers, organogenesis, and limb development.

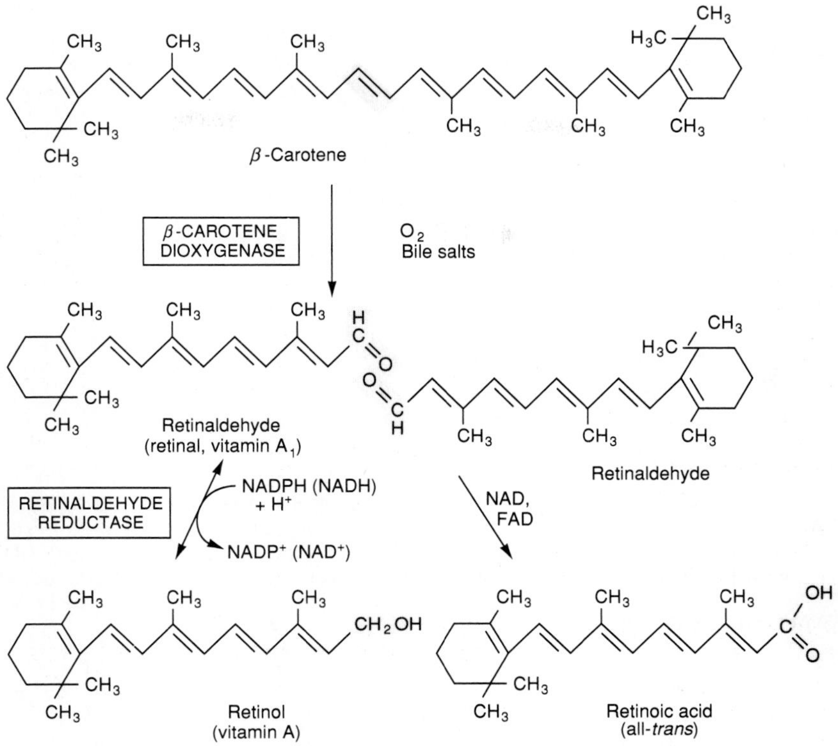

Figure 36–12. β-Carotene and its cleavage to retinaldehyde. The reduction of retinaldehyde to retinol and the oxidation of retinaldehyde to retinoic acid are also shown. 13-*cis*-Retinoic acid is also formed. (Reproduced, with permission, from Murray, RK: *Harper's Biochemistry*, 23rd ed. Appleton & Lange, 1993.)

B. Vision and the Role of Vitamin A: Photoreception in the eye is the function of two specialized cell types located in the retina: rods and cones. Both rod and cone cells contain a photoreceptor pigment in their membranes. The photosensitive compound of most mammalian eyes is a protein called **opsin**, to which is covalently coupled an aldehyde of vitamin A. The opsin of rod cells is called **scotopsin**; the photoreceptor itself is **rhodopsin**, or **visual purple**. This compound is a complex between scotopsin and the **11-*cis*-retinal** (also called 11-*cis*-retinene) form of vitamin A. Rhodopsin is a serpentine receptor embedded in the membrane of the rod cell. Coupling of 11-*cis*-retinal occurs at three of the transmembrane domains of rhodopsin. Intracellularly, rhodopsin is coupled to a specific G-protein called **transducin**.

When rhodopsin is exposed to light it is **bleached**, releasing the 11-*cis*-retinal from opsin. The absorption of photons by 11-cis-retinal triggers a series of conformational changes, which culminate in the conversion to **all-*trans*-retinal** (Figure 36–13). One important conformational intermediate in this pathway is **metarhodopsin II**. The photon-induced release of opsin brings about a conformational change in the photoreceptor. This change activates transducin, leading to increased binding of GTP by the α-subunit of transducin. The binding of GTP releases the α-subunit from the inhibitory β and γ-subunits. The GTP-activated α-subunit in turn activates an associated **phosphodiesterase**, an enzyme that hydrolyzes cyclic-GMP (cGMP) to GMP. Cyclic GMP is required to maintain the Na⁺ channels of the rod cell in the open conformation. The drop in cGMP concentration, however, results in complete closure of the Na⁺ channels, and it appears that metarhodopsin II is responsible for initiating the closure of the channels. This closure leads to hyperpolarization of the rod cell, with a concomitant propagation of nerve impulses to the brain (Figure 36–14).

C. Additional Role of Retinol: Retinol also functions in the synthesis of certain glycoproteins and mucopolysaccharides necessary for mucous production and normal growth regulation. This is accomplished by the phosphorylation of retinol to **retinyl phosphate**, which then functions similarly to dolichol phosphate (see Chapter 19).

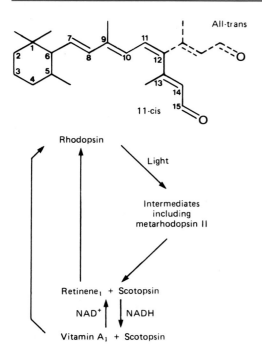

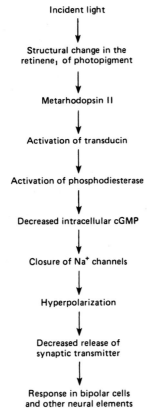

Figure 36–13. **Top:** Structure of retinene₁, showing the 11-*cis* configuration (unbroken lines) and the all-*trans* configuration produced by light (dashed lines). **Bottom:** Effects of light on rhodopsin. (Reproduced, with permission, from Ganong, WF: *Review of Medical Physiology*, 16th ed. Appleton & Lange, 1993.)

Figure 36–14. Sequence of events involved in phototransduction in rods and cones. (Reproduced, with permission, from Ganong, WF: *Review of Medical Physiology*, 16th ed. Appleton & Lange, 1993.)

Vitamin D

Vitamin D is a steroid hormone that regulates specific gene expression following interaction with its intracellular receptor. The biologically active form of the hormone is 1,25-dihydroxy vitamin D_3 (1,25-$(OH)_2D_3$, also known as **calcitriol**. The main function of calcitriol is to regulate calcium and phosphorous homeostasis.

Active calcitriol is derived from **ergosterol** (produced in plants) and from **7-dehydrocholesterol** (produced in the skin). **Ergocalciferol (vitamin D_2)** is formed by UV irradiation of ergosterol (Figure 36–15). In the skin, 7-dehydrocholesterol is converted to **cholecalciferol (vitamin D_3)** following uv irradiation (Figure 36–15).

Vitamins D_2 and D_3 are processed to D_2-calcitriol and D_3-calcitriol, respectively, by the same enzymatic pathways in the body. Cholecalciferol (or ergocalciferol) is absorbed from the intestine and transported to the liver bound to a specific **vitamin D-binding protein**. In the liver, cholecalciferol is hydroxylated at the 25 position by a specific **D_3-25-hydroxylase**, generating 25-hydroxy-D_3 [25-$(OH)D_3$] which is the major circulating form of vitamin D. The conversion of 25-$(OH)D_3$ to its biologically active form, calcitriol, occurs through the activity of a specific **D_3-1-hydroxylase** present in the proximal convoluted tubules of the kidneys, and in bone and placenta (Figure 36–16). 25-$(OH)D_3$ can also be hydroxylated at the 24 position by a specific **D_3-24-hydroxylase** in the kidneys, intestine, placenta, and cartilage (Figure 36–16).

Calcitriol functions in concert with parathyroid hormone **(PTH)** and **calcitonin** to regulate serum calcium and phosphorous levels. PTH is released in response to low serum calcium and induces the production of calcitriol. In contrast, reduced levels of PTH stimulate synthesis of the inactive 24,25-$(OH)_2D_3$. In the intestinal epithelium, calcitriol functions as a steroid hormone in inducing the expression of **calbindin D_{28}**, a protein involved in intestinal calcium absorption. The increased absorption of calcium ions requires the concomitant absorption of a negatively charged counter ion to maintain electrical neutrality. The predominant counter ion is P_i. When plasma calcium levels fall, the major sites of action of calcitriol and PTH are the bone, where they stimulate bone resorption, and the kidneys, where they inhibit calcium excretion by stimulating reabsorption by the distal tubules. The role of calcitonin in calcium homeostasis is to decrease elevated serum calcium levels by inhibiting bone resorption.

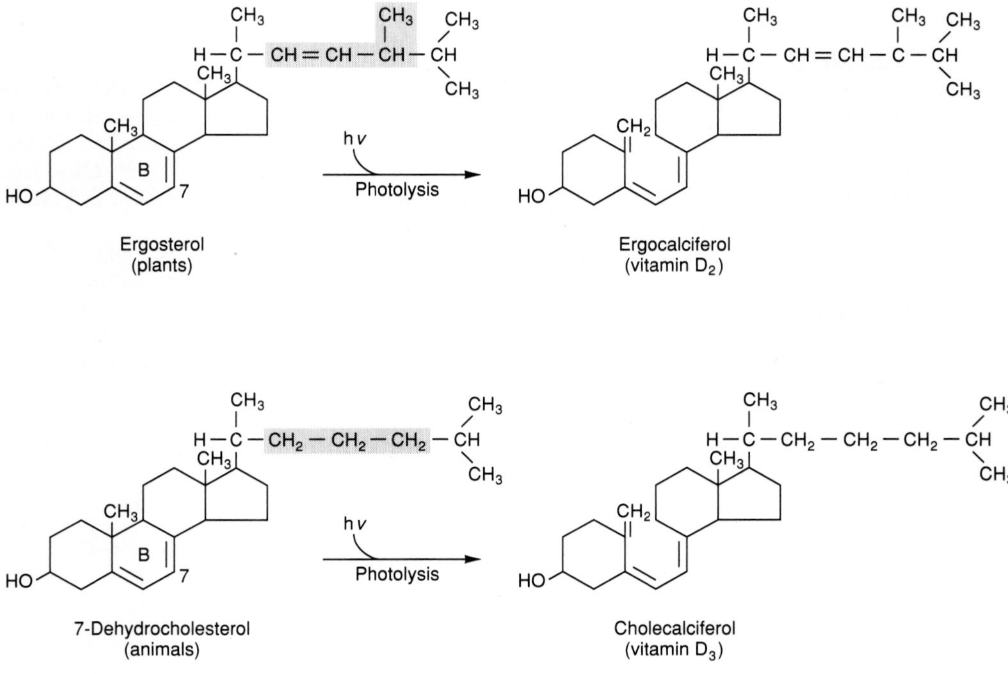

Figure 36–15. Ergosterol and 7-dehydrocholesterol and their conversion by photolysis to ergocalciferol and cholecalciferol, respectively. (Reproduced, with permission, from Murray, RK: *Harper's Biochemistry*, 23rd ed. Appleton & Lange, 1993.)

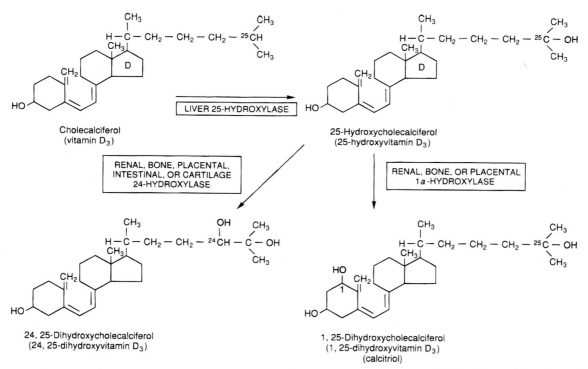

Figure 36–16. Cholecalciferol can be hydroxylated at the C_{25} position by a liver enzyme. The 25-hydroxycholecalciferol is further metabolized to 1α,25-dihydroxycholecalciferol or to 24,25-dihydroxycholecalciferol. The levels of 24,25-dihydroxycholecalciferol and 1,25-dihydroxycholecalciferol are regulated in a reciprocal manner. (Reproduced, with permission, from Murray, RK: *Harper's Biochemistry*, 23rd ed. Appleton & Lange, 1993.)

Vitamin E

Vitamin E is a mixture of several related compounds known as **tocopherols**. The α-tocopherol molecule (Figure 36–17) is the most potent of the tocopherols. Vitamin E is absorbed from the intestines packaged in chylomicrons. It is delivered to the tissues via chylomicron transport and then to the liver through chylomicron remnant uptake. The liver can export vitamin E in VLDLs. Because of its lipophilic nature, vitamin E accumulates in cellular membranes, fat deposits, and other circulating lipoproteins. The major site of vitamin E storage is adipose tissue.

The major function of vitamin E is to act as a natural **antioxidant** by scavenging free radicals and molecular oxygen. In particular, vitamin E is important for preventing peroxidation of polyunsaturated membrane fatty acids. The vitamins E and C are interrelated in their antioxidant capabilities. Active α-tocopherol can be regenerated by interaction with vitamin C following scavenge of a peroxy free radical (Figure 36–18). Alternatively, α-tocopherol can scavenge two peroxy free radicals and then be conjugated to glucuronate for excretion in the bile.

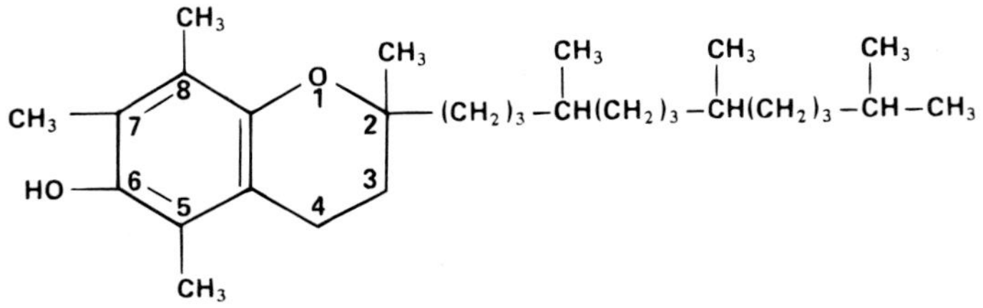

Figure 36–17. α-Tocopherol. (Reproduced, with permission, from Murray, RK: *Harper's Biochemistry*, 23rd ed. Appleton & Lange, 1993.)

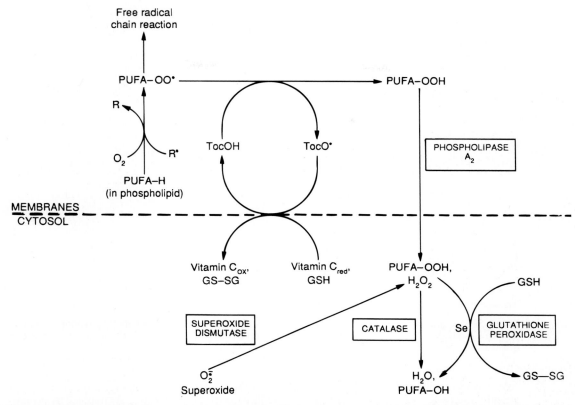

Figure 36–18. Interaction and synergism between antioxidant systems operating in the lipid phase (membranes) of the cell and the aqueous phase (cytosol). R•, free radical; PUFA-OO•, peroxyl free radical of polyunsaturated fatty acid in membrane phospholipid; PUFA-OOH, hydroperoxy polyunsaturated fatty acid in membrane phospholipid released as hydroperoxy free fatty acid into cytosol by the action of phospholipase A₂; PUFA-OH, hydroxy polyunsaturated fatty acid; TocOH, vitamin E (α-tocopherol); TocO•, free radical of α-tocopherol; Se, selenium; GSH, reduced glutathione; GS-SG, oxidized glutathione, which is returned to the reduced state after reaction with NADPH catalyzed by glutathione reductase. (Reproduced, with permission, from Murray, RK: *Harper's Biochemistry*, 23rd ed. Appleton & Lange, 1993.)

Vitamin K

The K vitamins exist naturally as **K₁** (phytylmenaquinone) in green vegetables and **K₂** (multiprenyl-menaquinone) in intestinal bacteria; **K₃** is synthetic menadione Figure 36–19).

The major function of the K vitamins is to serve as co-factors in the posttranslational modification of specific glutamatic acid residues within proteins. The modification is a carboxylation; generating γ-carboxyglutamate **(gla)** residues (Figure 36–20) and its extemely important for the proper function of several blood-clotting proteins, **factors II, VII, IX, and X** (see Chapter 37).

During the carboxylation reaction, the reduced **hydroquinone** form of vitamin K is converted to a **2,3-epoxide form**. The regeneration of the hydroquinone form requires an uncharacterized reductase. This latter reaction is the site of action of the **coumarin**-based anticoagulants, such as **warfarin** (see Chapter 37).

CLINICAL SIGNIFICANCE

Vitamin A

Vitamin A is stored in the liver, and deficiency of the vitamin occurs only after a prolonged lack in the diet. The earliest symptoms of vitamin A deficiency are **night blindness**. Additional early symptoms include follicular hyperkeratinosis, increased susceptibility to infection and cancer, and anemia equivalent to that of iron deficiency. Prolonged lack of vitamin A leads to deterioration of the eye tissue through progressive keratinization of the cornea, a condition known as **xerophthalmia**. The increased

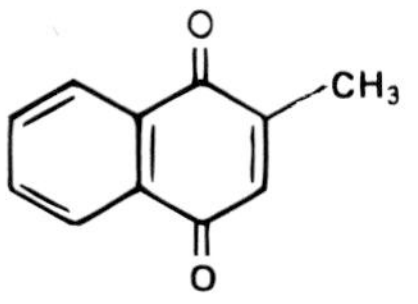

Menadione (vitamin K_3)

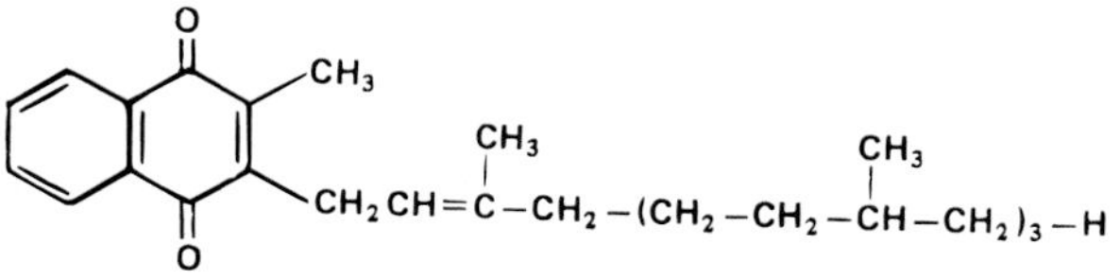

Phylloquinone (vitamin K_1, phytonadione, Mephyton)

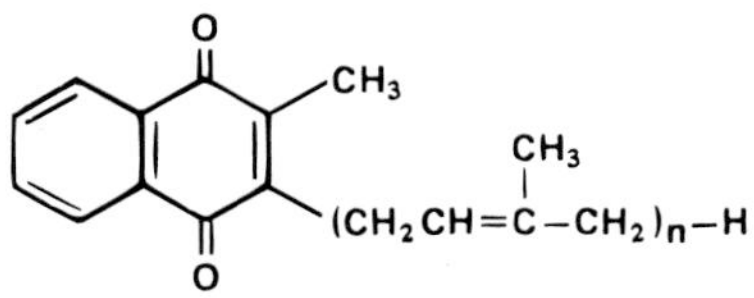

Menaquinone-n (vitamin K_2; n = 6, 7, or 9)

Figure 36–19. The naturally occurring vitamins K. Menadione is 2-methyl-1-1,4-napththoquinone. The other vitamin Ks are polyisoprenoid-substituted. (Reproduced with permission, from Murray, RK: *Harper's Biochemistry*, 23rd ed. Appleton & Lange, 1993.)

risk of cancer in vitamin A deficiency is thought to be the result of a depletion in β-carotene. A highly effective antioxidant, β-carotene is suspected to reduce the risk of cancers that are initiated by the production of free radicals. Of particular interest is the potential benefit of increased β-carotene intake to reduce the risk of lung cancer in smokers. However, caution is needed when increasing the intake of any lipid-soluble vitamin. Excess accumulation of vitamin A in the liver can lead to toxicity, which manifests as bone pain, hepatosplenomegaly, nausea, and diarrhea.

Vitamin D

As a result of the addition of vitamin D to milk, deficiencies in this vitamin are rare in this country. The main symptom of vitamin D deficiency in children is **rickets**, and in adults **osteomalacia**. Rickets is characterized by soft bones, a result of improper mineralization during the development of the bones. Osteomalacia is characterized by the demineralization of previously formed bone, leading to increased softness and susceptibility to fracture.

Vitamin E

No major disease states have been found to be associated with vitamin E deficiency, since the vitamin is supplied in adequate levels in the average American diet. The major symptom of vitamin E deficiency in humans is an increase in the fragility of red blood cells. Since vitamin E is absorbed from the intestines in chylomicrons, any fat malabsorption diseases can lead to deficiencies in vitamin E. Neurological disorders have been found to arise from vitamin E deficiencies associated with fat malabsorptive disorders. An increased intake of vitamin E is recommended for premature infants who receive formulas that are low in the vitamin, as well as for persons consuming a diet high in poly-

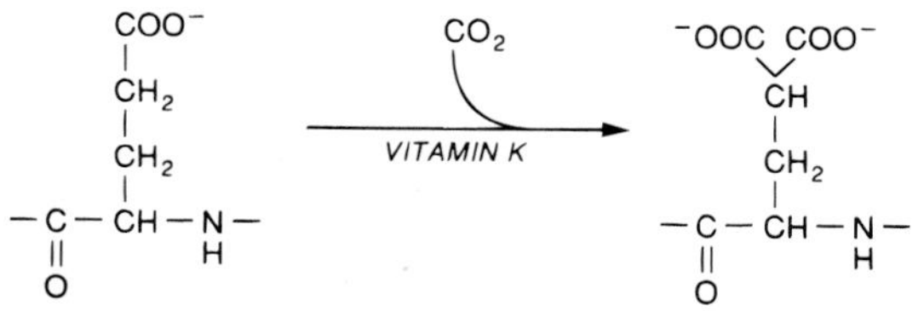

Figure 36–20. Carboxylation of a glutamate residue catalyzed by vitamin K-dependent carboxylase. (Reproduced, with permission, from Murray, RK: *Harper's Biochemistry*, 23rd ed. Appleton & Lange, 1993.)

Table 36–1. Essential macrominerals: Summary of major characteristics.

Elements	Functions	Metabolism[1]	Deficiency Disease or Symptoms	Toxicity Disease or Symptoms[2]	Sources[3]
Calcium	Constituent of bones, teeth; regulation of nerve, muscle function.	Absorption requires calcium-binding protein. Regulated by vitamin D, parathyroid hormone, calcitonin, etc.	Children: rickets. Adults: osteomalacia. May contribute to osteoporosis.	Occurs with excess absorption due to hypervitaminosis D or hypercalcemia due to hyperparathyroidism, or idiopathic hypercalcemia.	Dairy products, beans, leafy vegetables.
Phosphorus	Constituent of bones, teeth, ATP, phosphorylated metabolic intermediates. Nucleic acids.	Control of absorption unknown (vitamin D?). Serum levels regulated by kidney reabsorption.	Children: rickets. Adults: osteomalacia.	Low serum Ca^{2+}:P_i ratio stimulates secondary hyperparathyroidism; may lead to bone loss.	Phosphate food additives.
Sodium	Principal cation in extracellular fluid. Regulates plasma volume, acid-base balance, nerve and muscle function, Na^+/K^+-ATPase.	Regulated by aldosterone.	Unknown on normal diet; secondary to injury or illness.	Hypertension (in susceptible individuals).	Table salt; salt added to prepared food.
Potassium	Principal cation in intracellular fluid; nerve and muscle function, Na^+/K^+-ATPase.	Also regulated by aldosterone.	Occurs secondary to illness, injury, or diuretic therapy; muscular weakness, paralysis, mental confusion.	Cardiac arrest, small bowel ulcers.	Vegetables, fruit, nuts.
Chloride	Fluid and electrolyte balance; gastric fluid; chloride shift in HCO_3^- transport in erythrocytes.		Infants fed salt-free formula. Secondary to vomiting, diuretic therapy, renal disease.		Table salt.
Magnesium	Constituent of bones, teeth; enzyme cofactor (kinases, etc).		Secondary to malabsorption or diarrhea, alcoholism.	Depressed deep tendon reflexes and respiration.	Leafy green vegetables (containing chlorophyll).

[1]In general, minerals require carrier proteins for absorption. Absorption is rarely complete; it is affected by other nutrients and compounds in the diet (eg, oxalates and phytates that chelate divalent cations). Transport and storage also require special proteins. Excretion occurs in feces (unabsorbed minerals and from bile) and in urine and sweat.
[2]Excess mineral intake produces toxic symptoms. Unless otherwise specified, symptoms include nonspecific nausea, diarrhea, and irritability.
[3]Mineral requirements are met by a varied intake of adequate amounts of whole-grain cereals, legumes, leafy green vegetables, meat, and dairy products.

unsaturated fatty acids. The latter tend to form free radicals upon exposure to oxygen, and this may lead to an increased risk of certain cancers.

Vitamin K

Naturally occurring vitamin K is absorbed from the intestines only in the presence of bile salts and other lipids, through interaction with chylomicrons. Therefore, fat malabsorptive diseases can result in vitamin K deficiency. The synthetic vitamin K_3 is water soluble and absorbed irrespective of the presence of intestinal lipids and bile. Since the K_2 form is synthesized by intestinal bacteria, deficiency of the vitamin in adults is rare. However, long-term antibiotic treatment can produce this deficiency, The intestine of newborn infants is sterile; therefore, vitamin K deficiency in infants is possible if the vitamin is lacking from the early diet. The primary symptom of a deficiency in infants is a hemorrhagic syndrome.

Table 36–2. Essential microminerals (trace elements): Summary of major characteristics.

Elements	Functions	Metabolism[1]	Deficiency Disease or Symptoms	Toxicity Disease or Symptoms[1]	Good Sources[2]
Chromium	Trivalent chromium, a constituent of "glucose tolerance factor," which binds to and potentiates insulin.		Impaired glucose tolerance; secondary to parenteral nutrition.		Meat, liver, brewer's yeast, whole grains, nuts, cheese.
Cobalt	Required only as a constituent of vitamin B_{12}.	As for vitamin B_{12}.	Vitamin B_{12} deficiency.		Foods of animal origin.
Copper	Constituent of oxidase enzymes: cytochrome c oxidase, etc. Cystosolic superoxide dismutase. Role in iron absorption.	Transported by albumin; bound to ceruloplasmin.	Anemia (hypochromic, microcytic); secondary to malnutrition. Menke's syndrome.	Rare; secondary to Wilson's disease.	Liver.
Iodine	Constituent of thyroxine, triiodothyronine.	Stored in thyroid as thyroglobulin.	Children: cretinism. Adults: goiter and hypothyroidism, myxedema.	Thyrotoxicocis, goiter.	Iodized salt, seafood.
Iron	Constituent of heme enzymes (hemoglobin, cytochromes, etc).	Transported as transferrin; stored as ferritin or hemosiderin; lost in sloughed cells and by bleeding.	Anemia (hypochromic, microcytic).	Siderosis; hereditary hemochromatosis.	Red meat, liver, eggs. Iron cookware.
Manganese	Cofactor of hydrolase, decarboxylase, and transferase enzymes. Glycoprotein and proteoglycan synthesis. Mitochondrial superoxide dismutase.		Unknown in humans.	Inhalation poisoning produces psychotic symptoms and Parkinsonism.	
Molybdenum	Constituent of oxidase enzymes (xanthine oxidase).		Secondary to parenteral nutrition.		
Selenium	Constituent of glutathione peroxidase.	Synergistic antioxidant with vitamin E.	Marginal deficiency when soil content is low; secondary to parenteral nutrition, protein-energy malnutrition.	Megadose supplementation induces hair loss, dermatitis, and irritability.	Plants, but varies with soil content. Meat.
Silicon[3]	Role in calcification of bone and in glycosaminoglycan metabolism in cartilage and connective tissue.		Impairment of normal growth.	Silicosis due to long term inhalation of silica dust.	Plant foods.
Zinc	Cofactor of many enzymes: lactate dehydrogenase, alkaline phosphatase; carbonic anhydrase, etc.		Hypogonadism, growth failure, impaired wound healing, decreased taste and smell acuity; secondary to acrodermatitis enteropathica, parenteral nutrition.	Gastrointestinal irritation, vomiting.	
Fluoride[4]	Increases hardness of bones and teeth.		Dental caries; osteoporosis(?).	Dental fluorosis.	Drinking water.

[1]Excess mineral intake produces toxic symptoms. Unless otherwise specified, symptoms include nonspecific nausea, diarrhea, and irritability.
[2]Mineral requirements are met by a varied intake of adequate amounts of whole-grain cereals, legumes, leafy green vegetables, meat, and dairy products.
[3]Not yet demonstrated to be essential to humans but necessary in several animals.
[4]Fluoride is essential for rat growth. While not proved to be strictly essential for human nutrition, fluorides have a well-defined role in prevention and treatment of dental caries.

MINERALS

As discussed at the beginning of this chapter, the two broad classes of physiologically significant minerals are the **macrominerals**, required in quantities greater than 100 mg/day, and the **microminerals** (also known as the **trace minerals**), required in quantities less 100 mg/day. The functions of these minerals, and the clinical consequences of their deficiency or excess in the diet, are presented in Tables 36–1 and 36–2.

Questions

DIRECTIONS (items 1–10): Match the lettered list of vitamin names with the descriptions. Some vitamins may used more than once, others not at all.

a. Vitamin D.
b. Thiamin.
c. Vitamin A.
d. Folic acid.
e. Vitamin K.
f. Riboflavin.
g. Pantothenic acid.
h. Biotin.
i. Vitamin E.
j. Vitamin B_{12}.
k. Niacin.
l. Vitamin B_6.
m. Vitamin C.

1. This vitamin is a cofactor required for the activity of succinate dehydrogenase.

2. Deficiency in this vitamin leads to defective collagen processing and consequent disruptions in connective tissue production.

3. Deficiency of this vitamin leads to severe neurological complications that result from reduced ATP production in neural cells.

4. This vitamin is a cofactor required by malate dehydrogenase.

5. This vitamin is required for maintenance of active clotting factors.

6. Enzymes that perform carboxylation reactions require this vitamin as a cofactor.

7. Deficiencies in vitamin B_{12} are highly related to deficiencies in this vitamin.

8. This vitamin is required for the maintenance of mineralized bone.

9. One of the enzymes required for the utilization of an oxidation product of odd-chain fatty acids requires this vitamin as a cofactor.

10. The cancer-chemotherapeutic drug, methotrexate, inhibits the regeneration of the cellular pool of this vitamin.

DIRECTIONS (items 11–17): Each numbered item or incomplete statement is followed by answers or by completions of the statement. Select ONE lettered answer or completion that is BEST in each case.

11. Appetite suppression, constipation, nausea, mental depression, loss of eye coordination, and ataxia are symptoms of a deficiency of which vitamin?
a. B_6.
b. B_{12}.

    c. B_5.
    d. B_3.
    e. B_1.

12. Which of the following minerals is required as a constituent of vitamin B_{12}?
    a. Iron.
    b. Calcium.
    c. Cobalt.
    d. Manganese.
    e. Magnesium.

13. Hypochromic and microcytic anemias result from a deficiency of which mineral?
    a. Calcium.
    b. Iron.
    c. Magnesium.
    d. Selenium.
    e. Chromium.

14. Which of the following minerals is principally involved in the maintenance of acid-base balance?
    a. Sodium.
    b. Calcium.
    c. Phosphorus.
    d. Chlorine.
    e. Magnesium.

15. Which of the following minerals is important in the antioxidant capabilities of vitamin E?
    a. Iron.
    b. Zinc.
    c. Chromium.
    d. Selenium.
    e. Silicon.

16. The visual pigment of human rod cells is:
    a. Opsin.
    b. Metarhodopsin.
    c. Rhodopsin.
    d. Transducin.
    e. Scotopsin.

17. Propagation of visual signals from the eye to the brain requires:
    a. Opsin-induced opening of a K^+ channel in rod cells.
    b. cGMP-induced closing of a Na^+ channel in rod cells.
    c. Transducin-induced closing of a Na^+ channel in rod cells.
    d. Metarhodopsin-induced closure of a K^+ channel in rod cells.
    e. Phosphodiesterase-mediated hydrolysis of cAMP.

Answers

1. f	**7.** d	**13.** b
2. m	**8.** a	**14.** a
3. b	**9.** j	**15.** d
4. k	**10.** d	**16.** c
5. e	**11.** e	**17.** b
6. h	**12.** c	

Part X: Special Topics

Blood Coagulation

37

Objectives

- To describe the mechanisms of platelet activation.
- To describe the intrinsic and extrinsic pathways of blood coagulation.
- To describe the activation of thrombin.
- To describe the factors that control the level of circulating active thrombin.
- To describe the activation of fibrin.
- To describe the dissolution of fibrin clots.
- To describe the clinical significances of normal and aberrant blood coagulation.

Concepts

The ability of the body to control the flow of blood following vascular injury is paramount to continued survival. The process of blood clotting and then the subsequent dissolution of the clot, following repair of the injured tissue, is termed **hemostasis**. Hemostasis is composed of four major events that occur in a set order following the loss of vascular integrity:

1. The initial phase of the process is vascular **constriction**. This limits the flow of blood to the area of injury.
2. Next, platelets become activated by **thrombin** and aggregate at the site of injury, forming a temporary, loose **platelet plug**. The protein **fibrinogen** is primarily responsible for stimulating platelet clumping. Platelets clump by binding to collagen that becomes exposed following rupture of the endothelial lining of vessels. Upon activation, platelets release **ADP** and **TXA$_2$** (which activate additional platelets), **serotonin, phospholipids, lipoproteins**, and other proteins important for the coagulation cascade. In addition to induced secretion, activated platelets change their shape to accommodate the formation of the plug.
3. To insure stability of the initially loose platelet plug, a **fibrin mesh** (also called the **clot**) forms and entraps the plug. If the plug contains only platelets it is termed a **white thrombus**; if red blood cells are present it is called a **red thrombus**.
4. Finally, the clot must be dissolved in order for normal blood flow to resume following tissue repair. The dissolution of the clot occurs through the action of **plasmin**.

Platelet Activation

In order for hemostasis to occur, platelets must adhere to exposed collagen, release the contents of their granules, and aggregate. The adhesion of platelets to the collagen exposed on endothelial cell surfaces is mediated by **von Willebrand's factor (vWF)**. The function of vWF is to act as a bridge between a specific glycoprotein on the surface of platelets (GpIb/IX) and collagen fibrils.

The initial activation of platelets is induced by thrombin binding to specific receptors on the surface of platelets, thereby initiating a signal transduction cascade (Figure 37–1). The thrombin receptor is coupled to a G-protein that, in turn, activates phospholipase C-γ (PLCγ). PLCγ hydrolyzes phosphatidylinositol-4,5-bisphosphate (PIP$_2$), leading to the formation of inositol trisphosphate (IP$_3$) and diacylglycerol (DAG). IP$_3$ induces the release of intracellular C^{2+} stores, and DAG activates protein kinase C (PKC).

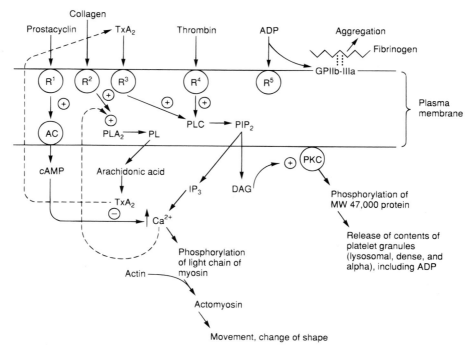

Figure 37–1. Diagrammatic representation of platelet activation. The external environment, the plasma membrane, and the inside of a platelet are depicted from top to bottom. Thrombin and collagen are the 2 most important platelet activators. ADP causes aggregation but not granule release. GP, glycoprotein; R^1, R^2, R^3, R^4, various receptors; AC, adenylate cyclase; PLA_2, phospholipase A_2: PL, phospholipids; PLC, phospholipase C; PIP_2, phosphatidylinositol 4,5-bisphosphate; cAMP, cyclic AMP; PKC, protein kinase C; TxA_2; IP_3, inositol triphosphate; DAG, diacylglycerol. (Reproduced, with permission, from Murray, RK: *Harper's Biochemistry*, 23rd ed. Appleton & Lange, 1993.)

The collagen to which platelets adhere, as well as the release of intracellular Ca^{2+}, leads to the activation of **phospholipase A_2 (PLA_2)**, which then hydrolyzes membrane phospholipids, leading to the liberation of **arachidonic acid**. The arachidonic acid release leads to an increase in the production, and subsequent release, of **thromboxane A_2 (TXA_2)**. This is another platelet activator that functions through the PLCγ pathway (Figure 37–1). Another enzyme activated by the released intracellular Ca^{2+} stores is **myosin light chain kinase (MLCK)**. Activated MLCK phosphorylates the light chain of myosin which then interacts with actin, resulting in altered platelet morphology and motility.

One of the many effects of PKC is the phosphorylation and activation of a specific 47,000-dalton platelet protein. This activated protein induces the release of platelet granule contents, one of which is **ADP**. ADP further stimulates platelets, increasing the overall activation cascade; it also modifies the platelet membrane in such a way as to allow **fibrinogen** to adhere to 2 platelet surface glycoproteins, GPIIb and GPIIIa (Figure 37–1), resulting in fibrinogen-induced platelet **aggregation**.

The activation of platelets is required for their consequent aggregation to a platelet plug. However, equally significant is the role of activated platelet surface phospholipids in the activation of the coagulation cascade (see below and Figure 37–2).

Two different pathways lead to the formation of a fibrin clot: the **intrinsic** and the **extrinsic** pathway. Although they are initiated by distinct mechanisms, the two converge on a common pathway that leads to clot formation. The formation of a red thrombus or a clot in response to an abnormal vessel wall in the absence of tissue injury is the result of the **intrinsic pathway** (Figure 37–2). Fibrin clot formation in response to tissue injury is the result of the **extrinsic pathway** (Figure 37–2). Both pathways are complex and involve numerous different proteins (Table 37–1). The proteins of the clotting cascade, in particular, can be divided into 5 different classes (Table 37–2).

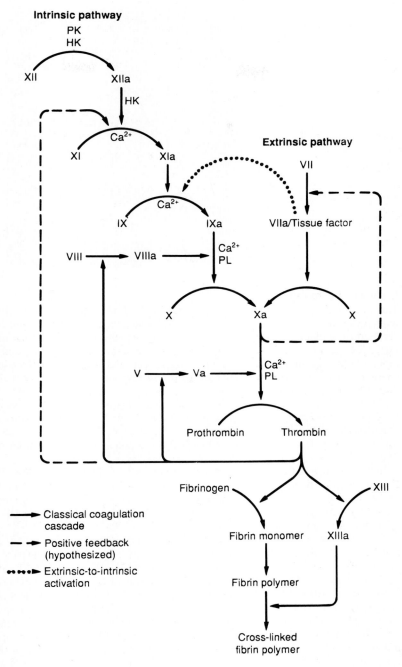

Figure 37–2. The pathways of blood coagulation. The intrinsic and extrinsic pathways are indicated. The events depicted below factor Xa are designated the final common pathway, culminating in the formation of cross-linked fibrin. New observations (. . .) include the finding that complexes of tissue factor and factor VIIa activate not only factor X (in the classical extrinsic pathway) but also factor IX in the intrinsic pathway. In addition, thrombin and factor Xa feedback-activate at the 2 sites indicated (−). PK, prekallikrein; HK, high-molecular-weight kininogen; PL, phospholipids. (Reproduced, with permission, from Roberts HR, Lozier JN: New perspectives on the coagulation cascade. Hosp Pract [Jan] 1992;27:97.)

Table 37–1. Numerical system for nomenclature of blood clotting factors. The numbers indicate the order in which the factors have been discovered and bear no relationship to the order in which they act.

Factor	Common Name
I	Fibrinogen ⎱ These factors are usually referred
II	Prothrombin ⎰ to by their common names
III	Tissue factor ⎱ These factors are usually not
IV	Ca^{2+} ⎰ referred to as coagulation factors
V	Proaccelerin, labile factor, accelerator (Ac-) globulin
VII[1]	Proconvertin, serum prothrombin conversion accelerator (SPCA), cothromboplastin
VIII	Antihemophilic factor A, antihemophilic globulin (AHG)
IX	Antihemophilic factor B, Christmas factor, plasma thromboplastin component (PTC)
X	Stuart-Prower factor
XI	Plasma thromboplastin antecedent (PTA)
XII	Hageman factor
XIII	Fibrin stabilizing factor (FSF), fibrinoligase

[1]There is no factor VI.

Intrinsic Pathway

The intrinsic pathway requires the clotting factors VIII, IX, X, XI, and XII. Also required are the proteins prekallikrein and high-molecular-weight kininogen, as well as calcium ions and phospholipids secreted from platelets (Figure 37–2). Each of these pathway constituents leads to the conversion of factor X (inactive) to factor Xa ("a" signifies "active"). Initiation of the intrinsic pathway occurs when prekallikrein, high-molecular-weight kininogen, factor XI, and factor XII are exposed to a negatively charged surface. This is termed the **contact phase**. Exposure of collagen to a vessel surface is the primary stimulus for the contact phase.

The assemblage of contact phase components results in conversion of prekallikrein to **kallikrein**, which in turn activates factor XII to **factor XIIa**. Factor XIIa can then hydrolyze more prekallikrein to kallikrein, establishing a reciprocal activation cascade. Factor XIIa also activates factor XI to **factor XIa** and leads to the release of **bradykinin**, a potent vasodilator, from high-molecular-weight kininogen.

In the presence of Ca^{2+}, factor XIa activates factor IX to **factor IXa**. Factor IX is a **proenzyme** that contains vitamin K-dependent γ-carboxyglutamate (gla) residues, whose serine protease activity is activated following Ca^{2+} binding to these gla residues. Several of the serine proteases of the cascade (II, VII, IX, and X) are gla-containing proenzymes (Table 37–2). Active factor IXa cleaves factor X at an internal arg-ile bond, leading to its activation to **factor Xa**.

The activation of factor Xa requires assemblage of the **tenase complex** (Ca^{2+} and factors VIIIa, IXa, and X) on the surface of activated platelets. One of the responses of platelets to activation is the presentation of phosphotidylserine and phosphotidylinositol on their surfaces. The exposure of these phospholipids allows the tenase complex to form. The role of factor VIII in this process is to act as a receptor, in the form of factor VIIIa, for factors IXa and X. Factor VIIIa is termed a **cofactor** in the clotting cascade. The activation of factor VIII to **factor VIIIa** (the actual receptor) occurs in the presence of minute quantities of thrombin. As the concentration of thrombin increases, factor VIIIa is ultimately cleaved by thrombin and inactivated. This dual action of thrombin, upon factor VIII, acts to limit the extent of tenase complex formation and thus the extent of the coagulation cascade.

Extrinsic Pathway

Activated factor Xa is the site at which the intrinsic and extrinsic coagulation cascades converge. The extrinsic pathway is initiated at the site of injury in response to the release of **tissue factor** (factor III) (Figure 37–2). Tissue factor is a **cofactor** in the factor VIIa-catalyzed activation of factor X. Factor VIIa, a gla residue containing serine protease, cleaves factor X to factor Xa in a manner identical to that of factor IXa of the intrinsic pathway. The activation of factor VII occurs by means of thrombin or factor Xa.

The ability of factor Xa to activate factor VII creates a link between the intrinsic and extrinsic pathways. An additional link between the two pathways exists through the ability of tissue factor and factor VIIa to activate factor IX. The formation of complex between factor VIIa and tissue factor is be-

Table 37–2. The functions of the proteins involved in blood coagulation.

Zymogens of serine proteases

Factor XII	Binds to exposed collagen at site of vessel wall injury; activated by high-MW kininogen and kallikrein.
Factor XI	Activated by factor XIIa.
Factor IX	Activated by factor XIa in presence of Ca^{2+}.
Factor VII	Activated by thrombin in presence of Ca^{2+}.
Factor X	Activated on surface of activated platelets by tenase complex (Ca^{2+}, factors VIIIa and IXa) and by factor VIIa in presence of tissue factor and Ca^{2+}.
Factor II	Activated on surface of activated platelets by pro-thrombinase complex (Ca^{2+}, factors Va and Xa). [Factors II, VII, IX, and X are Gla-containing zymogens.]

Cofactors

Factor VIII	Activated by thrombin; factor VIIIa is a cofactor in the activation of factor X by factor IXa.
Factor V	Activated by thrombin; factor Va is a cofactor in the activation of prothrombin by factor Xa.
Tissue factor (factor III)	A lipoprotein found on the surfaces of extravascular cells that acts as a cofactor for factor VII.

Fibrinogen

Factor I	Cleaved by thrombin to form fibrin clot.

Thiol-dependent transglutaminase

Factor XIII	Activated by thrombin in presence of Ca^{2+}; stabilizes fibrin clot by covalent cross-linking.

Regulatory and other proteins

Protein C	Activated to protein Ca by thrombin bound to thrombomodulin; then degrades factors VIIIa and Va.
Protein S	Acts as a cofactor of protein C; both proteins contain Gla residues.
Thrombo-modulin	Protein on the surface of endothelial cells; binds thrombin, which then activates protein C.

Gla = γ-carboxyglutamate.

lieved to be a principal step in the overall clotting cascade. Evidence for this stems from the fact that persons with hereditary deficiencies in the components of the **contact phase** of the intrinsic pathway do not exhibit clotting problems.

A major mechanism for the inhibition of the extrinsic pathway occurs at the tissue factor–factor VIIa–Ca^{2+}-Xa complex. The protein, **lipoprotein-associated coagulation inhibitor (LACI**, formerly named "anticonvertin") specifically binds to this complex. LACI is composed of 3 tandem protease inhibitor domains. Domain 1 binds to factor Xa and domain 2 binds to factor VIIa only in the presence of factor Xa.

Activation of Prothrombin to Thrombin

The common point in both pathways is the activation of factor X to factor Xa (Figure 37–2). Factor Xa activates **prothrombin** (factor II) to **thrombin** (factor IIa). Thrombin, in turn, converts fibrinogen to fibrin. The activation of thrombin occurs on the surface of activated platelets and requires formation of a **prothrombinase complex**. This complex is composed of the platelet phospholipids, phosphatidylinositol and phosphatidylserine, Ca^{2+}, factors Va and Xa, and prothrombin. Factor V is a **cofactor** in the formation of the prothrombinase complex, similar to the role of factor VIII in tenase complex formation. Like factor VIII activation, factor V is activated to **factor Va** by means of minute amounts and is inactivated by increased levels of thrombin. Factor Va binds to specific receptors on the surfaces of activated platelets and forms a complex with prothrombin and factor Xa (Figure 37–3).

Prothrombin is a 72,000-dalton, single-chain protein containing 10 gla residues in its N-terminal region (Figure 37–4). Within the prothrombinase complex, prothrombin is cleaved at 2 sites by factor

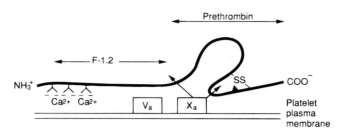

Figure 37–3. Diagrammatic representation of the binding of factors V_a, X_a, Ca^{2+}, and prothrombin to the plasma membrane of the activated platelet. The sites of cleavage of prothrombin by factor X_a are indicated by 2 arrows. The part of prothrombin destined to form thrombin is labelled prethrombin. The Ca^{2+} is bound to phospholipids and other molecules of the plasma membrane of the platelet. (Reproduced, with permission, from Murray, RK: *Harper's Biochemistry*, 23rd ed. Appleton & Lange, 1993.)

X_a. This cleavage generates a 2-chain active thrombin molecule containing an A and a B chain, which are held together by a single disulfide bond (Figure 37–4).

In addition to its role in activation of fibrin clot formation, thrombin plays an important regulatory role in coagulation. Thrombin combines with **thrombomodulin** present on the surface of endothelial cells, forming a complex that converts **protein C** to **protein Ca**. The cofactor **protein S** and protein Ca degrade factors Va and VIIIa, thereby limiting the activity of these 2 factors in the coagulation cascade.

Control of Thrombin Levels

The inability of the body to control the circulating level of active thrombin would lead to dire consequences. There are 2 principal mechanisms by which thrombin activity is regulated. The predominant form of thrombin in the circulation is the inactive **prothrombin**, whose activation requires the pathways of proenzyme activation described above for the coagulation cascade. At each step in the cascade, feedback mechanisms regulate the balance between active and inactive enzymes.

The activation of thrombin is also regulated by 4 specific **thrombin inhibitors**. **Antithrombin III** is the most important, since it can also inhibit the activities of factors IXa, Xa, XIa, and XIIa. The activity of antithrombin III is potentiated in the presence of heparin by the followings means: heparin binds to a specific site on antithrombin III, producing an altered conformation of the protein, and the new conformation has a higher affinity for thrombin as well as its other substrates. This effect of heparin is the basis for its clinical use as an anticoagulant. The naturally occurring heparin activator of antithrombin III is present as heparan and heparan sulfate on the surface of vessel endothelial cells. It is this feature that controls the activation of the intrinsic coagulation cascade.

However, thrombin activity is also inhibited by **α_2-macroglobulin, heparin cofactor II, and α_1-antitrypsin**. Although a minor player in thrombin regulation α_1-antitrypsin is the primary **serine protease inhibitor** of human plasma. Its physiological significance is demonstrated by the fact that lack of this protein plays a causative role in the development of emphysema.

Activation of Fibrinogen to Fibrin

Fibrinogen (factor I) consists of 3 pairs of polypeptides $(A\alpha B\beta,\gamma)_2$. The 6 chains are covalently linked near their N-terminals through disulfide bonds. The A and B portions of the $A\alpha$ and $B\beta$ chains comprise the **fibrinopeptides, A and B**, respectively. The fibrinopeptide regions of fibrinogen contain several glutamate and asparate residues, which impart a high negative charge to this region and aid in the

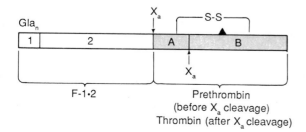

Figure 37–4. Diagrammatic representation of prothrombin. The amino terminus is to the left; region 1 contains all 10 Gla residues. The sites of cleavage by factor X_a are shown and the products named. The site of the catalytically active serine residue is indicated by ▲. The A and B chains of active thrombin (shaded) are held together by the disulfide bridge. (Reproduced, with permission, from Murray, RK: *Harper's Biochemistry*, 23rd ed. Appleton & Lange, 1993.)

solubility of fibrinogen in plasma. Active thrombin is a serine protease that hydrolyses fibrinogen at four arg-gly bonds between the fibrinopeptide and the α and β portions of the proteins (Figure 37–5).

Thrombin-mediated release of the fibrinopeptides generates fibrin monomers with a subunit structure $(\alpha\beta\gamma)_2$. These monomers spontaneously aggregate in a regular array, forming a somewhat weak fibrin clot. In addition to fibrin activation, thrombin converts **factor XIII** to **factor XIIIa**, a highly specific **transglutaminase** that introduces cross-links composed of covalent bonds between the amide nitrogen of glutamines and the ϵ-amino group of lysines in the fibrin monomers.

Dissolution of Fibrin Clots

Degradation of fibrin clots is the function of **plasmin**, a serine protease that circulates as the inactive proenzyme **plasminogen**. Any free circulating plasmin is rapidly inhibited by α_2-**antiplasmin**. Plasminogen binds to both fibrinogen and fibrin, thereby being incorporated into a clot as it is formed. **Tissue plasminogen activator (tPA)** and, to a lesser degree, **urokinase** are serine proteases that convert plasminogen to plasmin. Inactive tPA is released from vascular endothelial cells following injury; it binds to fibrin and is consequently activated. Urokinase is produced as the precursor **prourokinase** by epithelial cells lining excretory ducts. The role of urokinase is to activate the dissolution of fibrin clots that may be deposited in these ducts.

Active tPA cleaves plasminogen to plasmin (Figure 37–6), which then digests the fibrin; the result is a soluble degradation product to which neither plasmin nor plasminogen can bind. Following the release of plasminogen and plasmin, they are rapidly inactivated by their respective inhibitors. The inhibition of tPA activity results from binding to specific inhibitory proteins. At least 4 distinct inhibitors have been identified of which 2-**plasminogen activator-inhibitors type 1 (PAI-1) and type 2 (PAI-2)** are of greatest physiological significance.

CLINICAL SIGNIFICANCE

Deficiencies in proteins of the coagulation cascade can lead to **hemophilia,**, the inability to clot blood. **Hemophilia A** is the most common of these disorders and results from an inherited lack of **factor VIII**. **Hemophilia B** results from an inherited deficiency in **factor IX**. Treatment of these disorders currently consists of injections of factors that have been synthesized by means of recombinant DNA technology.

Coumarin drugs, such as **warfarin** and heparin, are useful as anticoagulants. The role of heparin in anticoagulation was discussed earlier. The coumarin drugs inhibit coagulation by inhibiting the vita-

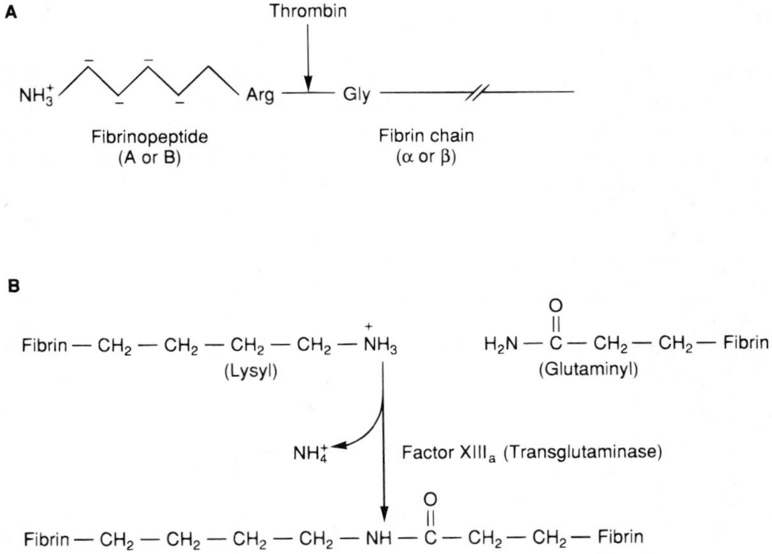

Figure 37–5. Formation of fibrin clot. **A:** Thrombin-induced cleavage of Arg-Gly bonds of the Aα and Bβ chains of fibrinogen to produce fibrinopeptides (left-hand side) and the α and β chains of fibrin monomer (right-hand side). **B:** Cross-linking of fibrin molecules by activated factor XIII (factor XIIa). (Reproduced, with permission, from Murray, RK: *Harper's Biochemistry*, 23rd ed. Appleton & Lange, 1993.)

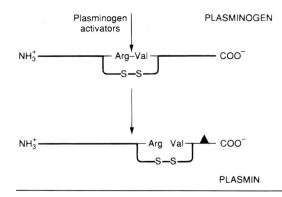

Figure 37–6. Activation of plasminogen. The same Arg-Val bond is cleaved by all plasminogen activators to give the 2-chain plasmin molecule. ▲ indicates the serine residue of the active site. The 2 chains of plasmin are held together by a disulfide bridge. (Reproduced, with permission, from Murray, RK: *Harper's Biochemistry*, 23rd ed. Appleton & Lange, 1993.)

min K-dependent γ-carboxylation reactions necessary to the function of factors II, VII, IX, and X, as well as proteins C and S. These drugs act by inhibiting the reduction of the quinone derivatives of vitamin K to their active hydroquinone forms (see Chapter 37). Because of the mode of action of coumarin drugs, it takes several days for their maximum effect to be realized. For this reason heparin is normally administered first, followed by warfarin.

The plasminogen activators also are useful for controlling coagulation. Because tPA is highly selective for the degradation of fibrin in clots, it is extremely useful in restoring the patency of the coronary arteries following thrombosis, in particular during the short period following myocardial infarct. **Streptokinase** is another plasminogen activator useful from a therapeutic standpoint. However, it is less selective than tPA, being able to activate circulating plasminogen as well as that bound to a fibrin clot.

Aspirin is an important inhibitor of platelet activation. By virtue of inhibiting the activity of cyclooxygenase (see Chapter 20), aspirin reduces the production of TXA_2. Aspirin also reduces the endothelial cells' production of **prostacyclin (PGI_2)**, an inhibitor of platelet aggregation and a vasodilator. Since endothelial cells regenerate active cyclooxygenase faster than platelets, the net effect of aspirin is more in favor of endothelial cell-mediated inhibition of the coagulation cascade.

Questions

DIRECTIONS (items): Each numbered item or incomplete statement is followed by answers or by completions of the statement. Select the ONE lettered answer or completion that is BEST in each case.

1. Control of the overall level of active circulating thrombin is exerted by:
   a. Factor Xa.
   b. Prothrombinase.
   c. Fibrin.
   d. Antithrombin III.
   e. Factor VIIIa.

2. Which of the following distinguishes red from white thrombi?
   a. The presence of erythrocytes.
   b. The presence of platelets.
   c. The presence of leukocytes.
   d. The presence of endothelial cells.
   e. The presence of epithelial cells.

3. Inducing the degradation of fibrin clots within excretory ducts is the function of:
   a. Streptokinase.
   b. Urokinase.
   c. tPA.
   d. Plasmin.
   e. Plasminogen.

4. Warfarin is useful as an anticoagulant because it inhibits:
   a. Thrombin activation.
   b. Platelet activation.
   c. Factor VII activation.
   d. Factor XII activation.
   e. Factor X activation.

DIRECTIONS (items 5–17): Match the following described properties with the right proteins.

   a. Contains factors VIIIa, IXa, X and Ca^{2+}.
   b. Hydrolyses prothrombin to thrombin.
   c. When activated is involved in degrading factors Va and VIIa.
   d. Caused by deficiency of factor IX.
   e. Contains factors Va, Xa and Ca^{2+}.
   f. Catalyzes conversion of soft fibrin clots to hard fibrin clots.
   g. Requires factor Xa for activation.
   h. Factor that activates factor XI to XIa.
   i. Factor XIIa releases this protein from high-molecular-weight kininogen.
   j. Caused by deficiency of factor VIII.
   k. Cleaves factor X to activate it.
   l. Activates antithrombin III.
   m. Protein bridge between GpIb/IX on platelets and collagen on endothelial cells.

5. Factor Xa.

6. Hageman factor.

7. Bradykinin.

8. Christmas factor.

9. von Willebrand factor.

10. Hemophilia A.

11. Protein C.

12. Heparin.

13. Tenase complex.

14. Factor XIIIa.

15. Factor II.

16. Prothrombinase complex.

17. Hemophilia B.

Answers

1. d	**7.** i	**13.** a
2. a	**8.** k	**14.** f
3. b	**9.** m	**15.** g
4. e	**10.** j	**16.** e
5. b	**11.** c	**17.** d
6. h	**12.** l	

38

Recombinant DNA Technologies in Medicine

Objectives

- To define the terms: cDNA, cloning, YAC, RFLP, PCR, RT-PCR, PCR-SSCP, LCR.
- To describe the function of restriction endonucleases.
- To describe cloning of cDNA and genomic DNA.
- To describe the use of YACs.
- To describe the techniques available to analyze cloned DNAs.
- To describe the polymerase chain reaction (PCR).
- To describe the RT-PCR technique.
- To describe the PCR-SSCP technique.
- To describe the ligase chain reaction (LCR).
- To describe transgenesis and the generation of transgenic animals.
- To describe the techniques of gene therapy.

Concepts

Modern medicine makes use of numerous molecular biological techniques in the analysis of disease, disease genes, and the functioning of these genes. The study of disease genes and their function in an unaffected individual has become possible only recently, with the development of **recombinant DNA** and **cloning** techniques: experimental manipulations made possible by the characterization of the activities of numerous DNA and RNA modifying enzymes (see Table 38–1). *Recombinant DNA* receives its name from the recombining of different segments of DNA. *Cloning* refers to the process of preparing multiple copies of an individual type of recombinant DNA molecule.

The classical process for producing recombinant molecules involves inserting exogenous fragments of DNA into a **vector,** or host chromosome. The vector may be derived from either a **plasmid** (a self-replicating structure of circular, double-stranded DNA, commonly found in bacteria), or a **bacteriophage** (a virus that infects bacteria).

Restriction Endonucleases

Restriction endonucleases are enzymes that can recognize, bind to, and hydrolyze specific sequences of nucleic acids in double-stranded DNA. The term *restriction endonuclease* was given to this class of bacterially derived enzymes when they were shown to be involved in restricting the growth of certain bacteriophages. Bacteria, meanwhile, protect themselves by modifying specific sequences within their genomes by methylation; this prevents their own DNA from being recognized by the restriction enzymes encoded by their genomes. The overall process is termed modification and restriction. Because the DNA of the infecting bacteriophage is not modified, it is ultimately digested by the restriction endonucleases present in the bacterium.

The key to the in vitro utilization of restriction endonucleases is their strict specificity: Each recognizes just one particular sequence of nucleotides. Restriction endonucleases are identified according to the bacteria from which they were isolated: for example, the enzyme **EcoRI**, which recognizes the sequence **5'-GAATTC-3'**, was isolated from *Escherichia coli*. A striking feature of restriction enzymes is that the nucleotide sequences they recognize are palindromic; that is, they are the same sequences in the 5' → 3' direction of both strands.

Some restriction endonucleases make staggered symmetrical cuts away from the center of their recognition site within the DNA duplex, others make symmetrical cuts in the middle of their recognition site, and still others cleave the DNA at a distance from the recognition sequence. Enzymes that make staggered cuts leave the resultant DNA with **cohesive** or **sticky ends** (Figure 38–1). Enzymes

Table 38–1. Enzymes used in molecular biology.

Enzyme(s)	Activity	Comments
Restriction endonucleases	Recognizes specific nucleotide sequences; cleaves the DNA within or near the recognition sequences	See Text and Table 38.2
Reverse transcriptase (RT)	RNA-dependent DNA polymerase; encoded by retrovirus	Used to convert mRNA into a complementary DNA (cDNA) copy for the purpose of cloning cDNAs
RNase H	Recognizes RNA-DNA duplexes; randomly cleaves the phosphodiester backbone of the RNA	Used primarily to cleave the mRNA strand that is annealed to the first strand of cDNA generated by reverse transcription
DNA polymerase	Synthesis of DNA	Used during most procedures where DNA synthesis is required; also used in in vitro mutagenesis
Klenow DNA polymerase	Proteolytic fragment of DNA polymerase; lacks the 5′→ 3′ exonuclease activity	Used to incorporate radioactive nucleotides DNA fragments generated by restriction enzymes; also can be used in place of DNA polymerase
DNA ligase	Covalently attaches a free 5′ phosphate to a 3′ hydroxyl	Used in all procedures in which two molecules of DNA need to be covalently attached
Alkaline phosphatase	Removes phosphates from 5′ ends of DNA molecules	Used to allow 5′ ends to be radiolabeled with the γ-phosphate of ATP in the presence of polynucleotide kinase; also used to prevent self-ligation of restriction enzyme–digested plasmids and lambda vectors
Polynucleotide kinase	Introduces γ-phosphate of ATP to 5′ ends of DNA	See above for alkaline phosphatase
DNase I	Randomly hydrolyzes the phosphodiester bonds of double-stranded DNA	Used in the identification of DNA regions that are bound by protein and thereby protected from DNase I digestion; also used to identify transcriptionally active regions of chromatin, which are more susceptible to DNase I digestion
S₁ Nuclease	Exonuclease that recognizes single-stranded regions of DNA	Used to remove regions of single-strandedness in DNA or RNA-DNA duplexes
Exonuclease III	Exonuclease that removes nucleotides from the 3′ end of DNAs	Used to generate deletions in DNA for sequencing or to map functional domains of DNA duplexes
Terminal transferase	DNA polymerase that requires only a 3′-OH; lengthens 3′ ends with any dNTP	Used to introduce homopolymeric (same dNTP) "tails" onto the 3′ ends of DNA duplexes; also used to introduce radiolabeled nucleotides on the 3′ ends of DNA
T3, T7, and SP6 RNA polymerases	RNA polymerase encoded by bacterial virus; recognizes specific nucleotide sequences for initiation of transcription	Used to synthesize RNA in vitro
Taq and Vent DNA polymerase	Thermostable DNA polymerases	Used in PCR (see Text)
Taq and Vent DNA ligases	Thermostable DNA ligases	Used in LCR (see Text)

that cleave the DNA at the center of the recognition sequence leave fragments of DNA with **blunt ends** (Figure 38–1). Any two pieces of DNA containing the same sequences within their sticky ends can anneal together and be covalently bound together in the presence of **DNA ligase** (Figure 38–2). Any two blunt-ended fragments of DNA can be bound together, irrespective of the sequences at the ends of the

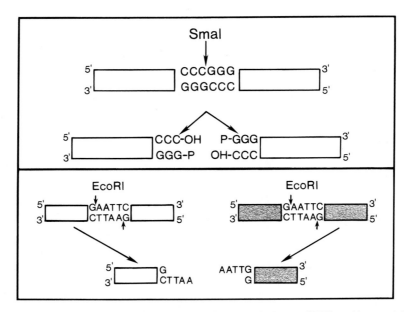

Figure 38–1. The restriction endonucleases EcoRI and SmaI cut duplexes of DNA and leave sticky and blunt-ended fragments of DNA, respectively. Any two DNA fragments harboring blunt ends (irrespective of the restriction enzyme used to generate the ends) can be bound together. Any two fragments of DNA harboring a 5′-AATT-3′ overhang (whether or not they were generated by EcoRI digestion) can be bound together, owing to the complementarity of the overhanging nucleotides.

duplexes. Table 38–2 shows the specificities and sites of phosphodiester bond cleavage that are catalyzed by a variety of restriction endonucleases.

Cloning

Any fragment of DNA can be cloned once it is introduced into a suitable vector for transforming a bacterial host. Cloning—the production of large quantities of identical DNA molecules—usually involves the use of a bacterial cell as a host for the DNA, although the process can be performed in eukaryotic cells as well. The term **cDNA cloning** refers to the production of a **library** of cloned DNAs that represent all the mRNAs present in a particular cell or tissue. **Genomic cloning** refers to the production of a library of cloned DNAs representing the entire genome of a particular organism. From either of these types of libraries, one can isolate (by a variety of screening protocols) a single cDNA or gene clone.

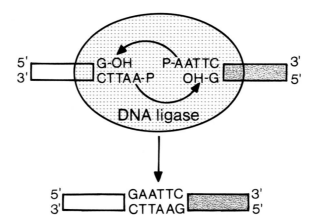

Figure 38–2. DNA ligase covalently attaches a free 5′ phosphate of one strand of DNA to the 3′-OH of an adjacent strand.

Table 38–2. Selected restriction endonucleases and their sequence specifications.[1]

Endonuclease	Sequence Cleaved	Bacterial Source
Bam HI	↓ G G A T C C C C T A G G 　　　　　↑	*Bacillus amyloliquefaciens* H
Bgl II	↓ A G A T C T T C T A G A 　　　　　↑	*Bacillus globigii*
Eco RI	↓ G A A T T C C T T A A G 　　　　　↑	*Escherichia coli* RY13
Eco RII	↓ C C T G G G G A C C 　　　　↑	*Escherichia coli* R245
Hind III	↓ A A G C T T T T C G A A 　　　　　↑	*Haemophilus influenzae* R_d
Hha I	↓ G C G C C G C G 　↑	*Haemophilus haemolyticus*
Hpa I	↓ G T T A A C C A A T T G 　　↑	*Haemophilus parainfluenzae*
Mst II	↓ C C T N A G G G G A N T C C 　　　　　↑	*Microcoleus* strain
Pst I	↓ C T G C A G G A C G T C ↑	*Providencia stuartii* 164
Taq I	↓ T C G A A G C T 　↑	*Thermus aquaticus* YTI

[1]A, adenine; C, cytosine; G, guanine; T, thymine. Arrows show the site of cleavage; depending on the site, sticky ends (Bam HI) or blunt ends (Hpa I) may result. The length of the recognition sequence can be 4 bp (Taq I), 5 bp (Eco RII), 6 bp (Eco RI), or 7 bp (Mst II). By convention, these are written in the 5′ to 3′ direction for the upper strand of each recognition sequence, and the lower strand is shown with the opposite (ie, 3′ to 5′) polarity. Note that most recognition sequences are palindromes (ie, the sequence reads the same in opposite directions on the 2 strands). A residue designated N means that any nucleotide is permitted.

To clone either cDNAs or copies of genes, a **vector** is required to "carry" the cloned DNA. As mentioned earlier, one main class of vectors is derived from bacterial **plasmids**: circular DNAs found in bacteria, able to replicate autonomously from the host genome. These DNAs were first identified because they harbored genes that conferred antibiotic resistance to the bacteria. These same genes are used in modern, in vitro-engineered plasmids to allow the selection of bacteria that have taken up plasmids containing the DNAs of interest. Unfortunately, the use of plasmids is somewhat limited: in theory, only fragments of DNA less than 10,000 base pairs (bp) in length can be cloned. In practice the limit is even tighter, around 5000 bp. The second class of vectors is derived from the bacteriophage (or bacterial virus) known as **lambda**. This virus is capable of both **lysogeny** (integration into the host genome) and **lysis** (infection followed by destruction of the infected host). The genes required for lysogeny have been removed from the lambda-based vectors, in order to allow only the lytic life cycle to take place. The advantage of using lambda-based vectors is that they can carry fragments of DNA up to 25,000 bp in length. In the analysis of the human genome, however, even lambda-based vectors are limiting, and a **yeast artificial chromosome (YAC)** vector system has been developed for the cloning of DNA fragments up to 500,000 bp (see below).

A. cDNA Cloning: Complementary DNAs are made from the mRNAs of a cell by any number of related techniques (Figure 38–3). Each technique begins with **reverse transcription** of the mRNA, followed by synthesis of the second strand of DNA and insertion of the double-stranded cDNA into either a plasmid or a lambda vector for cloning. Since all of the mRNAs are converted into cDNAs and are, in turn, inserted into a vector that is used to transform a host cell (for instance *E. coli*), the overall process creates a **"library"** of cloned cDNAs representing many copies of each mRNA species.

Screening of the library for a particular cDNA clone is accomplished by means of **probes**, which may be derived from nucleic acids or from proteins (either proteins or antibodies). cDNA libraries can also be screened by biological assay of the products synthesized by the cloned cDNAs. Screening with proteins, antibodies, or biological assay are all means of analyzing the expression of proteins from cloned cDNAs and are therefore known as **expression cloning**. Nucleic-acid probes can be generated from DNA (including synthetic oligonucleotides, oligos) or RNA. They can be labeled with a radioactive isotope or with modified nucleotides that are recognized by specific antibodies and detected by colorimetric or chemiluminescent assays.

B. Genomic Cloning: The majority of genomic cloning uses lambda-based vector systems, which are capable of carrying 15,000–25,000 bp of DNA. Cloning slightly larger fragments of genomic DNA can be accomplished with a chimeric plasmid-lambda vector system termed a **cosmid**. Cosmid vectors

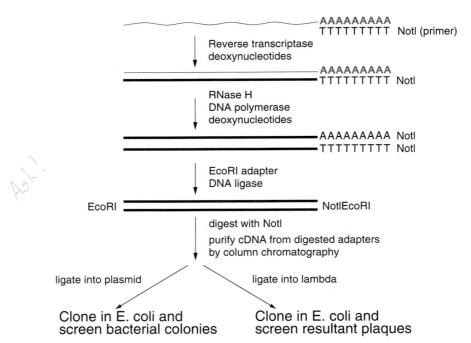

Figure 38–3. Typical process for production and cloning of cDNA. This example shows the use of a specific primer-adapter containing the sequences for the restriction enzyme NotI in addition to the poly(T) for annealing to the poly(A) tail of the RNA. It is possible to use only poly(T), or poly(T) with other restriction sites or random primers (a mixture of oligos that contain random sequences), to initiate the first strand cDNA reaction. In some cases poly(T) priming does not allow for extension of the cDNA to the 5'-end of the RNA; the use of random primers can overcome this problem since they will prime first-strand synthesis all along the mRNA. This technique shows the ligation of EcoRI adapters followed by EcoRI and NotI digestion. The process allows all the cDNAs to be cloned in one direction and therefore is termed **directional cloning**.

contain only the **"cos"** (cohesive) ends of the lambda genome (required for packaging the DNA into infectious virus particles), along with a plasmid antibiotic resistance gene and origin of DNA replication. Since approximately 30,000 bp of lambda DNA have been removed from cosmid vectors, larger genomic DNA fragments can be cloned. Still larger genomic DNA fragments can be cloned in **YAC** vectors (see below).

Genomic DNA can be isolated from any cell or tissue for cloning. As a first step, it must be digested with restriction enzymes to generate fragments in the optimal size range for the vector being utilized. Since some genes encompass many more base pairs than can be inserted into conventional lambda or cosmid vectors, it follows that the clones that are present in a genomic library must necessarily overlap. In order to generate overlapping clones, the DNA is only partially digested with restriction enzymes (Figure 38–4). This means that not every restriction site is cleaved. The partially digested DNA is then size-selected by a variety of techniques (eg, gel electrophoresis or gradient centrifugation) prior to cloning. The screening of genomic libraries is accomplished primarily with nucleic acid-based probes, but they can also be screened with proteins that are known to bind specific sequences of DNA (for instance, transcription factors).

C. Yeast Artificial Chromosome (YAC) Cloning: YAC vectors allow the cloning, within yeast cells, of fragments of genomic DNA that approach 500,000 bp. These vectors contain several elements of typical yeast chromosomes, hence the reference to yeast in their name (Figure 38–5). The YAC vectors contain a yeast **centromere (CEN)**, yeast **telomeres (TEL**, the specific sequences that are present at the ends of chromosomes and are necessary for replication) and a yeast **autonomously replicating sequence (ARS)**. As the name implies, a yeast ARS is like a bacterial plasmid in that it can be replicated autonomously from the yeast chromosomes. YAC vectors also contain genes (for instance, URA3, a gene involved in uracil synthesis) that allow the selection of yeast cells that have taken up the vector. In order that the vector can be propagated in bacterial cells, before the insertion of genomic DNA, YAC vectors also contain a bacterial replication origin and a bacterial selectable marker such as the gene for ampicillin resistance.

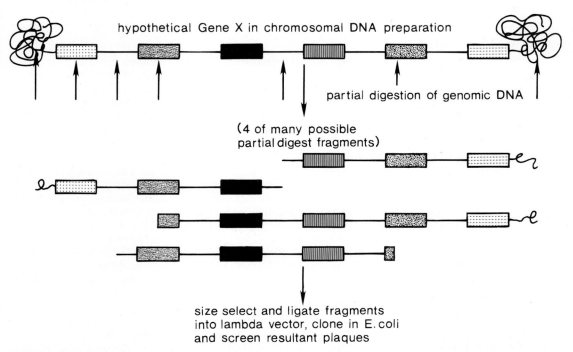

hypothetical Gene X in chromosomal DNA preparation

partial digestion of genomic DNA

(4 of many possible partial digest fragments)

size select and ligate fragments into lambda vector, clone in E. coli and screen resultant plaques

Figure 38–4. A hypothetical gene in a preparation of genomic DNA. The boxes indicate exons, and the lines separating them represent introns. Bold arrows indicate the sites where a particular restriction enzyme (such as Sau3AI) would cut this DNA. Partial digestion by a single enzymes generates a wide range of different fragments of the gene; 4 possible fragments are shown here. Fragments in the range of 15–25 kilobase pairs (kbp) are purified by gel electrophoresis or gradient centrifugation and ligated into a lambda vector. The DNA is then packaged into phage particles *in vitro* and used to infect *E. coli.*

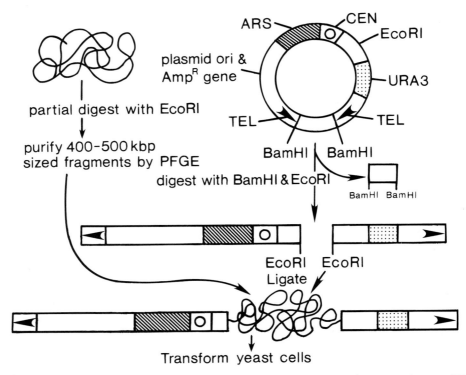

Figure 38–5. A typical YAC vector used to clone genomic DNA. The vector contains yeast telomeres (TEL), a centromere (CEN), a selectable marker (URA3), and autonomously replicating sequences (ARS) as well as bacterial plasmid sequences for antibiotic selection and replication in *E. coli*.

The process of cloning genomic DNA in a typical YAC vector is outlined in Figure 38–5. The genomic DNA is partially digested with EcoRI, and fragments in the range of 400–500 kilobase pairs (kbp) are purified by **pulsed field gel electrophoresis, PFGE** (see Chapter 11). The YAC vector is digested with EcoRI and BamHI, which places the telomere sequences at the ends of the linearized vector. The small BamHI fragment is separated from the rest of the YAC vector by standard gel electrophoresis. The genomic DNA is then ligated into the vector and used to transform yeast cells.

Analysis of Cloned Products

The analysis of cloned cDNAs and genes involves a number of techniques. The initial characterization usually requires mapping of the number and location of different restriction enzyme sites. This information is useful for DNA sequencing because it provides a means to digest the clone into specific fragments for sub-cloning, a process involving the cloning of fragments of a particular cloned DNA. Once the DNA is fully characterized, cDNA clones can be used to produce RNA *in vitro* and the RNA can be translated in vitro to characterize the protein. Clones of cDNAs also can be used as probes to analyze the structure of a gene by **Southern blotting** or to analyze the size of the RNA and pattern of its expression by **Northern blotting** (see below). Northern blotting is also a useful tool for analyzing the exon-intron organization of gene clones, since only fragments of a gene that contain exons will hybridize to the RNA on the blot.

A. DNA Sequencing: The sequencing of DNA can be accomplished by either chemical or enzymatic means. The original technique, **Maxam and Gilbert sequencing**, relies on the nucleotide-specific chemical cleavage of DNA and is not routinely used any more. The enzymatic technique, **Sanger sequencing**, involves the use of dideoxynucleotides (2′,3′-dideoxy) that terminate DNA synthesis and is therefore also called **dideoxy chain termination sequencing** (Figure 38–6).

B. Southern Blotting: Southern blotting is the analysis of DNA structure following its attachment to a solid support. The DNA to be analyzed is first digested with a given restriction enzyme; then the

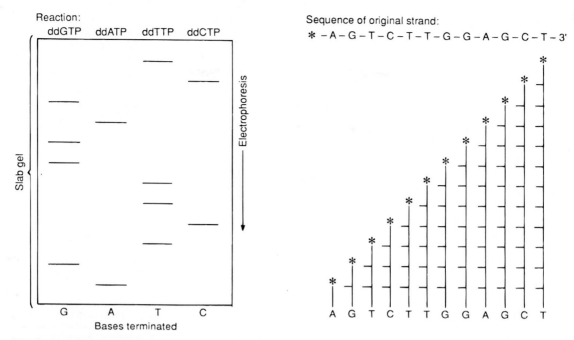

Figure 38–6. Sequencing of DNA by the method devised by Sanger. The ladderlike arrays represent from bottom to top all of the successively longer fragments of the original DNA strand. Knowing which specific dideoxynucleotide reaction was conducted to produce each mixture of fragments, one can determine the sequence of nucleotides from the labeled end toward the unlabeled end by reading up the gel. The base-pairing rules of Watson and Crick (A-T, G-C) dictate the sequence of the other (complementary) strand. (Reproduced, with permission, from Murray, RK: *Harper's Biochemistry*, 23rd ed. Appleton & Lange, 1993.)

resultant DNA fragments are separated in an agarose gel (see Chapter 11). The gel is treated with NaOH to denature the DNA, then the NaOH is neutralized. The DNA is transferred from the gel to nitrocellulose or nylon filter paper either by capillary diffusion or under electric current. The DNA is fixed to the filter by baking or by treatment with UV light. The filter can then be probed for the presence of a given fragment of DNA by various radioactive or nonradioactive means (Figure 38–7).

C. Northern Blotting: Northern blotting involves the analysis of RNA following its attachment to a solid support. The RNA is sized by gel electrophoresis, then transferred to nitrocellulose or nylon filter paper as for Southern blotting. Probing the filter for a particular RNA is carried out similarly to the probing of Southern blots (Figure 38–7).

D. Western Blotting: Completing the trio, Western blotting involves the analysis of proteins following their attachment to a solid support. The proteins are separated based upon size by SDS-PAGE (see Chapter 4) and electrophoretically transferred to nitrocellulose or nylon filters. The filter is then probed with antibodies raised against a particular protein (Figure 38–7).

Restriction Fragment Length Polymorphism (RFLP) Analysis

The genetic variability at a particular gene that may result even from minor base changes can alter the pattern of restriction enzyme digestion fragments. These alterations, which are detectable with an appropriate probe, may be due to deletions or insertions within the gene; even the substitution of a single nucleotide can change a restriction enzyme recognition site. The technique of RFLP analysis uses Southern blotting of genomic DNA that has been "digested" by restriction enzymes. The fragments of a given gene that are generated by digestion with a particular enzyme tend to fall into familial patterns, and these patterns can be detected by means of a probe corresponding to a gene of interest (Figure 38–8).

If the fragments within a family pedigree are of many different sizes, that is a sign that there are differences in the pattern of restriction sites within and around the gene being analyzed. RFLP patterns

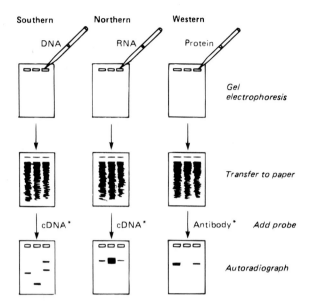

Figure 38–7. The blot transfer procedure. In a Southern, or DNA, blot transfer, DNA isolated from a cell line or tissue is digested with one or more restriction enzymes. This mixture is pipetted into a well in an agarose or polyacrylamide gel and exposed to a direct electrical current. DNA, being negatively charged, migrates toward the cathode; the smaller fragments move the most rapidly. After a suitable time, the DNA is denatured by exposure to mild alkali and transferred to nitrocellulose paper, in an exact replica of the pattern on the gel, by the blotting technique devised by Southern. The DNA is annealed to the paper by exposure to heat, and the paper is then exposed to the labeled cDNA probe, which hybridized to complementary fragments on the filter. After thorough washing, the paper is exposed to x-ray film, which is developed to reveal several specific bands corresponding to the DNA fragment that recognized the sequences in the cDNA probe. The RNA, or Northern, blot is conceptually similar. RNA is subjected to electrophoresis before blot transfer. This requires some different steps from those of DNA transfer, primarily to ensure that the RNA remains intact, and is generally somewhat more difficult. In the protein, or Western, blot, proteins are electrophoresed and transferred to nitrocellulose and then probed with a specific antibody or other probe molecule. (Reproduced, with permission, from Murray, RK: *Harper's Biochemistry*, 23rd ed. Appleton & Lange, 1993.)

are inherited. Moreover, because they segregate in a Mendelian pattern, they can be used for genotyping, such as in cases of paternity dispute or in criminal investigations.

Polymerase Chain Reaction (PCR)

PCR is a powerful technique that can amplify small quantities of DNA several millionfold in a short period of time. The process works by simple repeated replication of a template and uses sets of specific in vitro synthesized oligonucleotides to prime DNA synthesis. The design of the primers is dependent upon the DNA sequences that are to be analyzed. The technique is carried out through many cycles (usually 20–50) of melting the template at high temperature, allowing the primers to anneal to complementary sequences within the template, and then replicating the template with DNA polymerase (Figure 38–9).

PCR has been automated with the use of **thermostable** DNA polymerases, which are isolated from bacteria that grow in hot springs or in thermal vents in the ocean. During the first round of replication a single copy of DNA is converted to two copies; then two copies are converted to four; and so on, resulting in an exponential increase in the number of copies of the sequences targeted by the primers. After just 21 cycles, a single copy of DNA has been amplified over a millionfold.

The products of PCR reactions are separated in agarose or acrylamide gels, stained with ethidium bromide, and visualized with UV transillumination for analysis (see Chapter 10). Alternatively, radioactive dNTPs can be added to the PCR in order to incorporate a label into the products. In this case the products of the PCR are visualized by exposure of the gel to x-ray film. The added advantage of radiolabeling PCR products is that the levels of individual amplification products can thereby be measured.

A. Mst II restriction sites around and in the β-globin gene

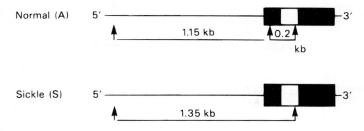

B. Pedigree analysis

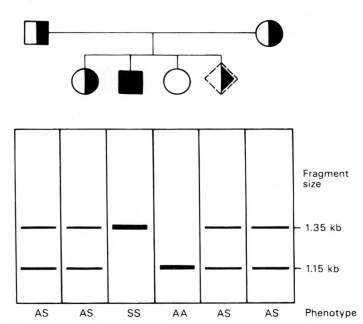

Figure 38–8. Pedigree analysis of sickle cell disease. The top part of the figure **(A)** shows the first part of the β-globin gene and the Mst II restriction enzyme sites ↑ in the normal (A) and sicle cell (S) β-globin genes. Digestion with the restriction enzyme Mst II results in DNA fragments 1.15 kb and 0.2 kb long in normal individuals. The T-to-A change in individuals with sickle cell disease abolishes one of the 3 Mst II sites around the β-globin gene; hence, a single restriction fragment 1.35 kb in length is generated in response to Mst II. This size difference is easily detected on a Southern blot **(B)**. (The 0.2kb fragment would run off the gel in this illustration.) Pedigree analysis shows 3 possibilities: AA = normal (○); AS = heterozygous (◐, ◻); SS = (■). This approach allows for prenatal diagnosis of sickle cell disease or trait (◆). (Reproduced, with permission, from Murray, RK: *Harper's Biochemistry*, 23rd ed. Appleton & Lange, 1993.)

Thanks to its ability to amplify specific fragments of DNA, PCR can be used in the analysis of disease genes (Table 38–3). The amplified fragments from disease genes may be larger (owing to insertions), or smaller (owing to deletions). Even extremely small samples of DNA are eligible for this technique. For example, in amniocentesis, only a few fetal cells need be extracted from the amniotic fluid in order to analyze for the presence of specific disease genes. Additionally, single point mutations can be detected by modified PCR techniques such as **LCR** and **PCR-SSCP** (see below). PCR also can be used to identify the level of expression of genes in other material, such as tissues or cells from the body. This technique is termed **RT-PCR** (see below).

Reverse Transcription-PCR (RT-PCR)

RT-PCR is a rapid procedure for the analysis of the level of expression of genes. This technique uses the ability of reverse transcriptase (RT) to convert RNA into single-stranded cDNA and couples it with

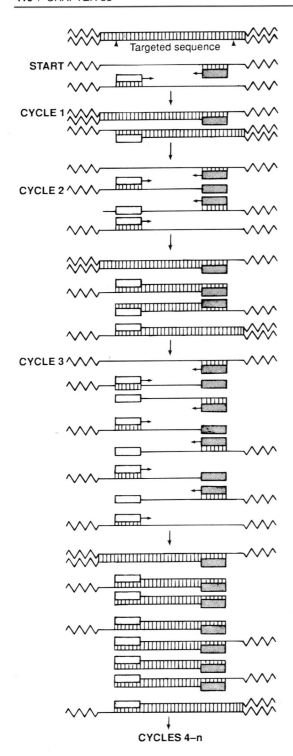

START

CYCLE 1

CYCLE 2

CYCLE 3

CYCLES 4–n

Targeted sequence

Figure 38–9. The polymerase chain reaction is used to amplify specific gene sequences. Double-stranded DNA is heated to separate it into individual strands. These bind 2 distinct primers that are directed at specific sequences on opposite strands and that define the segment to be amplified. DNA polymerase extends the primers in each direction and synthesizes 2 strands complementary to the original 2. This cycle is repeated several times, giving an amplified product of defined length and sequence. (Reproduced, with permission, from Murray, RK: *Harper's Biochemistry*, 23rd ed. Appleton & Lange, 1993.)

Table 38–3. Examples of inherited disorders detectable by PCR.

Disease	Affected Gene
Adenosine deaminase	Adenosine deaminase (see Chapter 27)
Lesch-Nyhan syndrome	HGPRT (see Chapter 27)
α_1-Antitrypsin deficiency	α_1-Antitrypsin
Cystic fibrosis	Cystic fibrosis transmembrane conductance (CFTR) protein
Fabry disease	α-Galactosidase (see Chapter 10)
Gaucher's disease	Glucocerebrosidase (see Chapter 10)
Sandhoff-Jatzkewitz disease	Hexosaminidase A and B (see Chapter 10)
Tay-Sachs disease	Hexosaminidase A (see Chapter 10)
Familial hypercholesterolemia	LDL receptor (see Chapter 21)
Glucose-6-phosphate dehydrogenase deficiency	Glucose-6-phosphate dehydrogenase
Maple syrup urine disease	α-Keto acid decarboxylase (see Chapter 25)
Phenylketonuria	Phenylalanine hydroxylase (see Chapter 24)
Ornithine transcarbamylase deficiency	Ornithine transcarbamylase
Retinoblastoma (Rb)	Rb gene product
Sickle-cell anemia	Point mutation in β-globin gene resulting in improper folding of protein
β-Thalassemia	Mutations in β-globin gene that result in loss of synthesis of protein
Hemophilia A	Factor VIII (see Chapter 37)
Hemophilia B	Factor IX (see Chapter 37)
Von Willebrand disease	Von Willebrand factor (see Chapter 37)

the PCR-mediated amplification of specific types of cDNAs present in the RT reaction. The cDNAs that are produced during the RT reaction represent a "window" into the pattern of genes that are being expressed at the time the RNA was extracted.

Total cellular RNA can be extracted from tissues or cells by any of several techniques and used as a template for RT. In most cases the RNA is primed using random primers. A small aliquot of the RT reaction is then added to a PCR reaction containing primers specific to the sequences of interest. The products of the RT-PCR can be then be visualized as described above for standard PCR.

PCR-Single Strand Conformational Polymorphism (PCR-SSCP)

Many inherited disorders are due to single nucleotide changes within critical regions of the affected gene. The PCR-SSCP technique can detect single mutations in genes, because these alter the mobility of the single strands of DNA harboring the mutation. Specific PCR primers are made that span both the sequences of a given disease gene where a mutation is known to exist and the region amplified by PCR. The same region of the wild-type gene is also amplified. For accurate visualization of the PCR products after gel electrophoresis, either the primers are radioactively labeled or radioactive nucleotides are incorporated into the PCR products.

The PCR products are separated in a polyacrylamide gel and visualized by exposure of the gel to x-ray film (Figure 38–10). Individuals who are homozygous wild-type at the locus being analyzed will exhibit two distinct bands in the gel, as will those who are homozygous mutant. However, because of the nucleotide change, the mutant PCR products will migrate with different mobilities in the gel. Individuals who are heterozygous will exhibit a pattern consisting of all four bands (Figure 38–10).

Ligase Chain Reaction

LCR is another technique for detecting single point mutations in disease genes. The technique uses a thermostable DNA ligase to bind together perfectly adjacent oligos. Figure 38–11 shows how this tech-

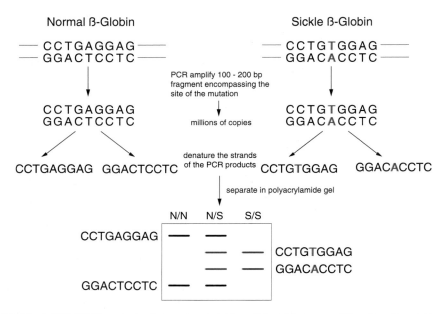

Figure 38–10. A PCR-SSCP analysis of normal and sickle-cell β-globin genes at the site of the sickle-cell mutation. The mutation in the sickle-cell gene is indicated by blue letters. Specific primers that span this mutation are utilized in a PCR. The products of the PCR are then separated due to conformational differences on a nondenaturing gel. The three possible results of this type of analysis for the sickle-cell gene are shown. An individual homozygous for wild-type will exhibit a distinct pattern of two bands, whereas a person homozygous for sickle-cell will exhibit a different pattern of two bands (in blue at the right). A person that is heterozygous for the wild-type and sickle cell genes will exhibit a pattern of all four bands.

nique can be used to detect the presence of the single point mutation in the sickle-cell β-globin gene. Two sets of oligos are designed to anneal to one strand of the gene at the site of the mutation, and a second set of two oligos anneals to the other strand. The oligos are designed in such a way that they will only completely anneal to the wild-type sequences. In the sickle-cell mutant the 3′ nucleotide of one oligo in each pair is mismatched. This mismatch prevents the annealing of the oligos directly adjacent to each other. Therefore, the DNA ligase will not bind the two oligos of each pair together.

With the wild-type sequence, the oligo pairs that are bound together become targets for annealing the oligos and, therefore, result in an exponential amplification of the wild-type target. Since prior knowledge of the sequence is required in order to detect point mutations in disease genes, LCR is used primarily to detect the presence of a mutant allele in high-risk patients.

Transgenesis

Transgenesis is the process of introducing exogenous genes into the germ line of an organism. The first successful transgenesis experiments were carried out in mice. For example, one well-known experiment introduced rat growth hormone gene into the germ line of mice. These transgenic mice grew to twice their normal size.

In the creation of a transgenic animal, the gene of interest must be passed from generation to generation, ie, it must be inherited in the germ line. To bring this about in mice or livestock animals, vectors containing the gene of interest with appropriate regulatory elements (eg, the β-lactoglobulin promoter, if expression of the transgene in the milk is desired) are injected into the nucleus of fertilized eggs. The eggs are then transplanted into the uterus of receptive females for development of the potential transgenic offspring. In order to test the resultant animal for germ-line transmission of the transgene, the chromosomal DNA of its *offspring* is tested. If the transgene is present, exhibiting Mendelian inheritance, then it has been transmitted in the germ line.

Currently, the process of transgenesis is used in both the plant and livestock industries. In the majority of these procedures, the aim is to generate plants and animals that are more resistant to diseases and infections. However, some transgenic farm animals such as sheep and cows are being developed

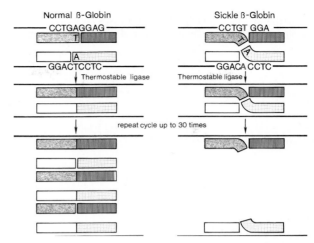

Figure 38–11. LCR analysis of normal and sickle cell β-globin genes. Genomic DNA is heat denatured and four oligonucleotides complementary to sequences of both strands are allowed to anneal. Only the wild-type globin sequence allows the two oligo pairs to anneal perfectly adjacent to each other. The ends of these perfectly matched oligos are a target for the thermal stable ligase. The mutant globin gene prevents complete adjacent annealing of the oligos and, thereby, prevents a target for ligase from forming. The ligated oligos themselves become targets for oligo annealing and ligation; as a result, after 30 cycles, detectable amounts of DNA are created.

in order to obtain high levels of expression of therapeutically important proteins in their milk. This allows large amounts of the protein of interest to be purified from the milk.

Gene Therapy

Transgenesis with humans would, in theory, allow for the elimination of disease genes in a population of offspring. However, technical as well as ethical issues likely will prevent any transgenic experiments with human eggs. Therefore, the ability to replace known disease genes with normal copies in afflicted humans is the ultimate goal of gene therapy. Human gene therapy protocols aim to introduce correcting copies of disease genes into somatic cells of an affected individual. The expression of a correct copy of an affected gene only in somatic cells prevents transmission through the germ line; in effect, it is a treatment analogous to organ or tissue transplantation. The most common techniques used in gene therapy include introducing the corrected gene into bone marrow cells, skin fibroblasts or hepatocytes. The most common vectors are derived from retroviruses and use only the transcriptional promoter regions of these viruses (the LTRs) to drive expression of the gene of interest. The advantage of retroviral-based vector systems is that expression can occur in most cell types. A number of human inherited disorders have been corrected in cultured cells (Table 38–4), and several diseases (eg, malignant melanoma and severe combined immunodeficiency disease, SCID) are currently being

Table 38–4. Human disorders treated in cultured cells by gene therapy.

Disorder	Affected Gene
SCID	Adenosine deaminase (see Chapter 27)
SCID	Purine nucleoside phosphorylase (see Chapter 27)
Lesch-Nyhan syndrome	HGPRT (see Chapter 27)
Gaucher disease	Glucocerebrosidase (see Chapter 10)
Familial hypercholesterolemia	LDL receptor (see Chapter 21)
Phenylketonuria	Phenylanine hydroxylase (see Chapter 24)
β-Thalassemia	β-Globin
Hemophilia B	Factor IX (see Chapter 37)

treated by gene therapy techniques. These successes so far indicate that gene therapy is likely to be a powerful therapeutic technique against a host of diseases in coming years.

Questions

DIRECTIONS (items 1–15): Match the following lettered list of enzymes or techniques with the the properties described. Some answers may be used more than once, others not at all.

    a. Reverse transcriptase.
    b. Taq DNA polymerase.
    c. RFLP.
    d. RT-PCR.
    e. T7 RNA polymerase.
    f. Bacteriophage lambda.
    g. Plasmid.
    h. DNA ligase.
    i. EcoRI.
    j. LCR.
    k. PCR-SSCP.
    l. Genomic library.
    m. Sticky ends.
    n. RNaseH.
    o. cDNA library.

1. Enzyme that digests any DNA at the same specific sequence.

2. Separation of the products of this assay by polyacrylamide gel electrophoresis is used to identify single mutations in DNA.

3. Enzyme that covalently connects together two fragments of DNA.

4. Enzymes used to convert RNA into DNA.

5. Enzyme used to generate RNA in vitro from a DNA template.

6. Technique used to analyze alterations at a specified genetic locus by restriction enzyme digestion, gel electrophoresis, and Southern blotting.

7. Enzyme required for PCR.

8. DNA molecule that can confer antibiotic resistance to bacteria.

9. Technique that uses a set of four distinct primers in a repetitive cycle to identify single mutations in a gene.

10. The use of reverse transcriptase is required for the generation of this type of library.

11. Technique used to measure the expression of a specific gene at any given time or in any specific tissue.

12. Vectors derived from this DNA are used in the cloning of genomic DNA.

13. This reaction requires a thermostable DNA ligase.

14. Enzyme used to remove the RNA from a cDNA synthesis reaction.

15. This reaction can be initiated with either oligo(dT) or random primers.

Answers

1. i	**6.** c	**11.** d
2. k	**7.** b	**12.** f
3. h	**8.** g	**13.** j
4. a	**9.** j	**14.** n
5. e	**10.** o	**15.** d

Index

Note: Page numbers followed by *t* and *f* indicate tables and figures, respectively. Page numbers in **boldface** indicate a major discussion.

Abetalipoproteinemia, 218*t*
Acanthocytosis, 218*t*
ACAT. *See* Acyl-CoA-cholesterol acyltransferase
ACC. *See* Acetyl-CoA carboxylase
Accessory proteins, processivity, 311
Acetoacetate, 199
    formation, 200*f*
Acetoacetate:succinyl-CoA transferase, 200
Acetoacetic acid, 6
Acetone, 199
Acetyl-CoA, 296
    origins for synthesis, 188–189, 189*f*
    threonine/glycine conversion, 257*f*
Acetyl-CoA carboxylase (ACC), 189, 374
    hormonal influences, 303*t*
N-Acetylgalactosamine, 178
N-Acetylglucosamine, 178
Acetyl lipoamide, formation, 155*f*
N-Acetylneuraminic acid, 87, 178
    structure, 75*f*
Acetyl-thiamin pyrophosphate, structure, 155*f*
Acid
    catalysis involving, 54
    conjugate base, 2
    excretion, 7
    secretion by proximal tubular cells, 7*f*
    titration curve, 3*f*
Acid anhydride bond, 94
Acid-base balance, 1–8
    kidneys and, 6
    plasma, 6
Acidosis, 8
    metabolic, 6
    respiratory, 6
Aconitase, 157
    production, 159*f*
ACP. *See* Acyl carrier protein
Active transport, 106, 107*f*, 116
Acute intermittent porphyria, 267

Acute lymphocytic leukemia (ALL), 363
Acute-phase response, 355
Acyl carrier protein (ACP), 189, 374
Acyl-CoA-cholesterol acyltransferase (ACAT), 214
Acyl-CoA dehydrogenase deficiency, 198–199
Acyl-CoA synthetase, 193, 201
ADA. *See* Adenine deaminase
Adapter protein, 361
Adenine, 92
    structure, 92*f*
Adenine deaminase (ADA), 283
Adenosine
    conformations, 93*f*
    derivatives, 94
    structure, 93*f*
Adenosine deaminase, polymerase chain reaction and, 411*t*
Adenosine diphosphate (ADP), 94, 392
    structure, 94*f*
Adenosine monophosphate (AMP), 94
    structure, 94*f*
Adenosine phosphoribosyltransferase (APRT), 280
Adenosine triphosphate (ATP), 94, 296
    formation, 44
    structure, 94*f*
S-Adenosylmethionine, 94
    formation, 245*f*
    structure, 95*f*
Adenylate cyclase, receptors modulating, 363
Adenylate kinase, 289
Adipose tissue, 296
    lipolysis, control, 197*f*
    metabolic functions, 297*t*, 300
    metabolism, 301*f*
ADP. *See* Adenosine diphosphate
ADP/ATP translocase, 171

Aerobic glycolysis, 121
Affinity chromatography, protein analysis and, 33
Agarose, 100
Alanine, 286, 298
    biosynthesis, 242
    isoelectric structure, 21*f*
    metabolic fate, 254–255
Alanine cycle, 242, 298, 300
ALA synthase. *See* Aminolevulinic acid synthase
Alcaptonuria, 24, 259
Aldolase, 124
Aldolase B, 147
Aldoses, structural relations, 72*f*
Alkalosis, 8
    metabolic, 6
    respiratory, 6
Alkylguanine transferase, 315
ALL. *See* Acute lymphocytic leukemia
Allopurinol, 95, 281–283
    structure, 96*f*
Allosteric effectors, 56
    interactions with monomeric enzymes, 56*f*
    interactions with oligomeric enzymes, 57*f*
    negative, 56
    positive, 56
Allosteric enzymes, 56–58
Allosteric sites, 56
Allysine, 62
    structure, 63*f*
Alpha-amanitin, 318
Alpha amylase, 121
Alpha-helix, 28–29
    right-handed, 30*f*
    view down axis, 31*f*
Alpha-hydroxylase, 194
Alpha-lipoprotein deficiency, familial, 218*t*
Alpha-tocopherol, antioxidant properties, 175
Amidotransferases, 278

Amino acid oxidases, 236
  oxidative deamination catalyzed, 236*f*
Amino acids, 18–24
  acid-base properties, 20–21
  activation, 329–336
  branched-chain, 258–259
    catabolism, 259*f*
  carbon skeletons, amphibolic intermediates formed, 254*f*
  chemical nature, 18–23
  classification, 18
    hydrophobicity/hydrophilicity and, 21*t*
  clinical significance, 23–24
  enzyme active-site, 55*f*
  essential, 233*t*
  glucogenic, 253–254
  ketogenic, 253–254
  metabolic fate, 253–265
  nitrogen removal, 234–236
  nonessential, 233*t*
    synthesis, 241–250
  optical properties, 22
  peptide bonds, 23
  present in proteins, 19–20*t*
  R-groups, 21, 55*f*
  synthesis and degradation, 296
  transport across intestinal wall, 232*f*
Aminoacidurias, 23–24
Aminoacyl-adenylate intermediate, 330
Aminoacyl-tRNA, formation, 331*f*
Aminoacyl-tRNA synthetases, 330–331
Aminoadipic semialdehyde synthase, 259–262
para-Aminobenzoic acid (PABA), 375
2-Amino-3-carboxymuconic acid, 262
Aminoisobutyrate, 286
Aminoisobutyric aciduria, defective enzyme, 288*t*
Aminolevulinic acid synthase (ALA synthase), 267–268
Aminopterin, 285, 380
Amino sugars, 178–186
  interrelationships, 179*f*
  N-linked, 182–183
  O-linked, 182
  physiologically important, 179–180
Aminotransferases, 234–235
  mechanism, 234*f*
Ammonia
  production, 230
  secretion, 7
Ammonium, formation, 8*f*
AMP. *See* Adenosine monophosphate
Amphibolic, 114
Ampholytes, 4, 33
Amylopectin, 76
Amylose, 76

Amytal, oxidative phosphorylation and, 172*t*
Anabolic reactions, 112–114
Anaerobic glycolysis, 121, 128
Anapleurotic, 114
Andersen's disease, deficiency, 137*t*
Anemia
  macrocytic, 380
  pernicious, 379
  sickle cell, 32
    pedigree analysis, 409*f*
    polymerase chain reaction and, 411*t*
Anion exchanger, 33
Annealing, 99
Anomeric carbon, 72
Anomers, 72
Antibiotics, translation and, 339*t*
Antibodies, 67
Anticoagulants, coumarin-based, 385
Antigens, tumor, 367
Antimetabolites, oxidative phosphorylation and, 172*t*
Antimycin A, oxidative phosphorylation and, 172*t*
Anti-oncogenes, 366
Antioxidants, 384
  interactions, 385*f*
alpha2-Antiplasmin, 397
Antiports, 107–108
Antithrombin III, 396
alpha1-Antitrypsin, 396
  deficiency, polymerase chain reaction and, 411*t*
Apo-A-1 deficiency, 218*t*
Apo-B-100, familial defective, 216*t*
Apo-C-II, 212
Apo-C-III deficiency, 218*t*
Apo-D, 215
Apo-E, 212
Apoenzymes, 37
Apoproteins, classifications, 213*t*
Apoptosis, 306
APRT. *See* Adenosine phosphoribosyltransferase
Arachidonic acid, 193
  conversion to leukotrienes, 207*f*
  conversion to prostaglandins and thromboxanes, 206*f*
  structure, 82*t*
Arginase, 236, 238
Arginine
  catabolism, 256*f*, 262
  metabolic fate, 255
  nitric oxide formation and, 273
Argininosuccinate lyase, 238
Argininosuccinate synthetase, 238
ARS. *See* Autonomously replicating sequence
Arthritis, rheumatoid, 109
Ascorbic acid, 377
  clinical significance, 380
  oxidation to dehydroascorbic acid, 379*f*
A-site, 334

Asparaginase, 254
  metabolic fate, 254–255
Asparagine, biosynthesis, 247–250
Asparagine synthetase reaction, 250*f*
Aspartate
  biosynthesis, 242, 247–250
  metabolic fate, 254–255
Aspartic acid, protonic equilibria, 22*f*
Aspartylglycosaminuria, enzyme deficiency, 185*t*
ATCase, 285
Atherosclerosis, 218
ATP. *See* Adenosine triphosphate
ATP-citrate lyase reaction, 188
ATP synthase, 167, 171
Attenuation, 344
Autonomously replicating sequence (ARS), 405
Axial, 73
Azide, oxidative phosphorylation and, 172*t*
Azidothymidine (AZT), 95
  structure, 96*f*
AZT. *See* Azidothymidine

Bacteria
  intestinal, products, 233*t*
  transcription cycle, 320*f*
  transcription termination signal, 321*f*
Bacteriophage, 400
Bacteriophage lambda, 403
Base-pairing, 98, 99*f*
  Watson-Crick, 98
Bases, catalysis involving, 54
Bassen-Kornzweig syndrome, 218*t*
Beriberi, 377
Beta-hydroxy-beta-methylglutaryl-CoA, 199
Beta-hydroxybutyrate, 199
Beta-hydroxybutyrate dehydrogenase, 199
Beta-oxidation, 194, 194*f*, 195*f*
Beta-sheets, 29
  antiparallel, silk fibroin as model, 61–62, 62*f*
  antiparallel/parallel, peptide bonds and R-groups, 31*f*
Bicarbonate
  formation in blood, 4
  reabsorption, 8*f*
Bicarbonate ion, formation, carbonic acid and, 5
Bile acids, 225–226
  biosynthesis and degradation, 227*f*
Bile pigments, 270–271
Bilirubin, 180, 270–271
  catabolism of heme and, 270*f*
Bilirubin diglucuronide
  conjugated, 180
  structure, 271*f*
Biliverdin, 270
Binding proteins, single-strand, 309
Biological catalysts, enzymes as, 44–45

Biological membranes. *See* Membranes
Biological oxidations, 164–175
Biotin, 374
  clinical significance, 379
  as coenzyme, 38*t*
  structure, 375*f*
2,3-Bisphosphoglycerate (BPG), 67
  pathway in erythrocytes, 127*f*
Blindness, night, 385
Blood, buffering, 4–6
Blood-clotting factors
  binding to plasma membrane, 396*f*
  numbering system, 394*t*
Blood-clotting proteins, 385
Blood coagulation, 391–398
  functions of proteins, 395*t*
  pathways, 393*f*
    extrinsic, 392, 394–395
    intrinsic, 392, 394
Blood group antigen
  Duffy, 185
  Lewis, 185
Blood group systems, MN, 185
Blot transfer procedure, 408*f*
Blunt end, 401
Boat form, 73
Bohr effect, 5, 67
BPG. *See* 2,3-Bisphosphoglycerate
Bradykinin, 394
Brain, 296
  metabolic functions, 297*t*, 300
Branched-chain amino acids,
  258–259
  catabolism, 259*f*
Branched-chain keto acid(s), genetic
  disease and, 259–265
Branched-chain keto acid dehydrogenase, 258
Branched chain keto acid oxidase, 24
Branching enzyme, 138
Breakpoint cluster region gene, 363
Buffering, 3–4
  blood, 4–6
ButyrylCoA, formation, 44

Cachectin, 357
Cafe-au-lait spots, 369
Calbindin D, 383
Calcitonin, 383
Calcitriol, 383
Calcium, characteristics, 387*t*
Calmodulin, 57, 142
cAMP, 94
  formation, 95*f*
Cancer, 314–315, 366–370
CAP. *See* Catabolite activator protein
Capsular space, 6
Carbamoyl aspartate, 285
Carbamoyl phosphate, 285
Carbohydrates, 71–78
  classification, 72*t*
  digestion, 120–121
  energy value, 115*t*

  linkage to proteins, 181–183
  specific dynamic action, 115*t*
Carbon, anomeric, 72
Carbon dioxide
  sources in tissues, 4
  transport from peripheral tissues to
    blood, 5*f*
Carbonic acid, 6
  bicarbonate ion formation and, 5
Carbonic anhydrase, 5, 6
Carbon monoxide, 270
  oxidative phosphorylation and,
    172*t*
Carboxyl proteases, proenzymes,
    activators, cleavage sites, 231*t*
Carboxypeptidases, 28
  proenzyme, activator, cleavage
    sites, 231*t*
  specificities, 28*t*
Carcinoma, colon, 370
Cardiolipin, 84
  biosythesis, 203*f*
  membrane, 104*t*
  metabolism, 203
  structure, 84*f*
Carnitine
  deficiencies, 198
  transport of long-chain fatty acids
    and, 193*f*
Carnitine acyltransferase I, 193, 198
Carnitine acyltransferase II, 193
  deficiencies, 198
beta-Carotene, 380
  cleavage to retinaldehyde, 381*f*
beta-Carotene dioxygenase, 380
Carotenoids, 380
Catabolic reactions, 112–114
Catabolism
  arginine, 256*f*, 262
  branched-chain amino acids, 259*f*
  cysteine, 258*f*
  glucose, 121
  heme, 269–271, 270*f*
  histidine, 262, 263*f*
  lysine, 259–262, 261*f*
  polyamine, 271, 273*f*
  proline, 256*f*
  purine nucleotide, 280–281
  pyrimidine nucleotide, 286–288,
    287*f*
  threonine, 255
  tryptophan, 264*f*
  tyrosine, 259
    intermediates, 260*f*
Catabolite activator protein (CAP),
  344
Catabolite-regulated operons, 343
Catabolite repression, 344
Catalase, 175
Catalysis
  by bond strain, 54
  covalent, 54–55
  proton donors and acceptors and, 54
  by proximity and orientation, 54
Catalytic site, 45–46

Catecholamines, 302
Cation exchanger, 33
CCAAT box, 346
CCK. *See* Cholecystokinin
CD45, 364
cDNA cloning, 402, 404, 404*f*
Cell cycle, eukaryotic, 307–308
  major phases, 308*f*
Cell death, 306
Cellular respiratory rate, 171
Cellular retinol binding protein
  (CRBP), 380
Centrifugation, protein analysis and,
  34
Centromere, 405
Ceramide, 204
  biosynthesis, 204*f*
Cerebellar ataxia, 23
Cerebrosides, 86–87
  metabolism, 204
CETP. *See* Cholesterol ester transfer
  protein
cGMP, 94–95
Chair form, 73
Chemical reactions
  conversion of reactants to products,
    energy changes during, 45*f*
  enzyme-catalyzed, effects of substrate concentration, 47*f*
  equilibrium states, 44
  order classifications, 44
  oxidation-reduction, 165–166
  rates, 43
  reaction order/number of reactants
    relationship, 45*t*
  thermodynamic potential, 166
Chemiosmotic potential, 170–171
Chemokines, 357
  receptor, 185
Chenodeoxycholic acid, 225
Chloramphenicol, translation and,
  339*t*
Chloride, characteristics, 387*t*
Chloride shift, 6
Cholecalciferol, 383
  formation, 383*f*
  hydroxylation, 384*f*
Cholecystokinin (CCK), 231
Cholesterol, 82, 87
  balance at cellular level, 224*f*
  biosynthesis, 222–223
  membrane, 104*t*
  metabolism, 221–227
    regulation, 223–225
  reverse transport, 215
  structure, 87*f*
  synthesis
    regulation by HMG-CoA reductase, 225*f*
    regulators, 298*t*
  synthesis and degradation, 296
  synthesis from mevalonic acid,
    223*f*
  transport between tissues, 226*f*
  utilization, 225

Cholesterol ester storage disease, 216t
Cholesterol ester transfer protein (CETP), 215
Cholestyramine, 219
Cholic acid, 225
Choline, 377
Choline plasmalogens, 203
Chondroitin sulfate, 76
  characteristics, 78t
  structure, 76f
Chromatin
  organization in cell, 308f
  structure, 306–307, 344
Chromatography
  DNA structure analysis and, 100
  protein analysis and, 33
Chromatosome, 307f
Chromium, characteristics, 388t
Chromosomes, recombinant, generation, 313f
Chronic myelogenous leukemia (CML), 363
Chylomicron remnant receptor, 212
Chylomicrons
  lipid transport and, 211–212
  metabolic fate, 214f
Chyme, 121
Chymotrypsin
  proenzyme, activator, cleavage sites, 231t
  specificities, 29t
Citrate, conversion to isocitrate, 159f
Citrate synthase, 157
  citric acid production and, 159f
Citric acid, production, 159f
Citric acid cycle, 158f
  enzymes, 157–162
  regulators, 298t
Citrulline, 235–236
Clofibrate, 219
Cloning, 400, 402–406
  analysis of products, 406–407
  cDNA, 402, 404, 404f
  expression, 404
  genomic, 402, 404–405
  yeast artificial chromosome, 405–406, 406f
CML. See Chronic myelogenous leukemia
CoA. See Coenzyme A
Coagulation. See Blood coagulation
Cobalamin, 374–375
  clinical significance, 379–380
  as coenzyme, 38t
  structure, 376f
Cobalt, characteristics, 388t
Coding strand, 318
Coenzyme(s), 37–38
  electron transport pathway, 168–169
  functional role, 38
  vitamins as, 38t
Coenzyme A (CoA), 374
  formation, 44

synthesis from pantothenic acid, 292f
Coenzyme Q, 169
Cofactor, 394
Cohesive end, 400
Colestipol, 219
Collagen, 62–64
  structure, 63f
  structure/function relationships, 61t
  unusual amino acids, 63f
Colon carcinoma, 370
Colony stimulating factors (CSFs), 357
Competitive inhibitors, 49
  characteristics, 49t
Complex enzymes, 37
Condensing enzyme, 157, 189
Cones, phototransduction, 382f
Conjugase, 375
Conjugate base, 2
Conjugated bilirubin diglucuronide, 180
Contact phase, 394, 395
Copper, characteristics, 388t
Coproporphyrinogen III, 268
Corepressor, 344
Cori cycle, 130, 131f, 298, 300
Cosmid, 404–405
Coumarin, 385, 397
Covalent catalysis, 54–55
c-ras, 33
CRBP. See Cellular retinol binding protein
Creatine, biosynthesis, 235f
Creatinine, 235
  biosynthesis, 235f
  urinary and plasma concentrations, 6t
Creatinine clearance rate, 235
Critical micellar concentration, 104
Crossing-over, 313f
CSFs. See Colony stimulating factors
CTP synthase, 285
Cyanide, oxidative phosphorylation and, 172t
Cyanogen bromide, 28
Cyclic pathway, 205, 206f
Cycloheximide, translation and, 339t
Cyclooxygenase, 205
Cystathionine synthase, 243
Cysteine
  biosynthesis, 242–243
    role of methionine, 244f
  catabolism, 258f
  metabolic fate, 255–256
Cysteine sulfinate pathway, 258f
Cystic fibrosis, polymerase chain reaction and, 411t
Cystine, 21
Cystine disulfide bond, reduction, 27f
Cystinuria, 24
Cytidine, structure, 93f
Cytochrome b5, 193
Cytochrome oxidase, 168
Cytokines, 355–357
  properties, 353–354t

Cytokinesis, 307
Cytosine, 92
  structure, 92f
Cytosolic NADH, energy conservation and, 172–173

DAG. See Diacylglycerol
dAMP, 94
Dansyl chloride method, 27
D-conformation, 71
ddI. See Dideoxyinosine
Debranching enzyme, 134–135
Decay-accelerating factor, 185
7-Dehydrocholesterol, 383
  conversion to cholecalciferol, 383f
Denaturation, 99
  thermal, 99
Deoxycytidine kinase, 287
Deoxyribonucleotides
  discontinuous polymerization, 311f
  formation, 288–289
Dermatan sulfate, 76
  characteristics, 78t
  structure, 76f
Dextrorotatory, 22
DHFR. See Dihydrofolate reductase
Diabetes
  insulin-dependent, 301–302
    interrelationships of organs, 302f
    metabolic effects, 301–302
  non-insulin dependent, 109, 302
Diabetic ketoacidosis (DKA), 201
Diacylglycerol (DAG), 391
1,2-Diacylglycerol, 201
1,2-Diacylglycerol phosphate, 201
Dideoxy chain termination sequencing, 406, 407f
Dideoxyinosine (ddI), 95
  structure, 96f
2,4-Dienoyl-CoA reductase, 194
Diffusion
  facilitated, 106–107
    passive diffusion compared, 107f
    "ping-pong" model, 109f
  simple, 106–107
Digestive tract, nitrogenous compounds, 231–234, 231t
Dihydrobiopterin reductase, 243
Dihydrofolate reductase (DHFR), 285, 375
Dihydroorotate, 285
Dihydroxyacetone, 71
  structure, 73f
Dimethylallyl pyrophosphate (DMPP), 223
2,4-Dinitrophenol, oxidative phosphorylation and, 172t
Dipeptide, 23
Diphosphatidylglycerol (DPG), 84
  metabolism, 203
  structure, 84f
Diphtheria toxin, translation and, 339t
Disaccharides, 73–74
  structures, 74f

Distal convoluted tubule, 6
Dithiothreitol, 27
DKA. *See* Diabetic ketoacidosis
DMPP. *See* Dimethylallyl pyrophosphate
DNA. *See also* Recombinant DNA technology
  analysis of structure, 100
  double helical structure, 97–98, 98*f*
  genomic, 405*f*
  helices
    parameters, 99*t*
    thermal properties, 99
  phosphodiester backbone, 97*f*
  postreplicative modification, 312
  recombination, 313, 314*f*
  repair of damaged, 314–315
  repetitive, 306
  replication, 306, 308–312, 310*f*
    model, 311*f*
  restructuring, 313–315
  synthesis, 306–315
    RNA-primed, 310*f*
DNA-binding proteins
  cancer and, 368
  single-strand, 309
  structural motifs, 346–347, 347*f*
DNA ligase, 312, 401, 402*f*
DNA photolyase, 315
DNA polymerase
  enzymatic activities, 312
  thermostable, 408
DNA polymerase holoenzyme, 309
DNA sequencing, 406
DNA topoisomerase, 311
DNA tumor viruses, 367
dNTP, regulation of formation, 289
Docking protein, 361
Dolichol, structure, 182*f*
Dolichol phosphate, 182
Dolichol-P-P-oligosaccharide, biosynthetic pathways, 183*f*
DOPA, 271
Dopamine, 271
DPG. *See* Diphosphatidylglycerol
Drug-induced orotic aciduria, defective enzyme, 288*t*
dTMP, 377
Duffy blood group antigen, 185
Dysbetalipoproteinemia, familial, 216*t*

Edman degradation, 27–28
eEFs. *See* Eukaryotic elongation factors
EFs. *See* Elongation factors
EGF. *See* Epidermal growth factor
Ehlers-Danlos syndrome, 32
Eicosanoids, 193
  properties, 205*t*
  synthesis, 204–205
eIFs. *See* Eukaryotic initiation factors
Elastase
  proenzyme, activator, cleavage sites, 231*t*
  specificities, 29*t*

Electron transport, 167
Electron transport pathway, coenzymes, 168–169
Electron transport proteins
  complexes I and II, 167–168
  complexes III and IV, 168
Electrophoresis
  DNA structure analysis and, 100
  protein analysis and, 33–34
Electrostatic forces, protein structure and, 31
Elongation factors (EFs), 334
Embden-Myerhoff pathway, 121
Enantiomers, 71
Endergonic reactions, thermodynamic principles, 14*t*
Endoglycosidases, 185
Endonucleases, 278
  restriction, 312, 400–402, 402*f*
    characteristics, 403*t*
Endopeptidases, 28
Endopeptidase V8, specificities, 29*t*
Endoproteases, specificities, 29*t*
Energy charge, 114
Enhancers, 319
Enhancer sequences, 345
Enolase, 127
Enoyl-CoA isomerase, 194
Enoyl-CoA reductase, 189
Enterohepatic circulation, 226, 226*f*
Enthalpy, 13–15
Entropy, 13–15
Enzymes, 36–40
  allosteric regulation, 56–58
  biochemical properties, 37*t*
  as biological catalysts, 44–45
  catalytic sites, 45–46
  classification, 37–39
  clinical significance, 39–40
  complex, 37
  inhibitors, 48–50
    competitive, noncompetitive, uncompetitive, 49
    interactions with substrates, 49
    reversible, 49–50, 49*t*
  kinetics, 42–50
    Michaelis-Menten analysis, 45–47
  Krebs cycle, 157–162
  mechanisms regulating activity, 55–56
  metabolic, hormonal influences, 303
  purification, 39–40
  reaction analysis using linear kinetic plots, 47–48
  simple, 37
  used in molecular biology, 401*t*
Enzyme-substrate (ES) complex, 46
Enzyme-substrate interactions, 54–55, 54*f*
Epidermal growth factor (EGF), 351
  properties, 352*t*
  receptor, 360

Epinephrine, 139, 302
  binding to plasma membrane receptors, 140*f*
  tyrosine conversion, 274*f*
Epitopes, 67
Epo. *See* Erythropoietin
2,3-Epoxide form, 385
Equatorial, 73
Equilibrium constant, 1, 43
eRFs. *See* Eukaryotic releasing factors
Ergocalciferol, 383
  formation, 383*f*
Ergosterol, 383
  conversion to ergocalciferol, 383*f*
Erythrocyte(s)
  2,3-bisphosphoglycerate pathway, 127*f*
  chemokine receptor, 185
Erythrocyte P antigen, 184
Erythromycin, translation and, 339*t*
Erythropoietin (Epo), 355, 357
  properties, 352*t*
  Escherichia coli, releasing factors, 334–336
Essential amino acids, 233*t*
Essential fatty acids, 82, 193
Essential macrominerals, characteristics, 387*t*
Essential microminerals, characteristics, 388*t*
Ethanol, energy value, 115*t*
Ethanolamine plasmalogens, 203
Eukaryotic cell cycle, 307–308
  major phases, 308*f*
Eukaryotic elongation factors (eEFs), 334
Eukaryotic gene control, 344–346
Eukaryotic genomes, 306
Eukaryotic initiation factors (eIFS), 332–334
  function in translation initiation, 332*t*
Eukaryotic mRNA, structure of cap, 322*f*
Eukaryotic mRNA gene
  processing, 321*f*
  structure, 346*f*
Eukaryotic pre-mRNA
  mechanism of splice site recognition, 326*f*
  nucleotide frequencies at boundaries, 326*f*
Eukaryotic releasing factors (eRFs), 334–336
Eukaryotic transcription, 319–320
Exergonic reactions, thermodynamic principles, 14*t*
Exoglycosidases, 185
Exons, 306, 323
Exopeptidases, 28
  specificities, 28*t*
Expression cloning, 404
Extracellular-signal regulated kinases, 362
Extrinsic pathway, 392, 393*f*, 394–395

Fabry's disease, 204
    enzyme deficiency, accumulating
        substance, symptoms, 89t
    polymerase chain reaction and, 411t
Facilitated diffusion, 106–107
    passive diffusion compared, 107f
    "ping-pong" model, 109f
Factor II, 385
Factor Va, 395
Factor VII, 385
Factor VIII, 397
Factor VIIIa, 394
Factor IX, 385, 397
Factor IXa, 394
Factor X, 385
Factor Xa, 394
Factor XIa, 394
Factor XIIa, 394
Factor XIII, 397
Factor XIIIa, 397
FAD. *See* Flavin adenine dinucleotide
Familial adenomatosis polyposis
    (FAP), 369
Familial alpha-lipoprotein deficiency,
    218t
Familial dysbetalipoproteinemia, 216t
Familial hyperalphalipoproteinemia,
    216t
Familial hyperapobetalipoproteine-
    mia, 216t
Familial hypercholesterolemia (FH),
    32–33, 218
    gene therapy, 413t
    polymerase chain reaction and,
        411t
Familial hyperchylomicronemia, 216t
Familial hypertriacylglycerolemia,
    216t
Familial hypobetalipoproteinemia,
    218t
Fanconi's syndrome, 23–24
FAP. *See* Familial adenomatosis
    polyposis
Farber's lipogranulomatosis, 89t
Farnesyl pyrophosphate (FPP), 223
FAS. *See* Fatty acid synthase
Fasting state
    interrelationships of organs during,
        299f
    metabolism and, 296–302
Fats
    energy value, 115t
    intestinal absorption, 211f
    specific dynamic action, 115t
Fatty acids, 82
    beta-oxidation, 194, 194f, 195f
    desaturation, 192–193
    elongation, 190–192
        microsomal system, 192f
    essential, 82, 193
    intake, transport, utilization, 83f
    long-chain
        biosynthesis, 191f
        transport through mitochondrial
            membrane, 193f

metabolism, regulation, 197–198
oxidation, 193–194
physiologically relevant, 82t
saturated, 82
    synthetic pathway, 189–190
synthesis, committed step, 189f
synthesis and degradation, 296
unsaturated, 82
    oxidation, 196f
Fatty acid synthase (FAS), 189–190
    multienzyme complex, 190f
Fatty acyl-CoA desaturase, 192
Fatty acyl-CoA ligase, 193
Fenofibrate, 219
Ferrochelatase, 268
FGF. *See* Fibroblast growth factor
FH. *See* Familial hypercholes-
    terolemia
Fibrillin, 32
Fibrin, 396–397
Fibrin clot
    dissolution, 397–398
    formation, 397f
Fibrin mesh, 391
Fibrinogen, 391, 392
    activation to fibrin, 396–397
Fibrinopeptides, 396
Fibroblast growth factor (FGF), 185,
    352
    properties, 352t
    receptor, 360
Fibrous proteins, 30
Fischer-style diagram, 72
Fish-eye disease, 218t
Flavin adenine dinucleotide (FAD),
    168–169, 290f, 373
Flavin mononucleotide (FMN),
    168–169, 373
Flavin nucleotide(s), oxidoreduction
    of isoalloxazine ring, 168f
Flavin nucleotide pyrophosphorylase,
    290
Flavokinase, 290
Flavoproteins, 373
Fluoride, characteristics, 388t
5-Fluorodeoxyuridine, 285
5-Fluorodeoxyuridylate, 285
5-Fluorouracil, 95, 285
    structure, 96f
FMN. *See* Flavin mononucleotide
Folic acid, 375–377
    clinical significance, 380
    as coenzyme, 38t
    structure, 377f
Foods
    energy values, 115t
    specific dynamic action, 115, 115t
Forbe's/Cori's disease, deficiency,
    137t
FPP. *See* Farnesyl pyrophosphate
Free energy, 13–15
Fructose, 74
    intolerance, 147
    metabolism, 147–148, 149f
    structure, 73f

Fructose-1,6-bisphosphate, glycoly-
    sis/gluconeogenesis and, 125f
Fructose-2,6-bisphosphatase, 303
    allosteric regulators, 125t
    hormonal influences, 303t
Fructose-6-phosphate, 179
Fructose-1-phosphate aldolase, 147
Fucose, 179, 181
    structure, 75f
Fucosidosis, enzyme deficiency, 89t,
    185t
Fumarase, 161
Fumarate hydratase, 161
Furan, 72
Furanoses, 72
Fusidic acid, translation and, 339t
Futile cycles, 114

GAGs. *See* **Glycosaminoglycans**
Galactose, 74, 179, 181
    conversion to glucose, 150f
    metabolism, 149–150
Galactose-1-P uridylyltransferase,
    150
Galactosemia, 150
Gangliosides, 87
    metabolism, 204
Gangliosidosis, enzyme deficiency,
    89t, 185t
GAPs. *See* GTPase activating pro-
    teins
Gaucher's disease, 204
    enzyme deficiency, accumulating
        substance, symptoms, 89t
    gene therapy, 413t
    polymerase chain reaction and,
        411t
G-CSF. *See* Granulocyte-CSF
Gemfibrozil, 219
Gene(s), nucleotide sequencing, 40
Gene expression, 342–348
    enzyme synthesis and, 55
Gene therapy, 413–414
Genetic code, 329, 330t
Genetic disease, branched-chain keto
    acid catabolism and, 259–
    265
Genomic cloning, 402, 404–405
Genomic DNA, 405f
Geranyl pyrophosphate (GPP), 223
Gibbs equation, 13
GlcNAc phosphotransferase, 183,
    186
GlcNAc-1-P-6-Man-protein, 183
GlcNAc-P-P-dolichol, 182–183
Globoid leukodystrophy, 89t
Globosides, 87
    metabolism, 204
Globular proteins, 30
Glomerular filtration, 6
Glucagon, 302
Glucan transferase, activity, 136f
Glucocorticoid receptor, 363
Glucogenic amino acids, 253–254

Glucokinase, 123
  glucose phosphorylating activity, 123*f*
Gluconeogenesis, 128, 296
  control in liver, 125*f*
  pathways in liver, 129*f*
  rate-limiting enzyme, 124
  regulators, 298*t*
Glucosamine, structure, 75*f*
Glucosamine-6-phosphate, 179
Glucose, 74, 179, 181
  conversion to lactose, 150*f*
  intracellular catabolism, 121
  monomers, 74
  structure, 73*f*
  triacylglycerol synthesis and, 300
  urinary and plasma concentrations, 6*t*
Glucose-alanine cycle, 130, 131*f*, 243*f*
Glucose-6-phosphatase, 130, 283
Glucose-6-phosphate, phospho-enolpyruvate conversion, 130
Glucose-6-phosphate dehydrogenase, 147
  deficiency, polymerase chain reaction and, 411*t*
Glucose-1-phosphate uridylyltransferase, 137–138
Glucuronate, 76, 180
  structure, 75*f*
  synthesis, 180*f*
Glucuronic acid, 180
Glutamate, 231
  biosynthesis, 242, 247–250
  metabolic fate, 254–255
  proline biosynthesis and, 247*f*
Glutamate dehydrogenase, 7, 230–236
Glutamate dehydrogenase reaction, 250*f*
Glutaminase, 7, 235–236, 254
Glutamine, 231
  biosynthesis, 247–250
  metabolic fate, 254–255
Glutaredoxin, 289
Glutathione, structure, 175*f*
Glyburide, 198
Glyceraldehyde, 71
L-Glyceraldehyde, 22
Glyceraldehyde-3-phosphate, mechanism of oxidation, 126*f*
Glyceraldehyde-3-phosphate dehydrogenase, 125–126
Glycerol, 298
Glycerol kinase, 201
Glycerol-3-phosphate dehydrogenase, 201
Glycerol phosphate shuttle, 172, 172*f*
Glycine, 377
  biosynthesis, 246–247
  conversion to serine, pyruvate, acetyl-CoA, 257*f*
  metabolic fate, 255

metabolites requiring as precursor, 248*t*
Glycine synthase, 255
Glycocholic acid, 226
Glycoconjugates, 178–186
Glycogen, 74–76, 296
  biosynthesis, 139*f*
  molecule, 75*f*, 135*f*
  "primer", 138
  reactions of synthesis, 139*t*
  regulation of metabolism, 139
  structure, 134–138
  synthesis and degradation, 296
Glycogenesis, 134, 137–138
  pathway in liver, 143*f*
  regulators, 298*t*
Glycogenin, 134, 138
Glycogenolysis, 134–136
  pathway in liver, 143*f*
  regulators, 298*t*
Glycogen phosphorylase, 303, 374
Glycogen storage disease, 77, 137
  categories, 137*t*
Glycogen synthase (GS), 138
  catalytic activity, 138*f*
  effects of effectors, 144*t*
  hormonal influences, 303*t*
  regulation, 142–143, 142*f*
Glycolipids, 71
Glycolysis, 120–130, 296
  aerobic, 121
  anaerobic, 121, 128
  control in liver, 125*f*
  energy yield, 121
  individual reactions, 121–128
  pathway, 122*f*
    in liver, 129*f*
  rate-limiting enzyme, 124
  regulators, 298*t*
Glycophorin, 185
Glycoprotein(s), 32, 71, 178, 180–181
  degradation, 185–186
  glycosidic bonds, 181*f*
  targeting defects, 186
Glycoprotein glycosyltransferases, 182
Glycosaminoglycans (GAGs), 76–77
  characteristics, 78*t*
  disaccharide repeating units, 76*f*
Glycosidases, 135, 185
Glycosides, 73
Glycosidic bonds, 73, 181
  types in glycoproteins, 181*f*
Glycosphingolipids, 82, 86–87, 204
  diagrammatic representation, 88*f*
Glypiated proteins, 185
Gly-X-Y, 62
GM-CSF. *See* Granulocyte-macrophage-CSF
Gout, 95, 281
  defect causing, 283*t*
G1 phase, 307
G2 phase, 308
GPP. *See* Geranyl pyrophosphate

G-proteins, 359
  cancer and, 367
  coupled receptors, 363
  regulators, 363
Granulocyte-CSF (G-CSF), 357
Granulocyte-macrophage-CSF (GM-CSF), 357
Growth factor(s), 351–355. *See also specific type*
  cancer and, 367
  properties, 352*t*
Growth factor receptor-binding protein 2, 361
Growth suppressors, 366
GS. *See* Glycogen synthase
GTPase activating proteins (GAPs), 363
Guanine, 92
  structure, 92*f*
Guanosine
  derivatives, 94–95
  structure, 93*f*
Guanylate cyclases, 94, 359
L-Gulonolactone oxidase, 377

Half cell, 165
Hartnup disease, 23, 379
Haworth-style diagram, 72
HCI. *See* Heme-controlled inhibitor
HDLs. *See* High-density lipoproteins
Heart, 296
  metabolic functions, 297*t*, 301
Helicases, 309
Helicobacter pylori, 185
Helix-loop-helix (HLH), 346
Heme
  biosynthesis, 269*f*
    characteristics, 267*t*
  catabolism, 269–271, 270*f*
  structure, 169*f*
  translation and, 336
Heme-controlled inhibitor (HCI), 336
  translation in reticulocytes and, 338*f*
Heme oxygenase, 269
  catabolism of heme to bilirubin and, 270*f*
Hemiacetals, 71
Hemiketals, 71
Hemoglobin, 64–67
  classification, 65*t*
  oxygenation, 65*f*
  oxygen binding curves, 66*f*
  structure/function relationships, 61*t*
Hemoglobin S, 32
Hemoglobinuria, paroxysmal nocturnal, 185
Hemophilia, 397
  gene therapy, 413*t*
  polymerase chain reaction and, 411*t*
Hemorrhagic syndrome, 387
Hemostasis, 391
HEMPAS, 184
Henderson-Hasselbalch equation, 3

Heparan sulfate, 76
  characteristics, 78*t*
Heparin, 76, 397–398
  characteristics, 78*t*
  structure, 76*f*
Heparin cofactor II, 396
Hepatic lipase deficiency, 216*t*
Hepatic remnant receptor, 212
Hepatocyte growth factor (HGF)
  receptor, 360
Hepatorenal tyrosinemia, 24
Her's disease, deficiency, 137*t*
Herpes virus type I, 185
Heterooligomers, 32
Heteropolysaccharides, 74
Heterotropic effectors, 56–57
Hexokinase
  characteristics, 123*t*
  glucose phosphorylating activity,
    123*f*
Hexokinase reaction, 123
HGPRT. *See* Hypoxanthine-guanine
  phosphoribosyltransferase
High-density lipoproteins (HDLs),
  214–215
  metabolism, 217*f*
High performance liquid chromatog-
  raphy (HPLC)
  protein analysis and, 33
  reversed-phase, 33
Histidase, 262
Histidine, catabolism, 262, 263*f*
Histidinemia, 262
Histone proteins, 307
HIV. *See* Human immunodeficiency
  virus
HLH. *See* Helix-loop-helix
HMG-CoA, 222–223
  acetoacetate formation and, 200*f*
HMG-CoA lyase, 199
HMG-CoA reductase, 199, 223
  cholesterol synthesis and, 225*f*
  hormonal influences, 303*t*
HMG-CoA synthase, 199, 223
Holoenzymes, 37
Homocystinuria, 243
Homogentisate oxidase, 24
Homogentisic acid oxidase, 259
Homologous recombination, 313
Homooligomers, 32
Homopolysaccharides, 74
Homotropic effectors, 57
Hormone(s), intracellular receptors,
  363–364
Hormone-releasable hepatic lipase
  deficiency, 216*t*
Hormone-sensitive lipase, 194, 198
  hormonal influences, 303*t*
HPLC. *See* High performance liquid
  chromatography
HTLVs. *See* Human T-cell leukemia
  viruses
Human immunodeficiency virus
  (HIV), 366
Human T-cell leukemia viruses
  (HTLVs), 366

Hunter's syndrome, 78
Hurler's syndrome, 78
Hyaline membrane disease, 87
Hyaluronate
  characteristics, 78*t*
  structure, 76*f*
Hyaluronic acid, 76
Hybridization, 99
Hydration, 4
Hydration shells, 4
Hydrazine, 28
Hydrazinolysis, 28
Hydrogen bond (H-bond), 1
Hydrogen bonding, protein structure
  and, 30
Hydrolases, biochemical properties,
  37*t*
Hydrophilicity, amino acid classifica-
  tion and, 21*t*
Hydrophobic forces, protein structure
  and, 30–31
Hydrophobicity, amino acid classifi-
  cation and, 21*t*
Hydroquinone, 385
Hydroxyapatite, 100
Hydroxylase P450, activity, 174*f*
Hydroxylysine, 62
  structure, 63*f*
Hydroxymethylbilane, 267–268
Hydroxyproline, 62
  metabolic fate, 255
  structure, 63*f*
4-Hydroxypyrazolopyrimidine. *See*
  Allopurinol
5-Hydroxytryptophan, structure, 275*f*
Hyperalphalipoproteinemia, familial,
  216*t*
Hyperaminoacidemia, 236
Hyperaminoaciduria, 236
Hyperapobetalipoproteinemia, famil-
  ial, 216*t*
Hyperbilirubinemia, 270
Hypercholesterolemia, familial,
  32–33, 218
  gene therapy, 413*t*
  polymerase chain reaction and,
    411*t*
Hyperchylomicronemia, familial,
  216*t*
Hypergalactosemia, 149–150
Hyperlipoproteinemias, 216*t*, 217
Hypertriacylglycerolemia, familial,
  216*t*
Hyperuricemia, 281
Hypobetalipoproteinemia, familial,
  218*t*
Hypolipoproteinemias, 217, 218*t*
Hypoxanthine, 95
Hypoxanthine-guanine phosphoribo-
  syltransferase (HGPRT), 281

ICAM-1. *See* **Intracellular adhesion**
  molecule-1
I-cell disease, 186

IDDM. *See* Insulin-dependent dia-
  betes
IDH. *See* Isocitrate dehydrogenase
IDLs. *See* Intermediate-density
  lipoproteins
Iduronate, 76
  structure, 75*f*
IFN-gamma. *See* Interferon-gamma
Ifosfamide, 23–24
IFs. *See* Initiation factors
IGF-I. *See* Insulin-like growth
  factor-I
IGF-II. *See* Insulin-like growth fac-
  tor-II
IgG. *See* Immunoglobulin G
ILs. *See* Interleukins
Immune interferon, 357
Immunodeficiency, defect causing,
  283*t*
Immunoglobulin G (IgG), 67
  schematic model, 68*f*
  structure/function relationships, 61*t*
IMP. *See* Inosine monophosphate
Induced fit model, 54
Initiation
  transcription, 319–320
    control, 345–346
  translation, 332–334
Initiation complex, 334
Initiation factors (IFs), 332
Inosine monophosphate (IMP), 278
  conversion to AMP and GMP, 280*f*
Inositol triphosphate (IP3), 391
Insulin, 301–302
Insulin-dependent diabetes (IDDM),
  301–302
  interrelationships of organs, 302*f*
Insulin-like growth factor-I (IGF-I),
  355
  properties, 352*t*
  receptor, 360
Insulin-like growth factor-II (IGF-II),
  183, 355
  properties, 352*t*
Interferon
  properties, 354*t*
  translation and, 336, 338*f*
Interferon-gamma (IFN-gamma), 357
Interleukin(s) (ILs), 351
  properties, 353–354*t*
Interleukin-1 (IL-1), hematopoietic
  and lymphoid tissues and, 355
Interleukin-3 (IL-3), 357
Intermediary metabolism, 112–117,
  113*f*
Intermediate-density lipoproteins
  (IDLs), 213
Internal energy, 13
International Union of Biochemists
  (I.U.B.), 37
Intestinal bacteria, products, 233*t*
Intestinal wall, amino acid transport
  and, 232*t*
Intracellular adhesion molecule-1
  (ICAM-1), 184

Intracellular hormone receptors, 363–364
Intracellular proteases, 234
Intracellular signaling molecules, 359f
Intrinsic factor, 375
Intrinsic pathway, 392, 393f, 394
Introns, 306, 323
    group I self-splicing, mechanism of splicing, 325f
    group II self-splicing, mechanism of splicing, 326f
Iodine, characteristics, 388t
Iodoacetic acid, 27
5-Iodo-2-deoxyuridine, 95
    structure, 96f
Ion exchange chromatography, protein analysis and, 33
Ionic equilibria, 1–8
Ionophores, 108–109
Ion product, 2
IP3. *See* Inositol triphosphate
IPP. *See* Isopentenyl pyrophosphate
Iron, characteristics, 388t
Isoaccepting tRNA, 332
Isocitrate, production, 159f
Isocitrate dehydrogenase (IDH), 157
    reaction catalyzed, 159f
Isoelectric pH, 4
Isoelectric point, 4
Isolectric focusing, 33
Isoleucine, metabolic fate, 258–259
Isomerases, biochemical properties, 37t
Isopentenyl pyrophosphate (IPP), 223
Isozymes, 39

Jaundice, 270

Kallikrein, 394
Keratan sulfate, 76
    characteristics, 78t
    structure, 76f
Keto acid(s), branched-chain, genetic disease and, 259–265
Keto acid dehydrogenase, branched-chain, 258
Ketoacidosis, 302
    diabetic, 201
3-Ketoacyl reductase, 189
3-Ketoacyl synthase, 189
Ketogenesis, 199–200
    pathways in liver, 199f
    regulation, 200
Ketogenic amino acids, 253–254
alpha-Ketoglutarate, 253
Ketoglutarate dehydrogenase, 373
alpha-Ketoglutarate dehydrogenase complex, 157–159, 160f
Ketone bodies, 199
    clinical significance, 200–201
Ketoses, physiologic significant, 73f
Kidneys, acid-base balance and, 6

Krabbe's disease, 89t
Krebs cycle, 158f
    enzymes, 157–162
Kwashiorkor, 234
Kynurenine, 262

LACI. *See* **Lipoprotein-associated coagulation inhibitor**
lac operon, 343–344
Lactase, 149
Lactate, 298
    conversion to phosphoenolpyruvate, 128–130
Lactate dehydrogenase (LDH), 39
Lactic acid, 6
Lactic acid cycle, 131f
Lactose, 74
    intolerance, 149
    structure, 74f
Lactose operon, repression and derepression, 343f
Lagging strand, 310
Lanosterol, 223
Lariat structure, 323
LATT. *See* Lysolecithin:lecithin acyltransferase
LCAD. *See* Long-chain acyl-CoA dehydrogenase
LCAT. *See* Lecithin-cholesterol acyl transferase
L-conformation, 71
LCR. *See* Ligase chain reaction
LDH. *See* Lactate dehydrogenase
LDLs. *See* Low-density lipoproteins
Leading strand, 310
Lecithin, 83
Lecithin-cholesterol acyl transferase (LCAT), 214–215
Lesch-Nyhan syndrome, 283
    defect causing, 283t
    gene therapy, 413t
    polymerase chain reaction and, 411t
Leucine, 253
    metabolic fate, 258–259
Leucine zipper, 346–347, 349f
Leukemia
    acute lymphocytic, 363
    chronic myelogenous, 363
    megaloblastic, 380
Leukocyte common antigen protein, 364
Leukotrienes (LTs), 204
    arachidonic acid conversion, 207f
    properties, 205t
Levorotatory, 22
Lewis blood group antigen, 185
Library, 402, 404
Ligand-receptor complex, 359
Ligase(s), biochemical properties, 37t
Ligase chain reaction (LCR), 411–412, 413f
Linear kinetic plots, 47–48
Linear pathway, 205, 207f

Lineweaver-Burk plots, effects of reversible inhibitors, 50f
Lingual amylase, 120–121
Linoleate, 193
Linoleic acid, 82
    structure, 82t
Linolenate, 193
Linolenic acid, 82
    structure, 82t
Lipase
    hepatic, deficiency, 216t
    hormone-sensitive, 194, 198
        hormonal influences, 303t
    lipoprotein, 211–212
    pancreatic, 211
Lipids, 81–88
    circulating carriers, 211–215
    complex, metabolism, 201–205
    intestinal uptake, 211–216, 211f
    membrane, 104, 104t
    metabolism, 188–205
    mobilization, 194–197
Lipid storage disease, 204
Lipocytes, 380
Lipogenesis, regulators, 298t
Lipoic acid, regeneration of oxidized form, 156f
Lipolysis, control, 197f
Lipoprotein(s), 32, 391
    composition, 212t
    high-density, 214–215
        metabolism, 217f
    intermediate-density, 213
    low-density, 213–214
        production, 215f
        receptor recycling, 218f
        receptors, 215–216
    structure, 212f
    very low-density, 212–213
        metabolic fate, 215f
Lipoprotein-associated coagulation inhibitor (LACI), 395
Lipoprotein lipase (LPL), 211–212
Lipoxygenase(s), 205
Lipoxygenase pathway, 207f
Liver, 296
    metabolic functions, 297–299, 297t
    role in nitrogen metabolism, 242t
Long-chain acyl-CoA dehydrogenase (LCAD), 198
Long terminal repeats (LTRs), 366
Long-term regulation, 197
Lovastatin, 219
Low-density lipoproteins (LDLs), 213–214
    production, 215f
    receptor recycling, 218f
    receptors, 215–216
LPL. *See* Lipoprotein lipase
LTRs. *See* Long terminal repeats
LTs. *See* Leukotrienes
Lyases, biochemical properties, 37t
Lymphokines, 351
Lymphotoxin, 357
Lysine, 253
    catabolism, 259–262, 261f

Lysinemia, 262
Lysinuria, 262
Lysis, 403
Lysogeny, 403
Lysolecithin:lecithin acyltransferase (LLAT), 203
Lysophospholipids, 83, 203
Lysosomal storage disease, 77, 87, 186
Lysosomes, 183
Lysyl oxidase, 63

Macrocytic anemia, 380
alpha2-Macroglobulin, 396
Macrominerals, 372, 389
  essential, characteristics, 387t
Macrophage-CSF (M-CSF), 357
Magnesium, characteristics, 387t
Maintenance methylase system, 312
Major groove, 97
Malate aspartate shuttle, 172, 173f
Malate dehydrogenase (MDH), 161–162, 188
Malic enzyme, 188
Malignant carcinoid syndrome, 379
Maltose, 74
  structure, 74f
Manganese, characteristics, 388t
Mannose, 179, 181
Mannosidosis, enzyme deficiency, 185t
MAP kinases. See Mitogen-activated protein kinases
Maple syrup urine disease, 24
  polymerase chain reaction and, 411t
Marfan's syndrome, 32
Maroteaux-Lamy syndrome, 78
Maxam and Gilbert sequencing, 406
MCAD. See Medium-chain acyl-CoA dehydrogenase
McArdle's syndrome, deficiency, 137t
M-CSF. See Macrophage-CSF
MDH. See Malate dehydrogenase
Medium-chain acyl-CoA dehydrogenase (MCAD), 198
Megaloblastic leukemia, 380
Melatonin
  structure, 275f
  synthesis, 271–273
Membranes
  abnormal functions, disease states associated, 109
  asymmetry, 106
  fluidity, lipid unsaturation and, 106t
  fluid properties, 104
  lipid composition, 104, 104t
  mobility of components, 106
  permeability, 106–109
  plasma, receptors, epinephrine binding, 140f
  proteins, 105–106
    integral, 105
    peripheral, 105

structure, 103–106
  fluid mosaic model, 105f
  transport systems, 108f
  water transport, 108
Menadione, structure, 386f
Menaquinone, structure, 386f
Mental retardation, cystathionine synthase and, 243
Mercaptoethanol, 27
6-Mercaptopurine, 95
  structure, 96f
3-Mercaptopyruvate pathway, 258f
Messenger RNA, 318
  eukaryotic, structure of cap, 322f
  polyadenylate tail added, 322f
Messenger RNA gene
  eukaryotic, structure, 346f
  processing, 321f
  transcription factors, characteristics, 348t
Metabolic acidosis, 6
Metabolic alkalosis, 6
Metabolic enzymes, hormonal influences, 303t
Metabolic pathways
  committed step, 116–117
  regulation, 117t
  regulators, 298t
Metabolic reactions, 114–115
Metabolic transport processes, factors regulating, 116t
Metabolism
  hormonal regulation, 302–303
  intermediary, 112–117, 113f
  regulation, 116
  well-fed versus fasting state, 296–302
Metabolites, transport, 117f
Metachromatic leukodystrophy, 204
  enzyme deficiency, accumulating substance, symptoms, 89t
Metalloenzymes, 38
Metalloflavoproteins, 373
Metarhodopsin II, 381
Methionine, 377
  metabolic fate, 256–258
  role in cysteine synthesis, 244f
Methionine synthase, 375
Methotrexate, 285, 380
Methylation, 312
5-Methylcytosine, 94
  structure, 312f
Methylmalonyl-CoA, 286
Methylmalonyl-CoA mutase, 375
Mevalonate
  activation, 223
  biosynthesis, 222f
Mevalonic acid, cholesterol synthesis and, 223f
Mevastatin, 219
Mevinolin, 219
Micelle, diagrammatic cross section, 105f
Michaelis-Menten analysis, 45–47

Michaelis-Menten equation, 46–47
  definition of symbols used, 46t
  limits of applicability, 48t
Microminerals, 372, 389
  essential, characteristics, 388t
Minerals, 389
  trace, 372, 389
Minor groove, 97
Mitochondrial respiratory rate, 171
Mitogen-activated protein kinases (MAP kinases), 362–363
Mitosis, 307
Mitotic phase, 307
MLCK. See Myosin light chain kinase
MN blood group system, 185
Mobile genetic elements, 314
Molybdenum, characteristics, 388t
Monobasic phosphate, formation, 8f
Monokines, 351
Monosaccharides, 71–73
Morquio's syndrome, 78
M phase, 307
MPS. See Mucopolysaccharidoses
mRNA, 318
  eukaryotic, structure of cap, 322f
  polyadenylate tail added, 322f
mRNA gene
  eukaryotic, structure, 346f
  processing, 321f
  transcription factors, characteristics, 348t
Mucolipidosis, enzyme deficiency, 185t
Mucolipidosis II, 186
Mucolipidosis III, 186
Mucopolysaccharides, 77
Mucopolysaccharidoses (MPS), 77–78, 78t
Multi-CSF, 357
Mutarotation, 72
Mutation
  transition, 315
  transversion, 315
Myoglobin, 64–67
  model, 64f
  oxygen binding curves, 66f
  structure/function relationships, 61t
Myosin light chain kinase (MLCK), 392
Myristic acid, structure, 82t

NAD. See Nicotinamide adenine dinucleotide
NADH
  cytosolic, energy conservation and, 172–173
  electron flow to oxygen, 167f
  electron flow to Coenzyme Q, 168f
  oxidation, 166–167
NADH-cytochrome b5 reductase, 193
NAD+ kinase, 293
NADP. See Nicotinamide adenine dinucleotide phosphate

Negative nitrogen balance, 233
Neomycin, translation and, 339*t*
Nerve growth factor (NGF), properties, 352*t*
Net charge, 21
Neurofibromatosis type-1 (NF1), 369
Neurofibromatosis type-1 gene, 363
Neurofibromin, 363
Neurotrophin receptor family, 360
NF1. *See* Neurofibromatosis type-1
NGF. *See* Nerve growth factor
Niacin, 373–374
    clinical significance, 378–379
    as coenzyme, 38*t*
Nicotinamide adenine dinucleotide (NAD), 373
    synthesis and breakdown, 291*f*
Nicotinamide adenine dinucleotide phosphate (NADP), 373
Nicotinic acid, 219, 293
NIDDM. *See* Non-insulin dependent diabetes
Niemann-Pick disease, 89*t*
Night blindness, 385
Nitric oxide (NO), formation from arginine, 273
Nitrogen
    metabolism, 229–238
        role of liver, 242*t*
    removal from amino acids, 234–236
Nitrogen balance, 233
    negative, 233
    positive, 233
Nitrogen compounds, biologically active, 266–274
Nitrogenous wastes, elimination, 6
NMDA, 18
NO. *See* Nitric oxide
Noncompetitive inhibitors, 49–50
    characteristics, 49*t*
Non-histone proteins, 307
Non-insulin dependent diabetes (NIDDM), 109, 302
Norepinephrine, tyrosine conversion, 274*f*
Northern blotting, 406, 407
Nucleic acids, 91–100
    purine/pyrimidine bases, 92*f*
    optical properties, 93
Nucleoside(s)
    base conformations, 92
    purine. *See* Purine nucleosides
    sugar conformations, 93–94
Nucleoside diphosphate kinase, 285, 288
Nucleoside kinases, 283, 289–290
Nucleosomes, core structure, 307*f*
Nucleotide(s)
    base conformations, 92
    co-factors, 290–293
    metabolism, 277–293
        clinical significance, 281–284
    pyrimidine. *See* Pyrimidine nucleotides

sugar conformations, 93–94
synthetic analogs, 95–96
Nucleotide sequencing, restriction endonucleases and, 40

Oculocutaneous tyrosinemia, 24
ODC. *See* Ornithine decarboxylase
3-OH ketoacyl dehydratase, 189
Okasaki fragments, 310, 311*f*
Oleic acid, structure, 82*t*
Oligomeric proteins, 32
Oligomycin, oxidative phosphorylation and, 172*t*
Oligonucleotides, 96–97
Oligosaccharides, 71, 178
    asparagine-linked, structures, 182*f*
    processing pathways, 184*f*
OMP decarboxylase, 285, 288
Oncogenes, 366, 367–368
    recessive, 366
Oncogenic form, 33
One-carbon metabolism, 246–247
Operators, 342
Operons, 342
    catabolite-regulated, 343
    lac, 343–344
        repression and derepression, 343*f*
    trp, 344
        modes of transcription, 345*f*
        transcription control, 344*f*
Opsin, 381
Ornithine, metabolic fate, 255
Ornithine decarboxylase (ODC), 271
Ornithine transcarbamoylase, 236
    deficiency, polymerase chain reaction and, 411*t*
Orotate phosphoribosyl transferase, 288
Orotic aciduria, defective enzyme, 288*t*
Osteogenesis imperfecta, 32
Osteomalacia, 386
Oxaloacetate, 253
Oxidases, 173–175
    amino acid, 236
        oxidative deamination catalyzed, 236*f*
Oxidation-reduction reactions, 165–166
Oxidative phosphorylation, 170–172, 170*f*, 296
    antimetabolites and, 172*t*
    regulation, 171–172
    stoichiometry, 171
Oxidoreductases, biochemical properties, 37*t*
Oxygen, 173–175
    binding curves, 66*f*
    partial pressure, 6
Oxygenases, 173–175
Oxygenation, addition of oxygen to heme iron, 65*f*

Oxygen free radicals, 174
    generation, 175*t*

p53, 369–370
PABA. *See* para-Aminobenzoic acid
PAF. *See* Platelet-activating factor
Palmitic acid, structure, 82*t*
Palmitoleic acid, structure, 82*t*
Palmitoyl thioesterase, 190
Pancreatic lipases, 211
Pantothenic acid, 293, 374
    clinical significance, 379
    as coenzyme, 38*t*
    coenzyme A synthesis and, 292*f*
    structure, 374*f*
PAPS. *See* 3-Phosphoadenosine 5-phosphosulfate
Parathyroid hormone (PTH), 383
Parkinson's disease, 271
Paroxysmal nocturnal hemoglobinuria (PNH), 185
Partial double-bond character, 23
Parvovirus B19, 184
Passive transport, 106, 107*f*, 116
PC. *See* Phosphatidylcholine
PCR. *See* Polymerase chain reaction
PCR-SSCP. *See* Polymerase chain reaction-single strand conformational polymorphism
PDGF. *See* Platelet-derived growth factor
PDH. *See* Pyruvate dehydrogenase
PE. *See* Phosphatidylethanolamine
Pedigree analysis, 409*f*
Pellagra, 23, 379
Pentachlorophenol, oxidative phosphorylation and, 172*t*
Pentose phosphate pathway, 147, 148*f*, 373
    regulators, 298*t*
PEP. *See* Phosphoenolpyruvate
Pepsin, specificities, 29*t*
Pepsin A, proenzyme, activator, cleavage sites, 231*t*
Peptide(s)
    C-terminal determination, 28
    N-terminal determination, 27–28
    signal, 336
    structural orientation and function, amino acid R-groups and, 21
Peptide bond, 23
    orientation in beta-sheets, 31*f*
    resonance stabilization, 23*f*
Peptidyltransferase, 334
Pernicious anemia, 379
Peroxidases, 175, 205
PFGE. *See* Pulsed-field gel electrophoresis
PFK. *See* Phosphofructokinase
PG. *See* Phosphatidylglycerol
PGs. *See* Prostaglandins
pH, isoelectric, 4
Phenylalanine
    catabolism in phenylketonuria, 246*f*

Phenylalanine (cont.)
  metabolic fate, 259
  ultraviolet absorption spectrum, 23f
Phenylalanine hydroxylase, 23, 243
Phenylalanine hydroxylase reaction, 245f
Phenylketonuria (PKU), 23, 243, 259
  gene therapy, 413t
  phenylalanine catabolism and, 246f
  polymerase chain reaction and, 411t
Phenylpyruvic acid, 244
Philadelphia+ chromosome, 363
Phosphatases, 57
  signal transduction and, 364
Phosphatidic acid, 82, 201
  structure, 83f
Phosphatidyl-4,5-bisphosphate (PIP), 391
Phosphatidylcholine (PC), 83
  membrane, 104t
  metabolism, 201
  structure, 83f
Phosphatidylethanolamine (PE), 83
  membrane, 104t
  metabolism, 201
  structure, 84f
Phosphatidylglycerol (PG), 84
  metabolism, 203
  structure, 84f
Phosphatidylinositol (PI), 84–85
  membrane, 104t
  metabolism, 201–203
  structure, 85f
Phosphatidylinositol-4,5-bisphosphate, 203
Phosphatidylinositol-3-kinase, 362
Phosphatidylserine (PS), 84
  membrane, 104t
  metabolism, 201
  structure, 84f
3-Phosphoadenosine 5-phosphosulfate (PAPS), 204
5-Phospho-alpha-ribosyl-1-pyrophosphate (PRPP), 278
Phosphocreatine, 300
Phosphodiesterase, 140, 278, 381
Phosphodiester bond, 96–97
Phosphoenolpyruvate (PEP)
  conversion to glucose-6-phosphate, 130
  lactate conversion, 128–130
Phosphofructokinase (PFK),
  allosteric regulators, 125t
Phosphofructokinase-1 (PFK-1), 124
Phosphoglucomutase, 135–136, 137
Phosphoglycerate kinase, 126
Phosphoglycerate mutase, 127
Phosphoglycerides, 83–85
  membrane, 104t
Phosphohexose isomerase, 124
Phospholipase(s), 203
  signal transduction and, 362
  sites of hydrolytic activity, 204f
Phospholipase A2 (PLA2), 392

Phospholipase C, 391
Phospholipids, 82–83, 391
  biosynthesis, 202f
  metabolism, 201–203
  signal transduction and, 362
Phosphopantetheine, 189
Phosphoprotein phosphatase-1 (PP-1), 224
Phosphoprotein phosphatase inhibitor-1 (PPI-1), 224, 302
  hormonal influences, 303t
Phosphoribosylation, 280
Phosphorus, characteristics, 387t
Phosphorylase(s), 134, 278
  activity, 136f
  control in muscle, 141f
  effects of effectors, 144t
  hormonal influences, 303t
  regulation of activity, 139–142
Phosphorylase kinase, 139–142, 303
  hormonal influences, 303t
Phosphorylation
  enzyme activity regulation and, 57–58
  oxidative, 170–172, 170f, 296
    antimetabolites and, 172t
    regulation, 171–172
    stoichiometry, 171
Phosphorylation potential, 114
Photoreceptors, 363
Phototransduction, 382f
Phylloquinone, structure, 386f
Phytanic acid, 194
pI, 4
PI. See Phosphatidylinositol
PK. See Pyruvate kinase
PKC. See Protein kinase C
PKU. See Phenylketonuria
PLA2. See Phospholipase A2
Plasma acid-base balance, 6
Plasmalogens, 84
  metabolism, 203
  structure, 85f
Plasma membrane, receptors, epinephrine binding, 140f
Plasmid, 400, 403
Plasmin, 391, 397
Plasminogen, 397
  activation, 398f
Plasminogen activator-inhibitors, 397
Plasmodium vivax, 185
Platelet(s)
  activation, 391–392, 392f
  aggregation, 392
Platelet-activating factor (PAF), 84–85, 203
  structure, 85f
Platelet-derived growth factor (PDGF), 351–352
  properties, 352t
  receptor, 360
Platelet plug, 391
PMF. See Proton motive force
PNH. See Paroxysmal nocturnal hemoglobinuria

PNP. See Purine nucleoside phosphorylase
Polyadenylate polymerase, 321
Polyamine(s), catabolism, 271, 273f
Polyamine oxidase, 271
Polyampholytes, 4
Polycistronic, 320, 342
Polyelectrolytes, 4
Polyhydroxyaldehydes, 71
Polyhydroxyketones, 71
Polymerase chain reaction (PCR), 45, 408–409, 410f
  inherited disorders detectable, 411t
Polymerase chain reaction-single strand conformational polymorphism (PCR-SSCP), 411, 412f
Polyphosphoinositides, 84
  hydrolysis, 363
Polysaccharides, 71, 74
Pompe's disease, deficiency, 137t
Porphobilinogen, biosynthesis, 267–268, 268f
Porphyria, acute intermittent, 267
Porphyrin, 180
Positive nitrogen balance, 233
Posttranscriptional processing, 320–322
Potassium, characteristics, 387t
PP-1. See Phosphoprotein phosphatase-1
PPI-1. See Phosphoprotein phosphatase inhibitor-1
Preinitiation complex, 332
Prenylation, enzyme activity regulation and, 57–58
Prepro-opiomelanocortin, 336
Preproproteins, 336
Preproteins, 336
Preribosomal RNA, 321
Primase, 309
Primer, 309
"Primer" glycogen, 138
Probes, 404
Probucol, 219
Processivity, 311
Processivity accessory proteins, 311
Proenzymes, 394
  enzyme activity regulation and, 55–56
Prokaryotic gene control, 342–344
Prokaryotic transcription, 319–320
Proline, 62
  biosynthesis, 246, 247f
  catabolism, 256f
  metabolic fate, 255
Promoter(s), 319
Promoter sequences, 342
Proproteins, 336
Prostacyclin, 398
Prostaglandin endoperoxide synthetase, 205
Prostaglandins (PGs)
  arachidonic acid conversion, 206f
  biosynthesis, 204–205
  properties, 205t

Prostanoids, 204
Prosthetic groups, 37–38
Proteases
    digestive tract, 231*t*
    intracellular, 234
    proenzymes, activators, cleavage
        sites, 231*t*
Protein(s), 296
    accessory, processivity, 311
    analysis, 33–34
    binding, single-strand, 309
    blood-clotting, 385, 395*t*
    carbohydrates linked, 181–183
    clinical significance, 32–34
    common acid groups, 22*t*
    DNA-binding
        cancer and, 368
        structural motifs, 346–347, 347*f*
    energy value, 115*t*
    fibrous, 30
    globular, 30
    glypiated, 185
    histone, 307
    membrane, 105–106
        integral, 105
        peripheral, 105
    non-heme-iron, 157
    non-histone, 307
    oligomeric, 32
    specific dynamic action, 115*t*
    structural orientation and function,
        amino acid R-groups and, 21
    structure, 26–32
        complex, 32
        forces controlling, 30–32
        primary, 27–28
        quaternary, 32
        secondary, 28–29
        super-secondary, 29–32
        tertiary, 29–30
    structure/function relationships,
        60–68, 61*t*
    synthesis, 329–339
        inhibitors, 336–338
        initiation, 332–334, 333*f*
        peptide elongation process, 335*f*
        regulation, 336–338
        termination process, 337*f*
    vitamin D-binding, 383
Protein C, 396
Protein Ca, 396
Protein kinase(s), mitogen-activated,
    362–363
Protein kinase (PKA), cAMP activa-
    tion, 57, 58*f*
Protein kinase C (PKC), 362, 391
Protein S, 396
Protein serine phosphatases (PSPs),
    364
Protein tyrosine phosphatases (PTPs),
    364
Proteoglycans, 77
    structure, 77*f*
Proteolysis, enzyme degradation and,
    55

Prothrombin
    activation to thrombin, 395–396
    binding to plasma membrane, 396*f*
    diagrammatic representation, 396*f*
Prothrombinase complex, 395
Proton motive force (PMF), 170–171
Proto-oncogenes, 33, 347, 366,
    367–368
Prourokinase, 397
Proximal convoluted tubule, 6
PRPP. *See* 5-Phospho-alpha-ribosyl-
    1-pyrophosphate
PRPP amidotransferase, 278
PRPP synthetase, 278
PS. *See* Phosphatidylserine
Pseudogenes, 314
Pseudo-Hurler polydystrophy, 186
P-site, 334
PSPs. *See* Protein serine phosphatases
PTH. *See* Parathyroid hormone
PTPs. *See* Protein tyrosine phos-
    phatases
Pulmonary surfactant, 83, 84
Pulsed-field gel electrophoresis
    (PFGE), 100, 406
Purine
    biosynthesis
        pathway, 279*f*
        regulation, 278–280
    metabolism, disorders, 283*t*
    structure, 92*f*
    synthetic analogs, 96*f*
Purine nucleoside(s), uric acid forma-
    tion and, 282*f*
Purine nucleoside phosphorylase
    (PNP), 283
Purine nucleotide(s), 377
    biosynthesis, 278–281
    de novo synthesis, control of rate,
        281*f*
    salvage and catabolism, 280–281
Purine nucleotide cycle, 281, 283*f*
Purine ribonucleotides, regulation of
    reduction, 289*f*
Purine ring, sources of nitrogen and
    carbon atoms, 280*f*
Puromycin, translation and, 339*t*
Putrescine, 271
Pyran, 72
Pyranoses, 72
Pyridoxal, 374
    phosphorylation, 375*f*
    structure, 374*f*
Pyridoxal kinase, 374
Pyridoxal phosphate, 374
    formation, 375*f*
Pyridoxamine, 374
    structure, 374*f*
Pyridoxine, 374
    as coenzyme, 38*t*
    structure, 374*f*
Pyrimidine
    biosynthesis, regulation, 285
    metabolism, disorders, 288*t*
    structure, 92*f*
    synthetic analogs, 96*f*

Pyrimidine nucleotides
    biosynthesis, 284–285, 286*f*
        pathway, 284*f*
    salvage and catabolism, 286–288,
        287*f*
Pyrimidine ribonucleotides, regula-
    tion of reduction, 289*f*
5-Pyrophosphomevalonate, 223
Pyruvate, 235, 296
    oxidation, regulators, 298*t*
    pyruvate dehydrogenase kinase
        and, 155
    threonine/glycine conversion, 257*f*
Pyruvate carboxylase, 374
Pyruvate dehydrogenase (PDH), 153,
    373
Pyruvate dehydrogenase (PDH) com-
    plex, 153–154
    components, 154*t*
    enzymatic activities and cofactors,
        154*f*
    regulation, 155–157, 156*f*
Pyruvate kinase (PK), 127–128
    hormonal influences, 303*t*

Racemases, 38–39
Rate constants, 43
RB. *See* Retinoblastoma
RDS. *See* Respiratory distress syn-
    drome
Reactant(s)
    conversion to products, energy
        changes during, 45*f*
    relationship with reaction order,
        45*t*
Reactant concentration, 43
Reaction order, 44
    relationship with number of reac-
        tants, 45*t*
Reaction rate, 43
Receptor serine/threonine kinases,
    361
Receptor tyrosine kinases (RTKs),
    359–360
    cancer and, 367
Recessive oncogenes, 366
Recombinant chromosomes, genera-
    tion, 313*f*
Recombinant DNA technology,
    400–414
Recombination, 313, 314*f*
Red thrombus, 391
Reductase kinase (RK), 223
    hormonal influences, 303*t*
Reductase kinase kinase (RKK), 223
Refsum's disease, 199
Releasing factors (RFs), 334–336
Renal lithiasis, defect causing, 283*t*
Repetitive DNA, 306
Replication
    DNA, 308–312
    semiconservative, 309
Replication fork, 309
Replisome, 311

Respiratory acidosis, 6
Respiratory alkalosis, 6
Respiratory distress syndrome (RDS), 87
Respiratory quotient (RQ), 115–116
Restriction endonucleases, 40, 312, 400–402, 402f
  characteristics, 403t
Restriction fragment length polymorphism (RFLP) analysis, 407–408
Restriction sites, 40
Reticulocytes, control of translation, 338f
Retinal, 380
Retinaldehyde reductase, 380
Retinene, structure, 382f
Retinoblastoma (RB), 368
  polymerase chain reaction and, 411t
Retinoblastoma susceptibility gene, 368
Retinoic acid, 380
  receptor, 363
Retinol, 380, 381
Retinyl phosphate, 381
Retroviral integration-induced transformation, 367
Retroviruses, 366
  oncogenic, 347
Reverse cholesterol transport, 215
Reverse transcriptase, 366
Reverse transcription, 404
Reverse transcription-PCR (RT-PCR), 409–411
Reversible inhibitors
  effects on Lineweaver-Burk plots, 50f
  kinetic effects, 49–50
RFLP analysis. See Restriction fragment length polymorphism (RFLP) analysis
RFs. See Releasing factors
R-groups, 21, 55f
  orientation in beta-sheets, 31f
Rheumatoid arthritis, 109
Rhinoviruses, 184
Rhodopsin, 381
Riboflavin, 290, 290f, 373
  clinical significance, 377–378
  as coenzyme, 38t
  structure, 373f
Ribonucleoside(s), structures, 93f
Ribonucleoside diphosphates, reduction, 288f
Ribonucleotide(s), polymerization into RNA sequence, 319f
Ribonucleotide reductase (RR), 283, 288–289
Ribosomal RNA, 318
  processing, 324f
Ribosomes, 318
Ribozymes, 37
Ribulose, structure, 73f
Ricin, translation and, 339t

Rickets, 386
RK. See Reductase kinase
RKK. See Reductase kinase kinase
RNA
  messenger, 318
    polyadenylate tail added, 322f
    structure of cap, 322f
  posttranscriptional processing, 320–322
  preribosomal, 321
  ribosomal, 318
    processing, 324f
  synthesis, 318–326
  transcription, 319–320
  transfer, 318
    amino acid activation and, 329–332
    isoaccepting, 332
    processing events, 323f
    structure, 331f
RNA polymerase, classes, 318–319
RNA polymerase III, transcription and, 347–348
RNA polymerase holoenzyme, 319
RNA primer, 309
RNA splicing, 321, 323
  alternative and aberrant, clinical significance, 324–325
RNA tumor viruses, 366
Rods, phototransduction, 382f
Rotenone, oxidative phosphorylation and, 172t
RQ. See Respiratory quotient
RR. See Ribonucleotide reductase
rRNA, 318
  processing, 324f
RTKs. See Receptor tyrosine kinases
RT-PCR. See Reverse transcription-PCR

Saccharopine, 259
Salvage pathways, 280
Sandhoff-Jatzkewitz disease
  enzyme deficiency, accumulating substance, symptoms, 89t
  polymerase chain reaction and, 411t
Sanfilippo syndrome, 78
Sanger sequencing, 27, 406, 407f
Saturated fatty acids, 82
  synthetic pathway, 189–190
SCAD. See Short-chain acyl-CoA dehydrogenase
Scatter factor receptor, 360
SCID. See Severe combined immunodeficiency disease
Scotopsin, 381
Scurvy, 380
SDH. See Succinate dehydrogenase
SDS polyacrylamide gel electrophoresis (SDS-PAGE), 33
Second substrates, 38
Sedoheptulose, structure, 73f
Selenium, characteristics, 388t

Semiconservative replication, 309
Serine, 377
  biosynthesis, 246, 248f
  metabolic fate, 255
  threonine/glycine conversion, 257f
Serine hydroxymethyltransferase reaction, 257f
Serine plasmalogens, 203
Serine protease(s), 55
  proenzymes, activators, cleavage sites, 231t
Serine protease inhibitor, 396
Serine/threonine kinases, 359, 361–362
  cancer and, 368
  non-receptor, 362
  receptor, 361
Serotonin, 391
  structure, 275f
  synthesis, 271–273
Serpentine receptors, 359
Serum glutamate-oxaloacetate-aminotransferase, 235
Severe combined immunodeficiency disease (SCID), 283
  defect causing, 283t
  gene therapy, 413t
Shine-Delgarno element, 332
Short-chain acyl-CoA dehydrogenase (SCAD), 198
Short-term regulation, 197–198
Sialidosis, enzyme deficiency, 185t
Sickle cell anemia, 32
  pedigree analysis, 409f
  polymerase chain reaction and, 411t
Sickle-cell beta-globin genes
  LCR analysis, 413f
  PCR-SSCP analysis, 412f
Signaling molecules, intracellular, 359f
Signal peptidase, 336
Signal peptide, 336
Signal recognition particle (SRP), 336, 339f
Signal sequence, 336
Signal transduction, mechanisms, 358–364
Silicon, characteristics, 388t
Silk fibroin, 61–62
  structure/function relationships, 61t
Simple diffusion, 106–107
Simple enzymes, 37
Single-strand binding proteins, 309
Site-specific recombination, 313
Size exclusion chromatography, protein analysis and, 33
Skeletal muscle, 296
  metabolic functions, 297t, 299–300
Small nuclear ribonucleoprotein particles (snRNPs), 323
snRNPs. See Small nuclear ribonucleoprotein particles
SOD. See Superoxide dismutase

Sodium
  characteristics, 387t
  urinary and plasma concentrations, 6t
Sodium bicarbonate, reabsorption, 6–7
Sodium/potassium-ATPase pump, stoichiometry, 108f
Solvation, 4
Somatomedin C, 355
Southern blotting, 406–407
Specific dynamic action, 115, 115t
Spermidine, biosynthesis, 271
  intermediates and enzymes and, 272f
Spermine, biosynthesis, 271
  intermediates and enzymes and, 272f
S phase, 308
Sphingolipid(s), 85–86
  diagrammatic representation, 86f
  metabolism, 204
Sphingolipid storage disease, 87, 89t
Sphingomyelin, 85–86
  biosynthesis, 204f
  membrane, 104t
  metabolism, 204
  structure, 85f
Sphingomyelin synthase, 204
Sphingosine, 85, 204
  structure, 85f
Splice acceptor site, 323
Splice donor site, 323
Spliceosomes, 323
Squalene monooxygenase, 223
Squalene synthase, 223
SRP. *See* Signal recognition particle
Standard biological electrode potential, 166
Standard electrode potential, 166
Starch, 76
Stearic acid, structure, 82t
Sticky end, 400
Streptokinase, 398
Streptomycin, translation and, 339t
Substrates, 45
  effects of concentration on enzyme-catalyzed reactions, 47f
  enzymic activity and, 38–39
  interactions with enzyme inhibitors, 49
  second, 38
Succinate
  formation from succinyl-CoA, 160f
  oxidation to *trans*-fumarate, 161f
Succinate dehydrogenase (SDH), 161, 373
  characteristics, 161t
Succinyl-CoA synthetase, 159
  succinate formation and, 160f
Succinyl thiokinase, 159
Sucrose, 74
  structure, 74f
Sulfatide(s), metabolism, 204

Sulfatide lipidosis, 204
  enzyme deficiency, accumulating substance, symptoms, 89t
Sulfhydryl bond, 21
Superoxide dismutase (SOD), 174
Surfactant, 83, 84
Surfactant proteins, 87
Symports, 107–108
Systemic lupus erythematosus, 325

Tangier disease, 218t
T antigens, 367
TATA box, 346
Taun's disease, deficiency, 137t
Taurine, 256
Taurochenodeoxycholate, 256
Taurocholate, 256
Taurocholic acid, 226
Tay-Sachs disease
  enzyme deficiency, accumulating substance, symptoms, 89t
  polymerase chain reaction and, 411t
Telomere, 405
Tenase complex, 394
Terminal differentiation, 307
Tetracycline, translation and, 339t
Tetrahydrobiopterin, 243
Tetrahydrofolate (THF), 278, 285, 375
  carbon atoms carried, oxidation states, 248t
  enzymes interconverting derivatives, 250t
  one-carbon units attached, interconversions, 249f, 378f
Tetrahydrofolic acid, 246
  as one-carbon atom carrier, 247
TGF-alpha. *See* Transforming growth factor-alpha
TGF-beta. *See* Transforming growth factor-beta
beta-Thalassemia, 325
  gene therapy, 413t
  polymerase chain reaction and, 411t
Thermal denaturation, 99
Thermodynamics
  first law, 12
  second law, 12–13
Thermolysin, specificities, 29t
THF. *See* Tetrahydrofolate
Thiamine, 372–373
  clinical significance, 377
  as coenzyme, 38t
  structure, 373f
Thiamine diphosphotransferase, 373
Thiamine pyrophosphate (TPP), 373
Thioesterase, 190
6-Thioguanine, 95
  structure, 96f
Thiokinase, 193
Thiolase, 199, 223
Thioredoxin, 289

Threonine
  catabolism, 255
  conversion to serine, pyruvate, acetyl-CoA, 257f
Thrombin, 391, 395–396
  inhibitors, 396
Thrombomodulin, 396
Thromboxane(s) (TXs), 204
  arachidonic acid conversion, 206f
  properties, 205t
Thromboxane A2 (TXA2), 392
Thrombus
  red, 391
  white, 391
Thryoid hormone receptors, 363
Thymidine kinase, 285
Thymidylate synthase, 285
Thymine, 92
  structure, 92f
Thymine nucleotides, synthesis, 285
Thymine phosphorylase, 286
Tissue factor, 394
Tissue plasminogen activator (tPA), 397
Titration curve, 3f
TN. *See* Turnover number
TNF. *See* Tumor necrosis factor
TNF-alpha. *See* Tumor necrosis factor-alpha
TNF-beta. *See* Tumor necrosis factor-beta
Tocopherols, 384
  structure, 384f
Tolbutamide, 198
Toxins, translation and, 339t
tPA. *See* Tissue plasminogen activator
TPP. *See* Thiamine pyrophosphate
Trace minerals, 372, 389
Transamidinases, 235
Transcobalamin II, 375
Transcription
  cycle in bacteria, 320f
  initiation, 319–320
    control, 345–346
  prokaryotic/eukaryotic, 319–320
  reverse, 404
  RNA polymerase III and, 347–348
Transcription factors, 344
  cancer and, 368
  mRNA gene, characteristics, 348t
Transcription termination signal, 321f
Transducin, 381
Transduction, 367
Transferases, biochemical properties, 37t
Transfer RNA, 318
  amino acid activation and, 329–332
  isoaccepting, 332
  processing events, 323f
  structure, 331f
Transformation, retroviral integration-induced, 367
Transforming growth factor-alpha (TGF-alpha), 354–355
  properties, 352t, 354t

Transforming growth factor-beta (TGF-beta), 351, 354
  properties, 352*t*, 354*t*
Transformylation, 278
Transgenesis, 412–413
Transglutaminase, 397
Transition mutations, 315
Transition state complex, 45
Transketolase, 373
Translation
  inhibitors, antibiotic and toxin, 339*t*
  initiation, 332–334
  interferon-regulated, 338*f*
  secreted and membrane bound proteins, 339*f*
Translocation, 334
Transmembrane, 106
  receptors, 359*f*
Transpeptidation, 334
Transposition, 313, 314
Transversion mutations, 315
Triacylglycerols, 82, 194, 296
  biosynthesis, 202*f*
  structure, 83*f*
  synthesis, glucose and, 300
Tricarboxylate transport, 188, 189*f*
Tricarboxylic acid cycle, 296
Triglycerides, metabolism, 201
Trimethoprim, 285
Triose phosphate isomerase, 125
trk oncogene, 360
tRNA, 318
  amino acid activation and, 329–332
  isoaccepting, 332
  processing events, 323*f*
  structure, 331*f*
  trp operon, 344
  modes of transcription, 345*f*
  transcription control, 344*f*
Trypsin
  proenzyme, activator, cleavage sites, 231*t*
  specificities, 29*t*
Tryptophan, 293
  catabolism, 264*f*
  metabolism, 262–265
  metabolites, structures, 275*f*
  ultraviolet absorption spectrum, 23*f*
Tubule reabsorption, 6
Tumor antigens, 367
Tumor necrosis factor (TNF), hematopoietic and lymphoid tissues and, 355
Tumor necrosis factor-alpha (TNF-alpha), 351, 357
  properties, 353*t*
Tumor necrosis factor-beta (TNF-beta), 357
  properties, 353*t*
Tumor suppressor(s), 366
Tumor suppressor genes, 368
Turnover number (TN), 40, 48

TXA2. *See* Thromboxane A2
TXs. *See* Thromboxanes
Tyrosine
  active metabolites, 271
  biosynthesis, 243–246
  catabolism, 259
    intermediates, 260*f*
  conversion to epinephrine/norepinephrine, 274*f*
  ultraviolet absorption spectrum, 23*f*
Tyrosine kinases, 359–361
  non-receptor, 360–361
    cancer and, 367
  receptor, 359–360
    cancer and, 367
Tyrosinemias, 24
  hepatorenal, 24
  oculocutaneous, 24
Tyrosine phosphatases, 359

Ubiquinone, structure, 169*f*
UDP-GlcNAc, 179
UDP-glucose dehydrogenase, 180
UDP-glucose pyrophosphorylase, 180
UDP-glucuronate, 180
UDP-glucuronosyl transferase, 180
UDP glucuronyl transferase, 270
Uncompetitive inhibitors, 49
  characteristics, 49*t*
Uniports, 107–108
Unsaturated fatty acids, 82
  oxidation, 196*f*
Uracil, 92
  structure, 92*f*
Urea, urinary and plasma concentrations, 6*t*
Urea cycle, 236–238
  reactions and intermediates, 237*f*
  regulation, 238
  summed reactions, 238*t*
Uric acid, 280–281
  formation from purine nucleosides, 282*f*
Uridine, structure, 93*f*
Uridine kinase, 286
Uridine phosphorylase, 286
Uridylate kinase, 285
Urocanate, 262
Urokinase, 397
Uronic acid pathway, 180
Uroporphyrinogen synthase, 268
Uroporphyrinogen I synthase, 267
Uroporphyrinogen III cosynthase, 268

Valine, metabolic fate, 258–259
Van der Waals forces, protein structure and, 31–32
Vascular constriction, 391
Vascular endothelial cell growth factor (VEGF) receptor, 360
Vector, 400, 403

Very low-density lipoproteins (VLDLs), 212–213
  metabolic fate, 215*f*
Vision, vitamin A and, 381
Visual purple, 381
Vitamin(s)
  as coenzymes, 38*t*
  lipid-soluble, 380–385
    clinical significance, 385–387
  water-soluble, 372–377
    clinical significance, 377–380
  water-soluble derivatives, 38
Vitamin A, 380–381
  clinical significance, 385–386
Vitamin B1, 372–373
  clinical significance, 377
Vitamin B2, 373
  clinical significance, 377–378
Vitamin B3, 373–374
  clinical significance, 378–379
Vitamin B5, 374
  clinical significance, 379
Vitamin B6, 374
  clinical significance, 379
  naturally occurring forms, 374*f*
Vitamin B12, 374–375
  clinical significance, 379–380
  reactions catalyzed, 376*f*
  structure, 376*f*
Vitamin C, 377
  clinical significance, 380
Vitamin D, 383
  clinical significance, 386
  receptor, 363
Vitamin D-binding protein, 383
Vitamin E, 384
  antioxidant properties, 175
  clinical significance, 386–387
Vitamin K, 385
  clinical significance, 387
  naturally occurring, 386*f*
VLDLs. *See* Very low-density lipoproteins
Von Gierke's disease, 283
  defect causing, 283*t*
  deficiency, 137*t*
Von Willebrand disease, polymerase chain reaction and, 411*t*
Von Willebrand's factor (vWF), 391
vWF. *See* Von Willebrand's factor

Warfarin, 385, 397
Water, transport across membranes, 108
Watson-Crick base-pairing, 98
Well-fed state
  interrelationships of organs following, 299*f*
  metabolism and, 296–302
Wernicke-Korsakoff syndrome, 377
Western blotting, 407
White thrombus, 391

Wilms' tumor (WT1), 368–369
Wobble hypothesis, 332
Wolman's disease, 216*t*
WT1. *See* Wilms' tumor

Xanthine oxidase, 95, 283, 373
Xanthinuria, defect causing, 283*t*
Xanthomas, 218

Xeroderma pigmentosum, 315
Xerophthalmia, 385
Xylulose, structure, 73*f*

YAC. *See* **Yeast artificial chromosome**
Yeast artificial chromosome (YAC), 404

Yeast artificial chromosome (YAC) cloning, 405–406, 406*f*

Zinc, characteristics, 388*t*
Zinc fingers, 346
Zn-proteases, proenzymes, activators, cleavage sites, 231*t*
Zwitterion, 4, 21

Basic Science Textbooks

Biochemistry
Examination & Board Review
Balcavage & King
1995, ISBN 0-8385-0661-5, A0661-7
Color Atlas of Basic Histology
Berman
1993, ISBN 0-8385-0445-0, A0445-5
1995 First Aid for the USMLE Step 1
Bhushan
1995, ISBN 0-8385-2593-8, A2593-0
Jawetz, Melnick, & Adelberg's
Medical Microbiology, 20/e
Brooks, Butel, & Ornston
1995, ISBN 0-8385-6243-4, A6243-8
Manual for Human Dissection
Photographs with Clinical
Applications
Callas
1994, ISBN 0-8385-6133-0, A6133-1
Concise Pathology, 2/e
Chandrasoma & Taylor
1995, ISBN 0-8385-1229-1, A1229-2

Medical Biostatistics & Epidemiology
Examination & Board Review
Essex-Sorlie
1995, ISBN 0-8385-6219-1, A6219-8
Review of Medical Physiology, 17/e
Ganong
1995, ISBN 0-8385-8431-4, A8431-7
Basic Histology, 8/e
Junqueira, Carneiro, & Kelley
1995, ISBN 0-8385-0567-8, A0567-6
Basic & Clinical Pharmacology, 6/e
Katzung
1995, ISBN 0-8385-0619-4, A0619-5
Pharmacology
Examination & Board Review, 4/e
Katzung & Trevor
1995, ISBN 0-8385-8067-X, A8067-9
Medical Microbiology & Immunology
Examination & Board Review, 3/e
Levinson & Jawetz
1994, ISBN 0-8385-6242-6, A6242-0

Clinical Anatomy
Lindner
1989, ISBN 0-8385-1259-3, A1259-9
Pathophysiology of Disease
McPhee, Lingappa, Ganong, & Lange
1995, ISBN 0-8385-7815-2, A7815-2
Harper's Biochemistry, 23/e
Murray, Granner, Mayes, & Rodwell
1993, ISBN 0-8385-3562-3, A3562-4
Pathology
Examination & Board Review
Newland
1995, ISBN 0-8385-7719-9, A7719-6
Basic Histology
Examination & Board Review, 2/e
Paulsen
1993, ISBN 0-8385-0569-4, A0569-2
Basic & Clinical Immunology, 8/e
Stites, Terr, & Parslow
1994, ISBN 0-8385-0561-9, A0561-9
Correlative Neuroanatomy, 22/e
Waxman & deGroot
1995, ISBN 0-8385-1091-4, A1091-6

(more on reverse)

Textbooks

...n Policy:

...nbach
...678-6, A3678-8
...gy, 6/e
..., & McIlroy
...35-1093-0, A1093-2
...olytes
... Pathophysiology

0-8385-2546-6, A2546-8
...linical Biostatistics, 2/e
Saunders & Trapp
...3N 0-8385-0542-2, A0542-9
...iynecology and Obstetrics
& Cunningham
..., ISBN 0-8385-9633-9, A9633-7
...iew of General Psychiatry, 4/e
...ldman
...995, ISBN 0-8385-8421-7, A8421-8

Principles of Clinical Electrocardiography, 13/e
Goldschlager & Goldman
1990, ISBN 0-8385-7951-5, A7951-5

Clinical Neurology, 2/e
Greenberg, Aminoff, & Simon
1993, ISBN 0-8385-1311-5, A1311-8

Medical Epidemiology
Greenberg, Daniels, Flanders, Eley, & Boring
1993, ISBN 0-8385-6206-X, A6206-5

Basic & Clinical Endocrinology, 4/e
Greenspan & Baxter
1994, ISBN 0-8385-0560-0, A0560-1

Occupational Medicine
LaDou
1990, ISBN 0-8385-7207-3, A7207-2

Primary Care of Women
Lemcke, Pattison, Marshall, & Cowley
1995, ISBN 0-8385-9813-7, A9813-5

Clinical Anesthesiology, 2/e
Morgan & Mikhail
1995, ISBN 0-8385-1381-6, A1381-1

Dermatology
Orkin, Maibach, & Dahl
1991, ISBN 0-8385-1288-7, A1288-8

Rudolph's Fundamentals of Pediatrics
Rudolph & Kamei
1994, ISBN 0-8385-8233-8, A8233-7

Genetics in Clinical Medicine and Primary Care
Seashore
1995, ISBN 0-8385-3128-8, A3128-4

Smith's General Urology, 14/e
Tanagho & McAninch
1995, ISBN 0-8385-8612-0, A8612-2

Clinical Oncology
...eiss
...3, ISBN 0-8385-1325-5, A1325-8

General Opthalmology, 14/e
Vaughan, Asbury, & Riordan-Eva
1995, ISBN 0-8385-3127-X, A3127-6

CURRENT Clinical References

CURRENT Critical Care Diagnosis & Treatment, 2/e
Bongard & Sue
1995, ISBN 0-8385-1454-5, A1454-6

CURRENT Diagnosis & Treatment in Cardiology
Crawford
1995, ISBN 0-8385-1444-8, A1444-7

CURRENT Diagnosis & Treatment in Vascular Surgery
Dean, Yao, & Brewster
1995, ISBN 0-8385-1351-4, A1351-4

CURRENT Obstetric & Gynecologic Diagnosis & Treatment, 8/e
DeCherney & Pernoll
1994, ISBN 0-8385-1447-2, A1447-0

CURRENT Pediatric Diagnosis & Treatment, 12/e
Hay, Groothuis, Hayward, & Levin
1995, ISBN 0-8385-1446-4, A1446-2

CURRENT Emergency Diagnosis & Treatment, 5/e
Saunders & Ho
1995, ISBN 0-8385-1450-2, A1450-4

CURRENT Diagnosis & Treatment in Orthopedics
Skinner
1995, ISBN 0-8385-1009-4, A1009-8

CURRENT Medical Diagnosis & Treatment 1995
Tierney, McPhee, & Papadakis
1995, ISBN 0-8385-1449-9, A1449-6

CURRENT Surgical Diagnosis & Treatment, 10/e
Way
1994, ISBN 0-8385-1439-1, A1439-7

LANGE Clinical Manuals

Dermatology
Diagnosis and Therapy
Bondi, Jegasothy, & Lazarus
1991, ISBN 0-8385-1274-7, A1274-8

Practical Oncology
Cameron
1994, ISBN 0-8385-1326-3, A1326-6

Office & Bedside Procedures
Chesnutt, Dewar, Locksley, & Tureen
1993, ISBN 0-8385-1095-7, A1095-7

Psychiatry
Diagnosis & Therapy 2/e
Flaherty, Davis, & Janicak
1993, ISBN 0-8385-1267-4, A1267-2

Neonatology
Management, Procedures, On-Call Problems, Diseases and Drugs, 3/e
Gomella
1994, ISBN 0-8385-1331-X, A1331-6

Practical Gynecology
Jacobs & Gast
1994, ISBN 0-8385-1336-0, A1336-5

Drug Therapy, 2/e
Katzung
1991, ISBN 0-8385-1312-3, A1312-6

Ambulatory Medicine
The Primary Care of Families
Mengel & Schwiebert
1993, ISBN 0-8385-1294-1, A1294-6

Poisoning & Drug Overdose, 2/e
Olson
1994, ISBN 0-8385-1108-2, A1108-8

Internal Medicine
Diagnosis and Therapy, 3/e
Stein
1993, ISBN 0-8385-1112-0, A1112-0

Surgery
Diagnosis & Therapy
Stillman
1989, ISBN 0-8385-1283-6, A1283-9

Medical Perioperative Management
Wolfsthal
1989, ISBN 0-8385-1298-4, A1298-7

LANGE Handbooks

Handbook of Gynecology & Obstetrics
Brown & Crombleholme
1993, ISBN 0-8385-3608-5, A3608-5

HIV/AIDS Primary Care Handbook
Carmichael, Carmichael, & Fischl
1995, ISBN 0-8385-3557-7, A3557-4

Pocket Guide to Diagnostic Tests
Detmer, McPhee, Nicoll, & Chou
1992, ISBN 0-8385-8020-3, A8020-8

Handbook of Poisoning
Prevention, Diagnosis & Treatment, 12/e
Dreisbach & Robertson
1987, ISBN 0-8385-3643-3, A3643-2

Handbook of Clinical Endocrinology, 2/e
Fitzgerald
1992, ISBN 0-8385-3615-8, A3615-0

Clinician's Pocket Reference, 7/e
Gomella
1993, ISBN 0-8385-1222-4, A1222-7

Surgery on Call, 2/e
Gomella & Lefor
1995, ISBN 0-8385-8746-1, A8746-8

Pocket Guide to Commonly Prescribed Drugs
Levine
1993, ISBN 0-8385-8023-8, A8023-2

Handbook of Pediatrics, 17/e
Merenstein, Kaplan, & Rosenberg
1994, ISBN 0-8385-3657-3, A3657-2

...ppleton & Lange • 25 Van Zant Street • P.O. Box 5630 • Norwalk, CT • 06856